Advance Praise for *Retu*

"A marvelous compendium of indisput
of the existence of God."

—Dr. Marcos Eberlin, professor of chemistry, Mackenzie University, Brazilian Academy of Sciences, Thomson Medalist

"A state-of-the-art treatise on arguments and counterarguments about intelligent design. It performs a gigaton task of covering the origin of everything from molecular machinery to the entire universe. A much-needed book."

—Dr. Stuart Burgess, professor of engineering design, Bristol University; research fellow, Clare Hall, University of Cambridge

"With this book, Stephen Meyer earns a place in the pantheon of distinguished, non-reductive natural philosophers of the last 120 years, from the great French savant Pierre Duhem through A. N. Whitehead to Michael Polanyi. . . . He has written a profound, judicious book of great value combining both advanced, detailed scientific expertise and philosophical, integrative wisdom."

—Professor Michael D. Aeschliman, emeritus Boston University; author of *The Restoration of Man*

"A meticulously researched, lavishly illustrated, and thoroughly argued case against the new atheism. Even if your mind is made up—especially if it is—Meyer's refreshing take on humanity's most unbridgeable divide: between secular and divine accounts of origins of the Universe, is a joy to read. You may not come away convinced, but you'll be richer for the journey."

—Dr. Brian Keating, Chancellor's Distinguished Professor of Physics, UC–San Diego; author of *Losing the Nobel Prize*

"Meyer masterfully summarizes the current evidence from cosmology, physics, and biology showing that the more we learn about the universe and nature, the more relevant the 'God hypothesis' becomes."

— Dr. Anthony Futerman, Joseph Meyerhoff Professor of Biochemistry, Weizmann Institute, Israel

"Reviewing all relevant evidence from cosmology to molecular biology, Meyer builds an irrefutable 'case for God' while delivering an unanswerable set of logical and scientific broadsides against the currently fashionable materialistic/atheistic worldview. Meyer builds his argument relentlessly, omitting no significant area of debate. The logic throughout is compelling and the book almost impossible to put down. Meyer is a master at clarifying complex issues, making the text accessible to the widest possible audience. Readers will be struck by Meyer's extraordinary depth of knowledge in every relevant area. The book is a masterpiece and will be widely cited in years to come. The best, most lucid, comprehensive defense of the 'God hypothesis' in print. No other publication comes close. A unique *tour de force*."

—Michael Denton, MD, PhD, former senior research fellow, biochemistry, University of Otago; author of *Nature's Destiny*

"No one else in my experience can explicate such complex material with the grace and clarity that seem so effortless to Stephen Meyer. With cold logic and meticulous rational analysis of the latest discoveries in cosmology, physics, and biology, Meyer confirms a truth that the ideologues find too frightening even to consider. By the *ad hominem* nature of their attacks on his brilliant work, they confirm its importance and suggest an eventual end to the scientism that warps our culture."

—Dean Koontz, *New York Times* bestselling author

"This is a long overdue book that until now neither scientific experts nor religious believers had the courage to write. For those genuinely curious about the God hypothesis, this book provides the fairest, most comprehensive statement available."

—Dr. Steve Fuller, Auguste Comte Chair in Social Epistemology, University of Warwick; author of *Knowledge: The Philosophical Quest in History*

"This is a truly superb analysis of the relevant evidence. Stephen Meyer convincingly demonstrates that the God hypothesis is not just an adequate explanation for the origin of our fine-tuned universe and biosphere: it is the best explanation."

—Professor David Galloway, MD, DSc, FRCS, FRCP, College of Medical, Veterinary & Life Sciences, University of Glasgow; former president, Royal College of Physicians and Surgeons of Glasgow

RETURN *of the* GOD HYPOTHESIS

Also by Stephen C. Meyer

Signature in the Cell

Darwin's Doubt

RETURN *of the* GOD HYPOTHESIS

THREE SCIENTIFIC DISCOVERIES
THAT REVEAL THE MIND
BEHIND THE UNIVERSE

STEPHEN C. MEYER

HarperOne
An Imprint of HarperCollins*Publishers*

To my mother, Patricia Meyer,
who first affirmed my philosophical interests;
to my father, Charles Meyer, who challenged
me to develop greater mathematical and
scientific proficiency; and to my wife, Elaine,
whose willingness to step out on faith has
made the whole adventure possible.

 For information, address HarperCollins Publishers, 195 Broadway, New York, NY 10007.

HarperCollins books may be purchased for educational, business, or sales promotional use. For information, please email the Special Markets Department at SPsales@harpercollins.com.

FIRST HARPERCOLLINS PAPERBACK PUBLISHED IN 2023

Designed by Terry McGrath
All illustrations © 2020 Ray Braun Design

Library of Congress Cataloging-in-Publication Data is available upon request.

ISBN 978-0-06-207151-4

23 24 25 26 27 LBC 7 6 5 4 3

Contents

Prologue

It was a public speaker's nightmare unfolding at a most inauspicious time. Eighteen minutes into my opening statement in a debate with physicist Lawrence Krauss, America's most prominent scientific atheist, I suddenly found I could no longer read my own PowerPoint slides. The brightly colored swirls, or "auras"—for me a telltale sign of the onset of a debilitating migraine—had begun to fill my visual field as I looked out through the blaze of lights behind the video cameras in a packed auditorium at the University of Toronto.

Intense light had often been a common migraine trigger for me, and it certainly was on that night in March 2016. As the auras spread, I began having trouble seeing not only the quotations and scientific diagrams on my slides, but Professor Krauss himself and the audience as well. Other neurological symptoms—numbness in my fingers and tongue, my voice echoing in my own head, and a difficulty finding words (aphasia)—followed predictably in rapid succession.

I was able to make it through the remaining seven minutes of my presentation by speaking more slowly and deliberately than I usually do and in some cases by using less technical words. But as I descended from the podium and was taken to a dark room, I felt both disoriented and disappointed. I realized it would now be difficult for me to say much in the ensuing roundtable (following a third speaker) about the main question of the forum, the one I specifically came to discuss.

The organizers of the forum had chosen the topic: "What's Behind It All? God, Science, and the Universe." Professor Krauss, then of Arizona State University, and I were a logical match to discuss this question from opposing points of view. Indeed, he and I had debated twice before, and I had often debated other scientific atheists during the preceding decade.

Krauss, who spoke first, had a reputation not only as an accomplished physicist, but also as a bold and outspoken controversialist—one with a talent for explaining scientific ideas to popular audiences. He is also well known for his provocative thesis that quantum physics can explain how the universe came into being from nothing. But that evening he didn't begin with a defense of that position. Instead, he began by declaring the topic of the forum unworthy of reflection and by characterizing me as unworthy of engagement. Indeed, he began the debate indulging in nearly ten minutes of what his boisterous supporters clearly regarded as deliciously personal invective, denouncing both me and, by extension, the organizers of the forum.

"If you appear on stage with someone talking about these ideas, it gives the impression that the ideas are worth debating or that the person is worth debating," Professor Krauss declared. "In this case, neither is true."[1]

When a rival in debate descends to *ad hominem* argument, I usually find myself surprised at his willingness to waste allotted time. Audiences typically find insults masquerading as arguments unpersuasive. Moreover, in a debate it usually takes little to defang such tactics beyond pointing them out. That night, however, Krauss's celebrity status had attracted hundreds of raucous supporters who laughed loudly at his punch lines, leaving me with the impression that an appeal to reason alone might not win the evening. As I began to speak, I pointed out that Krauss had provided little evidence to support his critique of my views, and still less in support of his own. Ordinarily, I might have also made light of his use of the *ad hominem* tactic, but on that night humor escaped me as my neurological distress grew progressively more acute while standing before a large audience in the auditorium and an estimated sixty thousand people watching online.

I had accepted the challenge of the debate in part to explain my own position about what science can tell us about the existence of God. This is, needless to say, an ultimate question and a subject of urgent concern for many thoughtful people. It is an important topic, as even many atheists would agree, and deserves a serious response. And although I sought to offer one that night, after the migraine set in I knew my ability to do so would be significantly limited—though, as it turned out, the cloud of my diminished condition would come with a silver lining.

For the debate I had planned, first, to explain my core argument for the intelligent design of life and then, in the ensuing discussion, to address a question I am often asked: "*Who is* the intelligent designer that you think is responsible for life?" I also meant to address a closely related question: "What does scientific evidence imply about the existence of God?"—or as the organizers of the forum put it: "What lies behind it all?"

Krauss answers that question with an emphatic "*Nothing*"—or at least nothing but the laws of physics. Though he denounces philosophy as a vacuous enterprise, he publicly advocates a philosophy that scholars call scientific materialism—an atheistic worldview affirmed by those who claim that science undermines belief in God.

Like other worldviews, scientific materialism attempts to answer some basic questions about ultimate reality—questions about human nature, morality and ethics, the basis of human knowledge, and even what happens to human beings at death. Most fundamentally, scientific materialism offers an answer to the question, "What is the entity or the process from which everything else came?"

Scientific materialists have traditionally answered that question by affirming that matter, energy, and/or the laws of physics are the entities from which everything else came and that those entities have existed from eternity past *as the uncreated foundation of all that exists*. Matter, energy, and physical laws are, therefore, viewed by materialists as self-existent.

Similarly, materialists hold that matter and energy organized themselves by various strictly naturalistic processes to produce all the complex forms of life we see today. This means scientific materialists also deny that a creator or designing intelligence played any role in the origin of the universe or life. Because materialists think that matter and energy are the foundational realities from which all else comes,[2] they deny the existence of immaterial entities such as God, free will, the human soul, and even the human mind conceived as an entity in some way distinct from the physiological processes at work in the brain.

Materialism is a venerable worldview with a long history going back to ancient Greece. It has had many prominent intellectual proponents, including Democritus, Thomas Hobbes, Charles Darwin, Ernst Haeckel, Bertrand Russell, and Francis Crick.

In recent years, powerful voices have popularized scientific materialism. Beginning about 2006 a group of scientists and philosophers known as the New Atheists ignited a worldwide publishing sensation. A series of bestselling books, led by Richard Dawkins's *The God Delusion*, argued that science properly understood undermines belief in God. Other books—by Victor Stenger, Sam Harris, Christopher Hitchens, Daniel Dennett, Stephen Hawking, and Krauss himself—followed suit.

In 2014 the Fox and National Geographic television networks aired a revamped version of a famous 1980 series with physicist Carl Sagan, *Cosmos: A Personal Voyage*. The new series, *Cosmos: A Spacetime Odyssey*, hosted by astrophysicist Neil deGrasse Tyson, began by replaying the audio of Sagan's memorable materialistic creed from the original series: "The cosmos is all that is, or ever was, or ever will be."[3]

The New Atheists and other science popularizers have explained the basis of their skepticism about the existence of God with admirable clarity. According to Dawkins and others, the evidence of design in living organisms long provided the best reason to believe in the existence of God, because it appealed to publicly accessible scientific evidence. But since Darwin, Dawkins insists, scientists have known that there is no evidence of *actual* design, only the illusion or "appearance" of design in life. According to Dawkins and many other neo-Darwinian biologists, the evolutionary mechanism of mutation and natural selection has the power to *mimic* a designing intelligence without itself being designed or guided in any way. And since random mutation and natural selection—what Dawkins calls the "blind watchmaker" mechanism—can explain away all "appearances" of design in life, it follows that belief in a designing intelligence at work in the history of life is completely unnecessary.[4]

Although Dawkins allows that it is still possible that a deity *might* exist, he insists there is absolutely no evidence for the existence of such a being, rendering belief in God effectively "delusional." Popular TV figure Bill Nye, the "Science Guy," has echoed this perspective. In his book *Undeniable: Evolution and the Science of Creation*, he says, "Perhaps there is intelligence in charge of the universe, but Darwin's theory shows no sign of it, and has no need of it."[5] Consequently, as Dawkins concluded in an earlier work: "Darwin made it possible to be an intellectually fulfilled atheist."[6]

Another New Atheist, philosopher Daniel Dennett, gives an evolutionary account of the origin of religious belief in his book *Breaking the Spell*, one that ultimately attributes belief in God to a cognitive impulse programmed into us by the evolutionary process rather than a rational or evidentially based system of belief. Thus, for those who know this,

Darwinism functions as a "universal acid" eating away at any basis for religious belief and traditional religious-based morality.[7]

Other New Atheists, including Lawrence Krauss (see Fig. 1.1b), say that physics renders belief in God unnecessary. Krauss contends that the laws of quantum physics explain how the universe came into existence from literally nothing. Consequently, he argues, it is completely unnecessary, even irrational, to invoke a creator to explain the origin of the universe.[8]

Stephen Hawking, formerly of the University of Cambridge and until his death in 2018 the world's best-known scientist, made a similar argument. In his book *The Grand Design*, coauthored with Leonard Mlodinow, he argues that "because there is a law such as gravity, the universe can and will create itself from nothing. Spontaneous creation is the reason there is something rather than nothing, why the universe exists, why we exist." Thus, for Hawking, "it is not necessary to invoke God to light the blue touch paper and set the Universe going."[9] The late Victor Stenger made similar arguments in his poignantly titled book *God: The Failed Hypothesis*.

All this high-profile science-based skepticism about God has percolated into the popular consciousness. Recent polling data indicate that in North America and Europe, the perceived message of science has played an outsized role in the loss of belief in God. In one poll, more than two-thirds of self-described atheists and one-third of self-described agnostics affirm that "the findings of science make the existence of God less probable." According to the same survey, the two most influential scientific ideas that have affected people's loss of faith are unguided chemical evolution (of the origin of life) and unguided biological evolution (of the development of life). According to these surveys, these two ideas have led more people to reject faith in God than has suffering from disease or death.[10]

Other polls have shown a dramatic rise in the group pollsters call "the nones"—religiously unaffiliated, agnostic, or atheistic respondents—among college and postcollege young people in the eighteen to thirty-three age range.[11] The rapid growth of this group occurred precisely during the recent decade in which the New Atheists have gained prominence. Indeed, there are many indications—from personal interviews, public opinion polls, and website testimonials—that college students in particular have been deeply influenced by the message of the New Atheists; many of these students now cite arguments similar to those made by Dawkins, Krauss, Dennett, and Hitchens as their main reasons for rejecting faith in God.

These developments have a particular poignancy and interest for me for two reasons—both of which help to explain why I agreed to debate Krauss in 2016 and why I've chosen to write this book. First, I have long been interested in the question of biological origins. Over the last decade I have written two books arguing that living systems exhibit evidence of intelligent design. Whereas Richard Dawkins contends that living systems merely "*give the appearance* of having been designed for a purpose,"[12] I have argued that certain features of living systems—in particular, the digitally encoded information present in DNA and the complex circuitry and information-processing systems at work in living cells—are best explained by the activity of an *actual* designing intelligence. Just as the inscriptions on the Rosetta Stone point to the activity of an ancient scribe and the software in a computer program points to a programmer, I've argued that the digital code discovered within the DNA molecule suggests the activity of a designing mind in the origin of life.

Nevertheless, in making my case for intelligent design, I have been careful not to claim more than the biological evidence alone can justify. In my previous books, I did not attempt to identify the designing intelligence responsible for the origin of the information present in living organisms or to prove the existence of God. After all, though I don't hold this view, it is at least logically possible that a preexisting intelligent agent somewhere else *within* the cosmos (i.e., not God) might have designed life and "seeded" it here on earth, as scientists who advocate a view known as "panspermia" have suggested.

Instead, I have simply argued that the information present in DNA suggests the prior creative activity of an intelligent agent *of some kind*, as opposed to an exclusively blind or undirected natural process such as random mutation and natural selection. Despite this limited claim, my explanation for the appearance of design still places me at odds with the New Atheists. Even so, though they and I have adopted diametrically opposed explanations for the appearance of design, we have focused our explanatory efforts on the exact same phenomenon of interest.

And that leads to the second, and perhaps more important, reason for my interest in what I call the God hypothesis. The New Atheists pose the question of what the evidence from the natural world as a whole shows about the existence (or nonexistence) of God. My readers evidently share an interest in that question. Many, upon encountering my argument for the intelligent design of life, have written asking a series of questions of roughly the following form: "If there is scientific evidence of the activity of a designing intelligence, then what kind of a designing mind are we talking about? An intelligent agent within the cosmos or beyond? An immanent or a transcendent intelligence? A space alien? Or God?"

Since my previous two books led inevitably to such questions, it has increasingly seemed a natural next step for me to explore what science can tell us about them and about the possible existence of God.

Slow to Speak

The debate in Toronto and its aftermath sealed my decision to address this subject in a book-length treatment. In the debate, I was able to explain my basic case for intelligent design in biology. Nevertheless, my migraine-addled state made it difficult for me to say much about the larger question of what science could tell us about God, as I had hoped to do in the ensuing discussion.

Nevertheless, one advantage of not being able to speak well, or only being able to speak slowly and deliberately, is that it forces you to say the most important things and to do so succinctly. I have a friend with Tourette syndrome who stutters and sometimes finds it difficult to work his way into fast-moving conversations. As a result he often blurts out incredibly pithy insights that distill the essence of a topic in a few words, sometimes to the amazement of friends. Something similar happened for me that night.

During the last five or ten minutes of the debate, as my symptoms started to dissipate, but only just, the moderator asked us to summarize our perspective on what science could tell us about "what lies behind it all." I found myself briefly describing three key scientific discoveries that I thought jointly supported theistic belief—what I call "the return of the God hypothesis": (1) evidence from cosmology suggesting that the material universe had a beginning; (2) evidence from physics showing that *from the beginning* the universe has been "finely tuned" to allow for the possibility of life; and (3) evidence from biology establishing that *since the beginning* large amounts of new functional genetic information have arisen in our biosphere to make new forms of life possible—implying, as I had argued before, the activity of a designing intelligence.

After the debate I received sympathetic mail from many people who felt badly about my having to battle a migraine at such a public event. But many who wrote also told me that the one thing they remembered about the substance of the debate was my closing statement and the succinct description of the three scientific discoveries that together point not just to a designer, but to an intelligence with attributes that religious theists have long ascribed to God. I realized later that I had, perhaps without planning to do so, distilled in a few words a way of structuring a persuasive and accessible science-based argument for the God hypothesis. Perhaps, I thought, it was time to develop this case.

An Unexpected Discovery

Another unexpected benefit of participating in the debate occurred completely out of view of the audience. As I prepared for the night in the two weeks leading up to it, I studied Krauss's proposed explanation for the origin of the universe. I also pored over a key technical paper and book written by a Russian physicist, Alexander Vilenkin, whose ideas Krauss had popularized in his book *A Universe from Nothing*. I was stunned by what I found. Krauss used the work of Vilenkin in effect to refute what is called the cosmological, or "first-cause," argument for the existence of God—an argument that posits God as the cause of the beginning of the material universe. As I reflected on what Vilenkin wrote, however, I concluded that Krauss completely missed the real import of Vilenkin's work, which arguably implied the *need* for a preexisting mind (see Chapters 17–19 for more detail).

Over the preceding few years I had noticed a similar pattern in the writings of other scientific materialists as they responded to arguments for intelligent design in both physics and biology. As I show in later chapters of this book, the allegedly strongest counterarguments against the theory of intelligent design often inadvertently seemed to strengthen, rather than weaken, the case for design. For example, attempts to explain the origin of what's called the fine tuning of the universe by invoking a "multiverse" inevitably required invoking prior unexplained fine tuning. Attempts to explain the origin of the information necessary to produce new forms of life invariably either required prior unexplained information or involved simulations that required the intelligent guidance of a programmer, biochemist, or engineer as a condition of their success. Thus, common responses to the argument for intelligent design in physics and biology typically begged the question as to the origin of prior indicators of design and, consequently, strengthened those arguments.

I now discovered that a similar problem attended claims to have explained the origin of the universe "from nothing." Properly interpreted, the physics used this way only seemed to reinforce the conclusion of the cosmological argument.

So my difficult evening in Toronto had another unexpected benefit. Going into it, I knew the typical and strongest counterarguments to each of the three interrelated arguments that I had long wanted to make in support of the God hypothesis. I already knew that two of those counterarguments inadvertently reinforced my case. Now I came to suspect from my debate preparation and my interaction with Krauss that the main counterargument to the third line of evidence I intended to marshal—evidence from cosmology—did the same thing.

I realized it was time to write this book.

Part I

The Rise and Fall of Theistic Science

1

The Judeo-Christian Origins of Modern Science

I live and work in Seattle, where, a few years ago, a prominent professor of evolutionary psychology, David Barash of the University of Washington, authored a startling *New York Times* op-ed. He told of "the talk" he gives each year to his students flatly informing them that science has rendered belief in God implausible. Or as he explained, "As evolutionary science has progressed, the available space for religious belief has narrowed: It has demolished two previously potent pillars of religious faith and undermined belief in an omnipotent and omnibenevolent God."[1]

Barash follows in a long tradition. Since the late nineteenth century powerful voices in Western culture—philosophers, scientists, historians, artists, songwriters, and science popularizers—have attested to the "death of God." By this they of course do not mean that God once existed and has now passed away, but instead that any credible *basis* for belief in such a being has long since evaporated.

Those who tout the loss of a rational foundation for belief in God often cite the advance of modern science and the picture of reality it paints as the chief reason for this demise. The idea that science has buried God is pervasive in the media, in educational settings, and in our culture broadly. For example, Richard Dawkins (Fig. 1.1a) has claimed that the scientific picture of the universe—and particularly evolutionary accounts of the origin and development of life on earth—supports an atheistic or materialistic worldview. As he put it, "The universe we observe

FIGURE 1.1A

FIGURE 1.1B

The prominent New Atheists, evolutionary biologist Richard Dawkins and physicist Lawrence Krauss.

has precisely the properties we should expect if there is, at bottom, no design, no purpose, no evil, no good, nothing but blind, pitiless indifference."[2]

This book will show that reports of God's decease have "been grossly exaggerated," to appropriate a quote from Mark Twain.[3] Instead, the truth is just the opposite of what Dawkins, Barash, and numerous other popular spokespersons for science have insisted. The properties of the universe and of life—specifically as they pertain to understanding their origins—are just "what we should expect" if a transcendent and purposive intelligence has acted in the history of life and the cosmos. Such an intelligence coincides with what human beings have called God, and so I call this story of reversal the return of the God hypothesis.

Three Big Questions

My own interest in what scientific discoveries show about the possible existence of God germinated over thirty years ago when I attended an unusual conference. At the time, I was working as a geophysicist doing seismic digital signal processing for an oil company in Dallas, Texas. In February 1985, I learned of a Harvard historian of science and astrophysicist, Owen Gingerich, who was coming to town to talk about the unexpected convergence between modern cosmology and the biblical account of creation as well as the theistic implications of the big bang theory. I attended the talk on a Friday evening and found that Gingerich had come to Dallas mainly to speak to a much larger conference the next day featuring leading theistic and atheistic scientists. They would be discussing three big questions at the intersection of science and philosophy: the origin of the universe, the origin of life, and the origin and nature of human consciousness.

Fascinated, I attended the Saturday conference at the Dallas Hilton. The organizers had assembled a world-class lineup of scientists and philosophers representing two great but divergent systems of thought. I was not surprised to hear outspoken atheists or scientific materialists explaining why they doubted the existence of God. What shocked me was the persuasive talks by other leading scientists who thought that recent discoveries in their own fields had decidedly *theistic* implications.

On the first panel, not only Professor Gingerich, but also the famed astronomer Allan Sandage, of Caltech, explained how advances in astronomy and cosmology established that the material universe had a definite beginning in time and space, suggesting a cause beyond the physical or material universe. Gingerich and Sandage also discussed discoveries in physics showing how the universe had been finely tuned from the beginning of time—in its physical parameters and initial arrangements of matter—to allow for the existence of complex life. This suggested to them some prior intelligence responsible for the "fine tuning."

FIGURE 1.2
Former Caltech astronomer Allan Sandage.

Neither wanted to claim that these discoveries "proved" the existence of God. They cautioned that science cannot "prove" anything with absolute certainty. Both argued, however, that the discoveries seemed to fit much better with a theistic perspective than a materialistic one. Professor Sandage (Fig. 1.2) caused a stir at the conference just by sitting down on the theistic side of the panel. It turns out that he had been a lifelong agnostic and scientific materialist and had only recently embraced faith in God. And he had done so in part *because* of scientific evidence, not in spite of it.

The panel on the origin of the first life featured another similarly dramatic revelation. One of the leading origin-of-life researchers in attendance, biophysicist Dean Kenyon (Fig. 1.3), announced that he had repudiated his own cutting-edge evolutionary theory of life's origin. Kenyon's theory—developed in a bestselling advanced text-

FIGURE 1.3
Dean Kenyon, biophysicist and origin-of-life researcher.

book titled *Biochemical Predestination*—articulated what was then arguably the most plausible evolutionary account of how a living cell might have "self-organized" from simpler chemicals in a "prebiotic soup."

But as Kenyon explained at the conference, he had come to doubt his own theory. Origin-of-life simulation experiments increasingly suggested that simple chemicals do not arrange themselves into complex information-bearing molecules, nor do they move in life-relevant directions—unless, that is, biochemists actively and intelligently guide the process. But if undirected chemical processes cannot account for the encoded information found in even the simplest cells, might a directing intelligence have played a role in the origin of life? Kenyon announced that he now held that view.

After the conference, I met one of Kenyon's colleagues on the origin-of-life panel, a chemist named Charles Thaxton. Thaxton, like Kenyon, thought that the information present in DNA pointed to the past activity of a designing intelligence—to an "intelligent cause," as he put it. As I talked more with him over the ensuing days and months, I became more intrigued with the question of the origin of life and whether a scientific case could be made for intelligent design based on the discovery of the digitally encoded information in DNA.

I decided to focus my own energies on assessing that possibility, eventually completing my PhD thesis at the University of Cambridge on the subject of origin-of-life biology. Much later, in 2009, I published *Signature in the Cell*. In that book, I made a case for intelligent design based upon the information stored in DNA, though, again, without attempting to identify the designing intelligence responsible for life. Even so, through those years I remained intrigued by the possibility that the evidence from cosmology and physics taken together with that of biology might provide the basis for a persuasive reformulation of a God hypothesis.

To say that the God hypothesis has returned implies that scientists must have previously rejected it and that, at some still earlier time, a theistic perspective reigned either as an inspiration for doing science, an explanation for specific scientific discoveries, or both. Yet few science popularizers today present the history of science and its relationship to religious belief this way. Instead, they not only assert that science and theistic belief currently conflict, but they also say that science and religion have nearly always been at war.[4] They describe the historical relationship between science and religion as one characterized by conflicting claims about reality and competing ways of knowing.[5]

This chapter challenges the New Atheist–favored narrative about the historical relationship between science and theistic belief. It does so by showing how Judeo-Christian ideas contributed crucially to the rise of modern science.

The History of Science (According to the New Atheists)

The standard story, advanced by New Atheists and more mainstream figures alike, asserts that science and religious belief have generally stood in direct opposition. Consider, for example, the revised, thirteen-part *Cosmos* series that aired in 2014. In the series, Neil deGrasse Tyson, a scientific materialist who dislikes the label "atheist," attributes a loss of belief in God during the seventeenth century to the triumph of Newtonian physics. In the third episode, Tyson gives a detailed account of the collaboration between astronomer Edmond Halley and Isaac Newton.[6] He recounts how this collaboration led to the publication of Newton's masterpiece the *Principia*, in which Newton developed his mathematically precise theory of gravity. Tyson claims that the applicability of Newton's theory of gravity to the motions of the planetary bodies undermined the "need for a master clockmaker to explain the precision and beauty of the solar system."[7]

Though Tyson acknowledged that Isaac Newton personally believed in God, calling him a "God-loving man," he assured his viewers that Newton's religious beliefs did nothing to advance his scientific endeavors. Instead, he insisted that Newton's religious study "never led anywhere" and that Newton's appeal to God represented "the closing of a door. It didn't lead to other questions."[8] Thus, according to Tyson, Newton's science liberated people from belief in God, even as his belief in God impeded his own scientific progress. Tyson's message was clear: to do good science, scientists must throw off the shackles of religion, and the advance of science has allowed people in Western culture to do just that.

The Warfare or Conflict Model

This perception of a perpetual and unavoidable conflict between science and faith arose relatively recently. During my first year of study at Cambridge, Professor Colin Russell, then the president of the British Society

FIGURE 1.4A

FIGURE 1.4B

John William Draper and Andrew Dickson White, the revisionist nineteenth-century historians who portrayed science and Christianity as at war with one another.

for the History of Science and a distinguished professor in the history of science at another British university (the Open University), gave a well-attended talk about the so-called warfare thesis. Professor Russell explained that the perception of a deep or inherent conflict between science and faith is a product of late nineteenth-century historical revisionism. Two such works helped give rise to this understanding. They are John William Draper's (Fig. 1.4a) *History of the Conflict Between Religion and Science* (1874) and Andrew Dickson White's (Fig. 1.4b) *A History of the Warfare of Science with Theology in Christendom* (1896).[9] These books appeared in the immediate aftermath of the publication of Darwin's *On the Origin of Species* in 1859. They "fostered the impression that religious critics of Darwinism threatened to rekindle the Inquisition,"[10] as historian Edward Larson has put it. Draper regarded organized religion as a direct and existential threat to the advancement of science. As he argued, ". . . Christianity and Science are recognized by their respective adherents as being absolutely incompatible; they cannot exist together; one must yield to the other; mankind must make its choice—it cannot have both."[11]

Another historian of science, Jeffrey B. Russell, observes that White's book was "of immense importance, because it . . . explicitly declared that science and religion were at war. It fixed in the educated mind the idea that 'science' stood for freedom and progress against the superstition and repression of 'religion.'"[12] In his Pulitzer Prize–winning history of the Scopes trial, *Summer for the Gods*, Edward Larson notes that in the decades following the publication of the *Origin of Species*, this "warfare model" of science and religion became "ingrained into the received wisdom of many secular Americans."[13]

In his talk at Cambridge and in his later writing, Colin Russell lamented how this model of the relationship between science and religion became "deeply embedded in the culture of the West" and "has proven extremely hard to dislodge."[14] He has noted that the "Draper-White thesis has been routinely employed in popular-science writing, by the media, and in a few older histories of science."[15] Indeed, though the New Atheists claim to advance a "new" perspective, their view of the relationship between science and faith merely echoes this late nineteenth-century historiography.

A Different Understanding

In truth, a chorus of twentieth- and twenty-first-century historians, philosophers, and sociologists of science tell a significantly different story. These scholars include Herbert Butterfield,[16] A. C. Crombie,[17] Michael B. Foster,[18] Loren Eiseley,[19] David Lindberg,[20] Owen Gingerich,[21] Reijer Hooykaas,[22] Robert Merton,[23] Pierre Duhem,[24] Colin Russell,[25] Alfred North Whitehead,[26] Peter Hodgson,[27] Ian Barbour,[28] Christopher Kaiser,[29] Holmes Rolston III,[30] Steve Fuller,[31] Peter Harrison[32] and Rodney Stark,[33] to name a handful. Although some scientific theories during the nineteenth century, particularly those concerning biological origins and geological history, did seem to challenge some traditional theistic ideas, these historians note that belief in a God—and Christianity specifically—played a decisive role in the rise of modern science during the sixteenth and seventeenth centuries.

Certainly, when Draper and White were writing, many scientists perceived a conflict between science and religious belief, but this was not always the case. Instead, almost all historians of science today offer a more nuanced view: they maintain that although some scientists or scientific theories challenged belief in God in some periods of scientific history, in others religious belief actually inspired scientific advance. In addition, many historians of science have shown that belief in God served both as an inspiration for doing science and as a framework for explaining scientific observations during the crucial period known as the scientific revolution (roughly between 1500 and 1750), in which modern science as a systematic endeavor first originated.

The Rise of Theistic Science

When I was a college professor, I used to team-teach a large core Western Civilization class on the history of science and technology. When I lectured about the scientific revolution, my students were always surprised to learn that so many historians and philosophers of science now identify Judeo-Christian ideas as crucial to the origin of modern science.

I first encountered this thesis myself during my first year of graduate school. That the thesis seemed so uncontentious among my Cambridge tutors also surprised me, given the atheistic or agnostic proclivities of most of the faculty in the department of the history and philosophy of science where I was a student.

One incident particularly confirmed my perception of the religious orientation of most of the faculty. After seminars, the faculty would often retire to a local pub for further discussion with students. On one occasion, a fellow graduate student from the United States, evidently feeling as insecure in these new surroundings as I was, decided to reveal—proudly—that he had rejected his former religious beliefs and now considered himself an atheist. This attempt to curry favor with the faculty members met with an unexpectedly dismissive response. One of the younger, hipper British lecturers in our department replied by saying, "Well, of course you're an atheist, but what else are you that makes you interesting?" I resolved at that point not to make a show of my own metaphysical opinions, as I doubted that they would have been even as well received as those of my classmate.

Despite such pervasive leanings, many of the historians, sociologists, and philosophers of science I encountered (and read) recognized the important role that belief in God had played in the scientific revolution. They did so in part because it has helped answer what might be called the "Why there? Why then?" question.

The X Factor: A Transposition in Thinking

The Cambridge University chemist and historian of science Joseph Needham first posed the "Why there? Why then?" question in his research on the scientific revolution. Needham observed that the material necessities for conducting science existed in many well-developed cultures. The Egyptians erected great pyramids, palaces, and funerary

FIGURE 1.5
Peter Hodgson, Oxford physicist and historian of science.

monuments. The Chinese invented the compass, block printing, and gunpowder. The Romans built great roads and aqueducts. And the Greeks had great philosophers, some of whom studied nature extensively. Yet none of these cultures developed the systematic methods for investigating nature that arose in western Europe between about 1500 and 1750.

The late Oxford physicist and historian of science Peter E. Hodgson (Fig. 1.5) made a similar observation. Many civilizations have had sophisticated material cultures, or what Hodgson calls "the material requirements for the growth of science."[34] As Hodgson explained:

> If we think about what is needed for the viable birth of science, we see first of all that it needs a fairly well-developed society, so that some of its members can spend most of their time just thinking about the world, without the constant preoccupation of finding the next meal. It needs some simple technology, so that the apparatus required for experiments can be constructed. There must also be a system of writing, so that the results can be recorded and sent to other scientists, and a mathematical notation for the numerical results of measurements. These may be called the material necessities of science.[35]

Since many cultures had these necessities, Needham and Hodgson wondered why modern science arose so dramatically in Europe during a fairly narrow window of time and why these other cultures did not develop anything like Western science with its formal methods for the study of nature. Why did human beings begin to unlock nature's secrets in such a revolutionary and systematic way in western Europe during the sixteenth and seventeenth centuries?

Needham, Hodgson, and many other historians, such as Herbert Butterfield and Ian Barbour, identify the missing "X factor" in the realm of ideas. They point to Judeo-Christian ideas prevalent in Europe before the sixteenth and seventeenth centuries. Barbour argues that "science in its modern form" arose "in Western civilization alone, among all the cultures of the world," because only the Christian West had the necessary "intellectual presuppositions underlying the rise of science."[36] As Butterfield, former professor of modern history at the University of Cambridge, explained, "It was supremely difficult to escape from the [previous] Aristotelian doctrine by merely observing things more closely. . . . It required a different kind of thinking cap, a transposition in the mind of the scientist himself."[37]

A Break with Greek Thought

These historians found that the reaffirmation of the Judeo-Christian doctrine of creation during the Catholic late Middle Ages and Protestant Reformation provided the needed transposition. That was the change that led to the break with the Greek thinking that had previously limited the advance of science.

The need for such a break puzzled me when I first learned of it. The Greeks are heroes of the Western intellectual tradition. Their "rationalist credo"—"The world is orderly, knowable and best known by human reason," as physicist Lois Kieffaber has formulated it[38]—made them the great paragons of reason in the ancient world.

Nevertheless, many Greek assumptions tended to impede the development of a rigorously empirical and observational approach to nature.[39] Though Greek, Jewish, and Christian philosophers all agreed about the rationality of nature, some Greek ideas induced a sterile armchair philosophizing unconstrained by actual observations.

Although the Greek philosophers thought that nature reflected an underlying order, they nevertheless believed that this order issued from an intrinsic self-existent logical principle called the *logos*, rather than from a mind or divine being with a will.[40] For this reason, many Greek thinkers assumed that they could deduce how nature ought to behave from first principles based upon only superficial observations of natural phenomena or without actually observing nature at all. In astronomy, for example, Aristotle (fourth century BCE) and Ptolemy (second century CE) both assumed that planets must move in circular orbits. Why? Because according to Greek cosmology, the planets moved in the "quintessential" realm of the crystalline spheres, a heavenly realm in which only perfection was possible. Since, they deduced, the most perfect form of motion was circular, the planets must move in circular orbits.[41] What could be more logical? As historian of science Reijer Hooykaas explained, when medieval Aristotelians said "things happened according to nature, this meant that they followed a pattern that seemed rational to the human mind, one which had been discovered by Aristotle."[42]

This overestimation of pure reason and the reliance upon logical necessity manifested itself in other ways. After the rediscovery of Aristotle's works in the West in the eleventh century,[43] Christian theologians were eager to synthesize their theological beliefs with the best of classical learning. They often adopted Greek assumptions about what nature must look like. Invoking considerations of logical necessity—and often Aristotle's authority—some medieval theologians and philosophers asserted that the universe must be eternal;[44] that God could not create new species;[45] that God could not have made more than one planetary system;[46] that God could not make an empty space;[47] that God could not give planets noncircular orbits; and many other such propositions.[48]

The Contingency of Nature

For science to advance, natural philosophers, or scientists, as we refer to them today, needed to develop a more empirical, evidence-based approach. This began to occur well before the scientific revolution because of a shift in thinking about the source of the order in the physical world. In 1277, Etienne Tempier, the bishop of Paris, writing with the support of Pope John XXI, condemned "necessarian theology" and 219 separate theses influenced by Greek philosophy about what God could or couldn't do.[49] Before the decree of 1277, Christian theologians and philosophers, particularly at the influential University of Paris, often assumed that nature must conform to seemingly obvious logical principles as exemplified by Aristotle's cosmological, physical, or biological theories.

The Judeo-Christian—indeed, biblical—doctrine of creation helped liberate Western science from such necessitarian thinking by asserting the contingency of nature upon the will of a rational God. Like the Greek philosophers, the early modern scientists thought that nature exhibited an underlying order. Nevertheless, they thought this natural order had been impressed on nature by a designing mind with a will—the mind and will of the Judeo-Christian God. For this reason, they thought that the order in nature was the product not of logical necessity, but of rational deliberation and choice, what the Scottish theologian Thomas Torrance calls "contingent rationality."[50] By reaffirming the doctrine of creation and the sovereignty and freedom of God to create as God saw fit, Tempier's decree in 1277 emphasized this principle.

This transposition in thinking led to a different approach to the study of nature in the centuries following Tempier's decree. Just as there are many ways to paint a picture or design a clock or organize the books in a library, there are many ways to design and organize a universe. Because it had been chosen by a rational mind, the order in nature could have been otherwise. Thus, the natural philosophers could not merely deduce the order of nature from logical first principles; they needed to observe nature carefully and systematically. As Robert Boyle, one of the most important figures of the scientific revolution and the founder of modern chemistry, explained, the job of the natural philosopher was not to ask what God must have done, but what God actually did.[51] Boyle argued that God's freedom required an empirical and observational approach, not just a deductive one.[52] Scientists needed to look, and to find out. As historian of science Ian Barbour explains, "The doctrine of creation implies that *the details of nature can be known only by observing them.*"[53]

The Intelligibility of Nature

Moreover, since nature had been designed by the same rational mind who had designed the human mind, the early modern scientists (or, again, "natural philosophers")[54] who began to investigate nature also assumed that nature was *intelligible*. It could be understood by the human intellect. The founders of modern science assumed that if they studied nature carefully, it would reveal its secrets. Their confidence in this assumption was grounded in both the Greek and the Judeo-Christian idea that the universe is an orderly system—a cosmos, not a chaos. As the British philosopher Alfred North Whitehead argued, "There can be no living science unless there is a widespread instinctive conviction in the existence of an *Order of Things*. And, in particular, of an *Order of Nature*."[55] Whitehead particularly attributed this conviction among the founders of modern science to the "medieval insistence upon the rationality of God."[56]

Other scholars have amplified this observation. They insist that modern science was specifically inspired by the conviction that the universe is the product of a rational mind who designed the universe to be understood and who also designed the human mind to understand it. As historian and philosopher of science Steve Fuller notes, Western science

FIGURE 1.6
The seventeenth-century astronomer Johannes Kepler, whose faith gave him confidence in the intelligibility of the universe.

is grounded in the "belief that the natural order is the product of a single intelligence from which our own intelligence descends."[57] Philosopher Holmes Rolston III puts the point this way: "It was monotheism that launched the coming of physical science, for it premised an intelligible world, sacred but disenchanted, a world with a blueprint, which was therefore open to the searches of the scientists. The great pioneers in physics—Newton, Galileo, Kepler, Copernicus—devoutly believed themselves called to find evidences of God in the physical world."[58] The astronomer Johannes Kepler (1571–1630) (Fig. 1.6), for example, exclaimed that "God wanted us to recognize" natural laws and that God made this possible "by creating us after his own image so that we could share in his own thoughts."[59]

Thus, the assumption that a rational mind with a will had created the universe gave rise to two ideas—contingency and intelligibility—which, in turn, provided a powerful impetus to study nature with confidence that such study would yield understanding.[60]

The Fallibility of Human Reasoning

Another biblical idea influenced the development of an observational and experimental approach. Even as scientists during the sixteenth and seventeenth centuries saw human reason as a gift of a rational God, these same scientists, many influenced by the Protestant Reformation and ideas from the church fathers recovered during the late Middle Ages, also recognized the fallibility of humans and, therefore, the fallibility of human ideas about nature. As Steve Fuller explains:

> Research in the history and philosophy of science suggests two biblical ideas as having been crucial to the rise of science, both of which can be attributed to the reading of Genesis provided by Augustine, an early church father, whose work became increasingly studied in the late Middle Ages and especially the Reformation. Augustine captured the two ideas in two

> Latin coinages, which *prima facie* cut against each other: *imago dei* and *peccatum originis*. The former says that humans are unique as a species in our having been created in the image and likeness of God, while the latter says that all humans are born having inherited the legacy of Adam's error, "original sin."[61]

Fuller goes on to argue that "once Christians began to read the Bible for themselves," as they did with the availability of printed books from the fifteenth century on, "they too picked out those ideas as salient in how they defined their relationship to God." This biblical understanding of human nature "extended to how they did science."[62]

Such a nuanced view of human nature implied, on the one hand, that human beings could attain insight into the workings of the natural world, but that, on the other, they were vulnerable to self-deception, flights of fancy, and prematurely jumping to conclusions. This composite view of reason—one that affirmed both its capability and fallibility—inspired confidence that the design and order of nature *could* be understood if scientists carefully studied the natural world, but also engendered caution about trusting human intuition, conjectures, and hypotheses unless they were carefully tested by experiment and observation.[63] Moreover, as the Australian historian of science Peter Harrison has noted, a renewed emphasis during the Protestant Reformation on the doctrine of the fall of humankind as well as the fallen state of nature meant that scientists should not take their initial observations of nature at face value. Instead they must "interrogate" nature using systematic experimental methods.[64]

This view of human reason has influenced science down to the present day. The methods for checking hypotheses developed by such figures as Francis Bacon during the seventeenth century and William Whewell during the nineteenth assumed both the capability and the fallibility of human reason. As Fuller notes, "This sensibility carried into the modern secular age, as perhaps best illustrated in our own day by Karl Popper's . . . method of 'conjectures and refutations,' the stronger the better in both cases. We should aspire to understand all of nature by proposing bold hypotheses (something of which we are capable because of the *imago dei*) but to expect and admit error (something to which we are inclined because of the *peccatum originis*) whenever we fall short in light of the evidence."[65]

In fact, one finds evidence of this concern to check human insight and intuition against empirical evidence going back farther, into the late

Middle Ages. As early as the thirteenth century, the Oxford theologian and philosopher Robert Grosseteste, along with his famed student Roger Bacon,[66] developed scientific methods of inference such as the method of "Resolution and Composition" and methods of testing causal hypotheses such as "Verification and Falsification." The latter closely resembled the modern scientific method of "isolation of variables."[67] These methods reflected both an early interest in the systematic study of natural phenomena and a recognition of the limits of human reason unaided by observation of the world. Scientists today use both these methods of hypothesis testing. Accordingly, the distinguished Oxford University historian of science Alistair C. Crombie calls Grosseteste "the real founder of the tradition of scientific thought in medieval Oxford, and in some ways, of the modern English intellectual tradition."[68] Grosseteste was also one of those theologians who helped to recover and reemphasize the doctrine of creation during the late medieval period. He lectured on numerous biblical books and especially on the book of Genesis.[69]

Ockham's Razor

Another thirteenth-century theologian who contributed to the development of scientific method was William of Ockham (Fig. 1.7). He is best known today for his famed "razor"—the methodological principle that encourages scientists to avoid multiplying unnecessary explanatory entities and, in that sense, to favor simpler hypotheses. William of Ockham also emphasized the contingency of creation and its dependence on the will of God, its creator.[70] Ockham's razor and his famed dictum—"Never posit pluralities [many explanatory entities] without necessity"[71]—helped to liberate science from appealing to what scholastic philosophers called Aristotelian "substantial forms."

FIGURE 1.7
William of Ockham, the thirteenth-century theologian who cautioned against multiplying theoretical entities and emphasized the "contingency" of creation.

According to Aristotle, all objects exemplify four different types of causes: material, formal, efficient, and final.[72] The material cause of an object is the material substance or stuff out of which the object is made—in the case of a chair, the wood. The formal cause represents either the form the entity exemplifies, for example, the shape of the chair, or the idea that produced that form, the blueprint in the mind of the designer of the chair. The efficient cause constitutes the means by which the object was made, the manufacturing process that produced the chair. And the final cause represents the purpose for which the entity was made—in the case of the chair, providing a place to sit.

Though this schema applied nicely to analyzing individual objects made by human agents, it did not apply so readily to understanding regularities or cause-and-effect patterns observed in nature. Instead, attempts to describe such repeatable phenomena by reference to formal and final causes often resulted in laughably vacuous or circular kinds of explanations, especially in physics and chemistry. Thus, when scholastics observed that bread always nourishes, or that a certain medicine ameliorates a condition, or that opium repeatedly puts people to sleep, they would invoke an abstract "formal cause" or "virtue" to explain the phenomenon. This practice often devolved into merely attributing causal powers to the name of the effect in question. For example, medieval scholastics would explain the ability of bread to nourish by citing its "nutritive virtue," the ability of a medicine to relieve discomfort by citing its "curative virtue," or the ability of opium to induce sleep by reference to its "dormitive virtue."[73]

Ockham regarded the existence of such "virtues" or formal causes (as well as universals such as Platonic forms) with skepticism. He suggested that they represented merely names or concepts in the human mind, not actual entities in reality.[74] He employed his razor to eliminate appeals to such explanatory entities, opening the door to simpler, or at least more illuminating, explanations.

Ockham's skepticism about the existence of scholastic substantial forms also reflected his belief in the fallibility of human reasoning and the vanity of human imagination.[75] Both, he thought, needed to be checked by observation and by the application of sound methodological principles such as his principle of parsimony (i.e., his "razor"). This thinking sprang from deep-seated theological convictions. Ockham's principle of parsimony came from his conviction in the underlying God-given order and elegance of natural phenomena. His emphasis on the need to check hypotheses against experience reflected his "theological voluntarism," the understanding that the natural world owes its orderly concourse to the *free* choice of an intelligent creator who could have made nature otherwise. Thus, in his own formulation of the principle of parsimony he expressed both the importance of experience (as well as reason) and his foundational theological convictions. As he put it: "For nothing ought to be posited without a reason given, unless it is self-evident or known by experience or proved by the authority of Sacred Scripture."[76]

Assessing the Warfare Model

I started my postgraduate work in the history and philosophy of science well before the New Atheists came on the scene. In 1986, my first year at Cambridge, Richard Dawkins had only just published *The Blind Watchmaker*, his popular treatise explaining how contemporary neo-Darwinian theory can explain the "appearance of design" in biological systems without invoking an actual designing intelligence. Later, in books such as *River Out of Eden* (1995) and *The God Delusion* (2006), Dawkins built on this argument to suggest that the absence of evidence for a designing intelligence in the history of life rendered the God hypothesis unnecessary (and even delusional).

These books argued that atheism (or scientific materialism) provided a better overall metaphysical explanation for the picture of the world revealed by science for two reasons. First, Dawkins claimed, as I noted earlier, that the universe "has precisely the properties we should expect if there is, at bottom, no design, no purpose," only "blind, pitiless indifference."[77] Second, since, in his view, neither life nor the universe revealed any evidence of actual (as opposed to apparent) design, atheism or materialism provided a more "parsimonious" explanation of the world. If there was no evidence of actual design in the universe and thus no evidence of design by God, why continue to believe in such a being?

In the late 1980s, aggressive atheistic polemics and historiography had not yet become a staple of our culture. Nevertheless, the idea that science and religion had long been at war had already deeply permeated scholarly thinking as well as popular culture and the assumptions of many in the media. Consequently, during my first few months at Cambridge as I was reading many historians of science who doubted the warfare model, I still wondered if *they* were the revisionists. Perhaps the contemporary historians had gotten it wrong. Maybe Draper and White and the popular spokespersons for science (such as the then popular Carl Sagan) were right after all. I decided to dig into the question of the historical relationship between science and theistic belief in more depth by reading the works of the founders of modern science for myself.

2

Three Metaphors and the Making of the Scientific World Picture

During my first year of postgraduate study in the history and philosophy of science, I encountered the works of numerous historians of science who argued that Judeo-Christian assumptions had influenced the rise of modern science. My program at Cambridge required students to write several long essays under the supervision of tutors before beginning work on a longer master's thesis, which, if approved, could lead to acceptance into the PhD program. I chose three topics exploring the ideological origins of modern science more deeply. My research covered books by historians of science about the ideological influences on the early modern scientists—but also the primary sources, the writings of the early modern scientists themselves. In this process, I repeatedly encountered three metaphorical ways of describing nature and nature's relationship to God. Time and time again scientists writing during the scientific revolution and the philosophers and theologians writing in the centuries leading up to it likened nature to a book, a clock, or a law-governed realm (Fig. 2.1).

Later when I was teaching, I highlighted these three metaphors. To show the importance of Judeo-Christian conceptions of God and nature to the rise of modern science, I described *what* scientists during the scientific revolution meant by these three figures of speech, *why* each reflected a Judeo-Christian worldview, and *how* each in turn inspired or guided scientific investigation.

FIGURE 2.1
Three metaphors used to describe nature during the scientific revolution: A book. A clock. A law-governed realm.

The Book of Nature

Early in the Christian era, theologians began referring to nature as a book, one that they likened to the Bible in its ability to reveal the attributes of God. Just as the book of scripture told of God's character and plan, so too did the book of nature reveal God's power and wisdom.[1] As early as the third century, the Christian monastic Anthony the Abbot referred to "created nature" as a "book," one always at his "disposal" whenever he wanted "to read God's words."[2] Another early church father, Basil the Great, similarly argued, "We were made in the image and likeness of our Creator, endowed with intellect and reason, so that our nature was complete and we could know God. In this way, continuously contemplating the beauty of creatures, through them as if they were letters and words, we could read God's wisdom and providence over all things."[3] Other influential Christian theologians including Augustine, Maximus the Confessor, and later Thomas Aquinas routinely employed the metaphor of two books.[4]

Moreover, scriptural texts such as Psalm 19, in the Old Testament, and Romans 1, in the New Testament, seemed to support this common usage. Psalm 19 affirms that "the heavens declare the Glory of God"[5] and even that "day after day they pour forth *speech*."[6] In Romans 1, St. Paul argues that "since the creation of the world God's invisible qualities—his eternal power and divine nature [sometimes translated 'wisdom']—have been clearly seen, being understood from what has been made."[7]

With the reaffirmation of the doctrine of creation during the late

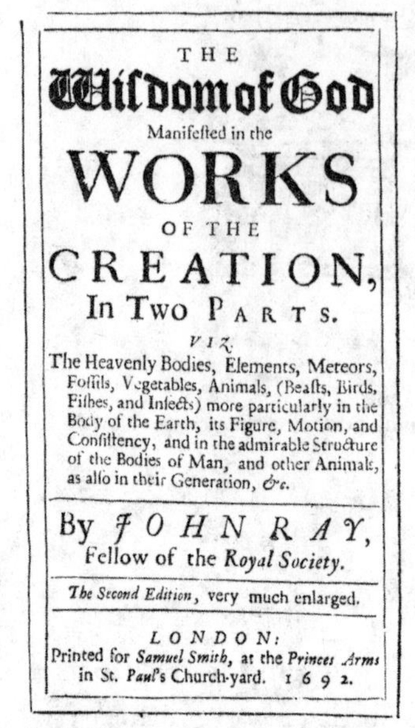

THE

Wiſdom of God

Manifeſted in the

WORKS

OF THE

CREATION,

In Two PARTS.

VIZ.

The Heavenly Bodies, Elements, Meteors, Foſſils, Vegetables, Animals, (Beaſts, Birds, Fiſhes, and Inſects) more particularly in the Body of the Earth, its Figure, Motion, and Conſiſtency, and in the admirable Structure of the Bodies of Man, and other Animals, as alſo in their Generation, &c.

By JOHN RAY,
Fellow of the *Royal Society*.

The Second Edition, very much enlarged.

LONDON:
Printed for *Samuel Smith*, at the *Princes Arms* in St. *Paul's* Church-yard. 1692.

FIGURE 2.2
The title page of biologist John Ray's 1692 book.

Middle Ages and the Reformation, these passages about the revelatory nature of the created order took on new significance. The founders of modern scientific disciplines would cite them as inspirations for the systematic study of nature. John Ray, a leading seventeenth-century biologist, published a massive two-volume study, *The Wisdom of God Manifested in the Works of the Creation*,[8] his title clearly referring to the passage from Romans just cited (Fig. 2.2). Robert Boyle (Fig. 2.3) not only referred to nature as a "booke," but went so far as to argue that "the study of the Booke of Nature is one of the Ends of the Institution of the Sabbath."[9] In other words, God had established a Sabbath in part to create time and leisure for scientific consideration of God's works of creation. He also explained in an essay titled "Of the Study of the Book of Nature" that his Sabbatarian convictions nevertheless allowed for the study of nature as an act of worship even on Sunday. As he put it, "I scruple not (when Opportunity invites) to spend some in Studying the Booke of the Creatures, either by instructing my selfe in the Theory of Nature; or trying those Experiments, that may improve my Acquaintance with her."[10]

Viewing the natural world as a book that reveals the character and nature of God provided a theological inspiration for the formal study of the natural world. It also reinforced conviction in the intelligibility of nature, because it implied that the divine author not only speaks through the book of nature, but that men and women made in his image and endowed with his rationality were equipped to read and understand it. The metaphor of the book of nature[11] also implied the legitimacy of scientific endeavor, since it affirmed that nature supplied a secondary source of authoritative revelation about the character and wisdom of the creator. This suggested that scientists could study nature and learn its secrets without needing to consult with theologians about whether their findings passed theological muster.

Nevertheless, since nature and scripture issued from the same source, namely, God, the early modern scientists assumed that both sources of revelation would ultimately align in either convergent or complementary testimony. Thus, Boyle, for example, never considered the possibility that the study of nature would undermine belief in God, but instead regarded devotion to the study of nature, like devotion to the study of scripture, as "an act of Piety,"[12] especially since he thought God desired "to have his Works regarded & taken Notice of."[13]

The Clockwork Nature

Leading natural philosophers during the scientific revolution commonly employed another metaphor. They often referred to nature as a clock—or more generally as a machine. This metaphor—also associated with Robert Boyle as well as Nicole Oresme,[14] a fourteenth-century philosopher and scientist at the University of Paris—implied both the contingency and intelligibility of nature. It implied the contingency of nature, since a good craftsman can make many different kinds of clockworks to accomplish the same end; it implied the intelligibility of nature, because, like the clocks in great medieval towers, the clockworks of nature were designed by a rational agent—thus making discernible the mechanisms upon which the orderly concourse of nature depends. As Boyle described nature: "'Tis like a rare Clock, such as may be that at *Strasbourg*, where all things are so skillfully contriv'd, that the Engine being once set a Moving, all things proceed according to the Artificer's first design."[15]

FIGURE 2.3
The seventeenth-century chemist and mechanical philosopher Robert Boyle, who likened nature to a clock and a book, metaphors that presupposed its divine origin.

This metaphor exercised considerable influence on how natural philosophers thought about their task. Many, including Boyle, referred to themselves as mechanical philosophers and, in so doing, explicitly broke with the Aristotelian scholastic practice of explaining natural phenomena by reference to insensible and, in Boyle's view, unintelligible substantial forms (or formal causes or "virtues").[16] Boyle and other mechanical philosophers rejected the "naming game" described in the previous chapter and instead insisted on looking for specific physical mechanisms—material interactions between corpuscles of matter or material structures—as explanations for the regularities of nature.[17]

Boyle himself did this in his work studying the properties of air, positing various mechanisms of interactions between the "corpuscles of air" to account for their springlike behavior under compression.[18] (The term "corpuscles" referred to small particles thought to be the most fundamental constituents of light or a material substance.) With his famous gas law, he also described the relationships between pressure, volume, and temperature of air.[19] In addition to rejecting appeals to substantial forms, Boyle discouraged appeals to the direct and singular activity of God to explain natural regularities. He did so because he thought such appeals subverted the attempt to understand the God-given causal powers of natural entities and thus the true, if only proximate, causal explanation for the regularities of nature.[20] In the same passage in which he likened God to the artificer of a great clock he also explained that his metaphor implied that God did not need to act discretely or specially many times to keep his original design working. Thus, he argued that the various natural and living systems that manifested design "do not require, like those of Puppets, the peculiar interposing of the Artificer, or any Intelligent Agent employed by him, but perform their functions upon particular occasions, by virtue of the General and Primitive Contrivance of the whole Engine."[21]

Though Boyle rejected appeals to formal causes and discrete and singular divine action to explain the regular motions or concourse of nature, he explicitly invoked the purposive or intelligent activity of God to explain the *original* construction of the universe, the mechanisms that made regularities possible and especially the diverse creatures of the living world. Indeed, as historian of science Edward Davis has pointed out, Boyle developed several design arguments to explain the origin of animals, the "Fabrick of the Universe,"[22] and the "First Formation of the Universe."[23]

For example, in an essay on "Final Causes" he argued: "The Wise Author of Nature has so excellently Contriv'd the Universe, that the more Clearly and Particularly we Discern, how Congruous the Means are to the Ends to be obtain'd by them, the more Plainly we Discern the Admirable Wisdom of the Omniscient Author of Things; of whom it is Truly said by a Prophet [Isaiah], that *He is Wonderful in Counsel, and Excellent in*

Working."[24] Consequently, he rejected "so Blind a Cause as *Chance*" as the explanation for both the orderly concourse of nature and the exquisite structures manifest in living things.[25]

Both Boyle's advocacy of the design argument and his insistence upon finding mechanistic explanations for those processes reflected the influence of the clockwork metaphor. The metaphor implied that the regular workings of nature could be explained by reference to the interactions of the material parts (or mechanisms) of the systems the divine clockmaker had put into nature. It also implied that merely naming those parts and asserting that they resulted from a substantial form or virtue with the same name did not yield any additional understanding of *how* these things work. Nor would invoking the activity of God at repeated instances to explain how the clock *normally* works afford deeper insight into the immediate causes of the regularities of the clockwork of nature. At the same time, the clockwork metaphor implied that the system of the universe as a whole as well as specific living systems owed their origin to the act of a designing intelligence, Boyle's great divine artificer God.

Edward Davis summarizes Boyle's methodological approach well: "Diligently pursue the physical causes of things, for that's how science is done; but, at the same time, [recognize that] design is sometimes evident in the whole contrivance one is studying."[26] Davis recounts his own experience of obtaining access to an original manuscript of Boyle's in the archives of the Royal Society in London and being "thunderstruck" by the similarity between one of Boyle's design arguments made in the seventeenth century and the later and more famous nineteenth-century design argument of William Paley.[27]

I had the same experience during my first year in graduate school when I read that same passage in another source. Paley famously asked his readers to imagine stumbling upon a watch on a heath and wondering "how the watch happened to be in that place." In such a case, Paley thought that any rational observer would conclude that an intelligent agent had fashioned the watch. Similarly, he argued, "Every indication of contrivance, every manifestation of design, which existed in the watch, exists in the works of nature."[28]

Now here is the similar passage from Boyle: "If an Indian or Chinois [Chinese] should have found a Watch cast on shore in some Trunke or Casket of some shipwrackt European vessel; by observing the motions and figure of it, he would quickly conclude that 'twas made by some intelligent & skillfull Being."[29] Clearly, Boyle not only *assumed* the intelligibility of nature; he also thought that he observed evidence in nature of an intelligent designer. Indeed, Davis goes so far as to call Boyle the father of the modern theory of intelligent design.[30]

The Laws of Nature

Natural philosophers during the scientific revolution often used another metaphor to characterize the workings of nature. Many began to characterize the natural world as a law-governed realm and sought to discover what they called the "laws of nature." This metaphor expressed—in part—an idea implicit in the one previously discussed. Since many natural philosophers thought that nature could be characterized as a kind of divine clockwork, they expected to observe *regularity* in the workings of nature. Boyle's analogy, likening God to a skilled watchmaker, expressed a common reverence for both the ingenuity *and* the regularity of divine activity as manifest in the natural world. The mechanical philosophers celebrated the regular concourse of nature as a consequence of God's ingenious design of various mechanistic processes.

For example, they thought that the predictable relationships between the temperature, pressure, and volume of gases described by Boyle's law derived more fundamentally from the mechanistic interactions of "corpuscles of air" and their God-given properties. For them, the regularity of natural phenomena derived from the original design of the mechanisms that made those phenomena possible—just as the regular movement of the hands of a clock derived from the underlying design of the arrangement of cogs and springs.

Nevertheless, many natural philosophers also conceived of the laws of nature as a more direct expression of God's orderly governance. To them, the metaphor of the "laws of nature" expressed a Judeo-Christian understanding of the relationship of nature to God, the divine legislator or governor, who sustained the existence of the natural world and ensured its orderly working. Thus, many founders of early modern science attributed the regularity of nature not only to God's original design of the natural world, but also to God's constant orderly supervision of it. As the Oxford University historian of science John Hedley Brooke has explained: "For Newton, as for Boyle and Descartes, there were laws of nature only because there had been a [Divine] Legislator."[31]

Numerous historians and philosophers of science, as well as a few scientists themselves, have made this connection. Some have even identified the Hebrew Bible as the ultimate source of the metaphor. As the Nobel laureate and University of California–Berkeley chemist Melvin Calvin argued, the notion of an "Order of Nature" was "discovered 2,000 or 3,000 years ago, and enunciated first in the Western world by the ancient Hebrews."[32] Calvin notes that the monotheistic worldview of the ancient Hebrews suggested a reason to expect a single coherent order in nature and thus a single, universally applicable set of laws governing the natural world.

By contrast, because animists, polytheists, and pantheists affirmed the existence of many spirits or gods, each possibly interacting with nature in different ways, they had no reason to think that natural phenomena would manifest uniformity and order. The ancient Hebrews, on the other hand, thought that, as Calvin put it, "the universe is governed by a single God, and is not the product of the whims of many gods, each governing his own province according to his own laws."[33] Calvin, like many historians and philosophers of science, identified this belief in an order-loving monotheistic God as "the historical foundation for modern science."[34]

During my studies, one of my supervisors directed me to a seminal article, "The Genesis of the Concept of Physical Law," by historian of science Edgar Zilsel. In it, Zilsel argues that the concept of the laws of nature expresses a "juridical metaphor" of theological origin.[35] He shows that the first people to conceive nature as an externally governed system were—in fact—the ancient Hebrews.[36] Zilsel notes that various passages from the Bible—from the book of Job, Proverbs, Psalms, and even the various prophetic books—implied that God had issued "laws" or "decrees" that set "boundaries" on the range of possible natural phenomena.[37]

These books of the Bible use the Hebrew word *chok*, translated variously as a "law," "decree," or "commandment," a word often used to describe a category of God's commandments for human beings. Zilsel points out that the Bible also uses this term to describe the limits that God placed upon the potential chaos of the material world. For example, the book of Job describes God as establishing by decree the exact strength of the force of the wind.[38] Other passages in Job and the book of Jeremiah describe God setting boundaries on the movements and extent of the sea.[39] As Zilsel explains, "The idea is distinctly implied that the sea, to which the divine command is addressed, wishes to offer resistance, but being too weak, is forced to bow before the supreme power of the Lord."[40] He argues that these biblical passages, as well as the idea of a single omnipotent and omniscient deity ruling over the nature, "decidedly contributed to the formation of" the modern concept of "the laws of nature during the seventeenth century."[41]

By contrast, Zilsel notes that ancient Greek or Roman philosophers, including those interested in nature, rarely used the term "laws of nature" or "natural law."[42] As he explains: "In the period of the sophists the terms 'law' and 'nature' . . . became even opposites, [with the term] 'law' designating [by convention] everything that is . . . artificially introduced by men." Similarly, he observes that the atomist philosopher Democritus "did not know anything of 'natural laws,' though he attempted to explain all physical phenomena by causes."

In addition, Zilsel notes that the most prominent Greek philosophers, Plato and Aristotle, simply didn't use the term "laws of nature" to describe natural phenomena.[43] As he explains, "Aristotle . . . never used the law-metaphor. Plato uses the term 'laws of nature' only once to characterize the behavior of the healthy in contrast to the sick human body." Indeed, neither Plato nor Aristotle attempted to identify universal laws that applied to all of nature. Plato sought to identify the immaterial forms revealed in specific material objects; Aristotle sought to identify the four causes—material, formal, efficient, and final—that he thought explained or described *specific* objects or organisms. Nevertheless, Aristotle did not attempt to relate these causes to universal and recurring phenomena in nature.

Later Greek and Roman philosophers, as well as Christian philosophers heavily influenced by the ancient Greeks, either lacked the concept of natural laws or did not use it to describe recurring physical phenomena. Epicurean philosophers, who were atomists and materialists, didn't conceive of nature as a law-governed system, because they thought the gods didn't care about nature.[44] The Roman Stoics, who recommended living in accord with Divine Reason as well as an attitude of indifference to pleasure and pain, did have a concept of natural law. Nevertheless, they thought of the underlying logos or logic of the universe, from which these laws derived, mainly as the source of moral principles that human beings could know by reason.[45] Similarly, many Christian Neoplatonists and Christian Aristotelians during the Middle Ages did use the term "natural law," but only to refer to the *moral* law knowable by reason, not to the regularities observed in the physical world.[46]

Only in the sixteenth and seventeenth centuries, after Western Christianity had begun to rely more heavily on the Bible as a source of authority and some of the passages discussed above came back into currency, did the term "laws of nature" begin to emerge in the writings of philosophers and scientists such as Bacon, Descartes, Kepler,[47] Christiaan Huygens, Richard Hooker, Boyle, and Newton to describe natural regularities.[48]

Zilsel acknowledges one early Greek scientist—Archimedes—who did describe certain phenomena in terms that we would today regard as laws of nature. Nevertheless, Archimedes himself conceived of these "laws" more as self-evident logical principles—principles that were intrinsic to nature and could not logically be otherwise.[49] For example, Archimedes discovered his buoyancy principle and what we now call his "law of the lever" and "law of optical reflection." Each of these principles would certainly qualify as a law in the modern sense. Archimedes, however, did not call them laws, but instead principles. Moreover, even though he discovered these principles by observation, he presented them as if they were deductions from self-evident logical postulates or axioms, as Euclid did in his works on geometry.[50]

Another historian of science, Francis Oakley, made a crucial distinction. He acknowledged that some ancient Greeks discovered principles that we would today regard as "laws" of nature. Yet whereas the Greeks conceived of these principles as *logically necessary* axioms inherent in (or internal to) nature itself, the scientists during the seventeenth century began to conceive of the laws of nature as contingent forms of order that were *impressed* upon nature from the outside by a creator.[51] Since the founders of modern science thought the laws of nature expressed the free will of the divine creator and sustainer of nature, they recognized that whatever order nature exhibits might well have been different had the creator chosen to create or order the natural world differently.[52]

This conception of natural law had several beneficial effects on the development of science as we know it today. First, natural philosophers no longer thought they could deduce how nature works from other axioms or first principles. They instead realized that they would need to observe nature in order to discover its lawlike regularities.[53] Thus, the concept of *impressed* or contingent laws of nature encouraged empirical observation. Second, since the metaphor of *laws* of nature implied reliable divine oversight, the metaphor encouraged the use of mathematics to describe natural regularities.[54] To employ philosophical terminology, the metaphor encouraged the development of a "rational empiricism" that combined mathematical description and deduction with careful inductive observations of nature. This mathematizing of the description of natural regularities enabled the prediction and control of nature and eventually fostered unprecedented technological advance. Only a few centuries after Newton characterized the universal law of gravity, human beings harnessed this knowledge to put men on the moon.

Action at a Distance and Constant Spirit Action

We've seen that many of the founders of modern science presupposed that the orderly concourse of nature reflected the sustaining power and free choice of a divine mind. Natural philosophers committed to such "theological voluntarism" thus saw the regularities of nature as a contingent expression of God's constant governance of the natural world. Newton's theory of universal gravitation and the controversy it provoked in his lifetime provide additional evidence in support of this thesis.

Though physicists today regard Newton (Fig. 2.4) as a giant in the field, some of his contemporaries regarded his theory of universal gravitation with considerable skepticism and even suspicion. According to the mechanical philosophers, all natural regularities were produced by underlying material mechanisms. Thus, all proper scientific explanations needed to invoke some kind of mechanical interactions between the material parts of objects or systems. For example, before Newton, physicists such as René Descartes[55] and Christiaan Huygens[56] proposed a theory of "traction" (gravitation), in which they envisioned swirling vortices of an invisible but material substance called "ether." The ether pushed heavenly bodies around the sun, as water pushes floating sticks in a spinning whirlpool.

FIGURE 2.4
A portrait of the young Isaac Newton.

Yet the law of gravity as proposed by Newton had no such mechanistic basis. Instead, it involved mysterious "action at a distance," in which the mass of one

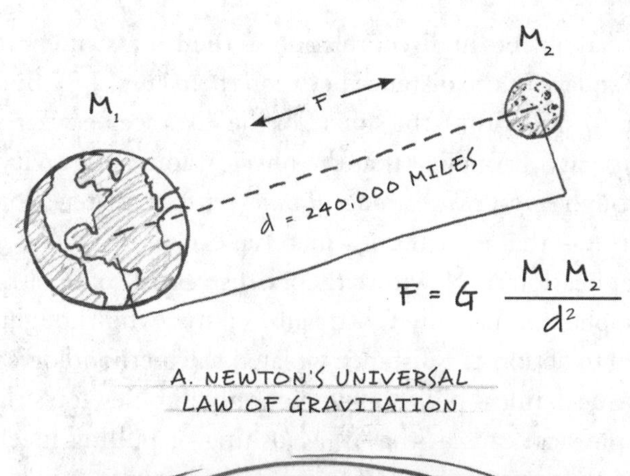

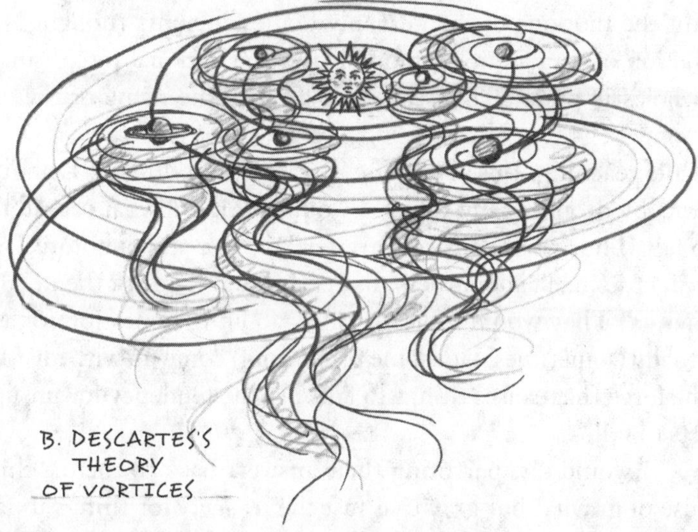

FIGURE 2.5
Comparison of Newton's and Descartes's view of gravitational attraction.
TOP: Newton's universal law of gravity states that massive bodies exert a force on each other that is proportional to the product of their masses and inversely proportional to the square of the distance between them. It envisioned action at a distance or force being transmitted through empty space.
BOTTOM: Descartes's theory of vortices postulated that space was filled entirely by an invisible material substance known as ether. As the ether whirled around the sun, it pushed the planetary bodies in orbit around it.

material body somehow transmitted a force through empty space attracting the mass of another material body without any physical contact between the two. Thus, Newton attributed—as we do today—the tidal action in the earth's oceans to the movement of the moon as it orbited around the earth, even though the moon and the earth do not have any direct physical contact.

According to Newton, all material objects exert a precise force on other material objects in direct relation to their mass and in inverse relation to the square of the distance between them (Fig. 2.5). But the second part of that formulation—the square of the distance between them—was the sticking part. It implied that the physical force of gravity was transmitted through *empty space across a distance* without material or mechanistic interaction—that is, without a material cause.[57]

This implication of Newton's theory distressed some of the mechanical philosophers, especially Gottfried Leibniz, who thought that Newton's appeal to action at a distance violated the methodological protocols of the new mechanical philosophy. Not only did Newton's theory fail to identify a physical cause—that is, a pushing or pulling mechanism—to explain the motions of the heavenly bodies, Leibniz thought Newton's attribution of "gravitational motion" to "gravitational force" smacked of the scholastic practice of treating as a cause the name of the effect in question.[58]

While teaching, I used to demonstrate what bothered Leibniz to my students by dropping my wallet. I would ask them what caused the wallet to fall. They would answer "gravity" or "gravitational force" without thinking too much about their answer. I would then ask them, "But what is gravity?" They would typically think about the question for a bit and eventually (sometimes with some prompting) come up with answers such as "the force that causes things to fall" or "the tendency for unsupported objects to fall."

"So," I would say, parroting their answers back to them, "things fall because of gravity, but gravity is just the *tendency* for things to fall. Isn't that circular and vacuous? What produces that tendency to fall? If we don't know, have we really identified the cause of gravity? Or have we just treated the name of the effect in question as its own cause?"

The puzzled looks on the faces of my students confirmed that they had begun to understand exactly why gravitational action at a distance bothered Leibniz. Newton only exacerbated Leibniz's concern by acknowledging explicitly that he "feigned" no knowledge of the *cause* of gravitational attraction—*hypotheses non fingo*, as he famously put it in Latin. But since Newton refused to propose a mechanism (such as swirling ether) to explain gravitational motion, Leibniz argued that Newton's law gave "refuge to ignorance and laziness by means of an irrational system which maintains not only that there are qualities which we do not understand—of which there are only too many—but further that there are some which could not be comprehended by the greatest intellect if God gave it every possible opportunity."[59]

Moreover, since Newton could not identify any *material* cause for gravitational action at a distance, Leibniz suspected that Newton secretly might be attributing the motion of the planets and other gravitating objects to the direct governance of an *immaterial* agent, namely, God.[60] Leibniz had good reasons for his suspicion. Since all matter "gravitated," or attracted other matter, the phenomenon could not be explained mechanistically as a consequence of the motion or interaction of some other matter that was itself unaffected by gravitational force—by a material force external to the bodies such as swirling ether, for example. If Newton had invoked some other mechanism to explain why all matter gravitates, the matter out of which that mechanism was composed would itself gravitate. In that case, the attractive capacity of that matter would itself need explanation, presumably by some other material mechanism. But that would raise the same question all over again, thus quickly degenerating into an infinite regress. To Leibniz this approach constituted a "recourse to absurdities."[61]

That left only two other options. Since Newton's system precluded an appeal to matter *external* to the bodies (such as swirling ether), his theory would need to refer to either some intrinsic property of matter or some *im*material entity as the cause of gravity. But Leibniz thought appealing to properties internal to the material bodies (a "gravitational virtue," an "attractive form," or even "gravitational force") constituted a return to the vacuous and unintelligible scholastic practice of attributing causal powers to the name of the effect. What if anything, he reasoned, was the difference between saying that "gravity is caused by a 'gravitational force'" and saying that "bread nourishes because of its 'nutritive virtue'"? Thus, Leibniz argued, the only alternative to such absurdities in the Newtonian system was a "perpetual miracle."[62] As he stated, if gravitational attraction "is not miraculous, it is false."[63]

Though a theist, Leibniz rejected attributing the regularities in nature to God's direct and constant governance. For Leibniz, a proper respect for the wisdom of God required seeing a preestablished design built into matter from the beginning, not a constant regulation of matter by the divine Spirit. As Leibniz argued, "God's excellency arises also from . . . his wisdom: whereby his machine lasts longer, moves more regularly, than those of any other artist whatsoever."[64] Leibniz saw such wisdom as a consequence of a "primitive active force" or "entelechy" embodied in all corporeal substances ensuring the orderly progression of the material world according to God's purposes.[65] In Leibniz's view, matter expresses purpose from within and "gives a physical [not a spiritual] basis for the regularities of nature."[66] Thus, whereas Newton perceived the ongoing and continuous activity of God in upholding the law of gravity, Leibniz thought of God's creative activity as a consequence of the infusion of his design into all matter at creation. He, therefore, expected gravity could be explained by reference to some mechanism.

Consequently, Leibniz believed that Newton's refusal to posit such a mechanism violated the division of labor that Boyle had established between natural philosophers and theologians. The former, whom we today call scientists, were to describe the ongoing *operation* and regularities of nature by reference to underlying physical mechanisms (such as swirling vortices or the interactions of "corpuscles of air"), while the theologians and natural theologians were to engage in identifying evidence of design in the *origin* of those mechanisms. Thus, Leibniz sought to place Newton's new theory on the horns of a dilemma: either universal gravitation represented a return to outdated scholastic practice or it brought divine action into science, where it did not, in Leibniz's view, belong. Either way, Leibniz asserted, Newton's theory relied on "occult" rather than mechanical causes. "Mr. Boyle," he observed, "would never have allowed such a chimerical notion."[67]

In his dispute with Leibniz, Newton defended the rigor and legitimacy of his new theory. He did so on the basis of its ability to describe with mathematic precision the motions of the heavenly bodies and the forces acting on falling bodies—however mysterious such action at a distance might be. Newton also thought that his use of mathematics to describe *universal* phenomena such as gravitational attraction represented an advance over the practice of trying to imagine a mechanical explanation for every *specific* class of phenomena. Indeed, mechanical models like the vortices theory of gravitation or the spring model of air pressure tended to claim applicability only to a specific process. Newton found this less intellectually satisfying than discovering laws that applied to all matter and the grand mathematical synthesis that he had advanced in the *Principia* (Fig. 2.6).[68]

Still, Newton likely did hold the view that Leibniz suspected. He seemed to affirm in private correspondence that he thought that an *im-*

Add. 3992

PHILOSOPHIÆ
NATURALIS
PRINCIPIA
MATHEMATICA.

AUCTORE
ISAACO NEWTONO,
EQUITE AURATO.

EDITIO SECUNDA AUCTIOR ET EMENDATIOR.

CANTABRIGIÆ, MDCCXIII.

FIGURE 2.6
The title page of Newton's *Principia* in which he developed his universal law of gravitation and the concept of action at a distance.

material agent, God, was responsible for the mysterious action that his law described. In a letter to Bishop Richard Bentley written in 1692, after his dispute with Leibniz, Newton implied that he thought that "gravity must be caused by *an agent* acting constantly according to certain laws."[69]

Newton did add the caveat: "Whether this agent be material or immaterial, I have left open to the consideration of my readers."[70] Nevertheless, it seems Newton was being coy. Earlier in the same letter Newton had already affirmed Bentley's explicitly theistic interpretation of Newton's view. Bishop Bentley had written to Newton as he was preparing to give lectures funded by a bequest from Robert Boyle.[71] The Boyle lectures were endowed to highlight the evidence for the existence of God from studies of nature. Bentley wrote to Newton asking him to clarify whether he thought God was responsible for the attraction between material bodies separated by empty space or whether gravitational attraction derived from an intrinsic material property of the material bodies themselves.[72]

In his letter, Bentley depicted Newton's view of the cause of gravity as follows: "'Tis unconceivable, that inanimate brute matter should (*without a divine impression*) operate upon & affect other matter without mutual contact."[73] Newton affirmed Bentley's interpretation and replied, "The last clause of your second position I like very well. 'Tis unconceivable that inanimate brute matter should (without the mediation of something else *which is not material*) operate upon & affect other matter without mutual contact; as it must if gravitation . . . be essential and inherent in it."[74]

Here Newton leaves no doubt as to his own view. He specifically rejects as "inconceivable" the idea that "brute matter" could cause action at a distance—i.e., move other matter "without mutual contact." Instead, he affirms that the constantly acting agent responsible for gravity must be "something else *which is not material*." This statement occurs in the context of Newton's response to Bentley's query about whether Newton thought that matter could "operate upon & affect" other matter at a distance "without a divine impression." Consequently, it clearly shows that Newton himself regarded the ultimate cause of gravity as "constant Spirit action," as one of my Cambridge supervisors characterized his view during a memorable tutorial. Newton also seemed to betray his own view of the cause of gravitational attraction publicly in a passage of the General Scholium to the *Principia*, an epilogue to the main work that he included in later editions partly as a reply to his critics. There he says explicitly: "In him [God] are all things contained *and moved*."[75]

These and many other passages also suggest that Newton affirmed the conception of natural law that historians have perceived in the writings of many founders of modern science.[76] Newton and his followers clearly regarded what we call the "laws of nature" as a mode of divine action and governance of the natural world.[77] The laws of nature not only reflected the past action of a divine creator who established the conditions necessary for orderly and regular natural processes, but the fundamental laws of nature also depend upon the ongoing and sustaining activity of a divine legislator.[78] As University of Cambridge historian of science Simon

Schaffer has explained, Newton thought what we call the laws of nature manifested "the essentially divine will evident in the common concourse of nature."[79]

The Importance of Design Arguments During the Scientific Revolution

Each of these three metaphors—the book of nature, the clockwork of nature, and the laws of nature—expressed and *presupposed* belief in a divine creator and/or sustainer of the natural world. Nevertheless, many of the founders of modern science did not just assume or assert by faith that the universe had been designed by an intelligent agent. They also argued for this hypothesis based on discoveries in their fields of study.[80] Johannes Kepler perceived intelligent design in the mathematical precision of planetary motion and in the three laws he discovered that describe that motion.[81] Robert Boyle insisted that the intricate clocklike regularity of physical laws and chemical mechanisms as well as the anatomical structures in living organisms suggested the activity of "a most intelligent and designing agent."[82] Carl Linnaeus later argued for design based upon the ease with which plants and animals fell into an orderly groups-within-groups system of classification.[83] Many other individual scientists made specific design arguments based upon empirical discoveries in their fields.

This tradition attained an almost majestic rhetorical quality in the writings of Newton. Newton not only viewed gravitational action at a distance as a manifestation of God's power, but he also made many powerful *design* arguments based upon other biological and astronomical discoveries. Newton viewed the order described by the laws of nature as a mode of divine action, but he also thought that many specific arrangements of matter (each subject to those laws) gave evidence of the design of an "intelligent and powerful being." For example, in the *Opticks*, his major treatise on light, Newton argued that the uncanny match between the optical properties of light and the structure of the mammalian eye suggested foresight and design. As he explained: "How came the Bodies of Animals to be contrived with so much Art, and for what ends were their several Parts? Was the Eye contrived without Skill in Opticks, and the Ear without Knowledge of Sounds? . . . And these things being rightly dispatch'd, does it not appear from Phænomena that there is a Being incorporeal, living, intelligent, omnipresent?"[84]

Writing in the General Scholium, the epilogue to the *Principia*, Newton suggested that the stability of the planetary system depended not only upon the regular action of universal gravitation, but also upon the precise initial positioning of the planets and comets in relation to the sun. As he explained: "though these bodies may indeed persevere in their orbits by the mere laws of gravity, yet they could by no means have at first deriv'd the regular position of the orbits themselves from those laws." Thus, "this most beautiful System of the Sun, Planets, and Comets, could only proceed from the counsel and dominion of an intelligent and powerful being."[85] As John Hedley Brooke has pointed out, Newton thought that "knowledge of God's power and wisdom could be inferred from the intelligence seemingly displayed in the designs of nature."[86]

Many histories of science—the kind you encounter in physics textbooks or in New Atheist books and videos—claim that Newton depicted a "mechanistic universe," an autonomous self-organizing and self-maintaining "world machine"—one that left no place for the activity of a divine creator, sustainer, or legislator of nature. This view misrepresents Newton in three ways. First, he rejected the idea that gravity—with its mysterious action at a distance—could be explained by any mechanistic cause. Second, Newton thought that laws of nature express God's way of ordering "brute matter" through the constant action of his will and spirit. Third, Newton saw evidence of initial acts of intelligent design in the complex configurations of matter in both the solar system and biological systems.

From Newton to Dawkins?

Presuppositions derived from a Judeo-Christian worldview helped to inspire and shape the foundation of modern science, and the founders of modern science themselves perceived evidence in support of those presuppositions, including the idea that life and the universe owe their origin to the activity of "a Being incorporeal, living, intelligent, [and] omnipresent."[87]

How, then, did we get from Newton to Dawkins? How did the theistic foundations and interpretation of the scientific enterprise in the sixteenth, seventeenth, and early eighteenth centuries give way by the nineteenth and twentieth centuries to the perception of conflict or even "warfare" between science and religious belief? How did the idea that nature displays the handiwork of the creator—central to the vigorous program of natural theology advanced by scientists such as Kepler, Boyle, and Newton—get completely overturned, so that the world's bestselling science popularizer in the twenty-first century could write a book claiming that science properly understood renders belief in God so "incredibly improbable" as to be effectively a "delusion"?

Clearly, an enormous intellectual shift has occurred in the West since the founding of modern science. But what caused it?

3

The Rise of Scientific Materialism and the Eclipse of Theistic Science

Belief in a rational and intelligent creator inspired the development of modern science. Yet after the eighteenth century, this understanding gave way to a perception, advanced by many powerful voices, that science and religious belief are at war. What produced this dramatic shift?

Early in my study of the history and philosophy of science, I found myself confronted with this question. During my first year at Cambridge, I developed a particular fascination with Newton and the way in which he so closely integrated his theistic and even biblical ideas into the development of the most foundational theories of physics. As one of my supervisors put it to me, "If you miss Newton's theism, you've missed everything." Newton not only had a profoundly theistic *philosophy* of nature, but he also developed several compelling (at least, at the time) arguments for natural *theology*—that is, arguments for the existence of God based upon observations of complex systems in the natural world.

I didn't have to go far, however, to find evidence of the opposite view of what science tells us. As I read the works of late nineteenth-century biologists, I came face to face with an entirely different philosophy of science than the one that had inspired the scientific revolution.

During this period, just as scientists were beginning to formulate theories of the origin and evolution of life on earth, the worldview known as scientific materialism, which I introduced in the Prologue, began to dominate thinking about the meaning of science. Its rise followed a movement in the history of philosophy during the eighteenth century that historians call the Enlightenment.

Intellectual historians and historians of science typically identify the shift away from the theistic foundations of modern science with three major developments in Western intellectual history: first, the Enlightenment idea that human reason could replace and function autonomously from religious belief;[1] second, the increasing skepticism about the existence of God, or at least about the soundness of arguments for God's existence, among many Enlightenment philosophers; and third, the rise of scientific materialism and with it both new norms of scientific practice and a worldview allegedly based on science that affirms matter and energy, rather than God, as the fundamental reality from which everything else comes.

Reason and Religion During the Enlightenment

Newton died in 1727. In the years following, scientists continued to demonstrate the power of the systematic investigation of nature. Increasingly, Enlightenment philosophers extolled the virtues of reason (and science) over religion as a source of authority and knowledge. Indeed, many philosophers viewed science and reason generally as sources of authority that could and should replace revealed religion. The idea seemed increasingly appealing in Europe after centuries of strife and warfare pitting Catholics and Protestants against each other, waged in part over competing claims about the source of religious authority. After the ravages of the Thirty Years' War (1618–48), many Europeans felt exhausted by religious conflict, leaving them open to new perspectives, even radically new ones.

In addition, French philosophers such as Voltaire and later Comte (writing after the Enlightenment) viewed science as perhaps the best example of the power of human reason and the most reliable source of knowledge. They assumed that science could function autonomously from religious belief—and thus independently from the presuppositions (such as the intelligibility, contingency, and uniformity of nature) that Christian theism supplied as part of the original epistemological foundation of modern science. Natural philosophers, or "scientists," as the English philosopher William Whewell dubbed them in 1833,[2] still sought to discover the "laws of nature" and the "mechanisms" of the great "clockwork" of nature. But these metaphors gradually lost their original theological connotations. At least, they did so for many Enlightenment philosophers who characterized science as a purely secular enterprise and depicted reason and revelation as opposites.

One prominent Enlightenment philosopher specifically appropriated the idea of the laws of nature as a reason to reject theistic belief. The skeptical empiricist David Hume (Fig. 3.1) argued that the lawful concourse of nature precluded the possibility of miraculous intervention by a transcendent God. Miracles, he said, are impossible because they *violate* the laws of nature. He depicted these laws as autonomous entities rather than descriptions of how God normally chooses to order the material world, as Newton and earlier scientists had believed.

FIGURE 3.1
The Scottish philosopher David Hume, whose radical empiricism led him to reject the design argument and the possibility of miracles.

Hume justified his rejection of the possibility of miracles by insisting *that uniform and repeated human experience* demonstrated that the natural laws could not be violated. As he explained: "A miracle is a violation of the laws of nature; and as firm and unalterable experience has established these laws, the proof against miracle, from the very nature of the fact, is as entire as any argument from experience can possibly be imagined."[3]

Hume advanced a theory of knowledge known as radical empiricism. Empiricism asserts that the observation of the natural world through the five senses offers the only sure path to knowledge. As such, it provides the only reliable source of ideas in our minds.[4] Hume's rejection of the possibility of miracles reflected this view, because it asserted that uniform human experience derived through the senses had established the laws of nature—and human beings had never observed any exceptions to them. As he noted: "Nothing is esteemed a miracle, if it ever happened in the common course of nature."[5] Thus, Hume not only argued that miracles violate the laws of nature, he also argued that our experience has established that what we call natural laws admit no exceptions. In short, miracles violate the laws of nature, and the laws of nature cannot be violated; therefore, miracles are impossible.

Hume's argument against miracles implied that the only kind of God that could exist—if any existed at all—would be a remote deistic being who never intervened or otherwise acted discretely or discernibly in nature. Hume's radical empiricism further implied, if subtly, that belief in God itself could not have a sound basis in reason, because God fails to qualify as the sort of entity that human beings could observe through the five senses.

The French philosopher Auguste Comte, another radical empiricist and the founder of a philosophy of science known as positivism, advanced an even more explicit dichotomy between science and religious belief. Writing in the 1840s, he maintained that the advance of knowledge occurred in three phases: a theological phase, a philosophical phase, and finally a "positive" or scientific phase. Comte argued that in the theological phase, people invoked the mysterious action of gods to explain natural phenomena, whether electrical storms or the spread of contagious disease. In a second, more advanced, metaphysical stage, philosophers offered abstract concepts like Plato's forms or Aristotle's final causes as explanations of natural phenomena. According to Comte, human beings only attained real, or "positive," knowledge when they replaced such superstitions and abstractions and explained natural phenomena by reference to natural laws or strictly material mechanisms.[6]

Thus, he secularized two of the theological metaphors—nature as mechanism and a lawful realm—that had previously expressed the theistic inspiration for doing science. Moreover, he insisted that science properly practiced could make no reference to divine action to explain any events or phenomena. Instead, explaining these scientifically, or "positively," required showing how such phenomena exemplified the laws of nature—now understood as entities that existed autonomously from the divine will or governance of nature.

The Demise of Theistic Arguments

A growing skepticism among Enlightenment philosophers about the classical arguments for God's existence—about the God hypothesis—provided an additional reason for the secularization of knowledge. During the late eighteenth century, leading philosophers such as David Hume and Immanuel Kant (Fig. 3.2) denied the soundness of two of the classical and most formidable arguments for God's existence from nature.

FIGURE 3.2
The German philosopher Immanuel Kant, who accepted a minimalist version of the design argument but rejected the cosmological argument for God's existence as inconclusive.

Hume raised powerful philosophical objections to the design argument; Kant expressed skepticism about the cosmological argument (though not necessarily the design argument or the existence of God).[7]

The classical design argument begins by noting certain highly ordered or complex features in nature, such as the configuration of planets or the architecture of the vertebrate eye. It proceeds to argue that such features must have arisen from the activity of a preexistent intelligence (typically equated with God). The cosmological argument assumes the principle of causality and/or the principle of sufficient reason (the idea that every event must have a cause, or reason for, its occurrence or existence). It seeks to deduce a necessary being—that is, God—as the first cause or sufficient reason for the universe's existence.[8]

Medieval Muslim scholars developed one of the most famous versions of the cosmological argument, known as the Kalām argument. It asserted that the universe had a temporal beginning—a proposition that philosophers typically sought to justify by showing the logical or mathematical absurdity of an infinite regress of cause and effect. The argument concluded that the beginning of the physical universe must have resulted from an uncaused first cause that exists independently of the universe.[9] The argument was typically expressed in a syllogism:

Whatever begins to exist must have a cause.
The universe began to exist.
The universe must have had a cause for its existence.

By further steps of reasoning, proponents of the Kalām argument deduced that the necessary first cause of the universe must transcend the physical universe (since a cause is necessarily separate from its effects)[10] and must be personal (since only a personal agent can act discretely to initiate a new line of causation without its action being caused by a prior set of necessary and sufficient material conditions). Finally, proponents of the Kalām argument equated that first transcendent and personal cause with God.

Throughout Western history, many philosophers and scientists have formulated arguments for the existence of God, some based upon observations of the natural world. Consequently, many Western thinkers also viewed science and theistic belief as mutually reinforcing. Yet skepticism about the most empirically based theistic arguments, such as the cosmological and design arguments, gradually became more pervasive after the end of the eighteenth century, largely because of developments within philosophy that were later reinforced by new scientific theories or interpretations of them.

The Demise of the Cosmological Argument

The philosopher Immanuel Kant, for example, undermined confidence in the Kalām cosmological argument. He did so by casting doubt on the validity of the second premise of the argument—the one affirming that the universe must have had a beginning. Instead, he argued that the question of whether the universe was finite or infinite in time could not be decided by reason. He thought that reason could lead to two equally rational but contradictory conclusions, or "antinomies," namely, that the universe did and did not have a beginning in time.[11] In his view, the universe might have had a beginning in time, but it could have also resulted from an unbroken line of effects and causes going back infinitely. In other words, Kant accepted the possibility that the universe might be eternal and self-existent.[12] He did not argue, as Aristotle had done, that the idea of *creatio ex nihilo*, creation from nothing, was logically incoherent. But he did regard the conclusion of the Kalām argument as uncertain.

Kant's philosophical skepticism about the cosmological first-cause argument was reinforced by the science of the day. Though Newton supported the design argument, one aspect of his physics—his postulation of infinite space—helped to undermine the classical Kalām argument.[13] According to Newton's theory of universal gravitation, all bodies attract one another with a force proportional to the product of their masses and inversely proportional to the square of the distance between them. His theory implied that *all* bodies of matter in the universe attract one another. Yet this created a puzzle. According to Newton's theory, every star should gravitate toward the center of the universe, until the whole universe collapses in on itself.

To explain the current stability of the universe, Newton proposed that "the matter was eavenly diffused through an infinite space," so that "it would never convene into one mass."[14] He thought that if there were an infinite number of stars scattered evenly throughout a universe of infinite space, then every star would attract every other star with equal force in all directions. Thus, the stars would remain forever suspended in a tension of balanced gravitational attraction.[15] Newton also found the infinite universe appealing for theological reasons. He thought of space as a "Divine Sensorium," a medium in which God perceived creation.[16] Since God was infinite, space had to be as well.

Physicists with a more materialistic outlook later found Newton's infinite universe philosophically agreeable. Some extended the infinite static-universe model by assuming that if space must be infinite, then time must also be infinite in both the forward and reverse directions. An infinite universe in a kind of steady-state or gravitational equilibrium—neither expanding nor contracting—lacked dynamic motion that would suggest either a beginning or an end. Thus, Newton's affirmation of a spatially infinite universe[17] seemed to many physicists to imply a temporally infinite universe as well.[18] Thus, skepticism about the idea that the universe had a beginning in time undermined support for, or interest in, the Kalām argument.

The Demise of the Design Argument

During the Enlightenment, the design argument also came under attack. Most design arguments then in currency had an analogical character. They likened, or analogized, living organisms to complex human artifacts (such as watches or clocks). Since such complex machines derived from the activity of intelligent agents, the much more complex machinery evident in living organisms must also have originated from a designing mind.

Hume took aim at this reasoning in his *Dialogues Concerning Natural Religion* (1779). He argued that the design argument depended upon a flawed analogy with human artifacts. He admitted that artifacts derive from intelligent artificers and that biological organisms have certain similarities to complex human artifacts.[19] Eyes and watches both depend upon the functional integration of many separate and specifically configured parts. Nevertheless, he argued, biological organisms also differ from human artifacts—they reproduce themselves, for example—and the advocates of the design argument fail to take these dissimilarities into account.[20] Since experience teaches that organisms always come from other organisms, Hume argued that an analogical argument really ought to suggest that organisms ultimately come from some primeval organism (perhaps a giant spider or vegetable), not a transcendent mind or spirit.[21]

Yet Hume's categorical rejection of the design argument did not prove entirely decisive with either theistic or secular philosophers. Thinkers as diverse as Kant,[22] the Scottish Presbyterian Thomas Reid,[23] the Enlightenment deist Thomas Paine,[24] and the English philosopher William Whewell continued to affirm various versions of the design argument. Indeed, science-based design arguments continued at least into the early nineteenth century, in works such as William Paley's *Natural Theology* (1802).[25]

Paley (Fig. 3.3) catalogued a host of biological systems that suggested the work of a superintending intelligence. Paley argued that the astonishing complexity and superb adaptation of means to ends in such systems could not originate strictly through the blind forces of nature. Paley also responded directly to Hume's claim that the design inference rested upon a faulty analogy. A watch that could reproduce itself, he argued, would constitute an even more marvelous effect than one that could not.[26] Thus, for Paley, the differences between artifacts and organisms only seemed to strengthen the conclusion of design. Moreover, well into the nineteenth century many scientists continued to find Paley's watch-to-watchmaker reasoning compelling. Other works of English natural theology during the first half of the nineteenth century, such as the famed *Bridgewater Treatises*,[27] also demonstrated the continuing popularity of design arguments, especially in Britain, despite Hume's well-known objections.

FIGURE 3.3
The early nineteenth-century proponent of natural theology, William Paley.

Indeed, it was not ultimately the arguments of the philosophers that began to erode the popularity of the design argument, but the emergence of increasingly powerful materialistic explanations of "apparent design." Charles Darwin's theory of evolution by natural selection stands out particularly.[28] In *On the Origin of Species*, published in 1859, Darwin (Fig. 3.4) argued that living organisms—previously considered the most obvious example of God's creative power—only appeared to be designed. Darwin proposed a concrete mechanism, natural selection acting on random variations, that could explain the adaptation of organisms to their environment (and other evidences of apparent design) without invoking an actual intelligent or directing agency. If the origin of biological organisms could be explained naturalistically, as Darwin argued, then explanations invoking a creative intelligence were unnecessary and even vacuous.[29]

FIGURE 3.4
Charles Darwin, British naturalist and author of *On the Origin of Species.*

The Rise of Scientific Materialism

The success of new scientific theories—about astronomical, geological, and biological origins—contributed to the rejection of theism as an explanatory framework for science. These new theories collectively seemed to support the alternative worldview of scientific materialism. For example, in 1796, the French physicist Pierre Laplace (Fig. 3.5) attempted to explain the origin of the solar system not as the product of the design of "an intelligent and powerful Being," as Newton had done, but as the result of purely natural gravitational forces. In his book *The System of the World* (*Exposition du Système du Monde*), Laplace outlined a scenario by which planets may have formed from the hot atmospheric gases surrounding the sun and other rotating stars.[30] Known as the nebular hypothesis, Laplace's theory provided an evolutionary and wholly materialistic account of the origin of the solar system. As he asked, "Could not this arrangement of the planets be itself an effect of the Laws of motion; and could not the supreme intelligence which Newton makes to interfere, make it to depend on a more general phenomenon? Such as . . . a nebulous matter distributed in various masses throughout the immensity of the heavens."[31]

FIGURE 3.5
Pierre Laplace, the French physicist and author of *A Treatise of Celestial Mechanics*.

Beginning in 1798, Laplace published *A Treatise of Celestial Mechanics* (*Traité de Mécanique Céleste*), a multivolume companion to his earlier work. He described the ongoing operation of the solar system in precise mathematical detail. In 1802, he presented some of the work to the new French emperor Napoleon Bonaparte. According to later (and possibly apocryphal) reports, when Napoleon eventually summoned Laplace to discuss his work, he questioned the scientist directly about the role of God in the origin of the solar system. The emperor asked, "They tell me you have written this large book on the system of the universe, and have never even mentioned its Creator." Laplace reportedly issued the now famous reply, "Sire, I had no need of that hypothesis."[32]

Historians are uncertain about whether Laplace actually said these words, in part because the first reported quotations of them don't appear until 1825.[33] Yet the earliest reports of the encounter certainly capture a shift in perspective about whether the God hypothesis could play an explanatory role in any aspect of the natural sciences. The English astronomer Sir William Herschel, the only eyewitness to the encounter between Laplace and Napoleon, reports that in response to "the first Consul" (Napoleon) asking Laplace "And who is the author of all this!" Laplace replied by indicating that (as Herschel paraphrases his response) "a chain of natural causes would account for the construction and preservation of the wonderful system."[34] And clearly Laplace's *Treatise* and his earlier *System of the World* provided a fully naturalistic account, not only of the ongoing operation of the celestial system, but also of its origin. Thus, Laplace's work marked a change in approach among many scientists.

Developments in other fields supported this trend. In geology, Charles Lyell explained the origin of the earth's most dramatic topographical features—mountain ranges and canyons—as the result of slow, gradual, and completely naturalistic processes of change.[35] In biology, as noted, Darwin's evolutionary theory sought to show that the blind process of natural selection acting on random variations accounted for the origin of new forms of life without any divine guidance. Living organisms only *appeared* to be designed.[36] As biologist Francisco Ayala has explained, "The functional design of organisms and their features would . . . seem to argue for the existence of a designer. It was Darwin's greatest accomplishment, [however,] to show that the directive organization of living beings can be explained as the result of a natural process, natural selection, without any need to resort to a Creator or other external agent."[37]

As I mentioned in the Prologue, every worldview or metaphysical system must address the question of origins. The writer James Sire has called this "the prime-reality question"[38]—the question "What is the thing or the entity or the process from which everything else comes?" By the end of the nineteenth century, scientists had formulated a reasonably comprehensive set of entirely materialistic theories sketching the origin and development of life, including human life, back to the origin of the earth and the solar system. Perhaps not surprisingly, then, a worldview arose at that time—scientific materialism—that claimed to offer a comprehensive materialistic answer to the prime-reality question based upon several then-popular scientific theories.

Changing Norms: The Role of Methodological Naturalism

It wasn't just the perceived success of materialistic theories of origins that banished God or the design hypothesis. New scientific norms and practices emerged during the nineteenth century making the God hypothesis seem, if not false, at least increasingly irrelevant. As historian of science Neil Gillespie argues in his classic work *Charles Darwin and the Problem of Creation*, the exclusion of the design hypothesis was reinforced by an emerging tradition that increasingly sought to exclude appeals to divine or intelligent causes from science *by definition*.[39] Gillespie attributes this emerging "positivistic episteme" in part to the influence of the philosopher Auguste Comte.

Other historians and philosophers see the emergence of strictly naturalistic methodological conventions more as a natural consequence of the practices of theistic scientists themselves. Steve Fuller has noted in correspondence with me that disagreements among theists about the explanatory role of God in science—such as the disagreement between Newton and Leibniz discussed in the previous chapter—contributed to the rise of a strictly materialistic approach to scientific explanation. Uncertainties among theists about how to invoke God appropriately as a scientific explanation left scientists increasingly inclined to consider only naturalistic explanations.

Either way, scientists increasingly understood that scientific theories should limit themselves to positing only materialistic entities or natural processes in scientific explanations. Positing creative intelligence violated such (spoken or unspoken) methodological norms. For example, in the *Origin of Species*, Darwin repeatedly argued against the scientific status of the received "theory of Creation." He often faulted his rivals not just for their inability to devise explanations for certain biological data, but for their inability to offer *scientific* explanations at all.[40] Indeed, some of Darwin's arguments for his theory of descent with modification depended not on newly discovered facts, but upon facts such as homology (similar structures in otherwise different animals) and fossil progression, that neither stymied nor puzzled contemporary palaeontologists such as Louis Agassiz and taxonomists such as Richard Owen. Both of these eminent scientists thought such phenomena reflected prior "acts of mind" or "the plan of creation."[41]

But in Darwin's view, explanations pointing to an immaterial mind, idea, or plan did not—in principle—qualify as proper scientific explanations. What he questioned in his attack against his rivals was not just their ability to explain the evidence, but rather the scientific *legitimacy* of any theory that failed to offer a materialistic cause for observable phenomena. Thus, Darwin dismissed Owen's explanation of the similarity of anatomical structures in different animals by reference to the "plan of creation" saying, "But that is not a *scientific* explanation."[42]

Darwin's assumptions about what scientific theories must look like influenced the way he made his case in the *Origin of Species*. The *Origin* also established new methodological norms that prohibited explanations from invoking creative intelligence or intelligent design in the history of life. Norms proscribing such appeals contributed to the repudiation of design arguments during and after the late nineteenth century.[43]

Of course, the perceived success of the fully naturalistic theories of origin in astronomy, geology, and especially biology reinforced this trend. These theories taken jointly suggested that the whole history of the universe could be told as a nearly seamless unfolding of the potentiality of *matter and energy.* The default cosmology of the nineteenth century—which assumed that matter and energy were eternal and self-existent—reinforced this materialistic perspective, since it seemed to eliminate any need to consider the question of the ultimate origin of matter.

Darwin, Marx, and Freud: A Comprehensive Materialism

In addition to Darwin, two other late nineteenth- and early twentieth-century figures contributed to this increasingly entrenched worldview. Both did so by developing influential social scientific theories. In economics and social philosophy, Karl Marx's (Fig. 3.6) dialectical materialism, and his utopian vision of the future based upon it, expressed a profoundly deterministic as well as materialistic understanding of human nature.[44] In psychology, Sigmund Freud (Fig. 3.7) formulated a complex characterization of the human psyche describing the different elements and the animating motivations or objectives of each part of the mind, in the process painting a strikingly deterministic vision of human nature.[45] Whereas Marx depicted human beings as determined by material needs and impersonal economic forces, Freud portrayed behavior as dictated by largely unconscious sexual desires.

FIGURE 3.6
Karl Marx, the founder of dialectical materialism.

Both Marx and Freud were atheists who expressed disdain for the God hypothesis.[46] Marx regarded religion as an opiate propagated by the bourgeois elite to anesthetize the working classes to their exploitation.[47] Freud also thought that belief in God served a utilitarian end. He thought human beings had invented the myth of a benevolent God to provide a comforting father

FIGURE 3.7
The famed psychologist Sigmund Freud.

figure as a psychological crutch to compensate for difficult relationships with their actual earthly fathers. Thus, his famous dictum reversing the Judeo-Christian creation story: "God did not create man; man created God."[48]

With these three great figures—Darwin, Marx, and Freud—science seemed to answer many of the deepest worldview questions that, heretofore, Judeo-Christian religion had answered for people in the West. As I've explained somewhat aphoristically in conference talks: "Darwin told us where we came from, Marx told us where we are going, and Freud told us about human nature and what to do about our guilt." All three claimed to base their theories on scientific evidence and analysis.

Thus, by the early twentieth century, science seemed to support, if it could be said to support anything, a materialistic worldview, not a theistic one. Science no longer needed to invoke a preexistent mind to shape matter in order to explain the evidence of nature. Matter had always existed and could—in effect—arrange itself without a preexistent designer or creator.

Scientific Materialism and the Relationship Between Science and Faith

Not surprisingly, the rise of scientific materialism altered the way many intellectuals conceptualized the relationship between science and theistic belief. Many twentieth-century scientists, philosophers, and theologians perceived science and theistic belief as standing in overt conflict with one another—the view promulgated by John William Draper and Andrew Dickson White in their late nineteenth-century revisionist histories[49] and later aggressively popularized by the New Atheists.

We have seen that most historians of science now regard as extremely simplistic attempts to cast the whole history of science as a battle between science and Christianity. Nevertheless, it does not follow that such a conflict has never existed. Many twentieth-century scientists and philosophers have advanced the conflict model as at least the correct *current*, if not past, understanding. Scientists and philosophers holding this view assert that even if science and religion were not in conflict in the past, they are now.

Many conflict advocates cite the ascendancy of Darwinian thinking—with its denial of actual as opposed to merely apparent design in living systems—as the principle and irreconcilable locus of this conflict. If theism asserts that creative intelligence played a key role in the origin of living forms, and if evolutionary biology can account for the origin of living organisms by reference to wholly undirected material processes, then one of these two views must be incorrect. Leading proponents of evolutionary theory such as Francisco Ayala,[50] the late William Provine,[51] Douglas Futuyma,[52] Richard Dawkins,[53] and the late George Gaylord Simpson[54] agree that neo-Darwinism denies any discernible evidence of design or guidance in the history of life. Neo-Darwinism teaches, as Simpson once put it, "that man is the result of a purposeless and natural process that did not have him in mind."[55] Or as Ayala avers, neo-Darwinism explains "design without designer."[56]

Of course, it does not follow from the truth of neo-Darwinism or some other materialistic evolutionary theory that a deity could not possibly exist. Even popular proponents of scientific atheism, including Dawkins and Nye, admit that science cannot categorically exclude that possibility.[57] They do not deny the possibility of a designer whose creative activity is so masked in apparently natural processes that it escapes scientific detection. Yet for most evolutionary biologists such an undetectable entity hardly seems worthy of consideration. As Dawkins has argued, if the appearance of design can now be fully explained by natural causes, then surely it is simpler to attribute the appearance of design directly to such causes rather than invoking an additional factor, an undetectable designer.[58] Although the existence of such a designer has remained a logical possibility, the vast majority of evolutionary biologists, and certainly New Atheists, have rejected the idea as an unnecessary and unparsimonious explanation. At the very least, neo-Darwinism makes "theological explanations" of life "superfluous,"[59] as evolutionary biologist Douglas Futuyma writes.

Other scientists have argued, however, that since science has not definitively disproved the existence of God, it does not necessarily contradict religious belief, even though explicitly theistic explanations for the origin and development of life may be unnecessary. Those who advance this view typically portray science and religion as such distinct enterprises that their teachings do not intersect in significant ways. They also usually deny that theistic religion properly understood makes any factual claims about human or natural history; its claims are only about morality and meaning.[60] For example, paleontologist Stephen Jay Gould in his book *Rocks of Ages* advanced the idea that science and religion exemplify completely separate forms of inquiry addressing completely different kinds of questions. He argued that they each have authority to speak only in their own separate realms, reflecting what he called their "non-overlapping magisteria," popularly abbreviated as NOMA.[61]

Though Gould, an agnostic with strong atheistic leanings, advanced this view in 1999 as an olive branch of sorts to religious believers, many theists had already advanced similar formulations. British neuroscientist Donald MacKay, physicist Howard Van Till, and biologist Jean Pond developed two such models known as complementarity (associated with MacKay and Van Till[62]) and compartmentalism or independence (associated with Pond).[63] Prominent theologians and philosophers such as Karl Barth, Søren Kierkegaard, and Martin Buber affirmed similar views. Like Gould's NOMA concept, these models asserted the religious and metaphysical neutrality of all scientific knowledge.

Advocates of compartmentalism and complementarity developed their models to defend theistic belief against the aggressive philosophical materialism of many conflict theorists. Even so, advocates of these models have generally conceded the failure of science-based theistic arguments. Instead, advocates of independence and complementarity (or "partnership" in Van Till's lexicon) have argued that materialist origins theories do not necessarily contradict theological accounts of creation, since God may have used Darwinian or other similarly materialistic processes to create new forms of life. In their view, statements by scientists about the *purposelessness* of evolution do not represent scientific statements per se, but instead represent "evolutionism"—an "extrascientific" or "pseudoscientific" apologetic for philosophical materialism. Even so, advocates of independence and complementarity generally have agreed with staunch Darwinists on one point. Both deny that evidence of intelligent design is scientifically detectable and both agree more generally that scientific evidence does not, and cannot, provide positive support for theistic belief.

Clearly, the very existence of the independence and complementarity perspectives shows that the demise of the theistic arguments did not eliminate theistic belief, even among scientists. Some prominent nineteenth-century scientists such as Michael Faraday,[64] Louis Agassiz,[65] and James Clerk Maxwell,[66] giants in their respective fields of chemistry, paleontology, and physics, even continued to advance the natural theological tradition inaugurated by Boyle and Newton. Nevertheless, many theists in the sciences since have adopted a more compartmentalized view of the relationship between science and religion.

Thus, the demise of theistic arguments and the rise of scientific materialism radically changed the terms of engagement between science and religion. Although many scientists came to regard the witness of science as hostile to a theistic worldview, others began to view it as entirely neutral. Few, however, have thought—in contrast to the founders of early modern science like Kepler, Boyle, and Newton—that the testimony of nature (or science) actually supports belief in God. By the beginning of the twentieth century, science—despite its theistic beginnings—seemed to have no need of the God hypothesis.

Part II

Return of the God Hypothesis

4

The Light from Distant Galaxies

For centuries, scientists and philosophers have wondered about the origin of the universe. Did it have a beginning? Or was it always here?

Perhaps they should have known, but they didn't. Philosophers and scientists alike had long tried to answer the question by reasoning from logical or theological principles alone. Yet in so doing they failed to take notice of a fact that had long stared everyone—at least everyone who has ever viewed and contemplated the night sky—in the face. Ironically, it was a poet who figured it out first.

Dating back to classical antiquity, most philosophers thought that the universe had existed forever. Aristotle, as I mentioned, affirmed an eternal universe without a beginning in time.[1] He argued that belief in a temporally finite universe entailed a logical contradiction. He thought of time as a series of connected moments, each with a beginning and ending, connecting to the beginning of the next moment and coming from the ending of the previous moment. If the universe began with the first moment in time, it must have come from the ending of an earlier moment. But to say that implied that a moment in time existed before the beginning of time—clearly an impossibility. What could be more contradictory than to talk about a time before time *first* started? For Aristotle, any consideration of a beginning of time led logically to just such an absurdity.[2]

During late antiquity and the Middle Ages, whether on theological or philosophical grounds, many Jewish and Christian philosophers broke with Aristotle's thinking. Augustine,[3] Thomas Aquinas,[4] Maimonides,[5]

Bonaventure,[6] and others reaffirmed the Judeo-Christian idea of creation *ex nihilo* and, with it, the idea that the universe had a definite temporal beginning.

Medieval proponents of the Kalām cosmological argument thought that they could prove this idea by showing the absurdity of what philosophers called "actual infinites."[7] If the past is infinitely old, then getting from the past to the present would be like trying to climb to the surface of the earth from a hole infinitely deep—from a bottomless pit. As one contemporary philosopher has characterized the problem, "one could get no foothold in . . . [an infinite temporal] series to even get started, for to get to any point, one *already* has to have crossed infinity."[8] Nevertheless, other medieval theologians and philosophers such as Augustine, Aquinas, and Maimonides[9] merely affirmed by faith that the universe had a beginning, convinced that reason could not decide the question one way or another.[10]

Philosophers and scientists also wondered about whether the universe contained a finite or an infinite amount of space. Though Aristotle thought that the universe had existed for an infinitely long time, he also thought that it contained only a finite volume of space extending to an outermost sphere, known as the quintessential realm. By contrast, the Stoics, a later philosophical movement in the ancient world, insisted that an infinite amount of space (or a void) surrounded the earth in every direction.[11] Though many thinkers during the Middle Ages and the scientific revolution affirmed that the universe *did* have a beginning a finite time ago, many of those same thinkers believed that the universe extended infinitely far in every direction of space. Descartes and Newton both held versions of this idea, and, as already noted, Newton justified his belief in the infinity of space on scientific and theological grounds.[12] Nevertheless, belief in a spatially infinite universe gave rise to a troubling paradox known after the 1820s as Olbers's paradox.

Hints of a Finite Universe

Olbers's paradox refers to the mystery of the dark night sky.[13] It is named after the German astronomer Heinrich Wilhelm Olbers (1758–1840), though awareness of the puzzle extended back almost 250 years before it was so named. Olbers, following many earlier astronomers, recognized that an infinitely large universe with a roughly uniform distribution

FIGURE 4.1A

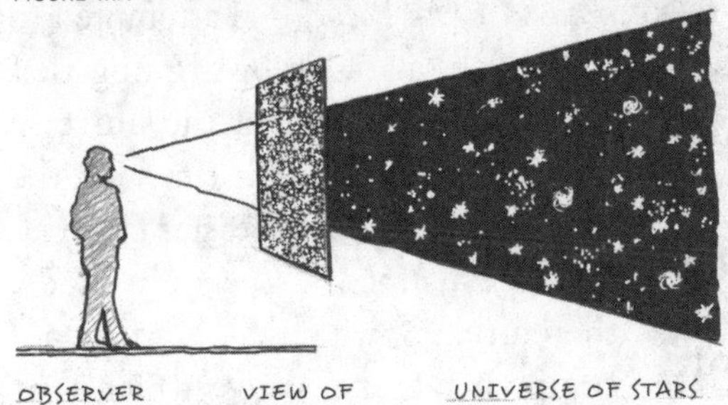

FIGURE 4.1B

Olbers's Paradox. The light from stars in the night sky at all distances appears to fill different parts of our visual field. If the universe were infinitely large, and stars or galaxies were distributed throughout it, every line of sight would terminate with a star or galaxy. In that case, the night sky would appear entirely illuminated and no dark regions would remain. That the night sky does not appear entirely white suggests that the universe is not infinitely large.

of stars or star clusters should exhibit a completely bright night sky.

Olbers had come to this conclusion because he realized that if space extended without end and if it contained a similarly infinite number of stars, then eventually any line of sight would terminate with a star (Fig. 4.1). Therefore, every possible point in the sky should shine brightly. To get a sense of why this follows, imagine standing in a forest surrounded by trees at various distances. If the forest continued forever in every direction, then however wide the trees were and however wide the gaps between the trees surrounding you were spaced, every line of sight would eventually terminate in a tree. And, consequently, the green of the forest would fill your entire field of view.

For centuries before Olbers, and ever since Copernican astronomer Thomas Digges first noted the puzzle in 1576, scientists and philosophers had tried to resolve the mystery of the dark night sky.[14] Various solutions were proposed. Digges argued that on its long trek across the universe the light from very distant stars might exhaust itself and thus never reach the earth. Others thought, as Newton initially did, that the universe had an infinite amount of space but not an infinite number of stars. Consequently, most quadrants of space would remain void of stellar material and the night sky would look mostly dark. Others argued that dark objects in the night sky—such as planets or the allegedly pervasive but invisible substance "ether"—would absorb light from stars before it ever reached the earth, causing the night sky to look dark even if stars were shining light along every line of sight toward earth.

Finally, in 1848, a poet, not an astronomer, formulated an explanation—one that anticipated by three-quarters of a century a later scientific discovery that would finally resolve Olbers's paradox to the satisfaction of scientists. The poet was an American known for his summoning of atmospheres of eeriness and dread: Edgar Allan Poe (Fig. 4.2).

FIGURE 4.2
The prescient poet Edgar Allan Poe.

In an extended essay entitled *Eureka: A Prose Poem*, Poe attempted to resolve the paradox by arguing that the immense extent of the universe did not afford enough time for the light to arrive from distant stars.[15] As he explained, "The only mode, therefore, in which . . . we could comprehend the voids which our telescopes find in innumerable directions, would be by supposing the distance of the invisible background so immense that no ray from it has yet been able to reach us at all."[16]

It's true that astronomers before Poe also proposed that the light from very distant stars couldn't reach us. But Poe had a different idea about the reason for this. Unlike the earlier thinkers, who thought that the light from distant objects might grow tired or that the ether might block it, Poe proposed that the universe was not old enough to give light sufficient time to get to the earth from the most remote regions of the vast night sky.[17] And if it was not old enough, that meant it was not of infinite age.

In Poe's view, if our present universe had existed for an infinitely long time, then the light would have had plenty of time to reach observers on earth, even if it traveled at a finite velocity across a great distance. But if our universe had only existed for a finite time, then the light from the extremely distant reaches of the universe might not have enough time to reach us here on earth. In this way, Poe resolved the paradox of the dark night—at least to his own satisfaction.

Poe postulated other ideas about the origin of the universe that have a strikingly modern feel. For example, he proposed that the universe started from a "primordial particle" and then expanded by "irradiating spherically" in all directions as new atoms were created.[18] This theory of the origin of the universe led him to propose what modern astronomers now think of as the correct solution to Olbers's paradox: the universe has a finite age, so light has only had time to reach us from a limited number of stars.

It would take nearly seventy-five years for most astronomers to catch up with the thinking of this visionary poet. And though the darkness of the night sky could have led astronomers to posit a temporally finite universe, in the end it was not the darkness but the light from distant stars—or associations of stars called galaxies—that finally tipped them off.

The Great Debate

Despite Poe's prescience, few physicists and astronomers at the beginning of the twentieth century doubted the infinite age of the universe for several reasons.[19] First, physicists during the nineteenth century accepted the framework of Newtonian physics, including Newton's affirmation of infinite space, which to some implied infinite time.[20] Further, during the late nineteenth century, as geologists used the measured rates of change of natural processes to establish the great antiquity of the earth, many scientists simply extrapolated these results and applied them to the universe. If the earth is extremely old, they reasoned, then perhaps the universe is infinitely old. As the theoretical physicist Simon Singh noted in his history of the formulation of the big bang theory, the nineteenth century's "growing uniformitarian movement came to the consensus that the earth is more than a billion years old, and that the universe must therefore be even older, perhaps even infinitely old." Why the jump from very old to infinitely old? Singh continues: "An eternal universe seemed to strike a chord with the scientific community, because the theory had a certain elegance, simplicity and completeness. If the universe has existed for eternity, then there was no need to explain how it was created, when it was created, why it was created or Who created it. Scientists were particularly proud that they had developed a theory of the universe that no longer relied on invoking God."[21]

Even though a consensus had formed by the turn of the twentieth century about the infinite age of the universe, considerable debate still raged among astronomers about its *size*. Harlow Shapley, a prominent astronomer at the Harvard College Observatory in the 1920s, maintained that our Milky Way galaxy contained the entire universe, and he estimated the diameter of the Milky Way (and thus the universe itself) at a relatively cozy 300,000 light-years across. Other astronomers at the time began to doubt that the whole of the universe was enclosed within the Milky Way, in part because of observations of structures in the night sky called spiral nebulae.[22]

Nebulae (singular: nebula) are gaseous regions in space visible either with the naked eye or through a telescope. The gaseous material in a nebula typically envelopes one or more stars, and the light from such stars illumines the surrounding gas, making a spectacular visual display.

In 1715 the famed British astronomer Edmond Halley described six prominent nebulae, then understood as any kind of cloudy-looking celestial object.[23] Over the next two hundred years, astronomers catalogued an increasing number of such astronomical phenomena and learned more about their structures and composition. With the invention of large telescopes capable of resolving fine details in distant objects as well as the increased use of photography, astronomers during the beginning of the twentieth century began to discern specific points of light and large-scale structures in many of the nebulae, suggesting that some contained many clusters of stars.

This discovery gave rise to the "Great Debate" between Shapley (Fig. 4.3a) and Heber Curtis (Fig. 4.3b),[24] the latter an astronomer at the Lick Observatory east of San Jose, California. Sponsored by the National Academy of Sciences, the debate took place in the spring of 1920 at the Smithsonian Institution in Washington, DC. At issue was whether the nebular "smudges" on astronomers' photographic plates represented indistinct gaseous material surrounding individual stars *within* the Milky Way or whether they represented distinct "island universes," or galaxies, *beyond* it.

FIGURE 4.3A

FIGURE 4.3B

The astronomers Harlow Shapley and Heber Curtis, opponents in "The Great Debate."

Astronomers regarded this question as immensely important. If the Milky Way encompassed all known nebulae, then its flattened shape, discovered by Herschel over a century before,[25] likely circumscribed the extent of the universe itself. But if the spiral nebulae represented galaxies in their own right—galaxies beyond the Milky Way—then of course the universe must extend far beyond our galaxy, and the Milky Way was but one of many galaxies.

At the debate, Shapley continued to maintain that the edge of the Milky Way galaxy defined the outer limits of the universe. To defend this view, he had to dispute that the spiral nebulae represented separate galaxies. Instead, he argued that they represented gaseous accumulations or clouds within the Milky Way itself. Curtis took the opposite view.

FIGURE 4.4
The astronomer Edwin Hubble.

Several years later Edwin Hubble, a young lawyer turned astronomer (Fig. 4.4), settled the question by decisively refuting Shapley. But his ability to do so depended upon an unsung hero of astronomy who previously had worked in relative obscurity during the first several years of the twentieth century.

The Cosmological Distance Ladder

Henrietta Leavitt (1868–1921) began examining photographic plates of stars at the Harvard College Observatory first as an unpaid volunteer, later as a "computer," as human tabulators were then called, and finally as an astronomer. The computers were often women, as you may know if you saw the 2016 movie *Hidden Figures*. Harvard College Observatory employed women to scan photographic plates and catalogue stars and other celestial phenomena at a time when few women participated in astronomical research.

The use of photographic plates had improved astronomy because it could catch more light over time than the human eye. With long exposures, stars too dim to be seen by the naked eye or even through a telescope would nonetheless leave distinct images on the plate. Leavitt

FIGURE 4.5
Henrietta Leavitt, the astronomer whose discoveries about Cepheid variable stars allowed astronomers to calculate distances to distant galaxies.

(Fig. 4.5), who was deaf from a disease suffered after she graduated from Radcliffe College,[26] had an extraordinary ability to analyze the telltale smudges of light on the plates and identify and catalogue stars. By doing so, she made a discovery that would help resolve the Great Debate.

Leavitt became interested in a type of pulsating star known as a Cepheid variable.[27] What she discovered about these stars enabled astronomers to determine—by a long chain of reasoning—the distances to the galaxies in which Cepheids could be observed. That, in turn, enabled them to determine that many galaxies were so far away that they could not reside in our Milky Way.

How Leavitt and other astronomers managed to do this stands as one of the great detective stories in the history of astronomy. Leavitt discovered that the brightness of Cepheid variables in a nebular structure called the Small Magellanic Cloud oscillates with a period that correlates with the magnitude of their brightness.[28] Leavitt noticed that the longer the *period of oscillation* from bright to dim to bright again (in days), the greater the *apparent brightness* of the Cepheid stars (Fig. 4.6).[29]

Technically speaking, "apparent brightness" is the brightness of a star as measured with a photometer, a device that measures the intensity of light. Photometers count the number of individual particles of light, called "photons," that arrive in a given area of a detector per second. Absolute brightness, on the other hand, is the brightness of a hypothetical star measured at some specific set distance from the earth (as it happens, 32.6 light-years, or 10 parsecs[30]). Since absolute brightness measures are calibrated using a standard distance measure, the absolute brightness varies only with pulsation, whereas apparent brightness varies with both pulsation rate and distance.

To see why this distinction between absolute and apparent brightness matters to an astronomer trying to determine the distance to luminous objects, imagine looking at a light coming from a lamppost through the fog while walking through a park at night. If you see a light in the park

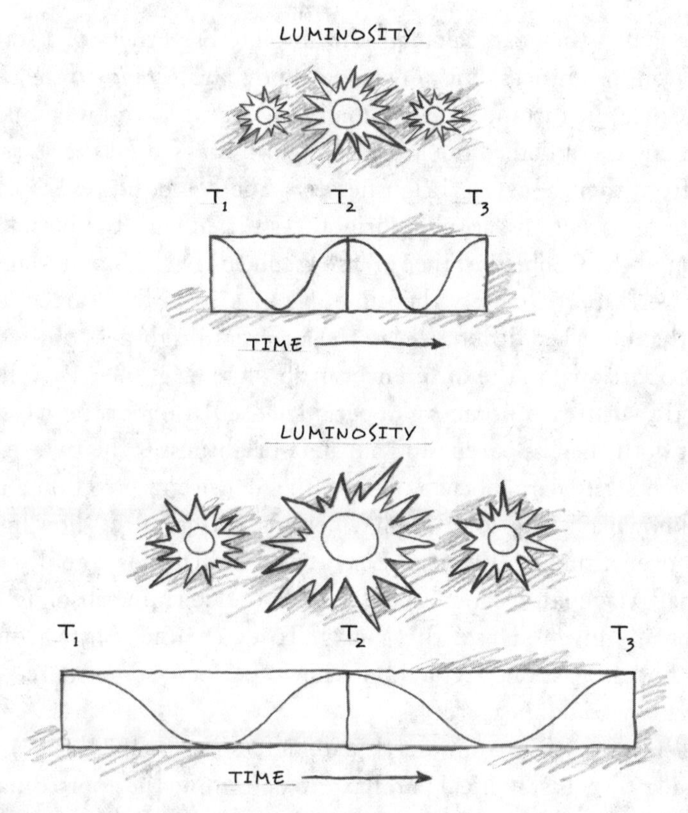

FIGURE 4.6

Cepheid variable stars. The brighter the star, the longer the period of oscillation of the light from bright to dark to bright again.

that looks extremely bright to you, you might attribute that apparent brightness to the light being extremely close at hand. Or you might attribute the brightness of the light to an extremely high-output lightbulb located on the other side of the park. In other words, the brightness you observe might be the result of either the close proximity or the extremely high luminosity of the light source. Unless you know the *distance* to the light, you cannot determine how brightly the light is shining at its source just from knowing how brightly it appears to you.

Astronomers have long faced a similar problem in using apparent brightness to calculate absolute brightness and distance. Astronomers can directly determine the apparent brightness of a Cepheid variable (or any) star by observation using a photometer. They also know that the intensity of light dissipates with distance. (Specifically, light intensity dissipates by the inverse of the square of the distance, or a $1/d^2$ factor.)

Consequently, they can calculate the absolute brightness of a star from its apparent brightness, but only if they know the *distance* to the star.[31]

Yet, since the distance to the Small Magellanic Cloud was unknown, Leavitt could not determine the absolute brightnesses of the stars in the cloud from their apparent brightnesses.[32] She could, however, plot how both the apparent and absolute brightness varied with the period of oscillation of the Cepheids. Since all the Cepheids in the Small Magellanic Cloud were approximately the same distance from the Earth, she reasoned that observed differences in the apparent brightness of these stars were proportional to the differences in absolute brightness as well. Consequently, when astronomers plot period of oscillation for those Cepheids against both their apparent and absolute brightnesses the two resulting lines necessarily parallel each other with a consistent offset on a type of graph known as a "log-log" graph (where the offset reflects how light dissipates over distance). Even so, Leavitt still didn't know the distance to the Small Magellanic Cloud, so she could not determine absolute brightness for any given star in the galaxy. To determine *that*, astronomers had to find at least one Cepheid of known period—somewhere—whose *distance* they could measure.

In 1913, Danish astronomer Ejnar Hertzsprung (1873–1967) used a method known as statistical parallax[33] to determine the approximate distance to a group of thirteen Cepheids relatively close to the sun. Since he could measure the *apparent* brightness and periods of pulsation of those Cepheids, he could calculate the *absolute* brightness of the Cepheids in that group using his newly calculated distance measurement.[34] He then used his knowledge of the average absolute brightness of the group as a whole to determine the distance to the Small Magellanic Cloud.

Here's how he did that. First, he found a Cepheid in the Small Magellanic Cloud with the same period of pulsation as the average of the group he studied nearer to the sun.[35] Since all Cepheids with the same period have the same absolute brightness, he could now determine the absolute brightness of that particular Cepheid in the Small Magellanic Cloud from its known period of pulsation. Indeed, that Cepheid would necessarily have the same absolute brightness as the average of the group near the Sun (since it had the same pulsation rate as that group). Since he also knew the corresponding apparent brightness of that Cepheid *in the Small Magellanic Cloud* (using the graph[36] based on Leavitt's measurements), he could now calculate the distance to that Cepheid in the Small Magellanic

Cloud—and, therefore, to the cloud itself.[37] Once Hertzsprung knew the distance to the Small Magellanic Cloud, he could calculate any Cepheid's absolute brightness in the cloud from its apparent brightness. Then, by matching the observed period of pulsation of other Cepheids in the universe to those on the same graph showing how absolute brightness varies with pulsation period, he could determine the distance to any Cepheid variable star in the universe (once he measured its apparent brightness).

This "bootstrapping" method formed part of what is now called the "cosmological distance ladder"—an overlapping system of different measurement techniques by which astronomers establish a known distance for a relatively nearby object and then use various methods to reason from known to unknown distances, step by logical step.

Even so, using these methods to calculate the distance to the Small Magellanic Cloud in 1913 still did not resolve the Great Debate. Hertzsprung calculated the distance to the Small Magellanic Cloud at about 30,000 light-years.[38] Yet at the time astronomers did not yet know the extent of the Milky Way, so they didn't know whether the Small Magellanic Cloud fell within the Milky Way or whether it might lie beyond it. In any case, few astronomers initially paid much attention to the significance of Hertzsprung's results.[39]

Galactic Distances and the Great Debate

Now at last Edwin Hubble enters the story. Using the most powerful telescope in the world at the time—an instrument 100 inches in diameter (Fig. 4.7) at Mt. Wilson in the San Gabriel Mountains above Pasadena, California—Hubble made a discovery that used Leavitt's findings and Hertzsprung's method of measuring distance to resolve the Great Debate. Hubble analyzed a forty-minute exposure of M31, the Andromeda nebula, which turned out to be much farther from the earth than the Small Magellanic Cloud. Hubble's photographic plates revealed what he first thought were three novae. (A nova is a star that displays an abrupt increase in brightness before gradually diminishing in brightness, usually over a period of months.) By analysis of variations on the photographic plate, he later realized that the three stars were in fact two novae and one Cepheid variable star.

On the plate next to the Cepheid, Hubble excitedly scratched out the "N" for nova and wrote "VAR!" (for variable).[40] He knew about

FIGURE 4.7
Edwin Hubble observing the night sky through the Hooker Telescope at Mt. Wilson Observatory in California.

Leavitt's findings and Hertzsprung's method showing how Cepheid variables could help calculate absolute brightness and thus distance. So after determining that the Cepheid in Andromeda brightened and dimmed over a 31.415-day period, he calculated its absolute brightness and then the difference between its absolute brightness and apparent brightness (as determined from his photographic plates). That in turn allowed him to determine the distance to the star in the Andromeda nebula. The result? His calculation showed that the star was approximately 900,000 light-years from the earth.[41] Yet Harlow Shapley had recently calculated that the Milky Way extended only about 300,000 light-years. That meant that the distance to the Andromeda nebula was three times greater than the total size of the Milky Way, the previously supposed maximum size of the universe as calculated by Shapley.

Hubble's conclusion: the Andromeda galaxy lies far beyond the Milky Way and the universe must be much bigger than Shapley and other astronomers had imagined. Thus, Andromeda was not a cloud of gas or a group of stars within the Milky Way, but a separate galaxy!

This discovery led to another discovery, one that would revolutionize cosmology. Like his first discovery, Hubble's second also rested on the work of another relatively unknown and less celebrated astronomer.

Spectroscopy and the Discovery of the Red Shift

Vesto Slipher (Fig. 4.8) was born ten years after the end of the Civil War and lived to see the first men walk on the moon. In 1912, before Hubble began to use the 100-inch telescope at Mt. Wilson and before he settled the Great Debate, Slipher used a smaller 24-inch telescope and new techniques in a field known as spectroscopy to document a curious feature of the light coming from faint nebulae. That work led to a discovery as

FIGURE 4.8
Vesto Slipher, the astronomer who first discovered the red shift of light coming from distant nebulae.

consequential in its own way as Neil Armstrong's "giant leap for mankind."

Spectroscopy is the study of the light emitted or absorbed by chemical elements and the characteristic wavelengths, frequencies, and colors of that light. When light passes through a prism, it separates into different colors, each of which has a different wavelength and corresponding frequency (longer wavelengths correspond to lower frequencies of oscillation of the light and shorter wavelengths to higher frequencies). Physicists refer to the full range of these wavelengths—including those not visible to the human eye—as the "electromagnetic spectrum."

They call the process by which a particular chemical element emits light of specific wavelengths a "spectroscopic emission." Physicists call the opposite process, where a chemical element (usually in gaseous form) absorbs specific frequencies or wavelengths of light, "spectroscopic absorption."

A short excursion into atomic theory and optics will help explain why the light coming from distant stars proved so important. When an atom gains energy from a collision with another atom, an electron, or a photon of light, the atom is said to become "excited." Upon excitation the electrons in the atom jump to higher energy levels. The "excited" electrons quickly drop down to lower energy levels, resulting in the emission of photons with energies equal to the differences between the energy levels (Fig. 4.9). Higher energy levels correspond to greater average distances between the electrons and the nucleus. Most important, the energy of an emitted photon

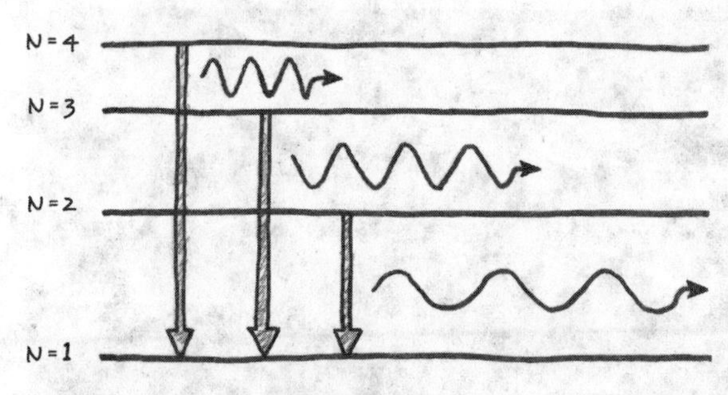

FIGURE 4.9
Spectral light emissions. When atoms gain energy from other atoms, electrons, or photons, they jump to higher energy levels. Such "excited" electrons then quickly drop down to lower energy levels, resulting in the emission of photons with energies equal to the differences between the energy levels. The energy of an emitted photon is directly proportional to its frequency and inversely proportional to its wavelength.

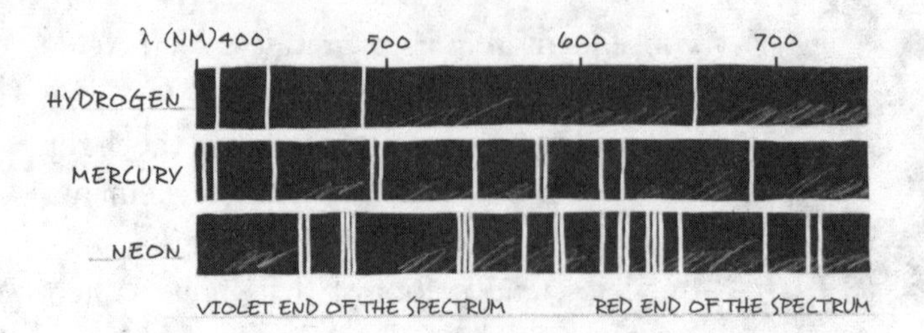

FIGURE 4.10
Different chemical elements emit a different combination of specific wavelengths of light in what is called an emission spectrum.

is directly proportional to its frequency and inversely proportional to its wavelength.

Each specific atomic element—whether hydrogen, helium, mercury, neon, or oxygen—has many unique energy levels. That means that when a particular element has excited electrons that drop to lower energy levels, the atom will emit photons with very specific frequencies and wavelengths corresponding to the energy differences between the levels in question. Since hydrogen, for example, has energy levels distinct from, say, helium or oxygen, it will emit a combination of photons with wavelengths different from those of other elements.

These patterns (Fig. 4.10) of specific emission wavelengths are known as "spectral lines," because they represent a discrete set of wavelengths within the spectrum or range of possible wavelengths of electromagnetic radiation. Because each chemical element has its own characteristic pattern (of emission or absorption lines and the spacing between them),[42] astronomers can use these patterns to determine the elemental composition of galaxies and stars based upon the light coming from them. Since

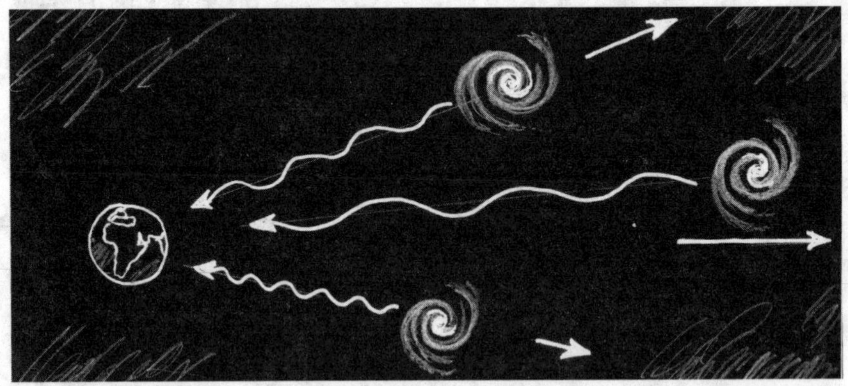

FIGURE 4.11A
Astronomers have discovered that distant galaxies are moving away from each other and from the earth. Consequently, light emitted at a given wavelength from distant stars will appear to be stretched out or "red shifted." Moreover, the farther galaxies are from the earth the faster they will recede from us and the more the wavelengths of light coming from them will be stretched out.

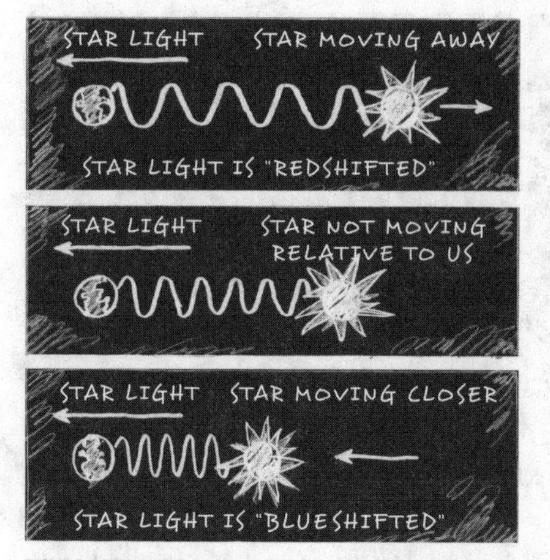

FIGURE 4.11B
The light coming from a galaxy moving away from the earth appears "red shifted" as the wavelengths of the light coming from that galaxy are stretched out or lengthened. The light coming from a galaxy moving toward the earth appears "blueshifted" as the wavelengths of the light coming from that galaxy are compressed or shortened.

the universe has more hydrogen than any other element, its lines are often the most dominant and easily identified.[43]

Using this knowledge, Vesto Slipher studied the spectra from planets, stars, and other bodies and determined their classification, elemental composition, and temperature. In 1912, he began to measure the light coming from spiral nebulae and discovered recognizable spectral-line patterns for each nebula, indicating the presence of specific elements.

He also discovered that the nebulae typically exhibited spectral lines that were shifted as a group *en masse* toward the red (longer wavelength) end of the electromagnetic spectrum. In other words, the spectral lines were at longer wavelengths than those of laboratory spectra for any given element, though the characteristic pattern of the spectral lines (the specific spacing between the lines) was roughly the same for each element.[44]

Slipher already knew what caused such shifts in wavelength. In 1848, the Austrian physicist and mathematician Christian Doppler discovered that the relative motion of an object affects the waves propagating from it. Sound waves from a train's whistle, for example, bunch up as the train approaches an observer and stretch out as the train recedes from that same observer. This phenomenon causes the sound of the train whistle to rise in pitch (shorten in wavelength) as it approaches and lower (lengthen in wavelength) as the train recedes.

This phenomenon is known as the Doppler effect. It also affects waves of light. With light, wavelengths are shortened if they issue from an object moving toward an observer and elongate if they are coming from an object moving away from an observer (Fig. 4.11). Since longer wavelengths correspond to redder light in the electromagnetic spectrum and shorter wavelengths correspond to bluer light, light from an approaching

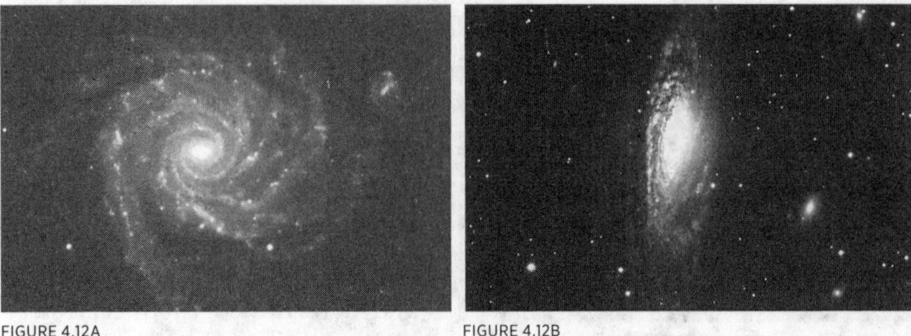

FIGURE 4.12A FIGURE 4.12B
Two of Edwin Hubble's original photographic plates showing spiral galaxies.

object will look bluer and light from a receding object redder than it otherwise would look.

As early as 1868, the British astronomers William and Margaret Huggins discovered that the light from stars exhibited such a Doppler shift. They made this discovery when analyzing the composition of Sirius, a nearby star in the Milky Way. Slipher discovered that individual stars that we now know are located within the Milky Way may exhibit either a blueshift or red shift, indicating that they may be moving either away from (red shifted) or toward (blue shifted) the earth. But he also discovered that the more indistinct spiral nebulae typically exhibited a much more substantial red shift, suggesting that they are receding extremely rapidly.[45]

As we saw, during the 1920s, Hubble determined that the spiral nebulae must lie well beyond the Milky Way, and therefore that they must represent separate galaxies. Moreover, the giant telescope through which he could now observe these galaxies allowed him to collect more light from them and record finer details on photographic plates. The resulting images confirmed that the spiral nebulae (Figs. 4.12a; 4.12b) were not individual nearby (perhaps forming) stars, but galaxies in their own right.

Hubble's next step was the revolutionary one. As he examined Slipher's red shift measurements as well as spectral studies of the red shift performed by fellow Mt. Wilson astronomer Milton Humason, he discovered an even more significant relationship.[46] The red shift associated with different galaxies at different distances revealed that more distant galaxies recede at faster rates than galaxies closer at hand. In fact, as Hubble plotted the recessional velocity and distance from earth of the different galaxies, he discovered a precise linear relationship between recessional velocity and distance.[47]

Hubble's plots showed a straight line roughly expressing a simple rule: the farther, the faster (Fig. 4.13). All other things being equal,[48] if one

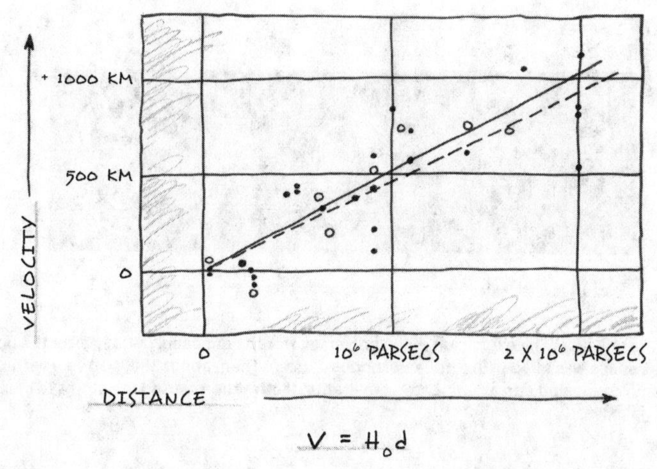

FIGURE 4.13

Galactic Recession and Hubble's Law. This chart shows the recessional velocity of several galaxies plotted against their distances from earth. It establishes that the farther galaxies are from earth, the faster they are receding from us. This linear relationship between recessional velocity and distance is known as Hubble's Law. (A parsec is a unit of distance used in astronomy.) See: Hubble, "A Relation Between Distance and Radial Velocity Among Extra-Galactic Nebula," 172.

galaxy is twice as far away from earth as another, it will be moving away twice as fast relative to the earth as that other, closer galaxy. In other words, Hubble discovered that the rate at which other galaxies retreat from ours correlates directly with their distance from us—just as if the universe were undergoing a spherical expansion like a balloon being blown up in all directions from a singular beginning.[49]

Hubble's discovery of an expanding universe (Fig. 4.14) was fraught with theoretical and philosophical significance. If the various galaxies are moving away from our galaxy and from each other in the forward direction of time, then at any time in the finite past the galaxies would have been closer together than they are today. As one extrapolates backward to determine the position of the galaxies at any given time in the past, not only would the galaxies have been closer and closer together, but eventually all the galaxies would have converged, bunching up on each other at some moment in the past. The moment where the galaxies converge marks the beginning of the expansion of the universe and, arguably, the beginning of the universe itself.[50]

When I explain the concept of the expanding universe to students or in public talks, I often illustrate the concept by blowing up a balloon with spiral galaxies drawn on the surface with a marking pen. As I inflate the balloon, the hand-drawn galaxies on the surface of the balloon get farther and farther away from each other. The galaxies in the universe do

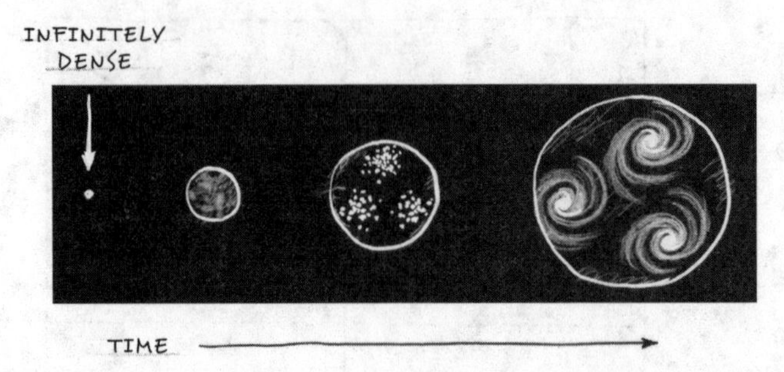

FIGURE 4.14
Expansion of the universe. The expansion of the universe after the big bang. Initially after the beginning of the universe, space was filled with a hot amorphous plasma. Then, about 380,000 years after the big bang, the plasma congealed into atoms. Later, gravitational attraction caused the atoms to coalesce into stars and galaxies.

the same as the result of the expansion of space itself. Moreover, the galaxies that are initially farther apart expand away from each other faster than the galaxies that are initially closer together, suggesting that the model of spherical or uniform expansion explains Hubble's observations well. By deflating the balloon, I can also illustrate how the material in the universe would have been closer and closer together at points farther and farther back in the past, suggesting that the matter making up the galaxies would have all come from the same initial point marking the beginning of the expansion of the universe.[51]

Of course, the actual universe contains galaxies *within* the expanding space as well as on "the edge" of space, and so the balloon analogy has its limits. For this reason, astronomers also sometimes think that a poppy-seed cake actually gives a better picture of the expansion of the universe. Before being baked, the poppy seeds are distributed throughout the cake. As the cake rises and expands during baking, each seed in the cake moves away from all the other seeds. The poppy-seed cake also illustrates why astronomers do not think that we can know where our Milky Way galaxy is located within the vast space of the universe itself. If we imagine ourselves sitting on any given poppy seed (or galaxy), all the other seeds (or galaxies) will appear to recede from us as the cake rises (or the universe expands), no matter where (or upon which seed or galaxy) we happen to be sitting.

In any case, Hubble's discovery implied an expanding universe in the forward direction of time and a finite universe with a definite beginning in the distant past. His discovery implied that the universe was expanding and had a beginning, just as Edgar Allan Poe's "Eureka" moment had anticipated three-quarters of a century before.

5

The Big Bang Theory

The evidence of galactic red shift, suggesting an expanding universe, was startling. But it became even more important, and revolutionary, when conjoined with Albert Einstein's new theory of gravity known as general relativity. Although Einstein's (Fig. 5.1) theory foreshadowed Hubble's discovery, Einstein himself initially rejected the proposal of an expanding universe. Even so, his work provided the framework in which Hubble's discovery would be understood. Eventually, the synthesis of Einstein's theory of gravity with the evidence for an expanding universe from observational astronomy came to be known as the big bang theory.

FIGURE 5.1
The physicist Albert Einstein who developed the theories of special and general relativity.

In 1915 Einstein shocked the scientific world with his theory of general relativity.[1] He built general relativity on his earlier theory of special relativity. Special relativity affirms the counterintuitive idea that distance and time are relative in the sense that two observers moving at different velocities will perceive time and space differently.[2]

Einstein came to this conclusion using a series of thought experiments (Fig. 5.2).[3] In one of these, he imagined himself sitting on a train traveling away from a clock tower approaching the speed of light. Since light

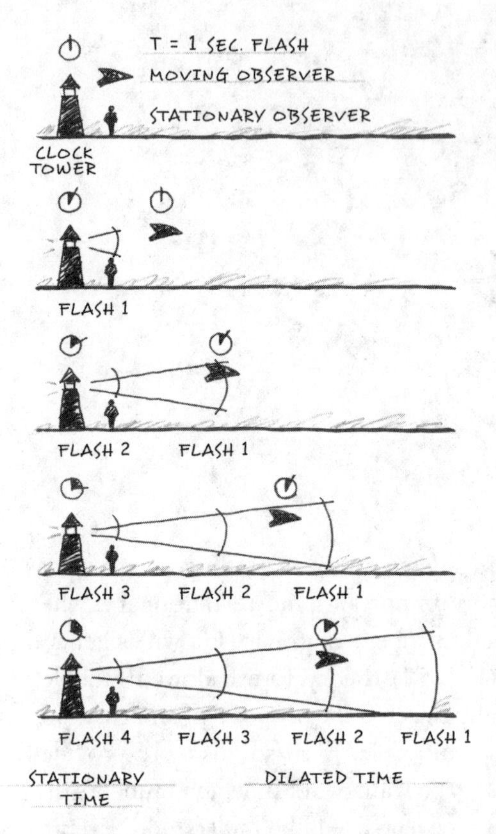

FIGURE 5.2
Time dilation. According to Einstein's theory of special relativity, time appears to slow down to an observer in a moving reference frame such as a spaceship as that moving object approaches the speed of light. This figure depicts the basis of Einstein's intuition by showing that as a spaceship moves away from a clocktower at high speed the information about the passage of time as conveyed by the moving hands on the clock (and successive flashes of light coming from the tower) will take longer to get to the spaceship than to a stationary observer closer to the tower. Thus, time near the clock will appear to move more slowly to the astronaut.

coming from the clock carried information about the change in position of the hands of the clock—that is, about the passage of time—he realized that if he traveled near the speed of light, information about the movement of the clock hands would take longer and longer to catch up to him as he moved away from the clock at higher and higher speeds and got closer and closer to keeping pace with the light.

Consequently, Einstein realized that he would observe time as measured near the clock (in its "frame of reference") as slowing down compared to time in his own frame of reference on board the speeding train. Yet a stationary observer near the clock (or one moving slowly relative to the speed of light) would perceive the clock ticking much more quickly (or "normally") in that reference frame.

Einstein's realization that the perception of time depends on the speed of the observer relative to the object observed is known as "time dilation." Physicists have actually measured time dilation by synchronizing highly accurate atomic clocks and comparing the passage of time on a supersonic airplane with time as measured on a stationary clock.

Through a similar analysis, Einstein recognized that spatial measurements must also dilate (or contract) at speeds approaching the speed of light. A spaceship passing an observer at speeds approaching that of light will appear shorter along the direction of motion than it would to an observer on board or to any observer not moving with respect to the spaceship.

Einstein's thought experiments showed that our measurements of space and time are fundamentally linked. Our perception of time depends on how fast we are moving *through* space; our perception of space depends upon how fast we are moving *over* time. That linkage suggested to him a new entity—*spacetime*. Spacetime combines the time variable (t) with the three spatial variables (x, y, z) in a four-dimensional continuum (x, y, z, ct) where c represents the speed of light.

Spacetime lay at the heart of his theory of general relativity, a novel view of how gravitational attraction works. Whereas Newton viewed gravity as a force between objects having mass, Einstein reconceived gravity as a geometric property of spacetime, something he saw as a multidimensional "fabric" that objects having mass could warp.[4]

Just as a bowling ball set down on a large trampoline makes a depression in its surface, a large mass such as the sun will curve or depress the fabric of spacetime. The more mass an object has, the larger the warp or depression. Objects having less mass "fall into" the depression in spacetime caused by objects with larger mass, just as tennis balls at the edge of a trampoline will roll into the depression created by a bowling ball placed in its center. Thus, general relativity, and Einstein's field equations expressing the theory mathematically, describe how curved space affects the movements of massive objects and how massive objects curve space. Or as the physicist John Archibald Wheeler cleverly summarized the theory, "Space tells matter how to move, and matter tells space how to curve."[5]

Astronomers confirmed Einstein's theory experimentally. They did so first by showing that it better accounted for a previously unexplained shift in the orientation of Mercury's orbit than Newton's theory did. Then, in 1919, an ingenious experiment showed that, just as general relativity predicts, light itself bends when it passes by massive objects. The British astrophysicist Sir Arthur Eddington demonstrated this by carefully observing starlight passing by our own sun during a total solar eclipse (Fig. 5.3).[6] The gravity of the sun had indeed bent the light, moving the apparent star positions farther from the sun.

Einstein's theory challenged Newton's theory of gravity in many important respects. But as Newton's theory of universal gravitation had

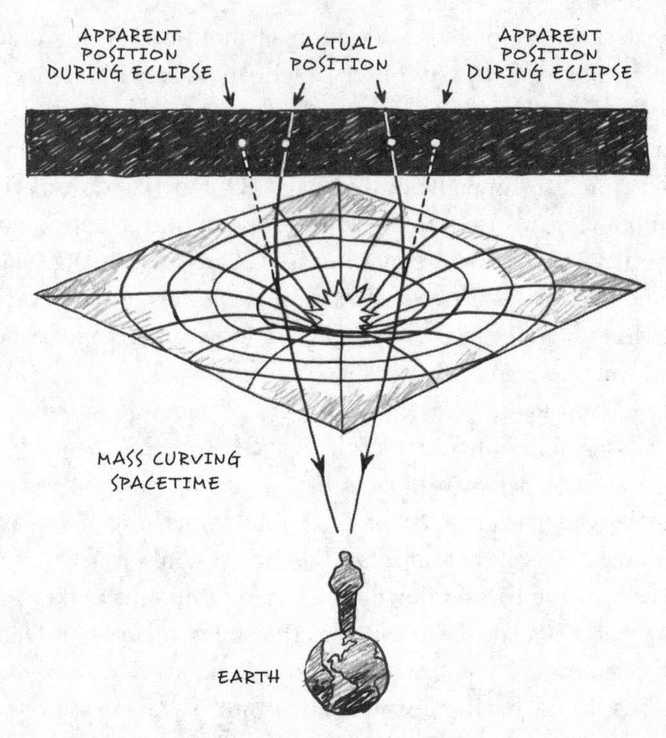

FIGURE 5.3

According to Einstein's theory of general relativity, massive bodies curve space. This curvature bends the path of light toward a massive body as it moves past. The diagram shows light coming from two distant stars passing through the gravitational field of the sun. The curvature of space around the solar mass alters the path, so the light curves around the sun. As a consequence, the apparent positions of the stars in the night sky appear to an earthbound observer to have shifted from their true position. Note how the stars' apparent positions in the diagram are shifted more to the left or more to the right of their actual positions. This effect is only observable on earth during a solar eclipse when the light coming from the sun is blocked by the moon. A famous experiment was performed in 1919 by Sir Arthur Eddington during a solar eclipse. He identified the predicted light-bending by observing the resulting shift in the apparent position of a particular star as the moon passed in front of the sun during the eclipse.

done before, it too implied that gravitational action would—in the absence of other counteracting forces—cause the matter in the universe to congeal into one place. Since according to Newton's theory "all matter gravitates"—all masses attract other masses—his theory implied that all matter would collapse in on itself into one great dense concentration of mass as the result of each massive body exerting an attractive force on every other massive body.

Since clearly all of the massive bodies in the universe have not congealed in this way, Newton attempted to solve the problem by positing an infinite amount of matter distributed throughout an infinite space. In so doing, he envisioned a balance of forces in which every massive body distributed throughout infinite space would attract every other massive body in every direction simultaneously, resulting in the various gravitational forces balancing each other (or, mathematically speaking, canceling each other out), thus putatively preventing the collapse.[7]

Einstein, however, realized that he could not avail himself of this solution. In his theory matter actually bends *space itself*. It does not just cause one mass to attract another mass *within* space. Consequently, according to general relativity, even within an infinite space, massive bodies would still cause space to curve in on itself, eventually resulting in the gravitational compaction of all matter and spacetime. In other words, Einstein realized that if gravity were the only force acting in the universe, it would necessarily cause matter to congeal and spacetime to contract in on itself. Yet since such a contraction hasn't happened (at least not yet) and since, further, the universe we observe today contains matter surrounded by empty space, Einstein thought something—some outward-pushing force of expansion—must be counteracting the effect of gravitation to account for the empty space between massive bodies in the universe.

The Cosmological Constant and the Static Universe

Thus, in his famous 1917 paper "Cosmological Considerations in the General Theory of Relativity,"[8] Einstein posited what he called the "cosmological constant" to describe a constantly acting repulsive force to counter the effects of gravitational contraction.[9] He further assigned a precise value to the cosmological constant to ensure that the strength of gravity and the repulsive force described by this constant exactly balanced, so as to sustain the universe in a kind of equipoised static state.[10]

Einstein's choice of the value for the cosmological constant had no physical justification. Instead, it followed from his assumption of an eternal, steady-state universe—an assumption he favored for explicitly philosophical reasons.[11] He also assumed that neither the density of mass-energy in the universe nor the radius of curvature of the universe[12]—two key terms in his equations—changed with time, though his initial equations treated these terms as potentially variable.[13] In so doing, Einstein depicted the universe as having always been static, neither expanding from a beginning nor contracting toward an end.[14] This allowed him to conceive of the universe as eternal and self-existent.[15]

Immediately after Einstein published his cosmology paper, a series of mathematical results challenged his static universe.[16] The Dutch mathematician and physicist Willem de Sitter solved Einstein's field equations for the special case of a universe without any matter in it. Since de Sitter's model also assumed the outward-pushing action of the cosmological constant, it necessarily implied an expanding universe. Nevertheless, Einstein rejected de Sitter's model and its implication of an expanding universe as unrealistic and theoretically inconsequential, since our universe obviously does contain matter.

A more significant theoretical challenge soon followed. In 1922, the Russian physicist Aleksandr Friedmann (Fig. 5.4) also solved Einstein's gravitational field equations, but he did so making more realistic assumptions about the universe. Unlike de Sitter, he assumed a universe with matter and energy in it as well as a roughly uniform distribution of that mass-energy. Friedmann's solutions and resulting equations included terms that allowed the density and radius of the universe to change or vary with time—a possibility that Einstein's arbitrary choice of the cosmological constant and initial conditions foreclosed.

FIGURE 5.4
The Russian physicist Aleksandr Friedmann, who solved Einstein's gravitational field equations.

Though Einstein initially expressed disapproval of Friedmann's decision to allow for the possibility of a dynamic universe, Friedmann's assumption followed logically and mathematically from the most basic physical principle of Einstein's theory of gravitation itself, namely, that massive bodies cause space to contract and therefore *to change*. Indeed, if mass causes space to curve or contract, then both the radius of curvature of space and the density of mass-energy within space could—depending upon the value of the cosmological constant—conceivably change over time. Friedmann himself did not attempt to decide whether the universe was static, expanding, or contracting, but he showed mathematically how different values of the cosmological constant could result in any one of those three possibilities.

Moreover, Friedmann's equations—his solutions to Einstein's field equations describing how matter bends space—implied a dynamic universe for *almost all* values of the cosmological constant and *almost all* choices of initial conditions. Indeed, his solutions implied that even for the exact value of the cosmological constant that Einstein had chosen—and except for Einstein's equally arbitrary assumption of a universe with unchanging radius and density—the universe would necessarily either expand or contract.[17]

Consequently, though Friedmann did not disprove Einstein's static universe, his solutions to the field equations implied the need for an implausible degree of fine tuning in both the value of the cosmological constant and the initial conditions of the universe in order to maintain a balance between the pressure of cosmic expansion and gravitational attraction. Friedmann thus amplified the tension between Einstein's preferred static-universe concept and the most natural implications for cosmology of Einstein's own theory of gravitation.

Other developments only deepened this tension. In 1927, the Belgian priest and physicist Georges Lemaître (Fig. 5.5) independently produced the same solutions to the field equations as Friedmann had done. Lemaître, however, not only showed that the field equations implied that the radius of curvature of space will change as time progresses; he also used observational data about the distant spiral nebulae (by now known as galaxies) to formulate a definite cosmological model of the universe.[18]

FIGURE 5.5
The Belgian priest and physicist Georges Lemaître, the father of the big bang theory.

Specifically, he incorporated Vesto Slipher's data about the Doppler shifts of the light from distant galaxies into his model and correlated that data with Hubble's 1924 measurements of the distances to other galaxies. These two data sets taken jointly implied that the galaxies were receding and that the galaxies that were farther away were receding faster than those close at hand. Though Hubble later formulated this relationship with more precision based on more observational data, Lemaître formulated it independently and before Hubble. This "the farther, the faster" relationship, later called Hubble's Law, suggested a spherical expansion of the universe in all directions of space.

Unlike Friedmann, whose equations merely implied that the universe *could* change in size over time, Lemaître cited evidence to show that it had changed—and was, in fact, expanding. Since Lemaître integrated the observational evidence of red shift into a cosmological model based upon general relativity (and his solutions to Einstein's field equations), his model implied that *space itself* was expanding, not just that the galaxies were receding into preexisting space (something that Hubble likely had not grasped). That, in turn, implied that the universe would have been much smaller in the past. More startlingly, it also implied, in the words of British physicist Stephen Hawking, that "at some time in the past . . . the distance between neighboring galaxies must have been zero."[19]

Thus, Lemaître both solved the field equations (as Friedmann had done) and used the red shift evidence (anticipating much of Hubble's later work) to develop a comprehensive cosmological model. His model implied an expanding universe in which space itself was expanding and, consequently, also implied a beginning to the expansion starting from what he described as a "primeval atom" or "cosmic egg." His model formed the foundation of the theory to which astrophysicist Fred Hoyle, a steady-state proponent, later applied the derisive label "the big bang."[20]

For philosophical reasons, Einstein disliked both Friedmann's and Lemaître's solutions to his gravitational field equations and specifically their implication of a dynamic and expanding universe. In 1922, he wrote a brief rebuttal to Friedmann's analysis claiming that Friedmann had not correctly solved the field equations. Einstein insisted that "the results concerning the non-stationary world, contained in [Friedmann's] work, appear to me suspicious. In reality it turns out that the solution given in it does not satisfy the field equations."[21]

After receiving a letter from Friedmann in which Friedmann responded persuasively to Einstein's criticism, Einstein published a retraction, acknowledging an error in his own calculations. Einstein also acknowledged that Friedmann had correctly shown that "the field equations admit, for the structure of spherically symmetric space, in addition to static solutions, dynamical solutions."[22]

Nevertheless, in an unpublished version of that same manuscript he characterized Friedmann's solutions as "unrealistic"—mathematically interesting, perhaps, but not applicable to the real world. In 1927, he offered a similar critique of Lemaître's solutions and his cosmological model at a prominent conference of physicists in Brussels, Belgium, called the Solvay Conference.[23] There he famously told Lemaître, "Your calculations are correct, but your physical insight is abominable."[24] Else-

FIGURE 5.6
Albert Einstein viewing the heavens through Hubble's telescope at the Mt. Wilson Observatory with Hubble (middle) and astronomer Walter Adams (right) watching.

where he expressed disdain for Lemaître's primeval atom hypothesis as an idea "inspired by the Christian dogma of creation, and totally unjustified from the physical point of view."[25]

But the heavens would soon talk back. In 1931 Einstein visited Hubble at Mt. Wilson and saw the astronomical evidence in support of the expanding universe through the great 100-inch telescope there. In the adjoining photograph of that visit (Fig. 5.6), you can see a famous picture of Einstein looking through the telescope, with Hubble in the background smoking his pipe. Soon after visiting Hubble at Mt. Wilson, Einstein publicly acknowledged that he recognized the necessity of a "beginning."[26]

This story, as commonly told, often places too much emphasis on Einstein's visit to Mt. Wilson and his interactions there with Hubble as the decisive event in his change of perspective. In truth, Einstein had probably come to accept the expanding-universe model more than a year earlier. He first learned about the red shift evidence from Lemaître in a taxicab ride during the Solvay Conference in 1927. In 1930, Sir Arthur Eddington (Fig. 5.7) also informed Einstein about the new developments in observational cosmology—including Hubble's 1929 paper establishing Hubble's Law—while Einstein was visiting Eddington at the University of Cambridge.[27]

FIGURE 5.7
The British astrophysicist Sir Arthur Eddington, who told Einstein about the evidence for galactic recession on a 1930 visit to Cambridge University, though Eddington himself did not much like its implications for a beginning to the universe.

Eddington also likely explained to Einstein why his static-universe model was unstable—and why Lemaître's equations, therefore, better represented the cosmological implications of general relativity than Einstein's own static-universe concept. Earlier that year, Eddington had shown that even for the value of the cosmological constant that Einstein chose, the universe would remain in static balance *only if* the mass and energy in the universe stayed evenly or homogeneously distributed. Even slight imbalances in the distribution of mass-energy would shift the universe toward a dynamic state in which pockets of space (or space as a whole) would either collapse or expand.

Eddington showed that the values for the cosmological constant and the curvature of the universe (as well as the mass-energy density of the universe) needed to be perfectly set and maintained. Even the slightest alteration in any of those values would cause the universe to either expand forever or contract back onto itself in a great cosmological "big crunch."[28]

So by the time Einstein arrived in Pasadena on January 29, 1931, he had already come to accept a dynamic and expanding universe and, with it, the implication of a beginning. As he explained in an interview with the *New York Times* published on January 3, "New observations by Hubble and Humason [astronomers at Mt. Wilson] concerning the red shift of light in distant nebulae" establish that "the general structure of the universe is not static."[29] He also stated in another *New York Times* interview on February 12: "The red shift of the distant nebulae have smashed my old construction like a hammer blow."[30]

Later Einstein said that his postulation of an arbitrary value for the cosmological constant—his cosmic fudge factor—was "the greatest blunder" of his life. Indeed, by seeking to preserve a static universe, Einstein inadvertently concealed an important cosmological reality implicit in his own theory of gravitation.

The Steady-State Cosmology

Einstein was not the only scientist who reacted reflexively against the idea of a beginning. Eddington himself found the metaphysical implications troubling. "Philosophically, the notion of a beginning of the present order is repugnant to me," he said. "I should like to find a genuine loophole. I simply do not believe the present order of things started off with a bang. The expanding Universe is preposterous. . . . It leaves me cold."[31]

Robert Dicke, a leading Princeton University physicist during the 1950s and 1960s, later explained why a finite universe elicited such knee-jerk philosophical opposition among so many scientists. An infinitely old universe "would relieve us," he said, "of the necessity of understanding

FIGURE 5.8
The three architects of the steady-state theory, Thomas Gold, Hermann Bondi, and Fred Hoyle.

the origin of matter at any finite time in the past."[32] A finite universe, by contrast, would force scientists to confront uncomfortable questions about the ultimate beginning of the material universe itself. It also raised the possibility that the universe had begun in something like a creation event produced by a cause that existed independently of matter, space, time, and energy.

Consequently, during the remainder of the twentieth century, physicists and cosmologists formulated many alternatives to the new big bang cosmology. Most of these attempted to restore the idea of an infinitely old universe. Some were formulated for explicitly philosophical reasons by scientists openly committed to a fully materialistic worldview.

For example, in 1948 three Cambridge researchers—Fred Hoyle, his fellow astrophysicist Thomas Gold, and mathematician Hermann Bondi (Fig. 5.8)—proposed the "steady-state" model to explain galactic recession without invoking the objectionable notion of a beginning.[33] Hoyle himself acknowledged that he proposed the steady-state model to circumvent what were to him the obvious theistic implications of the big bang theory.

According to the steady-state theory, as the universe expands, new matter is generated spontaneously in the space between expanding galaxies. For example, the matter of which the Milky Way galaxy is made would have popped into existence in between other galaxies that had in turn emerged from the empty space between other galaxies, and so on. Hoyle, Gold, and Bondi envisioned a universe of infinite extent in time and space—one that had always been expanding in the past and that would also always be expanding in the future.[34]

Oddly, the idea for the steady-state model came to Gold while he was watching a horror movie. The movie included a dream sequence in which the plot appeared to be changing, yet it always ended up exactly where it began.[35] Gold, with Hoyle and Bondi, proposed that the story of the universe might follow a similar script. Since the red shift evi-

FIGURE 5.9
According to the steady-state model, the universe must maintain a constant density of matter. But, as the universe expands, the density of the universe (i.e., the amount of matter per unit of volume) would begin to decrease. Consequently, to maintain a constant density, matter must be continually created throughout the universe. In effect, the stretching of space causes new matter to pop into existence. This figure depicts how steady-state proponents envision both the expansion of space and the continual creation of matter and energy.

dence supported an expanding universe, they proposed that the universe could endlessly double in size. But since doubling an infinite volume just generates another infinite volume, cosmic expansion would not actually change the measurable dimensions of the universe. As in the horror movie, the dream always returned to its starting place, an expanding but infinitely large universe.

So as long as some physical process, force, or field could continuously generate new matter from the empty but expanding space, the steady-state theory eliminated the need to posit a creation event at the beginning of time. Instead, the universe could simply continue to expand forever as it had been doing from eternity past. This idea was consistent with the red shift evidence, but it raised one obvious question: Where did the new matter come from?

Hoyle answered by postulating what he called a "C-field" or "creation field." To justify this proposal he somewhat arbitrarily asserted, as a fundamental physical principle, that the density of the universe must always remain constant (Fig. 5.9). It followed from that premise that, as the universe expands, it must produce a compensatory amount of new matter to maintain constant density. As astrophysicist Jean-Pierre Luminet notes, "Fred Hoyle demonstrated that the steady-state model was feasible on condition that a new field (which he called simply C for 'creation') was added to the equation; this ad hoc invention was envisaged as a reservoir of negative energy which had existed throughout the life of the universe—that is, forever."[36]

Evidential Challenges to Big Bang Cosmology

The steady-state theory remained the main competitor of the big bang model well into the 1960s. It was popular not only because it seemed philosophically less distasteful to many scientists, but also because the big bang theory had not yet explained several key classes of relevant evidence.

First, radiometric dating of terrestrial rock formations yielded an estimated age of the earth at 4.5 billion years.[37] Yet early versions of the big bang theory, which assumed an incorrect value for the constant in Hubble's Law, projected that only 1.9 billion years had elapsed from the beginning of the cosmic expansion till the present, implying that the earth was older than the universe that enclosed it—clearly an absurdity.[38]

Second, the big bang theory also offered no explanation for how, following the initial explosive beginning of the universe, lighter elements (such as hydrogen and helium, with just a few protons and neutrons) could have produced heavy elements (such as carbon and oxygen, with many more protons and neutrons).[39]

Finally, the big bang model predicted the presence of something that had remained undetected—a pervasive low-energy background radiation throughout the universe.[40] An analogy may help to illustrate why the theory made this prediction. Imagine roasting a turkey. When it's done, you take the dense, fully cooked bird out of the oven and place the turkey on the kitchen counter, being careful to then close the oven door. The turkey will radiate heat energy in all directions, raising the temperature of the room by a barely perceptible amount. In the same way, the dense concentration of mass-energy that, according to the big bang theory, existed after the beginning of the universe would have resulted in electromagnetic energy radiating throughout the universe as space expanded, leaving behind a background energy as a kind of signature of that initial hot, dense state.

Nevertheless, since the early universe contained not matter in a solid form as we know it (like a turkey), but hot plasma, the above analogy does not completely capture what big bang proponents envisioned. In their view, as the universe first began to expand, it would have had an incredibly tiny volume, with the mass and energy of the universe under extreme heat and pressure. In this state, known as a plasma state (a fourth phase of matter in addition to solid, liquid, and gaseous phases), electrons could not assume orbits around protons and neutrons to form stable atoms. Consequently, no light would radiate beyond the plasma. Instead, photons would scatter off electrons indiscriminately in all directions in much the same way that light scatters off the water droplets in a fog, making specific objects within the fog essentially invisible.

Next, according to the model, after 380,000 years the universe would have cooled to allow neutral hydrogen atoms to form, making it possible for light to travel freely. The light emitted from these first atoms would have then begun to bathe the expanding universe, moving through space in essentially straight paths in every direction.

In 1948, the physicists Robert Herman and Ralph Alpher predicted the existence of this light. They also predicted that the expansion of space would have gradually stretched out the light's wavelengths far toward the nonvisible end of the electromagnetic spectrum. Consequently, they expected that by the present day the light emanating from this original hot plasma would exhibit wavelengths of about 1 millimeter, corresponding to what physicists call the "microwave" portion of the electromagnetic spectrum.[41] Thus, Herman and Alpher dubbed the predicted energy the "cosmic microwave background radiation," or CMBR. This radiation would—if found—represent the afterglow of the big bang or, more specifically, that of the time just after the first atoms formed.

Herman and Alpher also calculated the temperature of a specific object, called a blackbody, that would today typically emit radiation with the same dominant wavelength as that of the predicted background radiation. Blackbodies are objects that absorb radiation of all wavelengths and that reemit radiation with a characteristic frequency distribution that depends only on the temperature of the blackbody,[42] kind of like the characteristic reddish glow from a cast-iron skillet when you heat it to a specific temperature. Herman and Alpher calculated that a specific blackbody with the same long-wavelength, low-energy radiation as that producing the radiation predicted on the basis of the big bang model would have a temperature of 5 degrees on the Kelvin scale (i.e., 5 degrees above absolute zero).

Here's how they did it. Herman and Alpher knew that since the time atoms first formed 380,000 years after the big bang, the distance across the universe had expanded by about 550 times. This expansion of space would have in turn caused the wavelengths of light coming from the initial plasma state to stretch out proportionately. They also estimated the temperature of the universe near the end of its early plasma stage at about 3000 degrees Kelvin. Since the characteristic temperature of a blackbody drops in proportion to an increase in wavelength, Herman and Alpher could calculate the temperature of a blackbody that would produce radiation equivalent to that present in the universe today. They did this by dividing 3000 degrees Kelvin (the temperature of the universe at the end of the plasma epoch) by 550 (the expansion factor of the universe), giving them a temperature equivalent of roughly 5 degrees Kelvin for the ubiquitous cosmic background radiation.

Their precise prediction constituted a solid piece of theoretical physics that researchers could in principle confirm by observation. Despite some early attempts, however, astronomers and astrophysicists in the early 1960s were unable to find any such low-energy, long-wavelength radiation. This presented big bang proponents with yet another anomaly to explain.

The Big Bang's Big Win

Thus, the big bang model was stymied for a time by evidential difficulties on three separate fronts. Yet new discoveries would soon resolve each of these.

First, there was the absurdity that the earth seemed to be older than the universe. In 1952 Walter Baade, of the California Institute of Technology, performed new studies on a particular class of Cepheid variables and discovered systematic errors in previous studies of these stars—errors that had the effect of underestimating the distances by a factor of two to faraway galaxies.[43] The need to recalibrate distances, in turn, implied that light coming from those galaxies would take longer to arrive. The result—by a new calculation—was a universe of 3.6 billion years in age.[44]

A few years later Caltech astronomer Allan Sandage, whom I mentioned in Chapter 1, demonstrated that the brightest stars in galaxies did not shine with approximately the same intensity, as astronomers had previously assumed. Correcting this assumption increased the estimated age of the universe to 5.5 billion years. Subsequent studies throughout the 1950s led Sandage to push back the estimated age of the universe to at least 10 billion years,[45] close to the current estimate of about 13.8 billion years.[46] These studies demonstrated the big bang occurred long enough ago to accommodate the ages of the astronomical objects contained in the universe, including a 4.5-billion-year-old earth.

Ironically, Fred Hoyle, a critic of the big bang theory, helped resolve the second problem facing the theory—that of explaining how heavy elements are produced. Hoyle took great interest in the issue because his steady-state theory also needed to account for the production of heavy elements.

Hoyle formulated a theory that showed how massive stars could synthesize carbon from lighter elements via a series of nuclear reactions known as the "triple-alpha process." (For more on this theory, see Chapter 7.) He thus inadvertently provided support for the big bang theory by removing one of the few remaining empirical obstacles to its acceptance.[47] Moreover, further studies of nucleosynthesis (how chemical elements are formed) continued to demonstrate the viability of fusion pathways in massive stars for elements heavier than helium in those stars.[48] These studies also implied a dynamic universe in which irreversible processes of change unfolded inexorably, leading to a present cosmos quite distinct from the cosmos of the distant past—hardly a picture naturally supportive of a static, steady-state model.

FIGURE 5.10
Physicists Robert Wilson and Arno Penzias, co-discoverers of the cosmic background radiation, standing in front of the Horn Antenna at the Bell Labs in 1965.

Last, the big bang theory faced the problem of the apparent absence of low-energy background radiation. But in 1965, two physicists, Arno Penzias and Robert Wilson (Fig. 5.10), at the Bell Telephone Laboratories in New Jersey, inadvertently discovered this remnant radiation. It turned up in the form of an annoying low hum in their highly sensitive large antennas at the Bell labs.

After trying to eliminate this effect by identifying many different possible sources of noise, including pigeons, Penzias and Wilson realized that the noise was coming from every direction and that it exhibited long, microwave-range

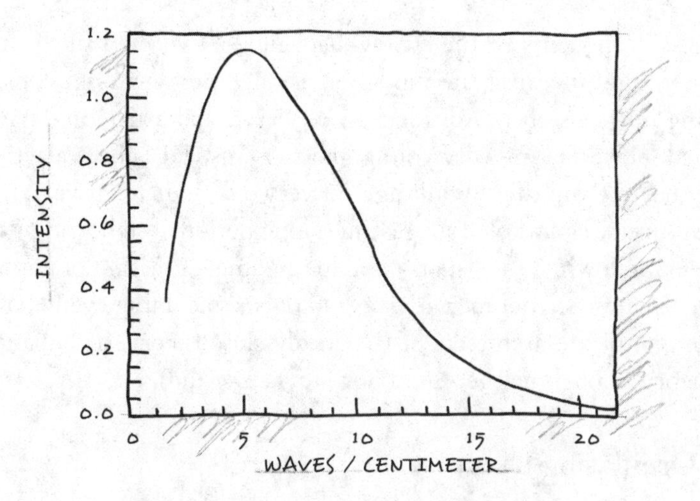

FIGURE 5.11
A perfectly opaque object in thermodynamic equilibrium, known as a "blackbody," exhibits a characteristic distribution of frequencies or wavelengths of radiation. This graph shows the distribution of wavelengths of the cosmic background radiation. It conforms beautifully to the curves characteristic of known blackbodies, suggesting that the cosmic background radiation issued from a relatively compact, opaque, early state of the universe.

wavelengths. When they found that the wavelength of this radiation computed to nearly the exact blackbody temperature equivalent that Alpher had predicted, they began to suspect that they had discovered something of cosmological importance. They contacted Robert Dicke at Princeton, who had himself been looking for the CMBR. After examining Penzias and Wilson's apparatus and data, he concluded that they had found what he had been looking for—the radiation left over from the hot, high-density plasma state postulated as a consequence of the big bang.[49]

The discovery of the cosmic background radiation, with almost the exact predicted wavelength and corresponding blackbody temperature (as later determined using a more accurate value of the Hubble constant), proved decisive (Fig. 5.11).[50] Whereas Alpher and Herman had predicted the existence of this microwave radiation as a consequence of the big bang model,[51] advocates of the steady-state model acknowledged that, given their model, no such radiation should exist.

Other evidence challenged the steady-state theory. For example, the steady state implied that galaxies should have a range of radically different ages, from extremely young, just-forming galaxies to extremely old galaxies. Proponents of the steady-state model expected such a distribution of ages, because the model envisioned new material constantly popping into existence. Advances in observational astronomy have not revealed, however, any very young galaxies. Instead, most galactic ages cluster narrowly in the "middle age" or very old range (relative to the age of the universe as a whole), suggesting a long period of stellar and galactic evolution following a singular—not an ongoing—creation of matter. By the 1970s, most astronomers and cosmologists, including even Hermann Bondi, one of the architects of the steady-state theory, had abandoned the theory (though neither Gold nor Hoyle ever did).[52]

The Oscillating Universe

Following the demise of the steady-state model in the mid-1960s, some physicists proposed an oscillating-universe model (Fig. 5.12) as an alternative to the finite universe suggested by the then ascendant big bang. Advocates of this oscillating model envisioned a universe that would expand, gradually decelerate, shrink back under the force of its own gravitation, and then, by some unknown mechanism, reinitiate its expansion, over and over, *ad infinitum*. For a time, the oscillating model preserved the notion of an eternal self-existing universe. But for several reasons physicists eventually rejected the model, its implication of an eternal universe, or both.

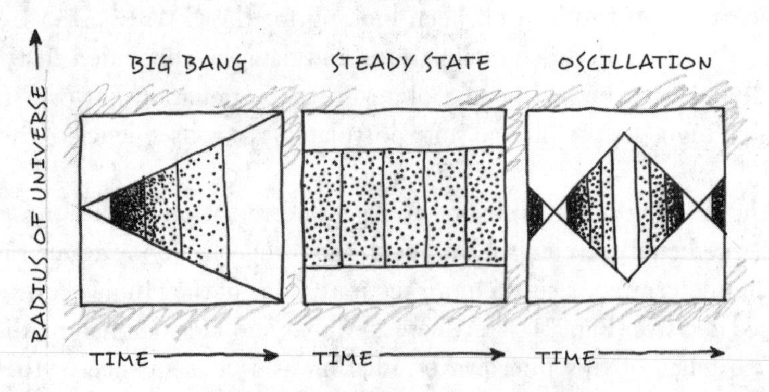

FIGURE 5.12
Three cosmological models: the big bang, steady state, and oscillating universe. The big bang model implies the universe had a beginning. The steady-state model implies that the universe has existed eternally and matter is being continuously created. The oscillating model depicts the universe expanding and collapsing an infinite number of times. All three models assume a presently expanding universe.

First, proponents could not devise a plausible mechanism to explain the successive reexpansions of the universe after the gravitational collapses they envisioned. Even on the somewhat implausible assumption that something like the expansion force of the cosmological constant would reinflate the universe after each collapse, the model ran into difficulties with the second law of thermodynamics, as MIT physicist Alan Guth demonstrated in 1984.[53] (The second law says that the disorder or entropy of an isolated system of matter and energy will increase over time.)

Guth showed that, according to the second law, the entropy (or disorder) of the matter and energy in the universe would increase over time in each cycle. But such increases in entropy (or the disorderly distribution of mass-energy) would result in less energy *available to do work* in each cycle. That would cause progressively longer and longer cycles of expansion and contraction, since increasing inhomogeneities in the mass-energy density throughout space would decrease the efficiency of gravitational contraction. Yet if the duration of each cycle necessarily increases as the universe moves forward in time, then it follows that each cycle in the past would have been progressively shorter. Since the periods of each cycle cannot decrease indefinitely, the universe—even in an oscillating model—would have had to have a beginning.

Similarly, if in every cycle mass and energy grow progressively more randomized, eventually—given infinite time—the universe would reach heat death in which *no* energy will be available to do work, like a rubber ball that bounces to a smaller and smaller height until finally it can bounce no more. Yet, if the universe was oscillating and infinitely old, it should have reached such a state an infinitely long time ago. But since we do not find ourselves in such a cold universe with maximally inhomogenous distributions of matter and energy, it follows—even assuming an oscillating universe—that the universe has not existed for an infinite amount of time.

In any case, recent astronomical measurements suggest that the universe has a mass density slightly less than the so-called critical density necessary to stop the expansion of the universe, thus ensuring that the universe will never recollapse.[54] Also, the expansion of the universe may actually be accelerating,[55] perhaps as the result of what astrophysicists call "dark energy," a postulated but unidentified form of energy that putatively permeates all of space and exerts an outward pressure on it.[56]

The Big Bang's Galaxy Problem

By the 1970s, most astronomers had come to accept the big bang over its rivals. Nevertheless, the discovery of the CMBR that effectively killed the steady-state theory left one mystery unsolved for proponents of the big bang. That mystery was galaxy formation.

For galaxies to form, the mass and energy just after the big bang must have exhibited fluctuations in density. This is necessary in order to account for the observed variations in the concentration of matter and energy throughout space today—as evidenced by, for example, galaxies and galaxy clusters surrounded by mostly empty space. In theory, these initial differences in the concentration of mass and energy would have affected the cosmic background radiation, since different concentrations of mass and energy would result in different characteristic wavelengths of light issuing from different places in the original hot, dense concentrations of matter and energy in the postplasma universe. For this reason, the big bang model implied that today's cosmic microwave background radiation (CMBR) ought to manifest small fluctuations in the intensity of the microwave radiation.

Using ground-based and airborne instruments, early attempts to locate these expected variations in the CMBR failed. Even tests using rockets launched above the atmosphere could not detect the predicted variations. In 1989, however, NASA launched a satellite known as the Cosmic Background Explorer, or COBE. As the COBE satellite swept the skies while

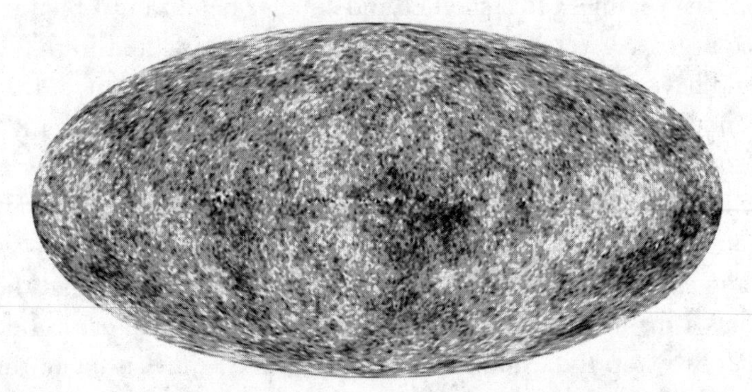

FIGURE 5.13

The big bang theory predicts the existence of a low-level cosmic background radiation. For the big bang to explain the origin of galaxies, there must have also been small variations in the intensity of this radiation from the earliest stages of the universe. As the Cosmic Background Explorer (COBE) satellite has scanned the night sky, it has detected these slight variations. This figure reproduces, in an enhanced black and white form, a famous color image of the night sky depicting these variations.

orbiting the earth above the atmospheric fray, it did indeed discover (Fig. 5.13) the predicted tiny variations in CMBR radiation.

These findings resolved one of the few remaining evidential challenges facing the big bang model and sealed the case from observational astronomy for a finite universe.[57] It gave a snapshot of the seeds of galaxies just after the creation of matter itself. For many scientists these images were startling in their significance. As George Smoot, the director of the COBE program, who eventually won the Nobel Prize for his discovery, put it: "If you're religious, it's like seeing God."[58]

The Final Rock

Clearly, Smoot spoke hyperbolically. But the discovery of a beginning to the universe has led many scientists to reflect seriously on the possible theistic implications of a finite universe.

I first encountered serious scientists doing just that at the Dallas conference I mentioned in Chapter 1. One of the first panels at that conference featured presentations on the evidence for the big bang theory and a temporally finite universe. A discussion followed of the philosophical implications of the theory. Panelists offered both theistic and materialistic perspectives. The panel included a veritable scientific "Who's Who" that included not only Harvard astrophysicist Owen Gingerich and Caltech astronomer Allan Sandage, but also Robert Jastrow, of the Goddard Space Institute, and Donald Goldsmith, the science adviser for the popular original *Cosmos* series hosted by Carl Sagan.

Of these luminaries, Sandage had perhaps the most profound effect on the audience. He described several of the lines of evidence supporting the big bang theory, including his own discoveries confirming the linear relationship between the distance to far-flung galaxies and their recessional velocities. After serving as a graduate assistant to Edwin Hubble and earning his PhD at Caltech under Walter Baade, Sandage continued the work of Hubble, refining the understanding of the Hubble relationship between recessional velocity and distance as it applied to galaxies in all quadrants of the night sky.[59]

By 1985, Sandage was widely respected as one of the great observational astronomers of the twentieth century. As I've noted already, he was also well known as an agnostic with a materialist philosophy of science and little interest in questions about the existence and nature of

God—or so many of the other panelists assumed that February morning. During his talk, however, he not only described the astronomical evidence for the beginning of the universe; he shocked many of his colleagues by announcing a recent religious conversion and then explaining how the scientific evidence of a "creation event" had contributed to a profound change in his worldview.

I recall his looking intently at the audience and gravely stating, "Here is evidence for what can only be described as a supernatural event. There is no way that this could have been predicted within the realm of physics as we know it." As he spoke, he paused between the words "super" and "natural," saying them separately for emphasis. He went on to explain that "science, until recently, has concerned itself not with primary causes but, essentially, with secondary causes. What has happened in the last fifty years is a remarkable event within astronomy and astrophysics. By looking up at the sky, some astronomers have come to the belief that there is evidence for a 'creation event.'"[60]

Sandage described his own internal struggle to reconcile his commitment to a reductionistic and materialistic philosophy of science with his growing convictions that something beyond the strictly material must have played a role in bringing the universe into existence. He explained that although he did not think that scientific evidence could *prove* God's existence, he did think that new discoveries in cosmology and physics had lent unexpected credibility and support to theistic belief.[61] He continued:

> I now have to go from a stance as a complete materialistic rational scientist and say this super natural event, to me, gives at least some credence to my belief that there is some design put in the universe. I cannot . . . with certainty say that. What now do I do? I am convinced that there is some order in the universe. I think all scientists, at the deepest level, are so startled by what they see in the miraculousness of the inner connection of things in their field . . . that they at least have wondered why it is this way.[62]

Listening to Sandage wrestle so honestly with the question of ultimate origins, with the implications of a theory that did not comport well with his previously long-held worldview, made a big impression on me. Could it be, I wondered, that scientific discoveries about the origin of the universe now challenged the long dominant materialism of the scientific establishment? Sandage seemed to be saying at least that much, and with good reason. If the material universe (of mass, energy, space, and time) itself came into existence a finite time ago, then matter and energy do not seem to be good candidates as explanations for the origin of the universe. Clearly, matter and energy could not cause themselves to come into existence before they themselves existed.

As it happens, Sandage was not the only astronomer at this time who perceived a convergence between the evidence for a beginning and a theistic perspective. Owen Gingerich, whose lecture the night before at Southern Methodist University had tipped me off about the conference, also made clear that he did not think that science could definitely *prove* the existence of God. Nevertheless, his popular lecture "Biblical Creation and Scientific Cosmogony" did explore what he called a "strange convergence" between the testimony of modern cosmology and the specifically biblical idea that the universe flashed instantly into existence a finite time ago.[63]

FIGURE 5.14
Astrophysicist Robert Jastrow of the Goddard Space Institute and author of *God and the Astronomers.*

Several years earlier, the late Robert Jastrow (Fig. 5.14), of the Goddard Space Institute, who also attended the Dallas conference, published a popular book, called *God and the Astronomers*, that made many of the same points. Jastrow, who was a religiously agnostic Jewish scientist, discussed the obvious theistic implications of the big bang theory. Though he acknowledged that these implications made him personally uncomfortable, he explained that the theory—with its affirmation of a beginning—seems to portray the origin of the universe in terms that closely match what a biblically informed theologian would expect.

In a memorable conclusion to his book, Jastrow observed that the discovery of a definite cosmic beginning:

> is an exceedingly strange development, unexpected by all but the theologians. They have always accepted the word of the Bible: In the beginning God created heaven and earth . . . . The development is unexpected because science has had such extraordinary success in tracing the chain of

> cause and effect backward in time. For the scientist who has lived by his faith in the power of reason, the story ends like a bad dream. He has scaled the mountains of ignorance; he is about to conquer the highest peak; as he pulls himself over the final rock, he is greeted by a band of theologians who have been sitting there for centuries.[64]

Smoot, Sandage, Gingerich, and Jastrow were each reflecting on the implications of a finite universe as revealed by discoveries in observational astronomy. Earlier developments in theoretical physics seemed to raise those same implications. Yet since Einstein, Friedmann, and Lemaître, significant new developments in theoretical physics have reinforced the conclusion of a cosmic beginning and done so in possibly an even more profound way.

6

The Curvature of Space and the Beginning of the Universe

The life of Stephen Hawking (Fig. 6.1) is a story of extraordinary scientific achievement in the face of acute physical challenge. From 1979 until 2009 Hawking held the prestigious Lucasian Chair of Mathematics at the University of Cambridge, a chair once held by Isaac Newton himself. As a graduate student at Cambridge, he first began to manifest troubling symptoms of the neuromuscular disease known as amyotrophic lateral sclerosis (or ALS), a disease that would eventually confine him to a wheelchair and force him to use a voice synthesizer for the rest of his life. Despite almost giving up on his PhD studies after receiving the ALS diagnosis, he eventually decided to press forward. As he did, he developed a profoundly suggestive insight about the origin of the universe.

FIGURE 6.1
A young Stephen Hawking, whose 1966 PhD thesis developed the first initial proof of a cosmological singularity theorem.

During his PhD research Hawking encountered the work of British physicist Roger Penrose. Penrose was working on the physics of black holes, locations in space where matter is so densely concentrated that even light cannot escape the gravitational pull of the mass. According to general relativity, the dense concentration of matter in a black hole will warp or bend the fabric of spacetime, creating a tightly curved, self-enclosed region of space. Such a dense concentration of matter makes a kind of gravitational trap that prevents anything on the inside of the tightly curved space from getting out—even light. Thus, the name "black hole."

Hawking realized that Penrose's work on black holes had implications for understanding the origin of the universe. He began to think about how the density and volume of the expanding universe would have changed over time. He realized that at every point in the past the mass of the universe would have been more densely concentrated. That meant that space would have been more tightly curved at every successive point farther and farther back in time. In his mind's eye, as he extrapolated backward in time, he saw that at some point the curvature of the universe would reach a limit—that is, it would attain an infinitely tight spatial curvature corresponding to zero spatial volume. This is called a "singularity," where the known laws of physics would break down and from which the universe would have begun its expansion.

In his PhD thesis, Hawking included one chapter about the implications of general relativity and the discovery of the expanding universe for our understanding of its origin. There he provided a preliminary mathematical proof for the occurrence of a spatial singularity at the beginning of the universe, "provided," he said, "certain very general conditions are satisfied." He first showed mathematically that any "time-like" or "light-like" path between two points in the curved space of the expanding universe must terminate at some finite point in the past. He then showed that, given such a finite termination point for light and time in the past, "there will be a physical singularity . . . where the density and hence the curvature [of the universe] are infinite."[1]

A 2014 film, *The Theory of Everything*, tells the story of Hawking's life and includes a memorable scene depicting his PhD examination. In it, the solitary postgraduate student Hawking stands at a plain wooden table. Across from him are the three distinguished physicists who will judge his thesis, Kip Thorne, Dennis Sciama, and Roger Penrose. As the examination begins, they critique his chapters. After finding each of his first three chapters deficient in some way, the three physicists begin to evaluate his fourth and culminating chapter, in which Hawking argues that increasingly dense concentrations of mass and energy in the reverse direction of time point to a singularity.

As the film depicts the scene, Thorne muses aloud over Hawking's main idea: "A black hole at the beginning of time?" Sciama then tersely expresses his understanding of Hawking's concept: "A spacetime singularity." The three physicists then exchange glances as Hawking worries about his fate. Then Sciama exclaims, "Brilliant. It's brilliant, Stephen. . . . Well done. Or should I say, well done, *Doctor.* An extraordinary theory." Hawking sighs in relief. As the scene concludes, Sciama, Hawking's supervisor, asks Hawking, "What comes next?" Hawking vows to develop further mathematical proof for the idea that "time has a beginning."

General Relativity and the Singularity Theorems

In the previous chapter, we saw how physicist Georges Lemaître incorporated evidence from observational astronomy into the structure of Einstein's new theory of gravity to develop the big bang theory. We also saw how Einstein's theory of general relativity seemed most naturally to suggest both a dynamic and finite universe. Since the late 1960s, further developments in theoretical physics have supplied additional support for the idea that the universe as well as space and time—or spacetime—had a beginning. (There are two types of singularities: spatial singularities, in which matter and space under the influence of gravity converge to a point of infinitely tight spatial curvature, and temporal singularities, in which light rays, particles, or events are traced back to an absolute starting point in time.)

Stephen Hawking, with two distinguished collaborators, including Roger Penrose, one of his PhD thesis examiners, played the central role in making these theoretical advances. Four years after being examined by Penrose, Hawking and Penrose together developed additional mathematical arguments for a spacetime singularity.[2] Then three years after that, in 1973, Hawking developed his case further with South African physicist George Ellis (Fig. 6.2).[3] Ellis had studied with Hawking in Cambridge under Dennis Sciama. After Ellis received his PhD, he stayed on as a research fellow and university lecturer until 1974 before taking a faculty position at the University of Cape Town. During the late 1960s and early 1970s, these three physicists produced a series of scientific publications that spelled out the implications of Einstein's theory of general relativity for the origin of space and

FIGURE 6.2
Physicist George Ellis, who collaborated with Stephen Hawking in their classic work *The Large Scale Structure of Space-Time* to prove cosmological singularity theorems based on general relativity.

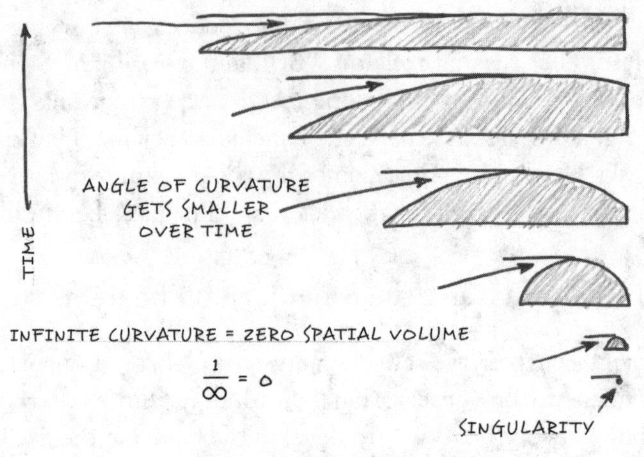

FIGURE 6.3
As the universe expands, space (or "spacetime") flattens and the curvature of space decreases and approaches zero. Curvature increases, however, in the reverse direction of time, eventually reaching a limit of infinite curvature. Infinite curvature corresponds to zero spatial volume, thus marking the beginning of the universe.

time.[4] Their solutions to Einstein's field equations implied a singularity at the beginning of the universe where the density of matter and the curvature of space would approach an infinite (Fig. 6.3).

Previously, during the 1920s, Lemaître had emphasized that the density of the universe would approach an infinite value as one extrapolated back to the earliest state of the universe. Thus, he portrayed the universe as beginning from a kind of cosmic egg or a primeval atom.[5] In fact, Friedmann's and Lemaître's solutions to the field equations also implied a singularity (including an infinitely tightly curved space in the finite past), but they made an unrealistic simplifying assumption about the early state of the universe.

They assumed a completely homogeneous distribution of matter and energy and a universe that was "isotropic." In a homogeneous universe the distribution of matter is the same everywhere, that is, *in all locations*; in an isotropic universe the distribution of matter and energy *looks* the same *in all directions* regardless of the vantage point of the observer. Lemaître later dropped the assumption about isotropy, but continued to presuppose homogeneity.[6] Yet cosmologists later realized that to explain the origin and evolution of galaxies required postulating slight differences in the density and distribution of matter and energy from the earliest stages of the universe.[7]

Since Friedmann and Lemaître ignored these differences by assuming perfect homogeneity—and since their assumption of homogeneity provided a plausible explanation of the emergence of the singularity in their solutions to the field equations—physicists disregarded the implication of the resulting temporal and spatial singularity in their solutions.[8] Indeed, many physicists suspected that the singularity in Lemaître's solutions was an artifact of his simplifying assumption of homogeneity, not a true picture of the origin of the universe.[9]

That's where Hawking, Penrose, and Ellis came in.[10] In works published between 1966 and 1973, they succeeded in solving the field equations without making the unrealistic assumption of perfect homogeneity. As Hawking and Ellis wrote in the preface to their 1973 book *The Large Scale Structure of Space-Time*: "For a long time it was thought that these singularities might simply be a result of the high degree of symmetry [i.e., homogeneity] and would not be present in more realistic models. It will be one of our main objects to show that this is not the case."[11]

The field equations allow physicists to describe differences in the spatial configuration of the universe (and coordinate systems describing them) that would derive from possible differences and irregularities in the initial (and present) distribution of matter. Taking these irregularities into account allows physicists to describe the initial state of the universe more accurately, but it makes solving the field equations more difficult mathematically. Nevertheless, Hawking and his colleagues solved the equations without assuming perfect homogeneity. In so doing, they demonstrated, based on general relativity, that the universe began in a "space-time" singularity of "infinite curvature."[12] Indeed, the theory of general relativity implies, as Hawking and Ellis wrote, "that there is a singularity in the past that constitutes, in some sense, a beginning of the universe."[13]

In 1973, Hawking and Ellis built on Hawking's dissertation and a 1970 paper by Hawking and Penrose to show, first, that the universe "is geodesically past incomplete."[14] "Geodesic" is a term from geometry designating the shortest distance between two points on a curved surface. Hawking and Ellis argued that the trajectory of any ray of light and/or timeline through curved space will necessarily terminate at some point in the finite past in an expanding universe. Or as they put it, "The gravitational effect of matter is always to tend to cause convergence of timelike and null [light-like] curves."[15] They then applied several powerful mathematical theorems that Hawking, Ellis, and Penrose had developed to show that in such "a geodesically past incomplete universe" certain mathematical contradictions[16] would result *if there were no singularities.* Since they had already proved past incompleteness (i.e., that the universe had a beginning in time), it followed that the universe began from a singularity in which the gravitational field would have been infinitely strong and the curvature of space infinitely tight. As they put it, "We show that in a generic space-time, an observer travelling on one of these incomplete [temporally finite] curves would experience infinite curvature forces."[17]

Oddly, however, an infinitely tightly curved space corresponds to a radius of curvature of zero units in length and thus to zero spatial volume. In 1978, the British physicist Paul Davies described the implications of the singularity theorems with great clarity:

> If we extrapolate this prediction to its extreme, we reach a point when all distances in the universe have shrunk to zero. An initial cosmological singularity therefore forms a past temporal extremity to the universe. We cannot continue physical reasoning, or even the concept of spacetime, through such an extremity. For this reason most cosmologists think of the initial singularity as the beginning of the universe. On this view the big bang represents the creation event; the creation not only of all the matter and energy in the universe, but also of space-time itself.[18]

To get my students to recognize the profundity of this result, I used to ask them, "How much stuff can you put in no space?" They would quickly realize that the answer to this question is: "None" or "No stuff." If, at some point in the past, space ceased to exist, then there would not at that point have been any place to put anything, whether matter or energy. Indeed, neither matter nor energy can exist in the absence of space and time. Thus, Hawking, Ellis, and Penrose's singularity proofs (interpreted as a realistic depiction of the history and spatial geometry of the universe) implied that a material universe of infinite density began to exist some finite time ago starting from nothing—or at least from nothing spatial, temporal, material, or physical.

Of course, thinking about the cosmological singularity can generate paradoxical or seemingly contradictory conclusions. One might also assert, for example, that the universe began from an enormous amount of mass-energy and an infinitely strong gravitational field since, at the singularity, the mass-energy density and the strength of the gravitational field would also have approached infinity. Even so, the singularity theorems do not permit one to posit mass-energy or a gravitational field as an eternal, self-existing entity, since "prior to" the singularity neither time nor space existed in our universe. And without space, mass-energy (and a corresponding gravitational field) would have no place to reside. In other words, however much mass-energy existed from the beginning of the universe, it had to arise *with* the beginning of time and space, both of which began a finite time ago. Thus, a spatial or temporal singularity prevents, as Davies noted, "any physical reasoning" about a prior state of the universe "through such an extremity," and thus that extremity (or singularity) does mark the beginning of the physical universe itself.

Taken at face value, the philosophical implications of a cosmological singularity are staggering. At the very least, a universe that begins in a spacetime singularity poses an acute challenge to any materialistic theory of the origin of the universe. Indeed, a singularity implies that not only space and time but also matter and energy *first* arose at the beginning of the universe, before which no such entities would have existed that could have caused the universe (of matter and energy) to originate.

Moreover, insofar as the spacetime singularity marks the point of origin of the universe from nothing physical, cosmological models based on solutions to the field equations of general relativity seem strangely reminiscent of what theologians long described in doctrinal terms as *creatio ex nihilo*—"creation out of nothing" (nothing physical, that is). Hawking and Ellis themselves addressed the issue of the creation of the universe in the conclusion of their 1973 book. As they reflected: "The creation of the Universe out of nothing has been argued, indecisively, from early times; see for example Kant's first Antinomy of Pure Reason. . . . The results we have obtained support the idea that the universe began a finite time ago. However, the actual point of creation, the singularity, is outside the presently known laws of physics."[19]

Conditions, Conditions, Conditions

But should we interpret the Hawking-Penrose-Ellis cosmological singularity as a realistic depiction of the spatial geometry of the universe all the way back to a temporal beginning? Hawking and Ellis themselves addressed this question in their 1973 work. They recognized that proofs of the spacetime singularity apply to our universe only if certain conditions are met. First, all singularity theorems presuppose general relativity as our best theory of gravity. And, indeed, numerous experimental confirmations of the predictions of general relativity have given physicists a high degree of confidence in the theory as it applies to the large-scale structure of the universe. These include increasingly accurate tests conducted with a hydrogen maser detector on a NASA rocket in 1980 and 1994. Such experiments have provided precise quantitative confirmation of the predictions of the theory, even out to the fifth decimal place.[20] Thus, general relativity now stands as one of the best confirmed theories of modern physics.

Questions about the applicability of the theory of general relativity arise, however, for extremely small subatomic and quantum-level phenomena. In the subatomic realm (10^{-12} cm or smaller), strange phenomena can occur, such as light or electrons acting like both waves and particles at the same time. Nondeterministic fluctuations in energy can occur at that scale as well. Since general relativity does not describe such effects, many physicists have proposed the need for a *quantum* theory of gravity, though no such theory has yet been definitively established (for an extensive discussion, see Chapters 17–19).

The inability of general relativity to describe gravitational phenomena on an extremely small subatomic scale has led to questions about the applicability of the theory during the earliest history of the universe, the first fractions of a second after the big bang. Indeed, when physicists extrapolate backward to envision the universe at different times in the remote past, they realize that at some point the universe would have been small enough that physicists would need to use quantum mechanics to describe the behavior of matter and energy in that subatomic realm.

Hawking, Penrose, and Ellis recognized this limitation in their proofs of the cosmological singularity. They acknowledged that, strictly speaking, they could only establish the contraction of spacetime back to the point in time where the universe would have had a radius of curvature of between 10^{-12} and 10^{-33} cm. Thus, strictly speaking, they could not prove that the universe began from an *absolute* spatial singularity, only a very, very, very tiny near singularity.[21] In that vanishingly small space, gravity might not function as described by general relativity.

Moreover, the possibility of fluctuations of energy in the tiny quantum realm posed a specific *technical* impediment to the singularity theorems. All singularity proofs are subject to so-called energy conditions of various kinds, including “weak,” “dominant,” and/or “strong” energy conditions. Different energy conditions describe various possible characteristics of the mass-energy of the universe. Assuming general relativity, the universe would need to meet different energy conditions in order to generate singularities, either at the beginning of the universe or as black holes within the universe. For example, the weak energy condition specifies that the mass-energy density of the universe must be positive or zero (i.e., nonnegative) for a singularity to occur. The dominant energy condition requires that criterion, but also holds that the pressure produced by the movement of energy in the universe must not exceed the value of the energy density. (The strong energy condition is defined in an even more technical mathematical way but also describes necessary properties of the energy in the universe.)[22]

For our purposes, the key point is this: meeting one or more of these energy conditions ensures the validity of the singularity proofs. But indeterministic quantum fluctuations in energy sometimes produce situations that violate one or more of these conditions, especially the strong energy condition. And, as it happens, the strong energy condition needs to be met to prove the universe began from an absolute spacetime singularity.[23] (A quantum fluctuation is a random local change in the energy associated with a particle or field occurring on a subatomic scale. In quantum physics such fluctuations can generate negative energy values, thus violating one or more of the energy conditions.)

In 1973, Hawking and Ellis acknowledged these limitations in the applicability of their proofs. They understood that the universe might not meet the strong energy condition “at every point” in its history, specifically during the earliest and tiniest fractions of a second after the presumed beginning of the universe—when the universe was vanishingly small. Thus, they acknowledged that general relativity only allowed them to extrapolate backward with absolute confidence to a point in the past where the universe had a radius of curvature of somewhere between a trillionth (10^{-12}) of a centimeter and a billion trillion trillionth (10^{-33}) of a centimeter. Understandably, however, they regarded a universe that tiny as effectively a spatial singularity. As they put it, “Such a curvature would be so extreme as that it might well count as a singularity.”[24]

Indeed, even if the singularity theorems did not prove a beginning of the universe from an absolute spatial zero point, these theorems did provide, for all practical purposes, a strong indicator of—or a pointer to—such a beginning. Thus, by the late 1970s and early 1980s, most astronomers, astrophysicists, and cosmologists came to regard the joint testimony of observational astronomy and theoretical physics as powerful confirmation of the big bang model.

Inflationary Cosmology

Nevertheless, other developments in theoretical physics and cosmology soon challenged the standard big bang model and reinforced concerns about the applicability of singularity theorems to the early universe. Later still, in an oddly unexpected reversal, these same developments eventually led theoretical physicists to discover that the universe must have had a beginning after all. Here's the story.

During the 1980s, the physicists Alan Guth, of MIT, Andrei Linde, of Stanford, and Paul Steinhardt, of Princeton, developed an alternative version of the big bang cosmology known as inflationary cosmology (or just "inflation"). Inflationary cosmology asserts that soon after the big bang, space experienced a short-lived but exponentially rapid expansion. This expansion was due to the negative gravitational pressure produced by a postulated "inflaton field"—a field that physicists conceived as generating an outward or repulsive pressure on space, causing the universe to expand. (Fields in physics are regions of space that are defined operationally by what they do—often by the forces they generate or the motions they induce in particles or radiation that come in contact with them.)

FIGURE 6.4
The MIT physicist Alan Guth, who developed the inflationary big bang model.

As originally proposed by Alan Guth (Fig. 6.4), inflation presupposed *a* beginning to the universe after which the space of the universe would rapidly expand for a brief period of time.[25] Subsequently, however, other physicists proposed "eternal chaotic inflation" models that envisioned not *a* beginning, but an infinite number of beginnings[26] (Fig. 6.5). These eternal chaotic inflation models gained in pop-

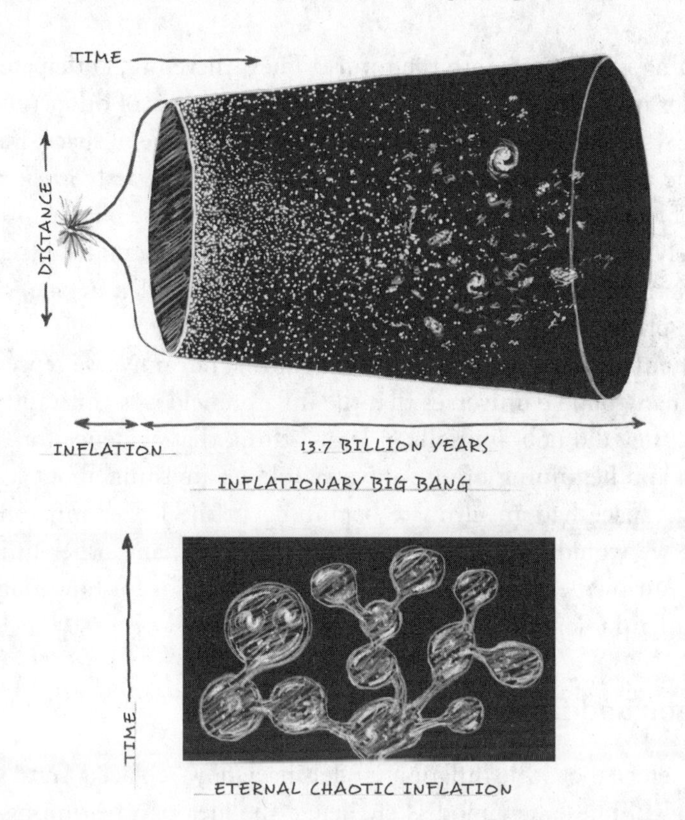

FIGURE 6.5

TOP: As first formulated, the inflationary cosmology model holds that the universe had a beginning, and it initially expanded extremely rapidly before slowing down to a more sedate pace of expansion. BOTTOM: Later cosmologists formulated the eternal chaotic inflation model. According to this model, as the universe expands, different regions of space will stop inflating, causing new bubble universes to emerge. This process would then continue indefinitely, producing an infinite number of "bubble universes" separated from each other by an inflating ocean of expanding space.

ularity among proponents of inflation, because many thought that the postulated "inflaton field" would be subject to *quantum fluctuations* in the energy of the field. As a result, they thought these fluctuations would necessarily produce causally disconnected regions of space—effectively, separate "bubble universes."

According to current eternal chaotic inflation models, after an initial phase of expansion, a quantum fluctuation in the energy of the inflaton field caused it to decay *locally* to produce our universe. The inflaton field also continued to operate outside our local area to produce a wider expansion of space into which other universes were birthed as the inflaton field decayed at other locations. Inflationary cosmologists envisioned inflation as having operated for an indefinitely long time in the past and as continuing indefinitely into the future. They, therefore, anticipated that the wider inflaton field will spawn an endless number of other universes as it decays in local pockets of an ever-growing volume of space. Further, since the inflaton field continues to expand at a rate vastly greater than the bubble universes expanding within it, none of these bubble universes will likely ever interfere with each other. The one inflaton field therefore gives birth to endless bubble universes—"many worlds in one," as the Russian physicist Alexander Vilenkin has described it.

Since inflationary cosmologists thought that our universe represented one of many bubble universes that the inflaton field was constantly generating, they did not—initially at least—think that evidence for the expansion and beginning of our universe told us anything about whether time and space had an ultimate beginning in the larger inflaton field. Indeed, we would have no way of knowing how many other universes beyond our own existed in the larger inflaton field or for how long they, or the inflaton field, had existed.

Inflation and Energy Conditions

Various eternal chaotic inflation models have now replaced Guth's original model. These new models challenge the idea of a beginning—and of a cosmological singularity—for yet another reason. Inflationary cosmological models affirm quantum fluctuations as the mechanism that produces bubble universes. Consequently, these models entail violations of the various energy conditions required to prove the cosmological singularity theorems.[27]

As envisioned by the architects of eternal chaotic inflationary cosmology, inflaton fields can experience random quantum fluctuations in net energy, sometimes resulting in short-lived but negative mass-energy densities, including temporary *negative* densities in the mass-energy of the universe. Such negative densities would violate the various energy conditions, including the strong energy condition, that the proofs of singularity theorems require. Consequently, inflationary cosmology has tended to undermine confidence in the relevance of singularity theorems for modeling the origin of the universe. As George Ellis told me in an interview recently, physicists who accept inflationary cosmology now typically see singularity theorems as an interesting piece of pure mathematics, but not as proofs of the beginning of our actual universe.[28]

The Explanatory Power of Inflation

What inflationary cosmology implies about singularity theorems might not seem significant, given its apparently speculative character. Nevertheless, despite the role that such hypothetical entities as "inflaton fields" and "bubble universes" play in inflationary cosmology, many cosmologists regard inflation as the best current cosmological model. Typically, they do so because of its ability to explain three main features of the universe—its homogeneity, its "flatness," and the absence of what are called "magnetic monopoles" in the visible universe.[29] These features were puzzling either from the perspective of standard big bang cosmology or, in the last case, from the standpoint of popular grand unified theories—theories that attempt to reduce the four fundamental forces of physics to one underlying physical law.

By homogeneity, cosmologists mean that the universe has the same composition and distribution of matter in all locations. One key aspect of this homogeneity is the uniformity of the cosmic background radiation, which has nearly the same temperature throughout the observable cosmos. This is a problem in standard big bang cosmology unless cosmologists postulate incredibly specific finely tuned initial conditions.[30]

As we saw in Chapter 5, this radiation has the same temperature in every direction to about 1 part in 100,000. This observed near uniformity can only be explained in standard big bang cosmology by postulating that the early (before about 380,000 years) plasma state of the universe was characterized by almost perfect uniformity in the temperature and distribution of mass-energy.

Inflationary cosmology attempts to explain the relative homogeneity of the background radiation not as the result of a finely tuned initial distribution of mass-energy (though it does invoke special conditions of its own), but instead as a consequence of an early, exponentially rapid rate of cosmic expansion. According to many inflationary models, during the first fractions of a second after the big bang, the temperature and density of the mass-energy in the tiny volume of space that would become our universe homogenized in a process of mixing, or "thermalization." Then the rapid expansion of the space that would become our visible universe distributed this homogeneous energy throughout it, ultimately resulting in the near homogeneity of the presently observable cosmic background radiation. Any remaining inhomogeneity—the expected remnant of the beginning of the universe according to the standard big bang model (absent extreme fine tuning)—would have been pushed *beyond* the edge of the visible universe as a result of the early inflationary expansion of space.

Inflationary cosmology also offers an explanation for the "flatness" of the universe. In a perfectly flat universe,[31] space would have no curvature such that two parallel beams of light would never converge or diverge. Such a universe will expand indefinitely, but its rate of expansion (due to its initial velocity and mass density[32]) will approach zero over time. Our universe is relatively flat because its initial rate of expansion has just barely overcome the gravitational attraction produced by its mass-energy density. In other words, many physicists think our universe has a mass density slightly less than the "critical mass density" necessary to halt its expansion. Consequently, space likely has a very slight overall curvature—that is, it is relatively flat. Proponents of the standard big bang theory would not expect this relative flatness unless the amount of mass-energy and the initial velocity of expansion had been precisely balanced (or fine-tuned) from the beginning of the universe.

Inflation explains the near flatness of the universe, like its homogeneity, as a consequence of the hyperexpansion of space during the early universe. Just as inflating a balloon to larger and larger sizes makes any small patch of it look flatter and flatter, so inflating the whole universe would make the spatial curvature of the universe and any smaller patch of spacetime (such as our observable universe) look flatter and flatter.[33]

In addition, inflation explains why physicists have not observed so-called magnetic monopoles. A magnetic monopole is a postulated (but not yet observed) elementary particle that would in theory act like a magnet with just one pole—one with either a north or a south pole, but not both. Magnetic monopoles are predicted in the visible universe based on popular (though inadequate) grand unified theories. Inflation ostensibly explains the apparent absence of magnetic monopoles by again invoking a rapid expansion of space that pushed the evidence of the monopoles (like that of inhomogeneity) beyond the visible universe.[34]

Inflation and the Borde-Guth-Vilenkin Theorem

Although the eternal chaotic inflation model prompted doubts about whether the universe did in fact have a beginning, it ultimately motivated another investigation in theoretical physics that led to a new and even more compelling proof of the beginning—indeed, one that holds *whether or not* inflationary cosmology turns out to be correct. (We'll see in Chapter 16 that some leading physicists, including Paul Steinhardt, one of the originators of inflationary cosmology, now think there are significant reasons for doubting all inflationary cosmological models.)

In any case, by the early 1990s, many physicists had embraced eternal chaotic inflation as the best model for the origin of the universe. The popularity of the model led two physicists, Arvind Borde and Alexander Vilenkin, of Tufts University, to investigate what inflation implied about whether the universe had a beginning. They sought to investigate whether the inflaton field could have been operating for an infinitely long time back into the past—that is, whether it could have been "past eternal," as they phrased it. Within a decade, Borde, Vilenkin, and a third physicist, Alan Guth, one of the original proponents of inflation, had come to a startling conclusion: the universe must have had a beginning, even if inflationary cosmology is correct.[35]

We've seen that previous attempts to prove a cosmological singularity at the beginning of the universe were based upon Einstein's theory of general relativity. This made sense given the strongly intuitive basis of Hawking's initial insight: if the universe is expanding, the density of mass and thus the curvature of the universe will eventually reach a limit in the reverse direction of time. Arguably, that insight and the mathematical arguments based on it remain strong *indicators* of a beginning, even if those arguments cannot conclusively prove the validity of extrapolating all the way back to an absolute spatial zero point.

Nevertheless, in 2003, Borde, Guth, and Vilenkin developed a proof for a beginning of the universe that did not depend on using Einstein's field equations of general relativity or on any energy condition.[36] Instead, the Borde-Guth-Vilenkin (BGV) theorem is based solely on geometric arguments and Einstein's theory of special relativity. Recall from Chapter 5 that special relativity addresses the relationship between the speed of light and time. The BGV theorem applies to any universe that meets very general conditions, including those implied by inflationary cosmological models. As Alexander Vilenkin explained, "A remarkable thing about this theorem is its sweeping generality. We made no assumptions about the material content of the universe. We did not even assume that gravity is described by Einstein's equations. So, if Einstein's gravity requires some modification, our conclusion will still hold. The only assumption that we made was that the expansion rate of the universe never gets below some nonzero value, no matter how small."[37]

Consequently, the theorem applies to nearly all plausible and realistic cosmological models. It states that any universe that is on average expanding is "past incomplete." In other words, if one follows any spacetime trajectory back in time, any expanding universe, including one expanding as a consequence of an "inflaton field," must have had a starting point to its expansion, indicating a beginning.

The theorem's conclusion of a beginning follows from a surprisingly intuitive set of considerations. Imagine a person moving toward you as space expands. For example, a football player might be running toward you on a football field as the field itself is being stretched and the yard lines on the field are receding from you. As you stand at one end of the field and try to estimate the speed at which the player is approaching, you will need to take into account how fast the yard lines are receding. If the player is running faster toward you than the yard lines are receding in the other direction, he will appear to be getting closer at some calculable rate. On the other hand, if the player is running more slowly than the yard lines are receding, he will appear to be getting farther away. Either way, the player will be approaching you more slowly than he would otherwise have been, had the yard lines on the football field not been moving away from you. In the language of physics, the "apparent" or "observed velocity" of the player will be slower than it would have been because of the "recessional velocity" of the yard lines.

This same logic applies in the cosmological case. If the universe is expanding, then any object, say, a spaceship (Fig. 6.6), moving toward an observer on earth will appear to be going more slowly than it otherwise would have were the universe not expanding.[38] Moreover, if the spaceship continues to fly at a constant velocity within the region of space in which it resides, but that region of space is itself receding from an earthbound observer because of the expansion of the universe, the velocity of the spaceship in relation to the earth will appear to get slower and slower as the universe continues to expand and the space around the spaceship recedes at a faster and faster rate. (Recall that in an expanding three-dimensional universe—think of the balloon analogy—the farther two objects are away from each other the faster they will recede from each other. Thus, if a spaceship is traveling toward the earth at a constant velocity in its frame of reference but it is also being simultaneously moved away from the earth

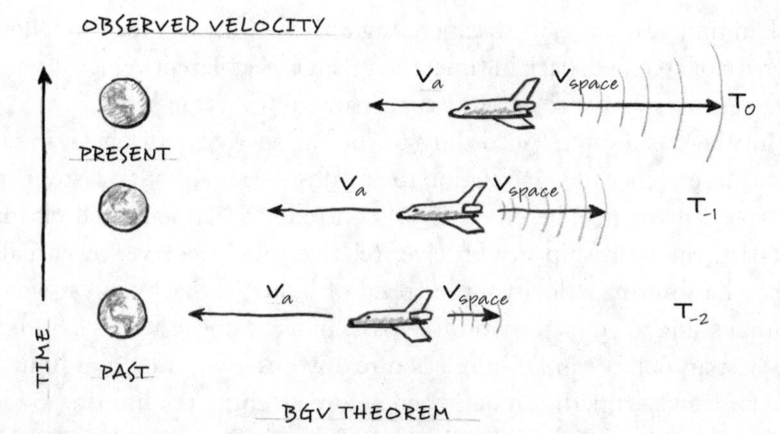

FIGURE 6.6
The BGV Theorem. The Borde-Guth-Vilenkin (BGV) theorem states that any universe that is on average expanding must have had a beginning. The theorem can be understood by imagining a spaceship traveling toward earth. The apparent velocity (V_a) for the spaceship as measured by an earthbound observer equals the actual velocity of the spaceship minus the velocity of the local space in which the spaceship resides as that space moves away from the earth due to the expansion of the universe. But what if we think about the apparent velocity of the spaceship in the past by back extrapolating in time? Since in the forward direction of time space is expanding and moving the spaceship farther from the earth (than it would otherwise be), if we extrapolate in the reverse direction of time, the spaceship would be closer to the earth (than it would otherwise be). The recessional velocity of space would have been smaller at that point in the past since recessional velocity increases with distance from the earth in an expanding universe but would have been slower in the past when the universe had not yet expanded as much. Consequently, V_a, the velocity of the spaceship relative to the earth, would be larger. Moving farther back in time still, the apparent velocity would increase again. With additional back extrapolations, the apparent velocity of the spaceship, V_a, would eventually equal the speed of light, which represents an absolute limit for the velocity of any object according to special relativity. At that point, no further back extrapolations in time would be possible (or physically meaningful), thus implying the universe and its expansion must have had a beginning.

because of the expansion of the universe, the recession rate of the whole region of space around the ship will be increasing. But if the velocity of the ship *within* that local space remains constant, the apparent velocity of the ship toward the earth will appear to be decreasing over time.)

Borde, Guth, and Vilenkin realized that these dynamics also work in reverse. In the reverse direction of time, the apparent velocity of the spaceship with respect to the observer (as measured at any point in the past) would progressively *increase* relative to its velocity in the present. That's because any two points in space would have been closer together than they are now, so we can think of space as effectively contracting in the past. Since the distance between our hypothetical spaceship moving toward the earth and the earth would have also been getting shorter and shorter, the apparent rate at which the spaceship approaches would progressively increase (as measured at successive points farther and farther back in time). In short, in an expanding universe, the farther one follows the path of an object back in time, the greater its apparent velocity would have been in relation to an observer separated from it in space.

But there is a catch. According to special relativity, an object in any frame of reference (i.e., in relation to any observer) cannot go faster than the speed of light. Therefore, if we continue to extrapolate back into the past, the spaceship would have, relative to an observer, eventually reached a limiting velocity—the speed of light. At that point, it would be impossible to go farther into the past. Indeed, since there is a limit to how fast an object can go in relation to any observer, there is a limit to how far back that path can be traced before reaching the limiting velocity of light. That point in the past would then represent an absolute beginning for the path of the spaceship and would mark the point at which space could not contract any further. Thus, it also would mark the point at which the expansion of space would have begun—in other words, the beginning of the universe.

Borde, Guth, and Vilenkin have shown that all cosmological models in which expansion occurs—including inflationary cosmology,[39] multiverses,[40] and the oscillating and cosmic egg models—are subject to the BGV theorem.[41] Consequently, Vilenkin argues that evidence for a beginning is now almost unavoidable. As he explains, "With the proof now in place, cosmologists can no longer hide behind the possibility of a past-eternal universe. There is no escape; they have to face the problem of a cosmic beginning."[42] Since our universe is expanding and the Borde-Guth-Vilenkin theorem does not depend upon any energy conditions, the theorem has reinforced one of the main conclusions of the original Hawking-Penrose-Ellis result (i.e., that the universe had a *temporal* beginning), albeit on different theoretical grounds.

Of Postulates and Proofs

Of course, all proofs, including those that support a cosmic beginning, depend upon some assumptions, postulates, axioms, and/or conditions. As noted, the Hawking-Penrose-Ellis singularity theorems depend upon different energy conditions. The Borde-Guth-Vilenkin theorem does not require any energy conditions, but does make an (albeit less restrictive) assumption as a condition of its validity—that is, the assumption that the universe is on average expanding. For this reason, Alan Guth has acknowledged the possibility of cosmological models that do not meet this condition. For example, he notes, "There may be models with regions of contraction [of space] embedded within the expanded region that could evade our theorem."[43] Even so, the models he cites that might evade the BGV theorem typically contradict the empirical evidence of cosmic expansion and/or they depend upon extremely complex and entirely hypothetical mathematical constructs. Thus, as a more plausible alternative, he allows that "some new physics (i.e., not inflation) would be needed to explain the past boundary of the inflating region. One possibility would be some kind of quantum creation event."[44]

Guth has in mind currently popular quantum cosmological ideas that portray the universe as arising not from a prior temporal state or material condition, but from a hypothetical "space of possibilities" described by the mathematics of quantum mechanics. In Chapters 17–19, I will examine these quantum cosmological models. In Chapter 17, I'll show that these models do not actually succeed in eliminating a cosmic beginning. Moreover, the speculative character of these alternative models and their inability to eliminate a temporal singularity only reinforce the sense that the singularity theorems do indeed provide at least a strong indicator of a beginning. In any case, I show that, if true, these models have unexpected theistic implications of their own.

No proof can establish any conclusion with certainty, since all proofs must make some assumptions. For now, though, it's worth noting that a proof (in the case of the BGV theorem) and a strong indicator (in the case of the Hawking-Penrose-Ellis singularity theorems) have reinforced the testimony of observational astronomy: as best we can tell, the universe did have a beginning.

7

The Goldilocks Universe

Astrophysicist Sir Fred Hoyle (Fig. 7.1) pioneered research on how the nuclear reactions in stars transform hydrogen into the many chemical elements, including carbon and oxygen, necessary for life.[1] He started his scientific career as a staunch atheist who saw no evidence of design in the universe. As he said in his early years as a scientist, "Religion is but a desperate attempt to find an escape from the truly dreadful situation in which we find ourselves. . . . No wonder then that many people feel the need for some belief that gives them a sense of security, and no wonder that they become very angry with people like me who say that this is illusory."[2]

FIGURE 7.1
Astrophysicist Sir Fred Hoyle. In the process of determining how the element carbon might have formed inside stars, Hoyle discovered many fine-tuning parameters.

His atheism played a major role in his approach to science, priming him to reject the idea that the universe had a beginning. In fact, as we saw, he coined the term "big bang" to ridicule the idea of a cosmic beginning and later developed the steady-state model as an alternative. Unfortunately for Hoyle, after the discovery of the cosmic microwave background radiation (CMBR), support for his steady-state model dwindled as more and more astronomers came to accept the big bang theory.

Nevertheless, it was not the discovery of the CMBR, but a different discovery that eventually shook Hoyle's atheism,[3] a discovery that Hoyle himself helped to make. Hoyle played an important role in uncovering one set of what physicists today call the "fine-tuning" parameters of the universe.

Fine tuning in physics refers to the discovery that many properties of the universe fall within extremely narrow and improbable ranges that turn out to be absolutely necessary for complex forms of life, or even complex chemistry, and thus any conceivable form of life, to exist.[4] Physicists now refer to the fortuitous values of these factors as "anthropic coincidences" (from *anthropos*, Greek for "human") and to the fortunate convergence of all these coincidences as the "anthropic fine tuning" of the universe.

Indeed, since the 1950s, physicists have discovered that life in the universe depends upon a highly improbable set of forces and features as well as an extremely improbable balance among many of them. The precise strengths of the fundamental forces of physics, the arrangement of matter and energy at the beginning of the universe, and many other specific features of the cosmos appear delicately balanced to allow for the possibility of life. If any one of these properties were altered ever so slightly, complex chemistry and life simply would not exist.

The fine tuning of these properties has puzzled physicists not only because of their extreme improbability, but also because there doesn't seem to be any necessary physical or logical reason why they have to be as they are. Philosophers of science call such fine-tuning features "contingent" properties, since they could conceivably have been different without violating either the fundamental laws of physics or any necessary principle of logic or mathematics.[5]

We apparently live in a kind of "Goldilocks universe," where the fundamental forces of physics have just the right strengths, the contingent properties of the universe have just the right characteristics, and the initial distribution of matter and energy at the beginning exhibited just the right configuration to make life possible. These facts taken together are so puzzling that physicists have given them a name—*the fine-tuning problem*.

The Mysterious Prevalence of Carbon in the Universe

Hoyle's contribution to the discovery of fine tuning began in the 1950s. What he discovered shocked him and eventually shook his atheism.

Hoyle knew that the universe contained a surprising abundance of carbon. He also knew the production of the element carbon was crucial to all known forms of life. Carbon forms long chain-like molecules that can carry information and store the energy that living cells need to survive.[6] People have speculated about life based on other elements, such as silicon, existing somewhere in the cosmos. But physicists have largely rejected this possibility for decades. As Robert Dicke, for one, wryly put it in 1961, "It is well known that carbon is required to make physicists."[7]

Indeed, carbon-based life is the only known form of life, and carbon has features that make it uniquely suitable as the basis for complex chemistry and life. For instance, carbon is essential for forming sufficiently stable, long, chain-like molecules capable of storing and processing genetic information. Carbon also combines with oxygen to form carbon dioxide in essential chemical reactions. Carbon dioxide is a gas, so it can easily escape cells as waste and readily mix throughout the biosphere. In contrast, silicon dioxide is a solid (familiar to us in the form of sand), and it cannot participate in biochemistry.

For a time Hoyle himself entertained the idea that other chemical elements might form the basis for life. At one point he wrote a novel speculating that cloud-like creatures might have self-organized from interstellar dust composed, presumably, of a variety of elements. Nevertheless, Hoyle later came to recognize the absolute necessity of carbon, making what he discovered about its synthesis all the more startling.

Hoyle knew that carbon is produced from the nuclear reactions taking place inside stars. He and other physicists thought that the most plausible pathway for building heavier elements (such as carbon) from lighter elements (such as hydrogen and helium) would require incremental accretion. In other words, they envisioned individual protons or neutrons (known collectively as "nucleons") colliding with lighter elements to produce successively heavier elements. They thought this process could build heavier elements one proton or neutron at a time, starting from the lightest element, hydrogen, with its one proton.[8]

Their models of how this might have occurred generated expected ratios of lighter elements to heavier elements—and these ratios matched the observed ratios in the universe, at least for the very light elements. For example, the nuclear physicists Ralph Alpher, Hans Bethe, and George Gamow demonstrated that fusion reactions in the early universe would result in the same relative abundances of the lightest elements as

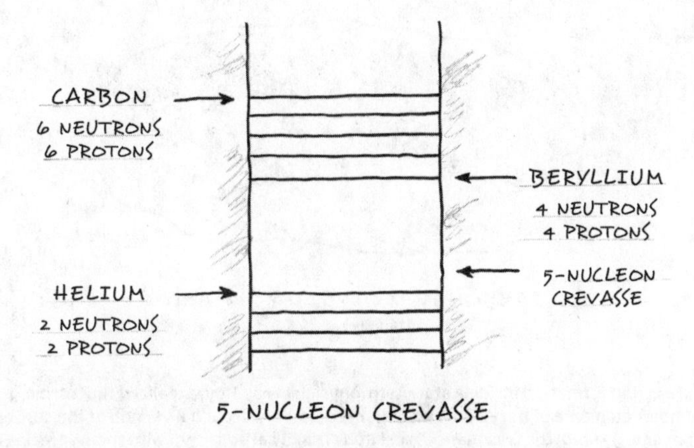

FIGURE 7.2
Astrophysicist Fred Hoyle initially thought that the most plausible pathway for building heavier elements (such as carbon) from lighter elements (such as hydrogen and helium) would occur as the result of incremental accretion of individual protons or neutrons (known collectively as "nucleons"). But Hoyle discovered that building elements heavier than helium in this manner required passing through atomic structures with five total protons and neutrons. Nuclear physicists know these five "nucleon" configurations to be unstable and call this barrier between lighter and heavier elements the "5-nucleon crevasse."

observed today—roughly 90 percent hydrogen and 10 percent helium by number of atoms (as opposed to mass).[9]

But fusing together these lighter elements to form elements heavier than helium requires passing through atomic structures with more than four protons and neutrons—in particular, nuclei with five total protons and neutrons. Nuclear physicists know these to be unstable and call this barrier between lighter and heavier elements the "5-nucleon crevasse" (Fig. 7.2).

This barrier results from the incredibly short half-lives—about one trillionth of a trillionth of a second ($1/10^{24}$ of a second)—of 5-nucleon configurations. These include lithium-5 (with three protons and two neutrons) and helium-5 (with two protons and three neutrons). What they had encountered was something like a 20-foot ladder with the rungs at the bottom and top but only one rung in the middle, making it impossible to climb. Except the situation was worse than that. Not only could the rung in the middle (representing the 5-nucleon state) not be reached, but if it could be reached it would vanish after only one trillionth of a trillionth of a second!

Gamow and Alpher in particular thought long and hard about this problem and considered various ideas about how to leapfrog the unstable 5-nucleon configurations of subatomic particles. They envisioned three

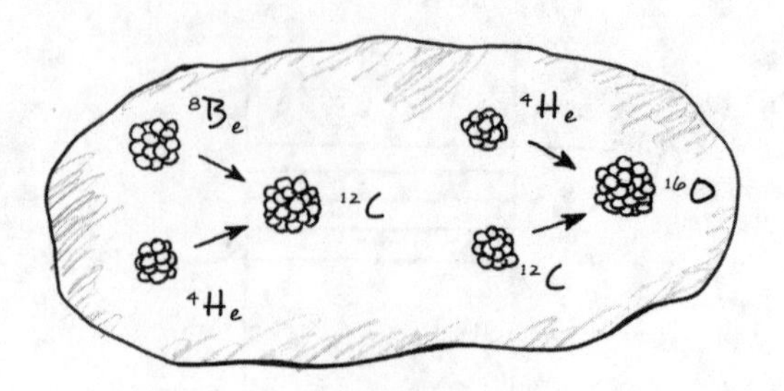

FIGURE 7.3
Carbon and oxygen formation inside a star. Astrophysicist Fred Hoyle realized that forming carbon from the simpler elements of beryllium and helium could only occur if a version of the carbon atom with a higher energy state (or "resonance") existed. That a carbon atom with such a precise resonance level does exist, implied a host of other prior finely tuned parameters in order for carbon formation to occur. Oxygen formation from carbon and helium also requires many prior finely tuned parameters.

helium atoms with two protons and two neutrons each (helium-4) coming together to make the most common form of carbon (carbon-12) with its characteristic six protons and six neutrons. They rejected this pathway as implausible, however, after they estimated the incredible improbability of three helium atoms colliding simultaneously.[10]

Gamow and Alpher discovered a kind of cosmic dilemma: collisions between smaller elements that skipped over the 5-nucleon step were incredibly unlikely; collisions that produced a 5-nucleon transition element would immediately disintegrate. There seemed no plausible path from the earlier conditions of the universe to the heavier elements capable of supporting life, whether one assumed a finite and dynamic universe or Hoyle's steady-state "continuous creation" concept.

Hoyle then considered a more radical alternate pathway. Based on quantum mechanical principles, he suggested that one nucleus of helium (with two neutrons and two protons) might combine readily with a beryllium-8 nucleus (containing four neutrons and four protons) to form carbon (which has six protons and six neutrons) (Fig. 7.3). Though beryllium-8 atoms are also highly unstable, they have half-lives just enough longer than elements with five nucleons to make a collision with a single helium atom likely enough to provide a plausible pathway for building carbon.

Or so it seemed at first. Hoyle soon recognized a problem that required significant fine tuning to solve. He calculated that the total energy of the beryllium-8 atom and the helium-4 atom exceeded the total energy of the carbon-12 atom. Consequently, the two smaller atoms would only fuse readily to form carbon if a higher-energy version of carbon existed, one with a precise excitation state (higher energy state) corresponding to a "resonance" matching the combined energies of beryllium-8 and helium-4 and the kinetic energy generated inside massive stars. A resonance is an energy level where two nuclei can—in accord with quantum mechanical principles—readily combine to form a new nucleus. Hoyle calculated the combined energy for helium and beryllium and determined that a carbon excitation state would need to have precisely 7.65 megaelectron volts (MeV) more energy than the "ground energy state" for carbon-12 (the common form of carbon).[11]

Since Hoyle knew that the universe contained large amounts of carbon, and since he could think of no other plausible pathway for its production, he predicted the existence of the precise excited energy state that he had calculated. Later, he visited the Kellogg Radiation Laboratory at Caltech and managed to convince an initially skeptical nuclear physicist named Willy Fowler to perform the required test to determine whether carbon with such an excitation state existed. In a striking example of theory leading to a specific empirical discovery, Fowler later confirmed the existence of carbon with an energy level with precisely the resonance that Hoyle had predicted. Thus, Hoyle demonstrated a plausible pathway to carbon from the lighter elements—one that could bypass the 5-nucleon crevasse.[12]

There was one problem, however. The carbon resonance level had to be just so or the whole process wouldn't work. And this raised the question of how it got that way.

The resonance levels of different elements are a consequence of many factors and can be calculated using the equations of quantum chromodynamics, a subdiscipline of quantum mechanics. Thus, the resonance levels of carbon would have been different if different factors had been in play. And if those resonance levels had been different, then beryllium-8 and helium-4 could not have combined to form carbon-12. Then life would likely not have arisen in our universe. All this led Hoyle to marvel that carbon did come in a form with the precise energy level needed to allow the smaller elements of beryllium and helium to combine to form it. But it also led him and other physicists to explore what conditions were needed to ensure that carbon would have the right resonance level, allowing it to form.[13]

The Discovery of the First Fine-Tuning Parameters

The question of how carbon acquired its precise, favorable resonance turned out to be just the tip of the iceberg. To answer this question, Hoyle formulated a theory about how collapsing, or "dying," stars could synthesize carbon from lighter elements under specific conditions.[14] His theory implied that whether beryllium and helium could combine readily to make carbon would in turn depend upon a multitude of contingent factors and forces, many of which—as it happens—had to be precisely fine-tuned and balanced.

The precise energy levels of beryllium and helium, on the one hand, and those of carbon, on the other, depend on several factors: the precise strength of two of the four fundamental forces of physics, the strong nuclear force and the electromagnetic force, as well as the masses of elementary particles called light quarks. In addition, another of the fundamental physical forces, gravitation, needs to have just the right strength to allow carbon and other elements to form inside stars.

There are four distinct fundamental forces in nature: gravitational force, electromagnetic force (EMF), the strong nuclear force (SNF), and the weak nuclear force (WNF). The weak nuclear force causes nuclear radiation (i.e., the radioactive decay of atoms). The strong nuclear force, an attractive force, holds protons and neutrons together; the electromagnetic force attracts particles with opposite charges and repels those with the same charge. The SNF operates at a short range, and the EMF operates at all distances.

In addition, the effects of these forces within specific atoms differ (e.g., their effects are different in beryllium than they are in helium). And these effects are related to the number and configuration of their elementary particles (protons and neutrons). Modern supercomputer calculations indicate that the EMF and the SNF must have precise strengths, within about .5–4 percent of their current levels, to make carbon production possible.[15]

Much more striking, the masses of "up quarks" and "down quarks," the constituent parts of protons and neutrons, must have precise values to allow for the production of the elements, including carbon, essential for a life-friendly universe. Indeed, the masses of these quarks must have simultaneously met nine different conditions for the right nuclear reactions to have occurred in the early universe[16] (Fig. 7.4). The "right" reac-

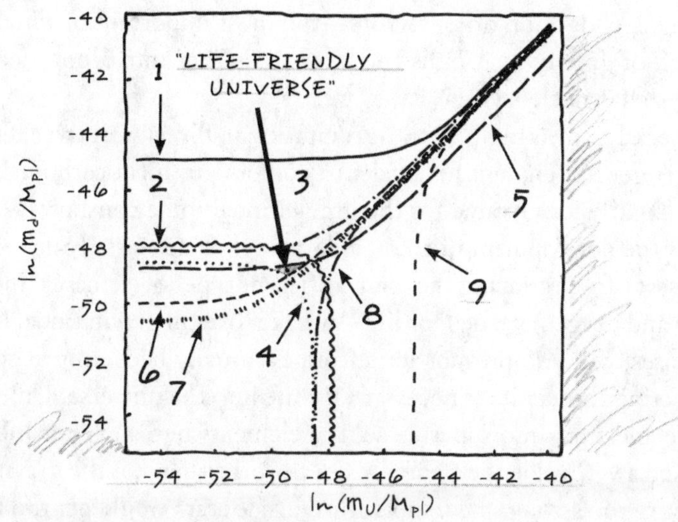

FIGURE 7.4

Each point on the graph corresponds to possible values for the masses of the up and down quarks (M_u, M_d). The masses are scaled by the Planck mass, M_{pl}, since Planck units are the most natural in cosmology. Each of the nine lines on the graph separates the regions corresponding to life-permitting and non-life-permitting universes for a specific criterion such as allowing for the existence of stable protons. In a universe capable of supporting life, all nine criteria must be met simultaneously, so the life-permitting region is the intersection of all nine life-permitting regions, marked in gray. That area corresponds to a minuscule proportion of all plausible values.

tions are ones that would produce the right elements (such as carbon and oxygen) in the right abundances necessary for life. The fine tuning of the masses of these two naturally occurring quarks in relation to the range of expected *possible* values for the mass of any fundamental particle is exquisite. Physicists conceive of that range as extending between a mass of zero and the so-called Planck mass, an important unit of measure in quantum physics. But the value of the "up quark" must have a precise mass of between zero and just one billion trillionth of the Planck mass, corresponding to a fine tuning of roughly 1 part in 10^{21}.[17] The mass of the "down quark" must have a similarly precise fine tuning.[18]

In addition, whether specific beryllium and helium atoms will combine to make carbon also depends on their kinetic (or thermal) energy, and this energy varies with the temperature of the stars in which these elements are forged. In order for carbon to form from the interaction of beryllium and helium, their nuclei must attain sufficiently high velocities to overcome the repulsive electromagnetic force between them. But that condition can only be met if the stars are hot enough to generate those critical atomic velocities. The ability to produce that much heat energy in turn depends upon other factors—the most important of which is the strength of gravity as it pulls the atoms together into a hot, dense ball during stellar nucleosynthesis.

Generally speaking,[19] if the gravitational force were weaker, stars wouldn't get hot enough for nuclei to combine to form carbon. In addition, a slightly lower value for the gravitational force constant (G) would prevent the development of thermal layering inside stars.[20] Such layering is necessary for producing the many different types of elements (including carbon and oxygen) needed for life. A weaker overall gravitational force, in most cases, will also prevent stars from eventually becoming supernovae and ejecting the elements necessary for life into the universe. Unless stars explode and turn into supernovae, the elements necessary for life would be locked away inside their cores.[21] On the other hand, if the gravitational force were too strong, the temperature inside stars would get too hot and nucleosynthesis would produce only elements heavier than carbon and oxygen. Nucleosynthesis would also proceed too quickly to allow for the formation of long-lived stars.[22] In that case, the stars would burn up too fast, thus depriving life of a fit place of habitation.[23]

Physicists have determined that the value of G is finely tuned to 1 part in 10^{35} in relation to a "natural" range of values that G could have (in possible alternate universes). Assuming that the SNF, the strongest of the four fundamental forces, establishes a reasonable upper limit for this range, the possible range of the four different fundamental forces can be conservatively set between zero and that of the SNF. The strength of gravity is about a factor of 10^{40} weaker than that of the SNF, so gravity could range between 0 and 10^{40} times G.[24] The value of the gravitational force constant could have been as much as 100,000 (or 10^5) times larger than its actual value without stars losing stability, though any further increases would produce such instability. Even so, since G could reasonably range from 0 to 10^{40} times its current value, the range of G consistent with stable stars still represents a small fraction of this range, 1 part in one hundred billion trillion trillion (1 in 10^{35}).[25]

Over the years, as Hoyle thought more about the discovery of the exact resonance level of carbon that he had predicted, and especially about all the factors that had to be just right to make carbon relatively easy to produce inside stars, he became convinced that some intelligence had orchestrated the precise balance of forces and factors in nature to make the universe life-permitting.[26] The strengths of the strong nuclear and electromagnetic forces, the ratio between the fundamental forces,[27] the exact kinetic energy of beryllium and helium, and thus the strength of gravitational forces inside stars as well as the excitation energy of carbon all had to be exquisitely tuned and coordinated within very narrow tolerances to promote the synthesis of large amounts of carbon inside stars. Yet without carbon life would be impossible.

Hoyle was stunned by these and other "cosmic coincidences" that physicists began to discover after the 1950s.[28] Whereas before he affirmed atheism and denied any evidence of design, he began to see fine tuning as obvious evidence of intelligent design. As he put it in 1981, "A common-sense interpretation of the facts suggests that a super-intellect has monkeyed with physics, as well as with chemistry and biology, and that there are no blind forces worth speaking about in nature. The numbers one calculates from the facts seem to me so overwhelming as to put this conclusion almost beyond question."[29]

The fine-tuning parameters that Hoyle discovered were by no means the only such parameters necessary to ensure a life-friendly universe. Indeed, cosmologists and physicists have found that the existence of life depends upon a dozen or so of these highly improbable finely tuned parameters.[30] Many have also noted that this fine tuning strongly suggests design by a preexistent intelligence. As the British physicist Paul Davies put it in 1988, "The impression of design is overwhelming."[31] Similarly, astrophysicist Luke Barnes notes: "Fine tuning suggests that, at the deepest level that physics has reached, the Universe is well put-together. . . . The whole system seems well thought out, something that someone planned and created."[32]

To see why, it helps to understand a bit more about the different types of fine tuning that physicists have discovered and the extraordinary extent of that fine tuning.

The Fine Tuning of the Laws and Constants of Physics

The most fundamental type of fine tuning pertains to the laws of physics and chemistry. Typically, when physicists say that the laws of physics exhibit fine tuning, they are referring to the constants within those laws.[33] But what exactly are the "constants" of the laws of physics?

The laws of physics usually relate one type of variable quantity to another. A physical law could tell us that as one variable (say, force) increases, another (say, acceleration) also increases proportionally by some factor. Physicists describe this type of relationship by saying that one variable quantity is *proportional* to another. Conversely, a physical law may stipulate that as one factor increases, another decreases by the same factor. Physicists describe this type of relationship by saying that the first variable quantity is *inversely proportional* to the other.

Newton's classical law of gravity, like most laws of physics, has a form expressing such relationships. The gravitational force equation asserts that the force of gravity between two bodies is proportional to the product of the masses of those bodies. It also stipulates that the force of gravity is inversely proportional to the distance between the bodies squared. Yet even if physicists know the exact masses of the bodies and the distance between their centers, that by itself doesn't allow them to compute the exact force of gravity.

Instead, an additional factor known as the gravitational force constant first has to be determined by careful experimental measurements. The gravitational force constant represents a kind of mysterious "X factor" that allows physicists to move beyond just knowing proportionality relationships—that is, that certain factors increase or decrease as other factors increase or decrease. Instead, it allows physicists to compute the force of gravity accurately if they know the values of those other variable quantities (mass and distance) *and* the value of the constant of proportionality. Physicists write Newton's gravitational force equation with the letter G representing the force constant, with *m* for mass and *d* for distance, as follows: $F = G\frac{M_1M_2}{d^2}.$

To explain the idea of a constant of proportionality, here's a thought experiment I used with my students. There was a Russian pole vaulter I admired named Sergey Bubka. In the 1980s and 1990s, Sergey set numerous pole-vaulting records. Now imagine you are a muscular vaulter like Sergey. You charge down the tarmac, plant your pole, and you begin to lift off, hoping to clear, say, 20 feet 3 inches, and set a new world record. Yet as you're about 10 feet in the air, some evil demon suddenly fiddles with the dials in the cosmic control room that sets the force constants for all the laws of physics. In the process, the demon changes the gravitational force constant. Your mass is still 100 kilograms, the earth still has the same mass (5.9736×10^{24} kilograms), and you are at that moment still roughly 10 feet away from the earth, as you were an instant before. But now, because the gravitational force constant has changed, the force of gravity acting on you has changed dramatically (Fig. 7.5).

On the basis of Newton's gravitational force equation, with the old

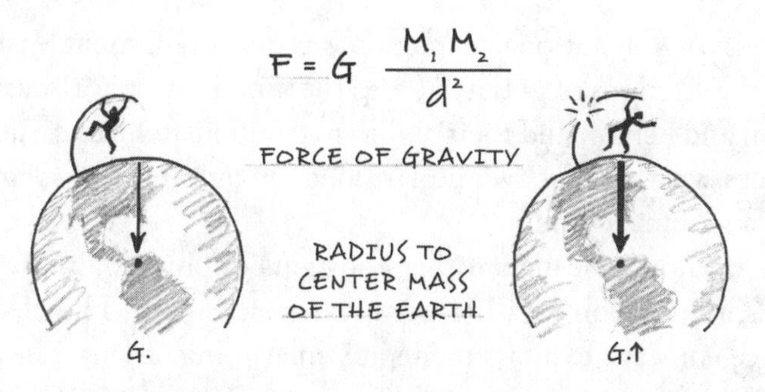

FIGURE 7.5
If the gravitational constant G were to increase dramatically during a pole vaulter's jump, the force of gravity exerted on the vaulter would increase proportionally, though the vaulter's mass, the earth's mass, and the distance from the center of earth (at that moment) would not have changed. Such a capricious change in the value of G could then result in the vaulter's pole snapping and the vaulter crashing to the earth.

gravitational force constant, you should clear the 20-foot, 3-inch bar with no trouble at all. (Yes, you're that good!) But now, due to that capricious cosmic fiddler, the force between you and the earth becomes much stronger. Consequently, your pole snaps and you crash to the earth.

All the equations describing the fundamental forces of physics include such force constants. Essentially, they measure or quantify the net effect of all the other factors (not specifically represented in the variables in the relevant equations) that affect the magnitude of the forces in question. And that's the curious thing. These force constants have one of the rare sets of values that make life in the universe possible. In other words, the constants in the equations describing the fundamental forces of physics turn out to be exquisitely finely tuned within extremely fine tolerances. As Paul Davies (Fig. 7.6) has marveled, "The really amazing thing is not that life on earth is balanced on a knife-edge, but that the entire universe is balanced on a knife-edge, and would be total chaos if any of the natural 'constants' were off even slightly."[34] Or as Stephen Hawking noted, "The remarkable fact is that the values of these numbers seem to have been very finely adjusted to make possible the development of life."[35]

FIGURE 7.6
In his many fine books, British physicist Paul Davies has explained and popularized the evidence for the fine tuning of the universe.

Recall from our discussion of Hoyle's discovery of the resonance level of carbon that the exact strength of gravitational attraction was important for at least two key reasons: the strength of gravity affects the kinetic energy associated with beryllium and helium and their ability to combine to form carbon, and it determines whether stars would last long enough to form solar systems capable of sustaining life.

We now know that many other constants of physics also exhibit fine tuning as a condition of a life-permitting universe.[36] The electromagnetic force constant exhibits moderate fine tuning of 1 part in 25.[37] The strong nuclear force constant is fine-tuned to 1 part in 200.[38] Moreover, the ratios of the values of the different force constants also require significant fine tuning. For example, the ratio of the weak nuclear force constant to the strong nuclear force constant had to have been set with a precision of 1 part in 10,000.[39] If the weak force had been weaker or stronger by that small fraction, stars powered by hydrogen fusion, required for life, would not have existed.

More impressively, the ratio of the electromagnetic force to gravity must be accurate to 1 part in 10^{40}. Were this ratio a bit lower, the gravitational attraction would be too strong in comparison to the contravening force of electromagnetism pushing nuclei apart. In that case, stars would, again, burn too quickly and unevenly to allow for the formation of long-lived stars and stable solar systems. Were this ratio a bit higher, gravitational attraction would be too weak in comparison to electromagnetism. That would have prevented stars from burning hot enough to produce the heavier elements needed for life.[40]

Indeed, slight differences in the strength of any of these constants or their ratios would preclude the possibility of life. Martin Rees, an emeritus professor of cosmology and astrophysics at the University of Cambridge, who is Astronomer Royal for the United Kingdom, summed up the matter with characteristic British understatement: "The possibility of life as we know it depends on the values of a few basic physical constants and is, in some respects, remarkably sensitive to their numerical values. Nature does exhibit remarkable coincidences."[41]

A Universe-Creating Machine?

When I was a PhD student, I took the opportunity to meet the Cambridge physicist John Polkinghorne. Sir John had retired as professor of mathematical physics a few years earlier and now held the position of president

FIGURE 7.7
Cambridge theoretical physicist Sir John Polkinghorne, who has argued that cosmic fine tuning now provides the basis for a revived program of natural theology.

of Queens' College, Cambridge (Fig. 7.7). I remember feeling a bit overawed on my way to meet him in his office, as I walked down the long hallway decorated with paintings of former presidents and other luminaries of the college. I had asked to meet in part because he had written extensively on the fine-tuning problem, and I wanted to ask if he would speak to a postgraduate student group to which I belonged. He agreed and later gave an excellent and remarkably clear talk to us about why he thought fine tuning provided persuasive evidence of design.

In his talks on the subject, he often used a memorable illustration to explain why he had come to that conclusion. Polkinghorne would ask students to imagine that they had traveled deep into space and had entered a space station in which they found a "universe-creating machine"—the very machine responsible for the fine tuning of the universe (Fig. 7.8). Polkinghorne would ask his audience to imagine that his hypothetical universe-generating machine displayed a number of different dials,

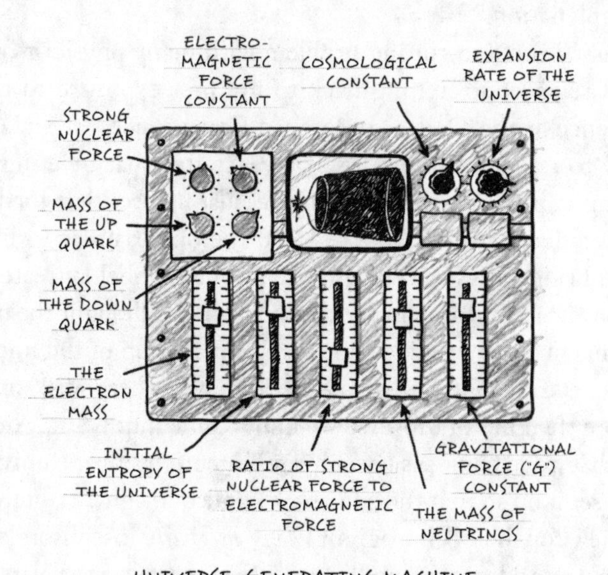

FIGURE 7.8
A hypothetical universe-generating machine illustrating the fine tuning of the laws and constants of physics and the initial conditions of the universe.

knobs, or adjustable sliders, each representing one of the many cosmological fine-tuning parameters. He also asked the audience to imagine that each of those dials was fixed to a specific setting out of a vast number of other possible settings.

I would later use an embellished version of Polkinghorne's illustration with my own students. I asked them to imagine that they were brilliant physicists who could—upon doing a few key calculations—quickly determine that the precise dial settings on the universe-creating machine, and nearly no other settings, allowed for the production of a life-friendly universe. "What if," I asked, "after making these calculations, you had discovered that even slight changes to the dial settings—by one click this way or that—resulted in catastrophic consequences that prevented the universe from sustaining life? What would you make of that?"

With somewhat less elaboration, Polkinghorne asked his audiences the same thing. Years later, I had the opportunity to interview him on camera about what *he* made of it. He answered by saying that although he did not think that the fine-tuning evidence *proved* the existence of God, he did think that a theistic designer provided a much better explanation of the fine tuning than any materialistic hypothesis. As he put it, "Well, I don't say that the atheist is stupid. I just say that theism provides a more satisfying explanation."[42]

I've discussed the fine-tuning problem with many physicists and philosophers of science, stretching back to my first exposure to the fine-tuning evidence at the Dallas conference I mentioned earlier. An early opportunity to do so presented itself a year after that meeting while I was attending a similar conference at Yale, just before I left for England for my postgraduate studies. There I had the opportunity to interview another Cambridge physicist, Brian Josephson, a Nobel laureate.

In our conversation, Josephson explained why he thought the choice of a prior intelligent mind provided a natural explanation of the fine-tuning evidence. As he explained, "It could have been [that there was] some mind around before the kind of universe we know came into being. And if that were right, that mind could, as it were, have intentions for the universe and been able to set it up so that the end result came out right."[43] Interestingly, in an interview for the PBS program *Closer to Truth*, Josephson later estimated his own confidence in intelligent design as the best explanation for the conditions that would make evolution possible at "about 80 percent."[44]

Another physicist I interviewed, the late Henry Margenau, a distinguished Yale professor of quantum physics, put the point even more emphatically. When I asked him how he explained the fine tuning of the laws and constants of physics, he simply said, "There is a mind which is responsible for the laws of nature and the existence of nature and the whole universe. And this is consistent with everything we know."[45]

As it happens, even physicists who have retained a materialistic perspective have found the fine-tuning evidence suggestive of intelligent design. Physicist and self-described atheist George Greenstein has confessed that, in the face of his materialistic predilections, "the thought insistently arises that some supernatural agency, or rather Agency, must be involved. Is it possible that, suddenly, without intending to, we have stumbled upon scientific proof for the existence of a supreme being? Was it a God who providentially stepped in and crafted the cosmos for our benefit?"[46]

Straws and Monkey Business

About the same time that I first met Polkinghorne, an article appeared in the *London Times* written by science journalist Clifford Longley. Longley, like Polkinghorne and Hoyle, argued that the theistic design hypothesis provided an obvious and commonsense explanation for the anthropic fine-tuning evidence. Attempts to explain the evidence by invoking chance alone or multiple other universes (more on that in Chapter 16) seemed to him to betray a kind of metaphysical special pleading, even desperation. As Longley explained, the anthropic design argument "is of such an order of certainty that in any other sphere of science, it would be regarded as settled." He continued: "To insist otherwise is like insisting that Shakespeare was not written by Shakespeare because it might have been written by a billion monkeys sitting at a billion keyboards typing for a billion years. So it might. But the sight of scientific atheists clutching at such desperate straws has put new spring in the step of theists."[47]

In the next chapter, I explain exactly what it is about the fine-tuning evidence that has led so many physicists to consider the design hypothesis. The question is not as obvious as it might at first seem. I also discuss other types of fine tuning that, in many cases, exhibit even more exquisite precision than the examples we have addressed up to this point. For now, it's worth noting that at least one version of the design argument has returned to scientific currency and that discoveries in physics, astronomy, cosmology, and chemistry have contributed to this unexpected development.

8

Extreme Fine Tuning—by Design?

"Intelligent design," as Nobel Prize–winning physicist Charles Townes has said, "as one sees it from a scientific point of view, seems to be quite real. This is a very special universe: it's remarkable that it came out just this way. If the laws of physics weren't just the way they are, we couldn't be here at all. The sun couldn't be there, the laws of gravity and nuclear laws and magnetic theory, quantum mechanics, and so on have to be just the way they are for us to be here."[1] Physicists such as Townes, who also taught at UC–Berkeley until his death in 2015, often speak of the fine tuning of the laws and constants of physics as the main type of fine tuning in the universe, as described in the previous chapter.

Even so, there are at least two other general types of fine tuning: the fine tuning of the initial conditions of the matter and energy at the beginning of the universe, and the fine tuning of other contingent features of the universe. These don't get mentioned quite as often as the fine tuning of the laws and constants of physics, but they are every bit as important for the existence and maintenance of life in the universe. In fact, some of these other features of the universe manifest an even more extreme degree of fine tuning than the examples we have already explored.

The Fine Tuning of the Initial Conditions of the Universe

Let's consider the initial fine tuning in the distribution or configuration of matter and energy at the beginning of the universe. You may remember that Einstein's famous equation $E = mc^2$ tells us that matter and energy are interconvertible; that is, matter has energy associated with it and energy can manifest itself as material particles.

Recall from Chapter 5 that the inhomogeneous distribution of mass-energy at the beginning of the universe accounted for the presence of galaxies in some parts of the universe and a dearth of matter in others. Physicists have determined that if the matter at the beginning of the universe had been configured even slightly differently, there would be either an extreme clumping of matter resulting in a universe in which only black holes would exist or, alternately, a highly diffuse arrangement of matter without any large-scale structures at all. Both of these alternatives would have prevented the formation of stable galaxies and stars in which life-friendly solar systems might later emerge.

Here is an illustration I have often used with students to explain the idea of initial-condition fine tuning. Before the invention of tunnel-boring machines, engineers who wanted to build a tunnel through a mountainside would use dynamite to blast a hole in its side. They would carefully plan how to "configure the charge," because they knew that tiny differences in the orientation and positioning of the explosives could make huge differences in the direction and shape of the hole left by the blast (Fig. 8.1). That is, they knew that the initial configuration of matter and energy together would determine the structure and shape of the resulting hole in the mountain.

In much the same way, the configuration of matter and energy at the beginning of the universe determined the distribution of matter and energy later in the history of the cosmos. Only the extreme fine tuning of that initial configuration enabled galaxies, stars, and planetary systems to form.

FIGURE 8.1

When creating a tunnel, the precise angle and force of dynamite charges will determine the outcome. In the same way, the initial configuration of matter and energy at the beginning of the universe will determine whether or not a life-permitting universe will result.

Initial-Entropy Fine Tuning

Physicists refer to the initial distribution of mass-energy as "entropy" (or "initial entropy") fine tuning.[2] Entropy measures the amount of disorder in a material system (of, e.g., molecules, atoms, or subatomic particles). Decreases in entropy correspond to increases in order. Increases in entropy correspond to increased *dis*order.[3] A universe in which ordered structures such as galaxies and solar systems can arise would require a low-entropy (highly specific) configuration of mass and energy at the beginning. In a universe with larger initial entropy, black holes would come to dominate.

Making an assessment of entropy requires determining the number of configurations of matter and energy that generate, or are consistent with, a particular state of affairs. If there are many configurations consistent with a given state of affairs, then physicists say that state has high entropy and is highly *dis*ordered. If there are only a few configurations consistent with a given state of affairs, then physicists say that state has low entropy and is highly ordered. For example, there are far fewer ways to arrange the books, papers, pencils, clothes, and furniture in a room that will result in it looking neat than there are ways of arranging those items that will result in it looking messy. We could say then that a tidy room represents a low-entropy, highly ordered state, whereas a messy room obviously represents a disordered, high-entropy state.

Or consider another illustration of these concepts. Liquid water exemplifies a high-entropy state. That's because, at temperatures between 32 and 212 degrees Fahrenheit, many different arrangements of water molecules are possible, all consistent with H_2O in a liquid state.[4] In other words there are lots of different ways (i.e., configurations of molecules) to have water as a liquid. Conversely, water in a solid state—namely, ice—exemplifies a low-entropy state, because ice has a rigidly ordered lattice structure. That structure restricts the number of ways of arranging water molecules. Consequently, there are relatively few ways (i.e., configurations) to have water in a solid state.

In the universe, a black hole represents a highly disordered (high-entropy) state, like an extremely messy room. That's because the intense gravitational forces at work in a black hole ensure that matter and energy may adopt many different chaotic configurations.[5] Yet regardless of which of those configurations result from the intense gravitational forces, the large-scale structure of the black hole will remain roughly the same.[6] Conversely, a galaxy represents a low-entropy state, like a tidy room, because there are relatively few ways to configure the elements out of which galaxies are made that will result in the orderly structures they exhibit. The universe as a whole also represents a lower-entropy system because the galaxies are uniformly distributed throughout space. On the other hand, if the universe were characterized by large irregularly distributed clumps of matter (e.g., in the form of lots of black holes), it would exhibit high entropy.

So how unlikely is it that our universe would have the low-entropy, highly ordered arrangement of matter that it has today? Stephen Hawking's colleague Roger Penrose (Fig. 8.2) knew that if he could answer that question, he would have a measure of the fine tuning of the initial arrangement of matter and energy at the beginning of the universe.

FIGURE 8.2
The Oxford physicist Sir Roger Penrose, who collaborated with Stephen Hawking in proving cosmological singularity theorems and later calculated the exquisite and hyper-exponential fine tuning of the initial entropy of the universe.

Penrose determined that getting a universe such as ours with highly ordered configurations of matter required an exquisite degree of initial fine tuning—an incredibly improbable low-entropy set of initial conditions.[7] His analysis began by assuming that neither our universe nor any other would likely exhibit more disorder (or entropy) than a black hole, the structure with the highest known entropy. He then calculated the entropy of a black hole using an equation based upon general relativity and quantum mechanics. The entropy value he calculated established a reasonable upper bound, or maximum possible entropy value, for the distribution of the mass-energy in our visible universe.[8]

Penrose then asked: Given the wide range of possible values for the entropy of the early universe, how likely is it that the universe would have the precise entropy that it does today? To answer that question, he needed to know the entropy of the present universe. Penrose made a quantitative estimate of that value.[9] He then assumed that the early universe would have had an entropy value no larger than the value of the present universe, since entropy (disorder) typically increases as energy moves through a system, which would have occurred as the universe expanded. (Think of a tornado moving through a junkyard or a toddler through a room.)

Then he compared the number of configurations of mass-energy consistent with an early black-hole universe to the number consistent with more orderly universes like ours. Mathematically, he was comparing the number of configurations associated with the maximum possible entropy state (a black hole) with the number associated with a low-entropy state (our observable universe). By comparing that maximum expected value of the entropy of the universe with the observed entropy, Penrose determined that the observed entropy was extremely improbable in relation to all the possible entropy values it could have had.[10] In particular, he showed that there were $10^{10^{101}}$ configurations of mass-energy—a vast number—that correspond to a highly ordered universe like ours. But he had also shown that there were vastly *more* configurations—$10^{10^{123}}$—that would generate black-hole dominated universes. And since $10^{10^{101}}$ is a minuscule fraction of $10^{10^{123}}$, he concluded that the conditions that could generate a life-friendly universe are extremely rare in comparison to the total number of possible configurations that could have existed at the beginning of the universe. Indeed, dividing $10^{10^{101}}$ by $10^{10^{123}}$ just yields the number $10^{10^{123}}$ all over again. Since the smaller exponential number represents such an incredibly small percentage of the larger exponential number, the smaller number can be ignored as the massively larger exponential number effectively swallows up the smaller one.

In any case, the number that Penrose calculated—1 in $10^{10^{123}}$—provides a quantitative measure of the unimaginably precise fine tuning of the initial conditions of the universe.[11] In other words, his calculated entropy implied that out of the many possible ways the available mass and energy of the universe could have been configured at the beginning, only a few configurations would result in a universe like ours. Thus, as Paul Davies observes, "The present arrangement of matter indicates a very special choice of initial conditions."[12]

That's putting it mildly. The mathematical expression $10^{10^{123}}$ represents what mathematicians call a hyper-exponential number—10 raised to the 10th power (or 10 billion) raised again to the 123rd power. To put that number in perspective, it might help to note that physicists have estimated that the whole universe contains "only" 10^{80} elementary particles (a huge number—1 followed by 80 zeroes). But that number nevertheless represents a minuscule fraction of $10^{10^{123}}$.[13] In fact, if we tried to write out this number with a 1 followed by all the zeros that would be needed to represent it accurately without the use of exponents, there would be more zeros in the resulting number than there are elementary particles in the entire universe. Penrose's calculation thus suggests an incredibly improbable arrangement of mass-energy—a degree of initial fine tuning that really is not adequately reflected by the word "exquisite." I'm not aware of a word in English that does justice to the kind of precision we are discussing.[14]

The Fine Tuning of the Expansion Rate of the Universe and/or the Cosmological Constant

In addition to the fine tuning of the laws and constants of physics and the arrangement of matter and energy at the beginning of the universe, physicists have discovered many other contingent, finely tuned features of the universe. For example, a life-permitting universe depends crucially on its precise expansion rate. Since the discovery of the red shift of the light coming from distant galaxies, astronomers have discovered that if the universe were initially expanding even a smidgeon faster or slower,[15] either stable galaxies would not have formed in the universe because matter would have dissipated too quickly for galaxies to congeal or else the universe would have quickly collapsed in on itself. Cosmologists refer to the first scenario as the "heat death of the universe" and the second scenario as the "big crunch." Neither outcome is friendly to life.

Though many leading physicists cite the expansion rate of the universe as a good example of fine tuning, some have questioned whether it should be considered an *independent* fine-tuning parameter, since the rate of expansion is a consequence of other physical factors.[16] Nevertheless, these physical factors are themselves independent of each other and probably finely tuned.[17] For example, the expansion rate in the earliest stages of the history of the universe would have depended upon the density of mass and energy at those early times. And the density of the universe one nanosecond (a billionth of a second) after the beginning had to have the precise value of 10^{24} kilograms per cubic meter. If the density were larger or smaller by only 1 kilogram per cubic meter, galaxies would never have developed.[18] This corresponds to a fine tuning of 1 part in 10^{24}.

The cosmological constant requires an even greater degree of fine tuning. (Remember that the cosmological constant is a constant in Einstein's field equations. It represents the energy density of space that contributes to the outward expansion of space in opposition to gravitational attraction.) The most conservative estimate for that fine tuning is 1 part in 10^{53}, but the number 1 part in 10^{120} is more frequently cited.[19] Physicists now commonly agree that the degree of fine tuning for the cosmological constant is *no less than* 1 part in 10^{90}.[20]

To get a sense of what this number means, imagine searching the vastness of the visible universe for one specially marked subatomic particle. Then consider that the visible universe contains about 200 billion galaxies each with about 100 billion stars[21] along with a panoply of asteroids, planets, moons, comets, and interstellar dust associated with each of those stars. Now assume that you have the special power to move instantaneously anywhere in the universe to select—blindfolded and at random—any subatomic particle you wish. The probability of your finding a specially marked subatomic particle—1 chance in 10^{80}—is still 10 billion times *better* than the probability—1 part in 10^{90}—that the universe would have happened upon a life-permitting strength for the cosmological constant.

Other Contingent Fine-Tuning Factors

Examples of the fine tuning of contingent features of the universe abound. For instance, to make life possible, the masses of the fundamental particles must meet an exacting combination of constraints. In the previous chapter, I discussed the fine tuning of the masses of the two naturally occurring quarks, the up quark and down quark, in relation to the range of expected *possible* values. Recall that the fine tuning of the masses of those quarks is considerable—1 part in 10^{21}. In addition, the *difference in masses* between the quarks cannot exceed one megaelectron volt, the equivalent of one-thousandth of 1 percent of the mass of the largest known quark, without producing either a neutron-only or a proton-only universe, both exceedingly boring and incompatible with life and even with simple chemistry.

Equally problematic, increasing the mass of electrons by a factor of 2.5 would result in all the protons in all the atoms capturing all the orbiting electrons and turning them into neutrons. In that case, neither atoms, nor chemistry, nor life could exist.[22] What's more, the mass of the electron has to be less than the difference between the masses of the neutron and the proton and that difference represents fine tuning of roughly 1 part in 1000.[23] In addition, if the mass of a special particle known as a neutrino were increased by a factor of 10, stars and galaxies would never have formed. The mass of a neutrino is about one-millionth that of an electron, so the allowable change is minuscule compared to its possible range.[24]

The combination of all these precisely fine-tuned conditions—including the fine tuning of the laws and constants of physics, the initial arrangement of matter and energy, and various other contingent features of the universe—presents a remarkably restrictive set of criteria. These requirements for the existence of life, again defying our ability to describe their extreme improbability, have seemed to many physicists to require *some* explanation.

The Weak Anthropic Principle

In 1974, the physicist Brandon Carter proposed what was initially the most popular naturalistic (or nontheistic) interpretation of the fine tuning. As Carter put it: "What we can expect to observe *must be* restricted by the conditions necessary for our presence as observers."[25]

Thus, Carter[26] and other proponents of what came to be commonly known as the "weak anthropic principle" (WAP) argued that we human beings should not be surprised to find ourselves living in a universe suited for life, because if the universe were otherwise, we wouldn't be here to observe it. And since there is nothing *surprising* about living in a universe that has the conditions necessary for our own existence, proponents of the WAP have argued that the fine tuning requires no explanation.

Nevertheless, the WAP encountered formidable criticism from philosophers of physics and cosmology. As philosopher John Leslie has argued,[27] contrary to what WAP advocates claim, the origin of the fine tuning *does* require explanation. He points out that though we humans should not be surprised to find ourselves living in a universe suited for life (since we are alive), we ought to be surprised to learn that the conditions necessary for life *are so extremely improbable.*

To illustrate the fallacy behind the WAP, Leslie likened our situation in our finely tuned universe to that of a blindfolded man who has discovered that, against all odds, he has survived a firing squad of one hundred expert marksmen. Though his finding himself still alive is certainly consistent with the fact that all the marksmen missed, it does not explain *why* the marksmen actually did miss. Instead, Leslie argues the prisoner should be surprised that he is still alive, because the marksmen are known to be excellent shots and the probability of all of them missing (if they had intended to kill him) is extremely small.

Clearly, the WAP makes a logical error. It treats stating a *necessary condition* of the occurrence of an event (in this case, our existence) as if it eliminated the need for a *causal explanation* of the conditions that make the event possible. To see part of the flaw in this reasoning, consider the following. Imagine that an insurance company sends an insurance investigator to find out why a warehouse burned down on a rainy night. After surveying the charred remains of the building, the investigator submits his report. The building burned down, he confidently explains, because of oxygen in the atmosphere.

The insurance company promptly fires the investigator and hires someone else. What was his mistake? The investigator confused a necessary condition of fire with the *cause* of the specific fire under investigation. In this case, the oxygen was not the "difference that made a difference"—that is to say, the *cause* of the fire. Oxygen had been in the air around the warehouse all along. Its presence was not by itself sufficient to produce the fire and thus does not explain why the building suddenly burned down. The gas cans and matchbooks left near the scene of the charred building would have provided more relevant information, had the investigator not been so literally "clueless."

Oxygen is indeed a necessary condition of fire, but saying so did not provide a *causal explanation* of the particular fire in question. Similarly, the fine tuning of the physical constants is a necessary condition for life, but that does not provide a causal explanation of, or eliminate the need to explain, the fine tuning itself.

Notice too that WAP advocates focus on the wrong phenomenon of interest. They think that what needs to be explained (or explained away) is why *we observe* a universe consistent with our existence. It's true that such an observation is not surprising. What needs explanation, though, is what caused the fine tuning of the universe in the first place—not our later observation of it. Thus, WAP advocates offer as a cause of the event that *does* need explanation a statement of a necessary condition of another event that *does not* need explanation.

The Strong Anthropic Principle

Another version of the anthropic principle known as the strong anthropic principle (SAP) or participatory anthropic principle (PAP) has attracted well-deserved disdain for its bizarre logic. Some physicists define the strong anthropic principle as simply the idea that "the Universe must have those properties which allow life to develop within it at some stage in its history,"[28] without saying what caused those finely tuned properties to develop. Nevertheless, a well-known extension of that idea also goes, somewhat confusingly, by the same name, the "strong anthropic principle." And it does attempt to explain the origin of the fine tuning.

According to this version of the SAP,[29] the need for observers to confer reality upon the universe means that the universe had to be fine-tuned to produce human observers to observe it. This explanation of the fine tuning is based on an interpretation of a strange phenomenon in the field of quantum physics. Physicists during the early part of the twentieth century discovered that electrons and photons of light will behave like either waves or particles depending upon how they are observed. In a now famous experiment called "the double-slit experiment" (see Chapter 17 for details), spatially extended waves of light producing characteristic interference patterns will—upon hitting a detection plate—manifest themselves as particles in spatially discrete locations. In quantum physics this phenomenon is known as the "collapse of the wave function." Some physicists have interpreted the phenomenon—of waves manifesting themselves as particles upon detection—as a consequence of the wave *being observed*.

Proponents of the strong anthropic principle have argued that just as the specific location of a photon of light or electron depends upon an observation and observer, the universe itself might depend for its existence upon an observer as well. But that means, they argue, that the universe would have had to have been finely tuned from the beginning in order for it to exist at all, since only a universe finely tuned for conscious life would produce observers capable of conferring existence upon the universe. Physicists John Barrow and Frank Tipler characterize this well-known version of the strong anthropic principle as affirming that: (1) "the Universe must have those properties which allow life to develop within it at some stage in its history"; (2) "there exists one possible Universe 'designed' with the goal of generating and sustaining 'observers'"; and (3) "observers are necessary to bring the Universe into being."[30]

This reasoning has an obviously problematic aspect. The observers allegedly causing the fine tuning of the universe make their observation of the fine tuning billions of years *after*, not *before*, the event that they allegedly cause. Yet clearly the very concept of "cause" implies an event that produces a *subsequent* effect. Even in the case of alleged observer-caused quantum phenomena, the effect—the collapse of the wave function—occurs after the cause—that is, the detection of the wave of light. The patent illogic of this formulation led the famous hard-nosed *Scientific American* writer Martin Gardner to express exasperation. He called the SAP and similar attempts to explain away the origin of the fine tuning "CRAP, the Completely Ridiculous Anthropic Principle."[31] Almost all physicists and philosophers now agree with this well-deserved, if intemperate, dismissal.

Chance and Natural Laws

The improbability of many of the individual fine-tuning parameters, to say nothing of the improbability of the whole ensemble, seems to preclude a straightforward appeal to chance. (I'll consider less straightforward appeals to chance involving the postulation of other universes in Chapter 16.) Yet the laws of physics do not seem capable of explaining the fine tuning either.

Indeed, the laws of physics do not account for why the laws of physics are as they are—that is, why they have the precisely fine-tuned features that they do. To see why, consider that several key fine-tuning parameters—in particular, the values of the constants of the fundamental laws of physics—are *intrinsic* to the structure of those laws. In other words, the precise "dial settings" of the different constants of physics represent specific features *of the laws of physics themselves*—just how strong gravitational attraction or electromagnetic attraction will be, for example. These specific and contingent values cannot be explained by the laws of physics because they *are part of* the logical structure of those laws. Scientists who say otherwise are just saying that the laws of physics explain themselves. But that is reasoning in a circle.

Similarly, no known law of physics can explain the initial distribution of matter and energy at the beginning of the universe, since the laws describe how various forces or fields act upon specific material conditions *once those conditions are present.* They do not explain how the conditions arose in the first place; they presuppose them. Indeed, the equations that express the fundamental laws of physics cannot be solved unless information about initial conditions is provided from some other source. The laws themselves neither furnish that information nor explain why the initial conditions have the values they do.

So, then, what does explain fine tuning?

Cosmic Clues

It may seem intuitively obvious that a finely tuned universe suggests a "fine tuner"—or a "superintellect" of some kind. But what exactly is it about the fine tuning that has suggested design—intelligent design—to so many physicists? Why have many thought that the theistic design hypothesis provides "a more satisfying explanation" of the fine tuning than various materialistic theories?

The mathematician and philosopher William Dembski (Fig. 8.3), has developed a theory about how we detect the activity of intelligent agents in the effects they leave behind. His theory helps explain why the fine tuning evidence suggests design to so many physicists. It also reinforces the conclusion that the fine tuning of the laws, constants, and initial conditions of the universe does indeed point to a designing mind.

In his groundbreaking book *The Design Inference*, Dembski explicated the criteria by which rational agents recognize the effects of other rational agents and distinguish them from the effects of unintelligent natural causes. According to Dembski, systems, sequences, or events that exhibit two characteristics at the same time—extreme improbability and a special kind of pattern called a "specification"—indicate prior intelligent activity. According to Dembski, extremely improbable events that also exhibit "an *independently recognizable* pattern" or set of functional

FIGURE 8.3
Mathematician and philosopher William Dembski. In his groundbreaking book *The Design Inference*, Dembski established a rigorous method of detecting the activity of intelligent agents and distinguishing such activity from purely natural causes.

requirements, what he calls a "specification," *invariably* result from intelligent causes, not chance or physical-chemical laws.[32]

I've often explained Dembski's theory by asking students to think about the faces on Mt. Rushmore in South Dakota. If you look at that famous mountain you will quickly recognize the faces of the American presidents inscribed there as the product of intelligent activity. Why? What about those faces indicates that an artisan or sculptor acted to produce them? You might want to say it's the improbability of the shapes. By contrast, we would not be inclined to infer that an intelligent agent had played a role in forming, for example, the common V-shaped erosional pattern between two mountains produced by large volumes of water. Instead, the faces on the mountain qualify as extremely improbable structures, since they contain many detailed features that natural processes do not generally produce. Certainly, wind and erosion, for example, would be unlikely to produce the recognizable faces of Washington, Jefferson, Lincoln, and Roosevelt.

Nevertheless, as Dembski points out, the precise arrangement of the rocks at the bottom of the mountain also represents an extremely improbable configuration, especially when one considers all the other possible ways those rocks might have settled. So in addition to the improbability of the shapes, what helps us to recognize that intelligent activity played a role in producing the faces?

The answer is the presence of a special kind of pattern. In addition to an improbable structure we see a shape or pattern that *matches* one we know from independent experience, namely, from seeing the human face and even the specific faces of the presidents on money or in history books. Thus, Dembski suggests that the improbability of the structure by itself does not trigger our awareness of prior intelligent design. Instead, intelligent agents recognize intelligent activity whenever they observe a highly improbable object or event that also matches an independently recognizable or meaningful pattern. The pile of stones at the bottom of the cliff does not form such a pattern, but the faces on the mountain do.

Consider another example. The cartoon in Figure 8.4 depicts a pattern of flowers on the hillside of the harbor in the city of Victoria on Canada's Vancouver Island. I've occasionally ridden a high-speed ferry from Seattle that docks in the Victoria harbor. Once, while standing on the bow of the ferry boat as it came into the harbor, I noticed this pattern of red and yellow flowers on the hillside. While I was still at

FIGURE 8.4
Specified complexity or functional information as an indicator or "signature" of intelligence. The inner harbor of Victoria, Canada houses flower beds that spell out the phrase "Welcome to Victoria." The arrangement of flowers conveys "specified" or functional information, an unmistakable sign of intelligence. No one, for example, would attribute this pattern of flowers to an undirected process such as birds flying over the harbor randomly dropping seeds.

some distance, the pattern caught my attention and made me curious, so I put on my glasses. When I did, I immediately made a design inference. Why? I realized the red and yellow flowers spelled the message "Welcome to Victoria"—clearly, the work of intelligent gardeners, not just natural processes.

Dembski's theory and his two criteria explain why I was right to make this inference. Given the many other ways the flowers might have been arranged and given how wind and rain and other natural forces would be expected to scatter the seeds for growing them, the specific arrangement of flowers qualified as an extremely improbable pattern. In addition, however, the arrangement exemplified several patterns that I recognized independently, namely, the shapes of several English letters. The arrangement of flowers also exhibited a *functionally significant* pattern in the sense that it met a set of requirements that enabled the arrangement as a whole to communicate a message. In this case, the arrangement of the flowers included words with known meanings arranged in accord with independent syntactic conventions and grammatical rules for conveying a message in the English language.

This example illustrates that Dembski's notion of specification subsumes the idea of "an independently given pattern" (a pattern match) *as well as* the idea of a "recognizable or significant outcome" or "set of functional requirements." One way of defining a specification helps to explain this equivalence. Indeed, we can think of a specification as a concise description of what something we recognize *is* or *does*. That can include a description of a recognizable pattern that we can describe concisely such as "the faces of the presidents." Or it can include *a set of functional requirements* such as those exemplified in a section of English text expressed in a flower arrangement, for example. In that case, we can concisely describe what the arrangement does: "the flowers convey a welcome message." In any case, in our experience a small-probability event that exhibits *a pattern recognized from independent experience or a set of functional requirements* reliably indicates intelligent design.

And that brings us back to fine tuning. Recall John Polkinghorne's illustration of a universe-creating machine. His hypothetical machine had lots of dials and knobs each with many possible settings. Yet each dial or knob was set precisely to make a significant outcome—i.e., life—possible. Polkinghorne used this illustration to elicit an awareness of design behind the fine tuning. Dembski's theory explains why it would. Polkinghorne's hypothetical machine illustrated how our universe exhibits (a) *an extremely improbable* ensemble of values and conditions that also (b) exemplify *a set of functional requirements*—those that we can concisely describe as "a set of parameters necessary for producing a life-sustaining universe."

Real-Time Design Detection: Improbability Plus

When I was teaching, I used to illustrate Dembski's theory and how it applied to the fine-tuning evidence by using a gag with a combination lock. I didn't initially tell my students the point I was trying to make. Instead, I started by telling them that I would use the lock to illustrate that chance alone was not a plausible way of explaining the fine tuning of the universe.

I would first ask students to try to open the lock by guessing the combination. I even told them that they needed to turn the dial first to the right, then to the left, and then back again to the right past the second number. As I passed the lock around the class, and as student after student failed to find the combination in three random trials, I acted increasingly smug, as if the demonstration was proving my point about the inadequacy of random processes as an explanation for extremely improbable outcomes.

Then, as if on cue and just as I was becoming insufferable, a student (we'll call her Paige) nonchalantly turned the dial three times—right, left, right—and popped open the lock. The class reacted predictably with laughter and taunting. Well, at least for a while.

At this point, I would feign shock at the outcome of the demonstration. I'd been proven wrong. A random search had produced an improbable and finely tuned outcome and had done so rather quickly.

Or had it? Invariably, someone would ask whether the student who opened the lock had really guessed the combination by chance. Then the accusations started.

"Was that for real?"

"Was she a plant?"

"Are you trying to trick us?"

"Who? Me?" I replied. "Why would I do something like that?"

As more and more students expressed skepticism, I asked why they suspected me. After all, even though it was improbable that the student would guess the combination, she still could've done it. "There was a chance," I said.

"I understand that," one student replied, "but it still seems much more likely that she knew the combination already." A consensus would form as other students began to suspect the same thing.

Eventually, I walked over to the student who had opened the lock and asked her to tell the truth. "Did I tell you the combination before class started?"

The student stood up, smiled, and then pulled a small slip of paper out of her pocket. More laughter would break out as she held up the combination for everyone to see.

After order was restored, I explained the real point of the demonstration. "As you thought about what you saw," I said, "most of you began to suspect something fishy. You rejected the chance hypothesis and instead began to suspect that intelligent design had played a role—both on the part of Paige, who used her knowledge of the combination to open the lock, and on my part in putting her up to it."

I then asked the class to think about what it was that they had just witnessed that made them suspect collusion—a type of design. Many first suggested that the improbability of Paige's opening the lock justified their suspicion. But I pointed out that all the other students had also spun the dial to three different settings (right-left-right) and that each three-number spin had the same probability of occurring: 1 in 64,000 (or $1/40 \times 1/40 \times 1/40$).

"So what was different that made you suspicious in this case?" I would ask. After a moment of perplexed silence, usually one or two students would say something like, "All the spins were equally improbable, but Paige just happened to turn the dial in just the right way to open the lock. That seemed pretty fishy since there are so many other combinations that wouldn't do that."

"Exactly," I would exclaim. I would then explain that in addition to witnessing an improbable occurrence, they had witnessed a specific pattern-matching event. Paige had spun the dial in just such a way as *to match* the independently established functional requirements—what we call the combination—for opening the lock. And, in our experience, events or systems that exhibit both extreme improbability and functional specificity invariably result from the activity of a designing mind.

Some students would probe my contention more deeply. They would ask *why* these joint criteria of improbability and specification indicate intelligent activity. "After all," they would argue, "every sequence of three spins *is* equally improbable. So why wouldn't we be justified in attributing what Paige did to chance?"

That would give me an opportunity to explain the probabilistic logic underlying design inferences more deeply. I would explain, first, that comparing the probability of any one arrangement to another is not the most important comparison. Instead, to assess the plausibility of a design hypothesis, we need to compare the probability of getting *any* non-functional outcome to the probability of getting *a specific* functional outcome (or recognizable pattern).

In the lock illustration, since only one of 64,000 possible sets of three turns matches the combination, the most likely outcome of a random or unguided trial is an *un*opened lock. Thus, an unopened lock would be the *expected outcome* of such a trial. Yet, if the dial spins were intelligently guided by knowledge of the combination, we would expect to see an event matching a recognizable pattern (i.e., the combination) and a functional outcome (i.e., opening the lock). We certainly would have more reason to expect an opened lock in the case of an intelligently guided try than we would if the try was unguided or random.

A Revived Natural Theology?

Dembski's theory—and the reasoning and repeated experience that underlies it—helps to explain why distinguished physicists have said things like a "superintelligence" provides a "commonsense interpretation" of the fine tuning; "the thought insistently arises that some supernatural agency . . . must be involved"; and theistic design provides "a more satisfying explanation" than competing materialistic hypotheses. With the extreme fine tuning of the fundamental physical parameters, physicists have discovered a phenomenon that exhibits precisely the two criteria—extreme improbability and functional specification—that in our experience invariably indicate the activity of a designing mind. Moreover, if the physical parameters of the universe were produced by a random or mindless process, we would expect to find a life-prohibiting (non-functional) universe since the overwhelming majority of the possible combinations of physical parameters preclude life. Alternatively, if a designing intelligence established the physical parameters of the universe, such an intelligence could well have selected a propitious, finely tuned set. Thus, the cosmological fine tuning seems more expected given the activity of a designing mind, than it does given a random or mindless process. The fine tuning does not seem, on its face at least, to be the kind of evidence we would expect if the universe had arisen from a completely purposeless process—from "blind, pitiless indifference," as Richard Dawkins has put it.

In later chapters, I show why this disparity in expectation concerning the fine-tuning evidence suggests intelligent design, indeed a transcendent intelligent designer, as a better explanation of the evidence than any purely materialistic or naturalistic hypothesis. In the process, I also evaluate recent attempts to explain the fine tuning without invoking a superintending intelligence, including a popular (if exotic) new idea known as the "multiverse hypothesis."[33]

For now, it's worth noting that scientists have discovered another class of unexpected evidence that suggests, to many leading physicists at least, the need to revive a God or design hypothesis. As John Polkinghorne notes, "We are living in an age where there is a great revival of natural theology taking place. That revival of natural theology is taking place not on the whole among theologians, who have lost their nerve in that area, but among the scientists."[34] Polkinghorne further observes that although this new natural theology generally has more modest ambitions than the natural theology of the Middle Ages, a profound intellectual shift has nevertheless begun to take place.

9

The Origin of Life and the DNA Enigma

Questions of origins are closely linked to ultimate questions of what, if anything, lies behind the natural world we see. From ancient times, observers of the heavens wondered at the spectacle of the turning stars, the wandering planets, the sun that warms us during the day, and the moon that provides a cold beacon at night. This spectacular order evoked the intuition of some guiding purpose, which modern science would begin to elucidate. But this was not all. Turning their gaze from the upper realms to the lower, human beings could not help noticing the organized structures in living organisms. These too looked as though they had been designed for a purpose—the elegant form and protective covering of the coiled nautilus; the interdependent parts of the vertebrate eye; the interlocking bones, muscles, and feathers of a bird wing. Our focus turns now, as well, to the appearance of design in life.

With the advent of Darwinism, and later of neo-Darwinism, evolutionary biologists claimed to explain that appearance in startling new terms. Rather than explaining that appearance as the product of design, they have contended that new forms of life arose from a purely undirected process. In *On the Origin of Species*, published in 1859, Darwin argued that the adaptation of organisms to their environments, previously considered a compelling evidence of design, could be explained by natural selection acting on random variations, a process that only mimics the powers of a designing intelligence.

Thus, since 1859 the appearance of design in living things has been understood by many biologists as a powerfully suggestive illusion, but an illusion nonetheless. As evolutionary biologist Francisco Ayala has argued, "Darwin's theory of natural selection accounts for the 'design' of organisms, and for their wondrous diversity, as the result of natural processes, the gradual accumulation of spontaneously arisen variations (mutations) sorted out by natural selection . . . a process that is creative, although not conscious."[1] Richard Dawkins is even more blunt, defining biology as "the study of complicated things that *give the appearance* of having been designed for a purpose."[2]

Thus, whereas many physicists have recently considered design by a "superintellect" as an explanation for the origin of the finely tuned features that make life possible in the universe, biologists have long resisted the design hypothesis. Ever since Darwin, they have assumed that they could, in the words of Ayala, explain "design without a designer."[3] Similarly, Francis Crick, who with James Watson in 1953 elucidated the structure of DNA, enjoined biologists to "constantly keep in mind that what they see was not designed, but rather evolved."[4]

It was, however, due in large measure to Crick and Watson's own discovery of the information-bearing properties of DNA that the materialist understanding of life has begun to unravel. Scientists have become increasingly, and in some quarters uncomfortably, aware that there is at least one appearance of design in biology that has not been explained by natural selection or any other purely naturalistic mechanism: the information present in even the simplest living cells.

The DNA Enigma

FIGURE 9.1
Watson and Crick presenting their model of the DNA double helix.

When Watson and Crick discovered the structure of DNA, they made a shocking discovery. DNA could store information in the form of a four-character digital code. Their structural model of DNA showed that strings of precisely sequenced chemical subunits called "nucleotide bases"—affixed along the interior of the DNA double-helix backbone—could store and transmit information (Fig. 9.1). Crick developed this idea further in 1958 with his now

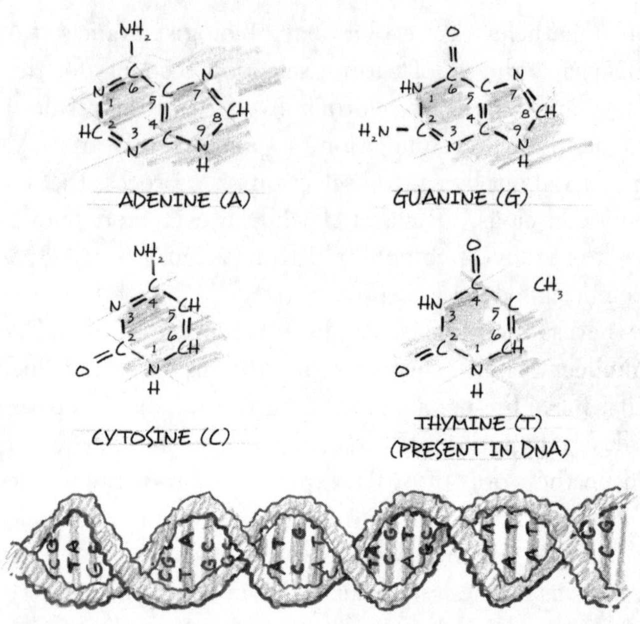

FIGURE 9.2
Francis Crick's sequence hypothesis. According to the sequence hypothesis, the four nucleotide bases of adenine, guanine, cytosine, and thymine function like alphabetic characters in a written text or digital characters in a section of machine code. In particular, their precise arrangement provides the instructions for building the proteins and protein machines that cells need to stay alive. The chemical formulas of these four bases are depicted at the top of the figure. Underneath them, a twisting DNA helix shows a series of these nucleotide bases (i.e., "genetic letters") conveying genetic assembly instructions.

famous "sequence hypothesis," according to which the chemical subunits of DNA (the nucleotide bases) *function* just as letters in a written text or digital characters in a section of computer software (Fig. 9.2). Just as letters or digital characters may convey information depending on their arrangement, so too do certain sequences of chemical bases along the spine of the DNA molecule convey precise instructions for arranging the amino acids from which proteins are made.

Crick knew that proteins perform most of the important life-maintaining functions in cells. For example, proteins function as enzymes catalyzing essential metabolic reactions at rates much faster than would otherwise occur (Fig. 9.3); they process genetic information; and they form the structural parts of tiny molecular machines in cells. Crick suspected that specific sequences of bases on the DNA strand provided the assembly instructions for building the protein molecules and protein machines that the cell needs to survive. By the mid-1960s, a series of brilliant studies and experiments in molecular biology had confirmed Crick's hypothesis (Fig. 9.4).

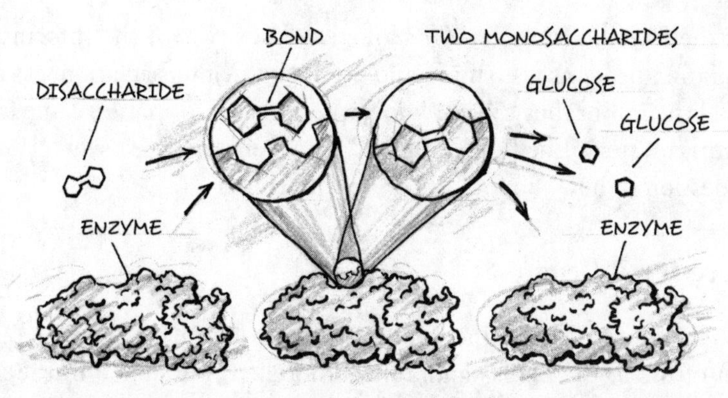

FIGURE 9.3

The three-dimensional specificity of proteins. The instructions in DNA direct the production of functional proteins, including enzymes. This diagram shows an enzyme breaking apart a two-part sugar molecule (a disaccharide). Notice the tight three-dimensional specificity of fit between the enzyme and the disaccharide at the active site where the reaction takes place.

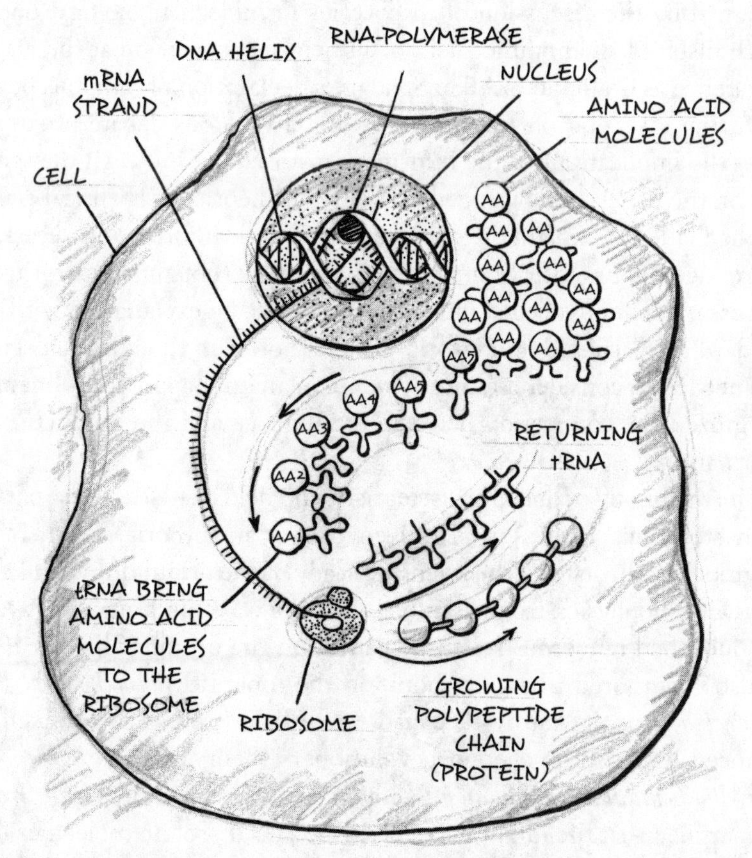

FIGURE 9.4

A simplified schematic of gene expression showing the process by which genetic information stored in DNA directs the production of proteins in the cell.

By the late 1960s, molecular biologists knew where the information for building proteins was stored and even how that information was used to build them. But they did not know where that information came from. I call this mystery "the DNA enigma," a mystery that is closely related to the question of how the first life on earth originated.

Where It All Began

I first encountered scientific doubts about evolutionary theories of the origin of life at the Dallas and Yale conferences I attended in the mid-1980s. Both events left me fascinated by discoveries in cosmology and physics—discoveries that had leading scientists considering anew the merits of a theistic perspective. Though my background up to that point had been mainly in physics (and geophysics), I found myself even more intrigued by the discussion of discoveries in molecular biology, one of which inspired an announcement of a scientific conversion at the Dallas conference as dramatic as Allan Sandage's declaration of religious belief.

During a session on the origin of life, a vigorous debate broke out about the implications of the information stored in DNA. All the scientists on the panel acknowledged that current theories of chemical evolution had failed to explain the origin of the genetic information necessary to produce the first life. Some of these scientists thought that origin-of-life research simply needed more time to devise an explanation within a standard materialistic framework. Others, however, thought that scientists needed to consider a radically new explanatory approach—one that recognized the connection between intelligence and the production of information.

The testimony of one of the scientists who held this latter view particularly stood out. Professor Dean Kenyon was an authority on chemical evolutionary theory and the scientific study of the origin of life. He held a PhD in biophysics from Stanford, had done research at NASA, and had published numerous scientific papers on the origin of life. In 1969, he also coauthored a seminal book on the topic titled *Biochemical Predestination*, a book that by 1985 had established itself as the bestselling advanced-level text on chemical evolutionary theory.

In *Biochemical Predestination*, Kenyon and his coauthor, Gary Steinman, argued that life might have arisen as crucial protein molecules first "self-organized" without assistance from DNA as the result of purely natural forces of chemical attraction between the smaller amino-acid subunits out of which proteins are made.[5] Many leading origin-of-life biochemists hailed this bold hypothesis as the most plausible chemical evolutionary approach to explaining the origin of life.

Yet by the late 1970s, Kenyon himself began to question the plausibility of his own theory. Experiments increasingly contradicted the idea that functional proteins could have assembled themselves from their amino-acid building blocks without preexisting genetic information in DNA directing the process. This forced Kenyon to reconsider the importance of DNA for building proteins and to search for an explanation for the origin of the information it contained. As he studied the structure of the DNA molecule more, Kenyon realized that the information in it could not have "self-organized." To say otherwise would be like saying a newspaper headline might arise as the result of the chemical attraction between ink and paper.

In Dallas, Kenyon publicly and dramatically repudiated his theory of "biochemical predestination." He also expressed misgivings about other chemical evolutionary theories and argued that the presence of information in the DNA molecule defied explanation by all current naturalistic theories of the origin of life, not just his own.[6]

Kenyon wasn't the only scientist on the panel who had come to this conclusion. A year before the conference, in 1984, chemist Charles Thaxton, polymer scientist Walter Bradley, and geochemist Roger Olsen published a book challenging the current chemical evolutionary theories of the origin of life. The book, titled *The Mystery of Life's Origin*, was published by Philosophical Library, at the time a prestigious New York scientific publisher that had previously published works by more than twenty Nobel laureates. In *Mystery*, Thaxton and his colleagues exposed many deficiencies in the various chemical evolutionary theories, but especially their inability to explain the origin of the genetic information necessary to produce the first living cell. They argued that, for various reasons, getting from chemistry to code unaided by an intelligence posed an insuperable difficulty.

After the conference, a mutual friend introduced me to Thaxton, who invited me to drop by his office to discuss the ideas presented in his book in more depth. Thaxton was living in Dallas at the same time that I was working as a geophysicist doing digital signal processing, an early form of applied information technology. It intrigued me to learn that a mystery as profound as the origin of the first life might turn on understanding the origin of the information and information-processing systems inside cells.

I began spending time at Thaxton's office regularly after work for long discussions. We talked about different chemical evolutionary theories, the molecular biology of the cell, the thermodynamics of living systems, the chemistry of the early earth's atmosphere, the logic of simulation experiments, and especially the application of information-theoretic concepts to DNA and molecular biology.

A year later, when I left for the University of Cambridge, I had a topic for a dissertation in mind. While in Cambridge, I eventually wrote both my MPhil and PhD theses on origin-of-life biology, devoting a portion of each to the critical question of the origin of genetic information. What I learned during my time in Cambridge about molecular biology and chemical evolutionary theories convinced me that origin-of-life biology had indeed reached a profound impasse.

Early Theories of the Origin of Life

Few people realize that Darwin's theory of biological evolution did not explain, or attempt to explain, how the first life—presumably a simple one-celled organism—might have first arisen. Instead, Darwin's theory sought to explain the origin of new forms of life from simpler *preexisting* forms. Nevertheless, in the 1860s and 1870s many biologists thought that they could devise a materialistic evolutionary explanation for the origin of the first life fairly easily. Why? They assumed that life was composed of a rather simple substance called protoplasm that could be easily constructed by combining and recombining simple chemicals such as carbon dioxide, oxygen, and nitrogen.

German evolutionary biologist Ernst Haeckel called this process cell "autogeny" and likened it to the process of inorganic crystallization. An English counterpart of Haeckel's, T. H. Huxley, proposed a simple two-step method of chemical recombination to explain the origin of the first cell. Just as salt could be produced spontaneously by adding sodium to chloride, so, Haeckel and Huxley thought, could a living cell be produced by combining several chemical constituents and then allowing spontaneous chemical reactions to produce the simple protoplasmic substance they assumed to be the essence of life.[7]

During the 1920s and 1930s a more sophisticated version of this "chemical evolutionary theory" was proposed by Russian biochemist Aleksandr I. Oparin. He too suggested that life could have first evolved as the result of a series of chemical reactions. But he envisioned many more chemical transformations and reactions over hundreds of millions of years. Oparin postulated these additional steps and additional time, because he understood more about the complexity of cellular metabolism than did Haeckel or Huxley.[8] Nevertheless, neither he nor anyone else in the 1930s fully appreciated the complexity—and information-bearing properties—of the DNA, RNA, and proteins that make life possible.

Though Oparin's theory appeared to receive experimental support in 1953 when Stanley Miller simulated the production of the amino-acid "building blocks" of proteins under ostensibly prebiotic atmospheric conditions (Fig. 9.5), his textbook version of chemical evolutionary theory soon encountered many difficulties. Origin-of-life researchers now know that Miller's simulation experiment has little, if any, relevance to explaining how amino acids—let alone their precise sequencing, necessary to produce proteins—could have arisen in the actual atmosphere

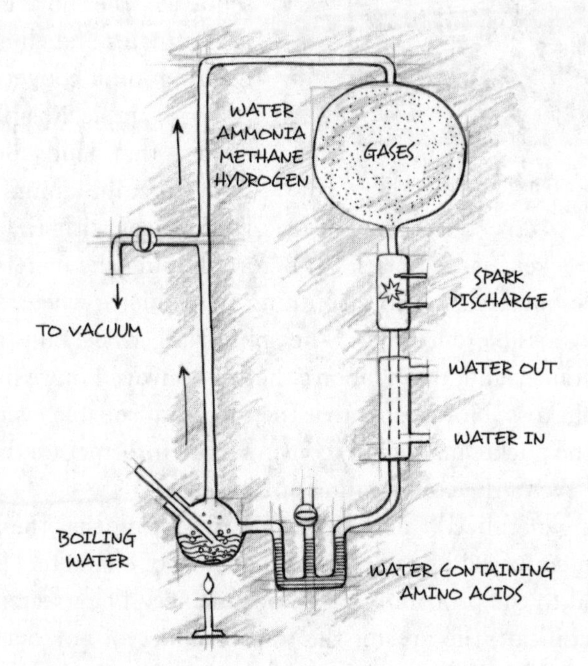

FIGURE 9.5

The Miller-Urey experiment simulating the production of amino acids from a mixture of gases that allegedly matched the prebiotic atmosphere.

of the early earth. Moreover, Oparin proposed no explanation for the origin of the information in DNA (or RNA) that present-day cells use to build proteins.

The Information in DNA: Shannon Plus

To understand why it has been so difficult to explain the origin of the information in DNA and other biomacromolecules in living cells, it's important to take a closer look at exactly what kind of information DNA, RNA, and proteins contain. In so doing, we'll see that DNA does not contain information in just the mathematical sense described by modern information theory as developed during the late 1940s by the MIT scientist Claude Shannon (Fig. 9.6).[9] Shannon's theory equated the amount of information with the amount of uncertainty that was reduced by a series of symbols or characters.[10] In Shannon's theory, the more improbable an event or sequence, the more uncertainty it eliminates and thus the more information it conveys.

FIGURE 9.6
Claude Shannon, MIT mathematician and information theorist.

For example, imagine flipping a coin that lands on "heads." Now imagine spinning a roulette wheel with the ball landing in the pocket marked black 33. Before flipping the coin, there were two possible outcomes. Before spinning the roulette wheel, there were thirty-eight possible outcomes. The spin of the wheel thus eliminated more uncertainty and, in Shannon's theory, conveyed more information than the coin toss. Notice also that the more improbable event (the ball landing in the pocket marked 33) conveys more information than the less improbable event (the coin turning up "heads").

Shannon generalized this relationship by stating that the amount of information conveyed by an event or sequence of characters is inversely proportional to the probability of its occurrence. The greater the number of possibilities, the greater the improbability of any one being actualized, and thus the more information transmitted when a particular possibility occurs.[11]

Nevertheless, as Shannon himself explained, his mathematical formalism could not detect whether a sequence of characters conveyed meaning or performed a communication *function*. To see the distinction between a merely improbable sequence of characters (one possessing Shannon information alone) and a sequence possessing *both* Shannon information and *functional* specificity, consider these two sequences:

inwehnsdysk]ifhsnmcpew,m.sa
Time and tide wait for no man.

Clearly, there is a qualitative difference between these two strings of characters. Whereas the bottom string performs a communication function, the top string does not. Thus, although the top string contains "Shannon information" and has a measurable improbability (or "complexity"), the bottom string contains both Shannon information and "functional" or "specified" information (sometimes called "specified complexity").[12]

It turns out that the specific arrangements of bases in DNA, like the arrangement of letters in an English sentence or digital characters in computer software, do not just exhibit a high degree of mathematical improbability. Instead, the *specific* arrangements of the nucleotide bases (especially in the coding regions of DNA) enable the DNA bases to perform a *function* in the cell. The bases in DNA convey instructions for building proteins—and do so in virtue of their *specificity* of arrangement. As Francis Crick explained in 1958, "Information means here the *precise* determination of sequence, either of bases in the nucleic acid [i.e., in the DNA] or of amino-acid residues in the protein."[13]

Thus, DNA not only has Shannon information; it also contains "specified" or "functional" information. Consequently, it also contains information in the ordinary sense of "alternative sequences or arrangements of characters *that produce a specific effect*," as the dictionary defines the term "information."

Thus Richard Dawkins notes that "the machine code of the genes is uncannily computer-like."[14] And software developer Bill Gates observes that "DNA is like a computer program."[15] Similarly, biotechnology specialist Leroy Hood describes the information stored in DNA simply as "digital code."[16] After the early 1960s, further discoveries made clear that the digital information in DNA and RNA represents only part of a complex information-transmission and -processing system—an advanced form of nanotechnology that both mirrors and exceeds our own in its complexity, design logic, and information-storage density.

But if this is true, how did the *functionally specified information* in DNA arise in the first place? And what produced the intricate information-processing systems in living cells that are essential for DNA to do its work? Are these striking appearances of design the product of actual design or of a natural process that merely mimics the powers of a designing intelligence?

The long-standing question of the origin of the first life turns on these questions. Since Watson and Crick's discovery, scientists have increasingly come to understand the centrality of information to even the simplest living systems. DNA stores the assembly instructions for building the many crucial proteins and protein machines that service and maintain even the most primitive one-celled organisms. It follows that building a living cell in the first place requires assembly instructions stored in DNA or some equivalent molecule.

In 1859, Darwin did not attempt to offer an explanation for the origin of the first life. Today the question of how life first originated is still widely regarded as a profound and unsolved scientific problem, largely because of the mystery surrounding the origin of functionally specified biological information. Indeed, since the 1950s, three broad types of naturalistic explanations have been proposed—those based on chance, those relying on the laws of physics and chemistry, and those combining natural law and chance. Each of these approaches has encountered severe difficulties.

Beyond the Reach of Chance

Initially, many scientists thought purely chance interactions between molecules in the earth's oceans or some favorable environment[17] could explain the origin of the information in DNA.[18] Since the late 1960s, however, few serious scientists have supported this view.[19] Since molecular biologists began to understand how the digital information in DNA directs the construction of proteins in the cell, many calculations have been made to determine the probability of formulating functional proteins or nucleic acids (DNA or RNA molecules) at random. Even assuming extremely favorable prebiotic conditions (whether realistic or not) and theoretically maximal reaction rates, such calculations have invariably underscored the implausibility of chance-based theories. These calculations have shown that the probability of obtaining functionally sequenced, information-rich biomacromolecules at random is, in the words of physicist Ilya Prigogine and his colleagues, "vanishingly small . . . even on the scale of . . . billions of years."[20]

In my book *Signature in the Cell*, I perform updated calculations of the probability of the origin of even a single *functional* protein or corresponding *functional* gene (the section of a DNA molecule that directs the synthesis of a particular protein) by chance alone. My calculations are based upon recent experiments in molecular biology establishing the extreme rarity of functional proteins in relation to the total number of possible arrangements of amino acids corresponding to a protein of a given length. Taking that and several other relevant independent factors into account, I show that the probability of producing even a single functional protein of modest length (150 amino acids) by chance alone in a prebiotic environment stands at no better than a "vanishingly small" 1 chance in 10^{164}, an inconceivably small probability. To put this number in perspective, recall that physicists estimate that there are only 10^{80} elementary particles in the entire universe.

In *Signature*, I also show that the probability of generating a single functional protein is extremely small *in relation to* all the opportunities for that event to occur since the beginning of time (what are called the "probabilistic resources" of the universe). Even if every event in the entire history of the universe (where an event is defined minimally as an interaction between elementary particles) were devoted to producing combinations of amino acids of a given length (an extravagantly generous assumption), the number of combinations thus produced would still represent only a tiny portion—less than one out of a trillion trillion—of the total number of possible amino-acid combinations corresponding to a functional protein—*any* functional protein—of that given length.

In short, it is extremely implausible to think that even a single protein would have arisen by chance on the early earth even taking into account the "probabilistic resources" of the entire universe over its 13.8-billion-year history. And a single protein, keep in mind, does not a living cell, with its many hundreds of specialized proteins, make.

For these and other reasons, serious origin-of-life researchers now consider "chance" an inadequate explanation for the origin of biological information.[21] Nobel laureate Christian de Duve, a leading origin-of-life biochemist until his death in 2013, categorically rejected the chance hypothesis precisely because he judged the necessary fortuitous convergence of events implausible in the extreme.[22] In a memorable passage in his 1995 article "The Beginnings of Life on Earth," de Duve made explicit the logic by which he rejected the chance hypothesis. As he put it, "A single, freak, highly improbable event can conceivably happen. Many highly improbable events—drawing a winning lottery number or the distribution of playing cards in a hand of bridge—happen all the time. But a string of improbable events—drawing the same lottery number twice or the same bridge hand twice in a row—does not happen naturally."[23]

Self-Organization Scenarios

Because of these difficulties, after the mid-1960s many origin-of-life theorists addressed the problem of the origin of biological information in a different way. Rather than invoking chance events or "frozen accidents," chemical evolutionary theorists such as Dean Kenyon suggested that the laws of nature or lawlike forces of chemical attraction might have generated the information in DNA and proteins. Proponents of such self-organizational models suggested that simple chemicals might possess properties capable of organizing the constituent parts of proteins, DNA, and RNA into the specific arrangements they now possess. Just as electrostatic forces draw sodium (Na+) and chloride (Cl–) ions together into a highly ordered pattern within a crystal of salt (NaCl), so too might amino acids with special affinities for each other arrange themselves to form proteins. This was Kenyon's thesis in his book *Biochemical Predestination*.[24]

For many current origin-of-life researchers, self-organizational models still seem to offer the most promising approach to explaining the origin of biological information. Nevertheless, there are serious scientific and conceptual reasons to doubt these models.

Kenyon's doubts about his self-organizational theory first surfaced in discussions with one of his students at San Francisco State University in an upper-division course on evolution. The student—ironically named Solomon Darwin—pressed Kenyon to examine whether his self-organizational model could explain the origin of the information in DNA. Kenyon might have deflected this criticism by asserting that his "protein-first" model of self-organization had circumvented the need

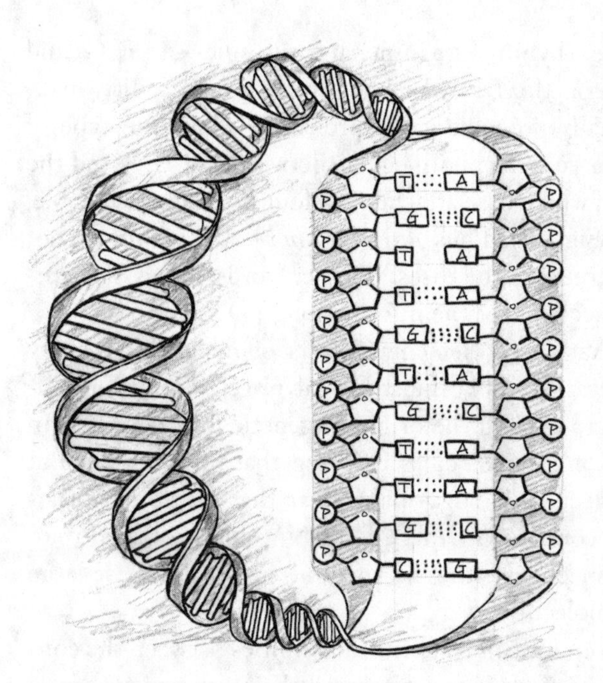

FIGURE 9.7
Model of the chemical structure of the DNA molecule depicting the main chemical bonds between its constituent molecules. Note that no chemical bonds link bases (designated by the letters in boxes) in the longitudinal message-bearing axis of the molecule. Note also that the same kind of chemical bonds link the different nucleotide bases to the sugar-phosphate backbone of the molecule (denoted by pentagons and circles). These two features of the molecule ensure that any nucleotide base can attach to the backbone at any site with equal ease, thus showing that the bonding properties of the chemical constituents of DNA do not determine its base sequences.

to explain the information in DNA. But Kenyon himself had begun to suspect that DNA needed to play a more central role in his account of the origin of life. At some point, DNA must have arisen as a carrier of the information for building proteins—and how that happened needed to be explained.

Yet explaining how the information in DNA arose posed a formidable difficulty for the self-organizational approach. This difficulty can be illustrated by examining the structure of the DNA molecule. Figure 9.7 shows that the structure of DNA depends upon several chemical bonds. There are bonds, for example, between the sugar molecules (designated by the pentagons) and the phosphate molecules (designated by the circled *P*s) that form the twisting backbones of the DNA helix. There are bonds fixing individual (nucleotide) bases to the sugar-phosphate backbones on each side of the molecule. Notice, however, that there are no chemical bonds, and thus no forces of attraction, between the bases that run along the spine of the helix. Yet it is precisely along this axis of the molecule that the genetic instructions in DNA are encoded.

Further, just as magnetic letters can be combined and recombined in any way to form various sequences on a metal surface, so too can each of the four bases adenine, thymine, guanine, and cytosine—A, T, G, and C—attach to any site on the DNA backbone with equal facility, making all sequences equally probable (or improbable). The same type of chemical bond (an N-glycosidic bond) occurs between the bases and the backbone regardless of which base attaches. All four bases are acceptable; none is preferred (see Fig. 9.7). Thus, *differences in bonding affinity* do not determine the arrangement of the bases. In other words, forces of chemical attraction do not account for the information in DNA.

Kenyon realized that these elementary facts of molecular biology had devastating implications. The most logical place to look for self-organizing properties to explain the origin of genetic information is in the constituent parts of the molecules carrying that information. But biochemistry and molecular biology make clear that the forces of attraction between the constituents in DNA, RNA, and protein do not explain the sequence specificity (the information) present in these large information-bearing molecules.[25]

There is a good reason for this. If chemical affinities between the constituents in the DNA message text determined the arrangement of the text, such affinities would dramatically diminish the capacity of DNA to carry information. Consider what would happen if the individual nucleotide "letters" (A, T, G, C) in a DNA molecule *did* interact by *chemical* necessity with each other. Suppose every time adenine (A) occurred in a growing genetic sequence, it would drag guanine (G) along with it. Or every time cytosine (C) appeared, thymine (T) would follow. In that case, the DNA message text would be peppered with repeating sequences of As followed by Gs and Cs followed by Ts.

Rather than having a genetic molecule capable of unlimited novelty, with all the unpredictable and aperiodic sequences that characterize informative texts, we would have a highly repetitive text awash in redundant sequences—much as happens in crystals. Indeed, in a crystal the forces of mutual chemical attraction *do* completely explain the sequential ordering of the constituent parts. Consequently crystals cannot convey novel information. Bonding affinities, to the extent they exist, cannot be used to explain the origin of information. Self-organizing chemical affinities generate highly repetitive "order," but not information; they create mantras, not messages (Fig. 9.8). What needs explaining is not the origin of order—whether in crystals, swirling tornadoes, or the "eyes" of hurricanes—but the origin of *information*.[26]

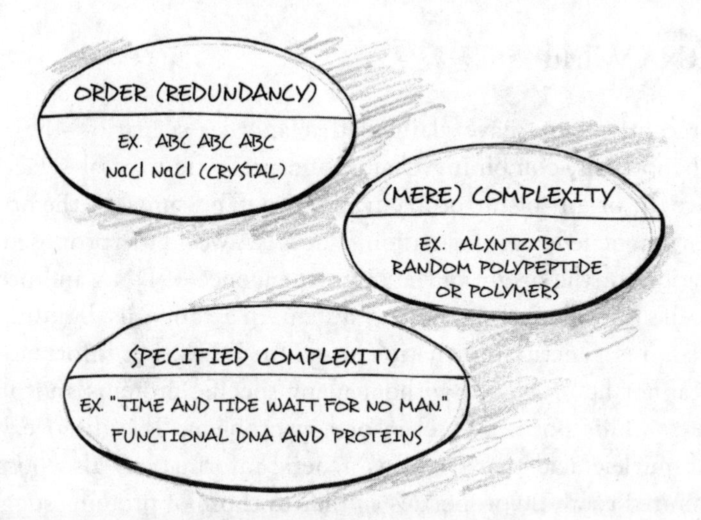

FIGURE 9.8
The concepts of order, complexity, and specified complexity are illustrated above. This figure shows three qualitatively different types of sequences as defined by the information sciences. Note that DNA contains sequences that exhibit specified complexity, not simple redundant order.

Chance and Necessity: Prebiotic Natural Selection

Other theories of chemical evolution have not relied exclusively on either chance or lawlike necessity alone. They have instead attempted to combine the two. For example, after 1953 Oparin revised his original theory of chemical evolution. He attempted to explain the origin of biological information as the product of the lawlike process of *natural selection* acting on the chance interactions of simple nonliving molecules. Yet Oparin's notion of prebiotic natural selection soon encountered obvious difficulties.

First, the process of natural selection presupposes the differential reproduction of *already* living organisms and thus a preexisting mechanism of self-replication. Yet self-replication in all extant cells depends upon functional (and therefore sequence-specific, information-rich) proteins and nucleic acids. And the origin of such information-rich molecules is precisely what Oparin needed to explain. Thus, many rejected his postulation of prebiotic natural selection as question-begging. As the evolutionary biologist Theodosius Dobzhansky insisted, "Pre-biological natural selection is a contradiction in terms."[27] Or as Christian de Duve explained, theories of prebiotic natural selection "need information which implies they have to presuppose what is to be explained in the first place."[28]

The RNA World

More recently, some have claimed that another scenario—the RNA-world hypothesis, combining chance and prebiotic natural selection—can solve the origin-of-life problem and with it, presumably, the problem of the origin of genetic information. The RNA world was proposed as an explanation for the origin of the interdependence of DNA and proteins in the cell's information-processing system. In extant cells, building proteins requires genetic information stored in DNA, but information in DNA cannot be processed without many specific proteins and protein complexes. This poses a chicken-or-egg problem. The discovery that RNA (a nucleic acid that contains genetic information) also possesses some limited catalytic properties similar to those of proteins suggested a way to solve that problem. "RNA-first" advocates proposed an early state in which RNA performed both the enzymatic functions of modern proteins and the information-storage function of modern DNA, thus allegedly making the interdependence of DNA and proteins unnecessary in the earliest living systems. They also envision primitive self-copying RNA molecules, or "RNA replicators," that can produce molecular offspring capable of competing for survival and, thus, inducing a process of prebiotic natural selection.

Nevertheless, many fundamental difficulties with the RNA-world scenario have emerged. First, synthesizing (or maintaining) many essential building blocks of RNA molecules under realistic conditions has proven either difficult or impossible.[29] Second, naturally occurring RNA possesses very few of the specific enzymatic properties of proteins necessary to extant cells. Indeed, RNA catalysts do not function as true enzyme catalysts. For instance, many enzymes are capable of coupling energetically favorable and energetically unfavorable reactions together (i.e., reactions that otherwise would not proceed spontaneously). RNA catalysts, so-called ribozymes, are not capable of doing this.

Third, attempts to enhance the limited catalytic properties of RNA molecules in "ribozyme-engineering" experiments have inevitably required extensive investigator manipulation, thus simulating, if anything, the need for intelligent design. Fourth, RNA-world advocates offer no plausible explanation for how primitive RNA replicators might have evolved into modern cells that rely almost exclusively on proteins to process and translate genetic information and regulate metabolism.[30]

Most important, the RNA-world hypothesis presupposes, but does not explain, the origin of sequence specificity or information in the original functional RNA replicators. To date, scientists have been able to design RNA catalysts that will copy only about 10 percent of themselves.[31] For strands of RNA to perform even this limited self-replication function, they must have very specific arrangements of their constituent nucleotide building blocks. Further, the strands must be long enough to fold into complex three-dimensional shapes (so-called tertiary structures). Thus, any RNA molecule capable of even limited function must have possessed considerable specified information content. Yet explaining how the building blocks of RNA arranged themselves into functionally specified sequences has proven no easier than explaining how the constituent parts of DNA might have done so. As de Duve noted in a critique of the RNA-world hypothesis, "Hitching the components together in the right manner raises additional problems of such magnitude that no one has yet attempted to do so in a prebiotic context."[32]

The Impasse Deepens

Since the 1980s, the crisis in origin-of-life research has only deepened. As Francis Crick lamented in 1981, "An honest man, armed with all the knowledge available to us now, could only state that in some sense, the origin of life appears at the moment to be almost a miracle, so many are the conditions which would have had to have been satisfied to get it going."[33] In 2008 in the film *Expelled*, Richard Dawkins publicly acknowledged that "we don't know" how life originated in the first place and even speculated that the information in DNA might represent a "signature of some kind of designer."[34] Not a divine designer, though. He proposed as an "intriguing possibility" that an alien civilization evolved elsewhere in the cosmos and then "designed" and "seeded" the first life on earth.[35]

Years earlier, in 1973, in a scientific paper in the astronomy journal *Icarus*, Francis Crick and his colleague Leslie Orgel advanced this same hypothesis, which they called "directed panspermia."[36] Later Crick revisited the hypothesis at greater length in the book *Life Itself.* That figures as prominent as Dawkins and Crick, ardent defenders of evolutionary theory and a materialistic approach to science, would posit such speculative hypotheses only underscores the depth of the origin-of-life problem and the closely associated enigma of the origin of genetic information.

The Mystery of Life's Origin and the "Intelligent Cause" Hypothesis

It was this growing crisis that led Charles Thaxton (Fig. 9.9), Walter Bradley, and Roger Olsen to write *The Mystery of Life's Origin*. In it, they not only critiqued then current evolutionary theories, but also proposed a radically new approach. In a philosophical epilogue, they proposed an "intelligent cause" as an explanation for the origin of the genetic information necessary to produce life in the first place.[37] They also argued that positing such a cause might constitute a completely legitimate and appropriate scientific hypothesis within the historical sciences, a mode of inquiry they called "origins science."

FIGURE 9.9
The American chemist and coauthor of *The Mystery of Life's Origin*, Charles Thaxton.

Drawing on the work of the physical chemist Michael Polanyi and others, they argued that chemistry and physics alone could not produce information any more than ink and paper could produce the information in a book. Instead, our uniform experience suggests that information always arises as the product of mind or what they called an "intelligent cause."[38]

By the mid-1980s, Dean Kenyon had come to consider this same possibility. For him, the digital information in DNA provided "evidence for intelligent purpose in the cosmos, or design" and he suggested as a result that "the natural theological question should now be reopened by the philosophers."[39]

Abductively, My Dear Watson

My exposure in 1985 to Kenyon and his ideas and my marathon discussions with Thaxton captured my philosophical and scientific interest and left me, though not yet fully convinced, deeply intrigued by their perspective. Consequently, after moving to Cambridge, I investigated questions that had emerged in my discussions with Thaxton. What kind of information is present in DNA? Do scientists use a distinctive method of historical-scientific inquiry to study biological and cosmological origins? After completing my PhD, I took up a closely related question:

Could the intuitive connection between information and the prior activity of designing intelligence justify a rigorous scientific argument for intelligent design based upon the presence of the functionally specified digital information in DNA?

As I began to study the reasoning that historical scientists use to identify causes responsible for events in the remote past, I discovered that scientists who use this reasoning often make inferences with a distinctive logical form, known technically as "abductive inferences."[40] Geologists, paleontologists, evolutionary biologists, and other historical scientists reason like detectives, inferring *past* conditions or causes from *present* clues. As Stephen Jay Gould noted, historical scientists typically "infer history from its results."[41]

Nevertheless, this kind of reasoning can be problematic. That's because more than one cause can often explain the same effect. To address this problem in geology, the nineteenth-century geologist Thomas Chamberlain developed a method of reasoning he called the "method of multiple working hypotheses."[42]

During my last year in Cambridge, I met a visiting American philosopher of science named Peter Lipton (Fig. 9.10) who had extensively characterized this method of reasoning during his doctoral studies at Oxford. Lipton called this method of reasoning "inference to the best explanation." He was visiting from Williams College in Massachusetts as a candidate for a position in the Department of the History and Philosophy of Science at Cambridge—a position he later accepted before eventually rising to the position of head of the department. In the spring of 1990, Lipton left me a copy of his soon to be published manuscript,

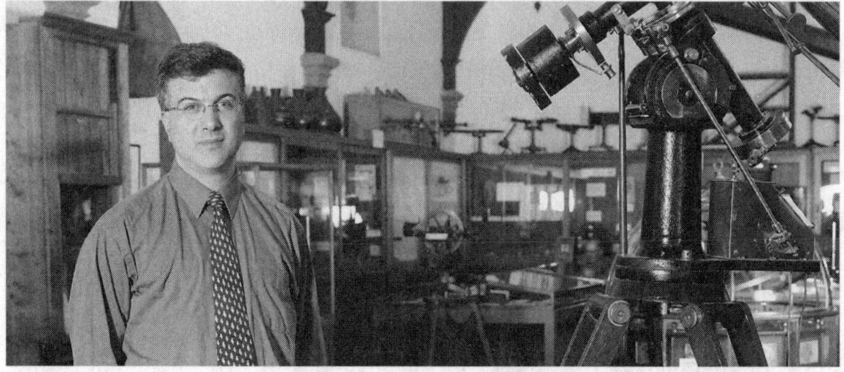

FIGURE 9.10
Peter Lipton, the Cambridge philosopher of science who authored *Inference to the Best Explanation*.

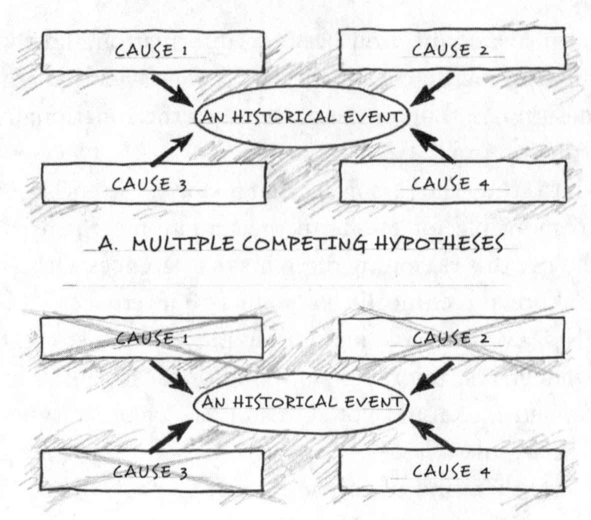

FIGURE 9.11

In the method of multiple competing hypotheses, or inference to the best explanation, scientists posit multiple possible hypotheses and then elect that hypothesis which, if true, would best explain the event or evidence in question. Historical scientists have identified causal adequacy as a key criterion for determining which hypothesis or explanation qualifies as the best. The above figure depicts a process of reasoning in which historical scientists have proposed four potential causal explanations, eliminated three from consideration, and elected a fourth. In the diagram, this fourth causal hypothesis would presumably represent a cause known to be sufficient to produce the event in question—in other words, a causally adequate hypothesis.

titled *Inference to the Best Explanation*. In it, he explained that when scientists use this method of reasoning to explain the occurrence of an event or structure (including one from the past), they often compare various hypotheses to see which would, if true, best explain the relevant evidence[43] (Fig. 9.11).

Lipton's detailed philosophical defense of this distinctive method broke new ground. But what exactly makes an explanation *best*? As it happens, I later learned that nineteenth-century historical scientists had already developed practical criteria for answering that question. The most important of these criteria is called "causal adequacy."

This criterion requires that historical scientists identify causes known to have the power to produce the kind of effect, feature, or event in question. In making these determinations, they evaluate hypotheses against their present knowledge of cause and effect. Causes known to produce the effect in question are judged better candidates than those that do not. For instance, a volcanic eruption provides a better explanation for a white ash layer in the earth than an earthquake or a flood, because volcanic eruptions have been observed to produce ash layers, whereas earthquakes and floods have not.

One of the first scientists to develop this principle was the geologist Charles Lyell, who influenced Charles Darwin. Darwin read Lyell's magnum opus, *Principles of Geology*, on the voyage of the *Beagle* and employed its principles of reasoning in the *Origin of Species*. The subtitle of Lyell's *Principles* summarized the geologist's methodology: *Being an Attempt to Explain the Former Changes of the Earth's Surface, by Reference to Causes Now in Operation*. Lyell argued that when scientists seek to explain events in the past, they should not invoke unknown types of causes. Instead, they should cite causes known from our uniform experience to have the power to produce the effect in question.[44] Historical scientists should cite "causes now in operation." This was the idea behind his famous "uniformitarian" dictum: "The present is the key to the past."

Darwin himself adopted this methodological principle as he sought to demonstrate that natural selection qualified as a *vera causa*, that is, a true, known, or actual cause of significant biological change.[45] He sought to show that natural selection was *causally adequate* to produce the effects he was trying to explain.

Philosophers of science have also noted that assessments of explanatory power lead to conclusive inferences only when it can be shown that there is *just one known cause* for the effect or evidence in question.[46] When scientists can infer a *uniquely* plausible cause, they can avoid the fallacy of affirming the consequent—the error of deciding on one causal explanation while ignoring other possible causes.[47] (See Chapter 11.)

What did all this have to do with the origin of biological information, what I have called "the DNA enigma"? I wondered if a case for an intelligent cause of the information in DNA could be formulated and justified in the same way that historical scientists would justify any other causal claim about an event in the past. I began to frame a series of questions. If neither chance, nor physical-chemical necessity, nor the two acting in combination produces specified information, what does? Do we know of any "cause now in operation" that has the power to create large amounts of specified information?

As I considered these questions, I came across a book written by Henry Quastler, one of the early scientists who first began to apply informational concepts to molecular biology. In it, Quastler made an almost offhand and seemingly obvious observation. As he put it, "The creation of new information is habitually associated with conscious activity."[48]

Quastler's remark hit me like a thunderbolt. It suggested a radical possibility, a way to formulate a rigorous scientific case for intelligent design as an inference to the best explanation—specifically, the best explanation for the origin of biological information.

The creative action of a conscious and intelligent agent clearly represents a known and adequate cause (one "now in operation") for the origin of specified information. Uniform and repeated experience affirms that intelligent agents can produce large amounts of functional or specified information, whether in software programs, ancient inscriptions, or Shakespearean sonnets.

The specified information in the cell also points to intelligent design not just as an adequate explanation, but as the *best* explanation (Fig. 9.12). Why? Experience shows that large amounts of specified information *invariably* originate from an intelligent source.

This is particularly apparent when the information is expressed in a digital or alphabetic form. A computer user who traces the information on a screen back to its source invariably comes to a mind, that of a software engineer. Similarly, the information in a book or newspaper article ultimately derives from a writer—from a mental, rather than a strictly material, cause.

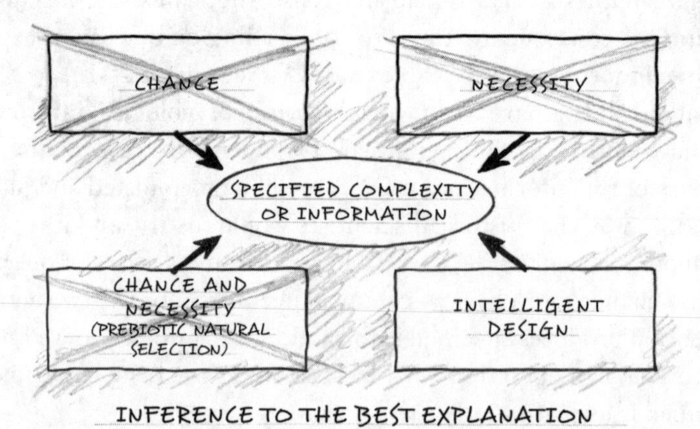

FIGURE 9.12
In *Signature in the Cell* and in this chapter, I infer intelligent agency or design as the best, most causally adequate explanation for the origin of the functional information or specified complexity necessary to produce the first living cell.

Ironically, the generalization that intelligence is the *only known cause* of specified information (at least starting from a nonbiological source) has received support from the specialized scientific discipline of origin-of-life research itself. During the last seventy years, every proposed materialistic model has failed to explain the origin of the functionally specified genetic information required to build a living cell.[49] Moreover, origin-of-life simulation experiments only succeed in producing life-relevant chemistry or information-rich molecules—in, for example, simulations of the RNA world—as the result of the interventions of *intelligent origin-of-life biochemists*, or "ribozyme engineers." Thus, intelligence or mind or what philosophers call "agent causation" now stands as the only known cause capable of generating large amounts of specified information,[50] in particular, the amounts necessary to produce a new protein fold, the minimal unit of biological innovation.[51] (See Chapter 10.)

Scientists in many fields recognize the connection between intelligence and information and make inferences accordingly. Archaeologists assume that a scribe produced the inscriptions on the Rosetta Stone. The search for extraterrestrial intelligence (SETI) presupposes that specified information imbedded in electromagnetic signals coming from space would indicate an intelligent source.[52] As yet radio astronomers have not found any such information-bearing signals. But closer to home, molecular biologists have identified specified information-rich sequences and systems in the cell, suggesting, by the same logic, the past existence of an intelligent cause for those effects.

Our uniform experience affirms that specified or functional information—whether inscribed in hieroglyphics, written in a book, encoded in a radio signal, or produced in an RNA-world "ribozyme-engineering" experiment—*always* arises from an intelligent source, from a mind, not a strictly material process. So the discovery of functional digital information in the DNA and RNA molecules in even the simplest living cells provides strong grounds for inferring that intelligence played a role in the origin of the information necessary to produce the first living organism.

In my book *Signature in the Cell*, I develop this argument further and respond in detail to various objections to the case for intelligent design as sketched briefly in this chapter. I address the objections that intelligent design is religion, is not science, is not testable, is based on flawed analogical reasoning, is a fallacious argument from ignorance, and others.

I also provide extensive documentation of the scientific discussion provided here. But the evidence and the logic supporting intelligent design can be grasped without all that technical detail.

Indeed, when I first noticed the subtitle of Lyell's book referring to "causes now in operation," a light came on for me. I asked myself a question: "What cause *now in operation* produces digital code or specified information?" Is there a known cause—a *vera causa*—of the origin of such information? What does our uniform experience tell us? It occurred to me that by Lyell's and Darwin's own criterion of a sound scientific explanation, intelligent design qualifies as the best explanation for the origin of biological information. Why? Because we have independent evidence—"uniform experience"—that intelligent agents can and do produce specified or functional information and we know of no other cause that does (at least starting from a purely physical or chemical state).

Late nineteenth-century scientists knew nothing, of course, about the importance of information to living systems. They assumed that the universe consisted of two fundamental entities: matter and energy. But during the 1950s and 1960s molecular biologists discovered a third fundamental entity at the foundation of life—information. Moreover, the functional digital information in the "machine code of the genes," as Dawkins put it, does not seem—based upon our experience—"to be the kind of quality we should expect to observe" if there was "no design, no purpose . . . nothing but blind, pitiless indifference"[53] at work in the origin of life.

Instead, our experience of the twenty-first-century information revolution, to say nothing of centuries of using and generating information, suggests that the presence of functional information—especially if in an alphabetic or digital form—is one of those qualities that we *should expect* only if intelligent design and purpose had played a role in the origin of life.

Yet the case for intelligent design in biology grows even stronger when we consider not just the information necessary to produce the first life, but also the information explosions that mark life's subsequent history. The next chapter addresses this subject.

10

The Cambrian and Other Information Explosions

The fossil record on our planet documents the origin of major innovations in biological form and function. These episodes—if we take the fossil record at face value—often occur abruptly or *discontinuously*, meaning that newly arising forms bear little resemblance to what existed earlier. In my book *Darwin's Doubt*—the sequel to *Signature in the Cell*—I wrote about one of the most dramatic of those discontinuous events, the Cambrian explosion. During this event, beginning about 530 million years ago, most major groups of animals first appear in the fossil record in a geologically abrupt fashion.

Although the Cambrian explosion of animals is especially striking, it is far from the only "explosion" of new living forms. The first winged insects, birds, flowering plants, mammals, and many other groups also appear abruptly in the fossil record, with no apparent connection to putative ancestors in the lower, older layers of fossil-bearing sedimentary rock. Evolutionary theorist Eugene Koonin describes this as a "biological big bang" pattern. As he notes, "Major transitions in biological evolution show the same pattern of sudden emergence of diverse forms at a new level of complexity. The relationships between major groups . . . do not seem to fit the pattern that, following Darwin's original proposal, remains the dominant description of biological evolution."[1]

In the *Origin of Species*, Darwin depicted the history of life as a *gradually* unfolding, branching tree, with the trunk representing the first one-celled organisms and the branches representing all the species that evolved from these first forms.[2] In this view, novel animal and plant species arose from a series of simpler precursors and intermediate forms over vast stretches of geologic time. Darwin argued vigorously for this view. At the same time, he acknowledged that the sudden appearance of many major groups of organisms in the fossil record did not fit easily into his picture of gradual evolutionary change.[3]

Instead, this pattern challenged Darwin's claim that natural selection acting on random variations had produced all the new forms of life. As Darwin understood it, the process of natural selection acting on random variations necessarily operated slowly and gradually—thus rendering any pattern of sudden appearance a puzzling anomaly.

Darwin saw natural selection as slow and gradual because of the intrinsic logic of the process. Significant biological changes in any population occur only when randomly arising variations in the features or traits of organisms confer functional advantages in the competition for survival and reproduction. Those organisms that acquire new advantageous traits tend to prevail in the competition, enabling them to pass on those traits to the next generation. As nature "selects" successful variations, the features of a population change.

Yet, as Darwin conceived of the process, the variations responsible for permanent changes would have to be relatively modest, or "slight," in any given generation. Major or large-scale variations—what evolu-

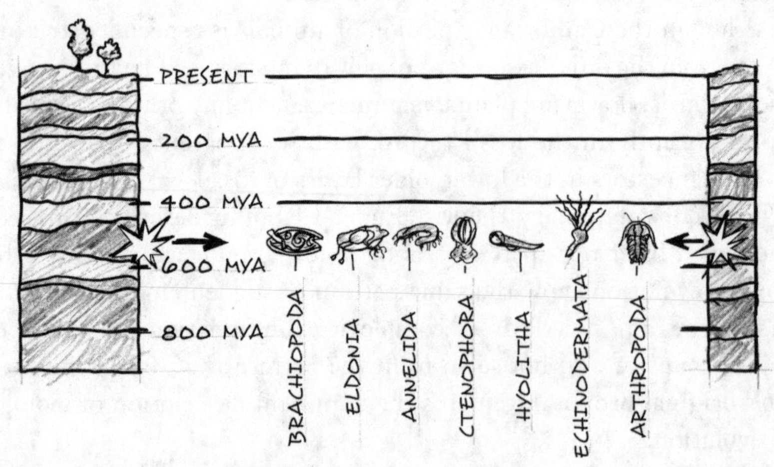

FIGURE 10.1

Representatives of some of the major animal groups that first appear abruptly in the sedimentary rock record during the Cambrian period.

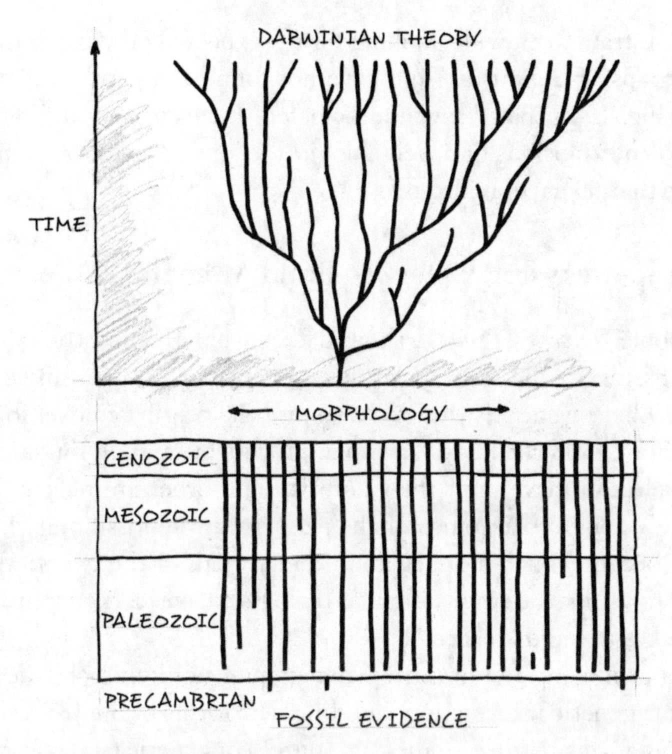

FIGURE 10.2
The origin of animals. Darwinian theory (top) predicts gradual evolutionary change in contrast to the fossil evidence (bottom), which shows the abrupt appearance of the major animal groups.

tionary biologists would later term "macromutations"—would inevitably produce dysfunction, deformities, or even death. Only minor variations would be viable and therefore heritable.

Any larger-scale changes would have to be built slowly from a long series of smaller-scale, heritable variations. Significant changes to organismal form and function would thus require many hundreds of millions of years. That is precisely what appears unavailable in the case of many salient episodes of evolutionary innovation, such as the Cambrian explosion (Fig. 10.1), the angiosperm (flowering-plant) "big bloom" during the Cretaceous (130 million years ago), and the mammalian radiation in the Eocene period (about 55 million years ago).

Darwin hoped the mystery of the missing ancestral fossils would be solved by future geological discoveries documenting the gradual transitions his theory predicted. But the opposite has occurred. In the 160 years since the publication of the *Origin*, paleontologists have combed geological strata worldwide, looking for the expected precursors to many major groups of organisms[4] and have not found the pattern of gradual change (Fig. 10.2) that Darwin anticipated. Instead, new findings have often shown explosions of novel biological form to have been even more dramatic than Darwin realized.[5]

The Deeper Mystery: How to Build Animals

In my book *Darwin's Doubt* I showed that what I called the "mystery of the missing fossils" betrays an even deeper and more fundamental mystery. That mystery is this: What *caused* the origin of novel forms of animal life? Could the neo-Darwinian process of random mutation and natural selection have built the Cambrian and other animals and done so quickly enough to account for the patterns in the fossil record? That question became much more acute in the last half of the twentieth century and now into the twenty-first, as biologists have learned more about what it takes to build an animal.

Just as producing the first life from simpler nonliving chemicals requires new genetic information, building new forms of life from simpler preexisting forms requires additional sources of genetic (and other types of biological) information. During the Cambrian explosion, an explosion of new animal body plans occurred. But building new body plans requires new organs, tissues, and cell types. And new types of cells require many kinds of specialized or dedicated proteins. Animals with gut cells, for example, require new digestive enzymes—a type of protein. But building new proteins requires genetic information stored on the DNA molecule.

Later many other such explosions of new biological form—the first insects, turtles, dinosaurs, sea reptiles, birds, flowering plants, mammals, and other major groups of organisms—also appeared abruptly in the fossil record.[6] In each case, building these new forms of life with their anatomical novelties would have required massive "explosions" of new genetic information.

We've seen that chemical evolutionary theory has not explained the origin of the functional digital information and information-processing systems in the simplest living cells—striking appearances of design, as even leading evolutionary proponents have acknowledged. Has neo-Darwinism, or any other evolutionary theory, explained the origin of the information necessary to produce *new* forms of life without invoking a designing intelligence?

In 1959 few people in evolutionary biology worried about this question.

Centennial Euphoria

The year 1959 witnessed the centennial celebration of *On the Origin of Species*. The modern version of Darwin's theory known as *neo-Darwinism* stood unrivaled—and unchallenged—in its scientific status. In preceding decades, evolutionary biologists had formulated this new theory of evolution by synthesizing classical Darwinism and Mendelian genetics. Whereas classical Darwinism proposed natural selection and random *variation* as the main mechanism of evolutionary change, the *neo-Darwinian* synthesis envisioned natural selection acting on a particular type of variation known as a "genetic mutation." These mutations were conceived as errors or alterations in hereditary material, supplying the minor variations upon which natural selection could act.

As we saw in the previous chapter, molecular biologists during the 1950s discovered that the DNA molecule stores genetic information as a linear array of precisely sequenced chemical subunits. This advance clarified where and how random mutations occurred and how they might produce new variations and genetic traits. Just as a fortuitous rearrangement of letters in an English text might generate new words or sentences, so too might random changes in, or rearrangements of, the genetic text in DNA produce new genes and, ultimately, new proteins. Such changes—sifted by natural selection—could provide the new genetic information to produce novel forms of life without any intelligent direction. For evolutionary biologists, these developments removed any lingering doubts about the creative power of natural selection working on random variation.[7] According to the neo-Darwinian synthesis, small-scale "microevolutionary" changes could be extrapolated indefinitely to account for large-scale "macroevolutionary" development.

Consequently, the tone of the main 1959 centennial meeting, held at the University of Chicago, was nothing if not triumphalist. As the British evolutionary biologist Julian Huxley proclaimed, "Future historians will perhaps take this Centennial Week as epitomizing an important critical period in the history of this earth . . . when the process of evolution, in the person of inquiring man, began to be truly conscious of itself."[8]

Does Not Compute

Nevertheless, doubts soon began to arise. These doubts came at first from an unexpected quarter—a group of mathematicians, physicists, and computer scientists, some of whom were faculty members at the Massachusetts Institute of Technology. In 1966, they joined with leading evolutionary biologists at a seminal conference, "Mathematical Challenges to the Neo-Darwinian Interpretation of Evolution." Held at the Wistar Institute in Philadelphia, the conference sought to evaluate the mathematical plausibility of natural selection and random mutation as a means of producing new genes and proteins—and thus new genetic information.

For the mathematically minded scientists at Wistar, doubts about the creative power of the mechanism of random mutation and natural selection stemmed from the elucidation of the nature of genetic information and the confirmation of Francis Crick's sequence hypothesis during the early 1960s. The discovery that DNA stores information as a four-character digital code raised questions about the efficacy of random mutational changes in producing such information—or at least enough of it to produce a novel protein structure and therefore any major innovation during the history of life.

Murray Eden (Fig. 10.3), one of the MIT professors who convened the event, emphasized that in all computer codes and written texts specificity of sequence determines function. Thus, random changes in sequence consistently degrade function or meaning. Indeed, no computer programmer wants random changes introduced into a program that he or she has written. Such changes will inevitably degrade and ultimately destroy the function of the existing program long before a new program would emerge through such a process. As Eden explained, "No currently existing formal language can tolerate random changes in the symbol sequences which express its sentences. Meaning is almost invariably destroyed."[9] He suspected that the need for specificity in the arrangement of DNA bases would

FIGURE 10.3
Former MIT computer engineering professor Murray Eden. Eden helped convene the now famed Wistar Institute conference "Mathematical Challenges to Neo-Darwinism" in 1966.

also render any random mutational "search" for a new functional gene or protein inevitably unsuccessful as well.

Later, worries about neo-Darwinism spread to evolutionary biologists themselves.[10] Over the past three decades, many evolutionary biologists have challenged a key tenet of the neo-Darwinian synthesis, namely, the idea that small-scale microevolutionary changes can be extrapolated to explain large-scale macroevolutionary innovations. For the most part, microevolutionary changes (such as variation in color) merely use or express existing genetic information, while the macroevolutionary change necessary to assemble new organs or whole body plans requires the production of new genetic information.

Recognizing this and other problems, in 2008 a group of sixteen evolutionary biologists met in Altenberg, Austria, to discuss their doubts about the creative power of the mechanism of random mutation and natural selection. Known as the "Altenberg 16," they and others have called for a new theory of evolution—one based on some mechanism other than—or in addition to—random mutation and natural selection.

FIGURE 10.4
Austrian evolutionary biologist Gerd Müller. At a 2016 Royal Society conference in London, Müller presented a notable talk on "The Explanatory Deficits of neo-Darwinism."

In November 2016, the Royal Society, the world's oldest and arguably most august scientific body, over which Isaac Newton once presided, hosted a similar conference in London to address perceived inadequacies in the standard neo-Darwinian theory of evolution. Austrian evolutionary biologist Gerd Müller (Fig. 10.4) opened the proceedings by outlining "the explanatory deficits" of neo-Darwinism, including its inability to explain the origin of "phenotypic complexity" and "anatomical novelty" in the history of life.[11]

I attended this 2016 meeting and it was clear to me that Müller's Royal Society audience understood the grave significance of his indictment, though the colorless technical terms "phenotypic complexity" and "anatomical novelty" might have obscured that significance for nonspecialists. What exactly does neo-Darwinism fail to explain? A phenotype refers to the visible form of an animal's or plant's anatomy. Müller was therefore saying that standard neo-Darwinian theory has failed to explain the origin of the new and complex anatomical features and structures that have arisen throughout the history of life. That would include novel animal architectures such as the arthropod, chordate, and molluscan body plans; new anatomical structures such as wings, limbs, eyes, nervous systems, and brains; and new specialized organs such as the vertebrate liver, digestive system, and kidneys. In short, neo-Darwinism fails to explain the origin of the most important defining features of living organisms, indeed, the very features that evolutionary theory has, since Darwin, claimed to explain.

Müller's talk echoed his earlier technical publications making the same points. In a provocative technical book, "Origination of Organismal Form," Müller and biologist Stuart Newman argued that neo-Darwinism has "no theory of the generative."[12] In other words, neo-Darwinism cannot explain what caused new forms of life to arise. In this book, published nearly 150 years after the *Origin of Species*, Müller and Newman characterized the "origination of organismal form" as an unsolved problem for evolutionary theory. Yet, again, the origin of biological form is precisely what Darwinism, and later neo-Darwinism, claimed to explain.

Other evolutionary biologists have echoed this concern. Many now repeat an old aphorism affirming that mutation and natural selection can account for "the survival of the fittest, but not the arrival of the fittest"[13]—that is, small-scale variations, but not large-scale innovations in biological form.

Evolutionary biologists, and especially public promoters of the theory such as Richard Dawkins and Eugenie Scott, formerly of the National Center for Science Education, sometimes acknowledge that *chemical* evolutionary theories have so far failed to account for the origin of the first living cell. Nevertheless, they typically treat the origin of the first life as a kind of isolated anomaly—an interesting puzzle that stands as an outlier against the otherwise comprehensive explanatory power of materialistic evolutionary theory. Consequently, they continue to affirm that evolutionary theory can (or eventually will) explain the origin of all new forms of life. In so doing, they gloss over the problem of the origin of genetic information as it confronts biological, as well as chemical, evolutionary theory.

The Problem of the Origin of Biological Information

During my PhD years, I came to appreciate the depth of the information problem confronting chemical evolutionary theory. But I soon learned that the problem goes much deeper. Specifically, it also poses a formidable challenge to neo-Darwinism and other theories of biological evolution—theories that attempt to explain not the origin of the first life, but the major innovations that have occurred during life's subsequent history.

In the summer of 1988, I came home from Cambridge to attend a conference in the Seattle area on the topic "Sources of Information Content in DNA." There I was introduced to a wider network of biologists and other scientists who doubted the creative power of the mechanism of random mutation and natural selection. These included Michael Denton (Fig. 10.5), a British-born Australian molecular biologist. I had read his groundbreaking book *Evolution: A Theory in Crisis* the year before. And now, in July 1988, he and I arrived in Seattle, jet-lagged, from two opposing points on the map—Australia and England. After we were introduced, and with both of our body clocks out of sorts, we ended up talking deep into the night before the first day of the conference. During our conversation, he told me more about why the mathematicians and computer scientists at Wistar were so deeply skeptical about the ability of random mutations and natural selection to produce new genetic information.

FIGURE 10.5
The Australian molecular biologist Michael Denton, whose 1985 book *Evolution: A Theory in Crisis* emboldened a number of younger scientists to express their growing doubts about neo-Darwinism.

According to neo-Darwinian theory, new genetic information arises as random mutations occur in the DNA. "Random" means that mutations occur without respect to the functional needs of the organism—mutations have no inherent directionality. Nevertheless, natural selection can only "select" what random mutations first generate.[14] And for the evolutionary process to produce new forms of life, random mutations must first generate—at the very least—new genetic information for building novel proteins.

And that, Denton told me, was the problem. When it comes to producing new genetic information, the neo-Darwinian mechanism, with its reliance on random mutations, faces a needle-in-a-haystack dilemma, or what mathematicians call a "combinatorial problem."

Proteins in a Combinatorial Haystack

In mathematics, the term "combinatorial" refers to the number of possible ways that a set of objects can be arranged or combined. Simple bike locks, for example, typically have four *dials* with ten *settings* on each dial. A bike thief encountering one of these locks (and lacking bolt cutters) faces a combinatorial problem since there are 10 × 10 × 10 × 10, or 10,000, possible ways of combining the possible settings on each of the four dials and only one combination that will open the lock. Unless the thief has a lot of time to spend trying various combinations, a random search of possible combinations is unlikely to yield the correct combination.

In a memorable chapter in his book (chap. 13), "Beyond the Reach of Chance," Denton explained why the mechanism of random mutation and natural selection also faces a combinatorial problem.[15] He did so by drawing an analogy to English text. As Denton noted, linguists have estimated that for every meaningful sequence of English characters 12 letters long there are *one hundred trillion* (i.e., 100,000,000,000,000, or 10^{14}) corresponding gibberish sequences of the same length—effectively a lock with fourteen dials and ten digits but only one combination.

Denton asked his readers to consider what would likely happen to the meaning of an English phrase or sentence if many of the letters were randomly altered. Because there are a number of similarly spelled words (sample, example, trample, apple, etc.), random changes might at first alter the meaning of the original sentence but maintain some meaning. As the random changes accumulated, however, they would not only alter the original message beyond recognition, but would eventually efface or destroy any meaning altogether.

Drawing on the insights of the Wistar conferees, Denton explained the mathematical reason for this. In English there are vastly more ways "to go wrong than to go right"—that is, for any sequence of any given length, there are more combinations of English letters that will *not* produce a meaningful phrase or sentence than combinations of those same 26 letters that *will* generate a meaningful sentence. Indeed, the number of nonfunctional gibberish sequences *dwarfs* the number of functional combinations. Consequently, random changes in letters are overwhelmingly more likely to "find the gibberish," or degrade meaning, than to generate a new meaningful sentence, especially as the number of changes to the original meaningful sequence increases.

FIGURE 10.6
The mathematician and philosopher David Berlinski, who has written persuasively about the problem of "combinatorial inflation" and the implausibility of a random mutational search producing novel functional genes and proteins.

Moreover, as the length of the required phrase or sentence grows, the number of possible letter sequences of that length grows exponentially, and grows much faster than the number of possible meaningful sequences, so that the probability of finding a functional sequence via a random search diminishes precipitously with necessary sequence length. Denton noted that whereas for every meaningful sequence of English letters 12 letters long there are one hundred trillion (or 10^{14}) corresponding gibberish sequences, for every meaningful sequence of English letters 100 letters long there are 10^{100} corresponding gibberish sequences, an unimaginably large number. Mathematician David Berlinski (Fig. 10.6) has dubbed this the problem of "combinatorial inflation" in his seminal 1996 essay titled "The Deniable Darwin."[16]

Like the Wistar mathematicians and computer scientists, Denton told me that he suspected (but could not prove in 1988) that the mechanism of random mutation and natural selection faced a similarly formidable combinatorial search problem. For a sequence of bases in DNA of any significant length, there likely were vastly more ways of arranging nucleotide bases that would *not* produce a functional protein than there were ways of arranging nucleotide bases that *would*. Since both English text and DNA store information in long sequences of alphabetic characters (or more precisely in the case of DNA, in long sequences of four distinct chemical subunits functioning as such), he suspected that both forms of information were subject to the problem of combinatorial inflation, making a random search for functional sequences a needle-in-a-haystack proposition. Consequently, random mutational changes were overwhelmingly more likely to degrade biological function than to generate a new functional gene or protein.

Nevertheless, when I first met Denton, he told me that it was not yet possible to make a conclusive mathematical determination of the plausibility of a random mutational search for new functional genes and proteins. Molecular biologists, he told me, could not yet quantify how rare functional DNA sequences (genes) and proteins were among all the possible sequences of nucleotide bases and amino acids of a given length.

Consequently, they couldn't yet calculate the relevant probabilities—and thus assess the plausibility of random mutation and natural selection as a means of producing new genetic information.

As far back as the 1950s and 1960s, molecular biologists understood that the size of the "sequence space" of possible arrangements of nucleotide bases and amino acids is extremely large. Indeed, as the required length of a gene or protein sequence grows, the number of possible base or amino-acid sequence combinations of that length grows exponentially.

For example, for protein chains, there are 20^2, or 400, ways to make a two-amino-acid combination, since each position could feature any one of 20 different amino acids. Similarly, there are 20^3, or 8,000, ways to make a three-amino-acid sequence, and 20^4, or 160,000, ways to make a sequence four amino acids long, and so on. Yet most functional proteins are made of *hundreds* of amino acids. Even a relatively short protein of, say, 150 amino acids represents one sequence among an astronomically large number of other possible sequence combinations—approximately 10^{195}. That is an enormous number, the digit 1 followed by 195 zeroes. Intuitively, this suggests that the odds of finding even a single functional sequence—a working gene or protein—as the result of random genetic mutations may be prohibitively small, even taking into account the time available to the evolutionary process.

But knowing the total number of alternative possible combinations associated with a sequence of any given length does not by itself allow scientists to make a definitive quantitative assessment of the plausibility of a random search for a new functional information-rich sequence. Instead, molecular biologists must also be able to determine two other variables. And the most important of those variables has only recently been determined.

Unlocking the Mystery of Information

In *Darwin's Doubt*, I used the example of a bike lock to illustrate why assessing the plausibility of a successful random search requires knowledge of the size of a sequence space, but also of two other variables.[17]

First, we must know how many opportunities there are for opening the lock. Remember the simple four-dial bike lock described above. A typical bike thief has a negligible chance of finding the right combination for such a lock if he has only one opportunity to open it. Now imagine that we encounter a really committed thief. This thief is willing to search the "sequence space" of possible combinations at a rate of about one new possible combination per ten seconds and to keep at it long after the typical bike thief has given up. If our committed bike thief had fifteen hours and took no breaks, he could generate more than half (about 5,400 of the 10,000) the total combinations of a four-dial bike lock. In that case, the probability that he will stumble upon the right combination exceeds the probability that he will fail. Given this, it would be *more likely than not* that he will succeed in opening the lock by random search. The chance hypothesis—that he will succeed in opening the lock via a random search—is therefore also more likely to be *true* than false.

But now imagine a much more complicated lock. Instead of four dials, this lock has ten dials. Instead of 10,000 possible combinations, this lock has 10^{10}, or 10 billion, possible combinations. With only one combination that will open the lock out of 10 billion—a prohibitively small ratio—it is much more likely that the thief will fail *even if he devotes his entire life to the task*.

A little math shows this to be true. If the thief did nothing but sample combinations at random, one every ten seconds for an entire one-hundred-year lifetime, he would still sample only about 3 percent of the total number of combinations on a lock that complex. In this admittedly contrived case, it would be *much more likely than not* that he would *fail* to open the lock by random search. In that case, the chance hypothesis—that the bike-obsessed thief will succeed in finding the combination by a random search—is much more likely to be *false* than true.

So what about relying on random mutations to "search" for a new DNA base sequence capable of directing the construction of a new protein? Would such a random search be more likely to succeed—or to fail—in the time available to the evolutionary process? Is a random mutational search for a new protein more like the case of our hypothetical thief searching for the combination on the four-dial or the ten-dial lock?

As our examples show, the ultimate probability of the success of a random search—and the plausibility of any hypothesis that affirms the success of such a search—depends upon both the *size of the space* that needs to be searched and the *number of opportunities* available to search it.

But scientists need to know something else to determine the probability of success in the case of genes and proteins. They also need to know how rare or common functional arrangements of DNA are among all the possible arrangements for a protein of a given length. That's because for genes and proteins, unlike in our bike-lock example, there are many functional combinations of bases and amino acids (as opposed to just one) among the vast number of total combinations. Thus, they need to know the overall ratio of functional to nonfunctional sequences in the DNA.

Imagine that our hypothetical thief must choose between cracking a lock with four dials and cracking one with ten dials. The four-dial lock, however, has only one combination that will open it, while on the ten-dial lock every other combination (50 percent of the combinations) will open the lock. If the thief knew this, which lock should he choose? He might be tempted to opt for the smaller lock. In fact, though, the larger lock gives him better odds (1 in 2 as opposed to 1 in 10,000).

Thus, to assess the difficulty of a random search, it's necessary to know how many of the combinations will open the lock. The key isn't just the number of total combinations that have to be searched, but the *ratio* of the number of combinations that will open the lock to the total number of combinations. In the same way, it isn't just the total number of possible combinations in the amino-acid sequence space that determines the difficulty of a random search for a new protein structure. Ultimately, it's the ratio of functional to nonfunctional sequences that determines the difficulty.

In 1966 at the Wistar conference—and in 1988 when I met Denton—molecular biologists knew that the combinatorial sequence space associated with even a protein of modest length was enormously and exponentially large. Yet they didn't know how many of those arrangements were functional. In effect, they didn't know how many of the possible combinations would "open the lock."

Determining the Golden Ratio

Two years after I met Denton and had begun teaching in the fall of 1990, I had to return to Cambridge to defend my PhD thesis. On that visit, a mutual friend introduced me to Douglas Axe (Fig. 10.7), a protein scientist with a recently minted PhD from Caltech. Axe had just begun to do experiments in Cambridge designed to answer questions about the rarity of functional genes and proteins in combinatorial sequence space—to determine, in other words, the relevant ratio of functional to nonfunctional sequences.

FIGURE 10.7
Molecular biologist Douglas Axe. While working at the University of Cambridge Medical Research Council Laboratory from 1990 to 2003, Axe established that DNA base sequences capable of making protein "folds" are extremely rare among the vast number of corresponding possible sequences.

While working at the University of Cambridge and the prestigious Medical Research Council Laboratory from 1990 to 2003, he ultimately established that DNA base sequences capable of making the complex three-dimensional structures called protein "folds" are extremely rare among the vast number of possible sequences. A protein fold is a distinctive, stable, complex, three-dimensional structure that enables proteins to perform specific biological functions. Since proteins are crucial to almost all biological functions and structures, protein folds represent the smallest unit of structural innovation in living systems (Fig. 10.8).

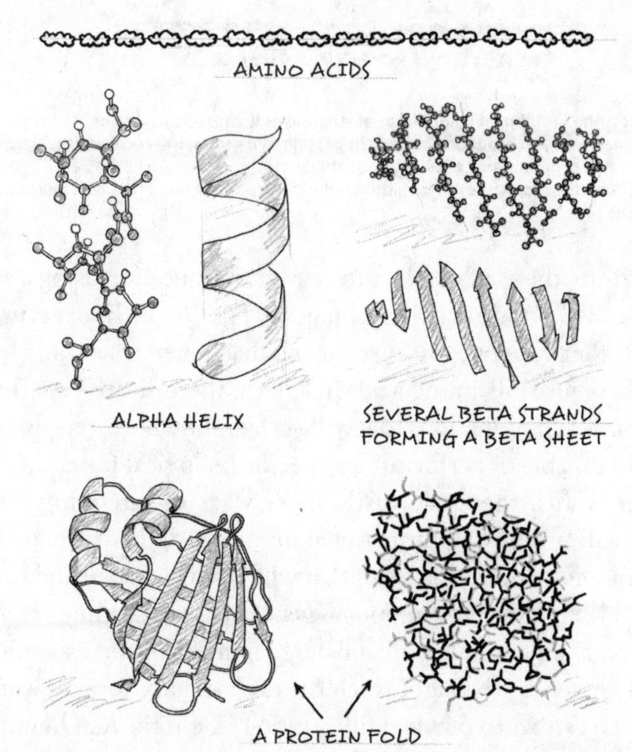

FIGURE 10.8
Different levels of protein structure. The first panel at the top shows the primary structure of a protein: a sequence of amino acids forming a polypeptide chain. The second panel depicts, in two different ways, two secondary structures: an alpha helix (left), and beta strands forming a beta sheet (right). The third panel at the bottom shows, in two different ways, a tertiary structure—that is, a protein fold. Protein folds represent the smallest unit of structural innovation in living systems.

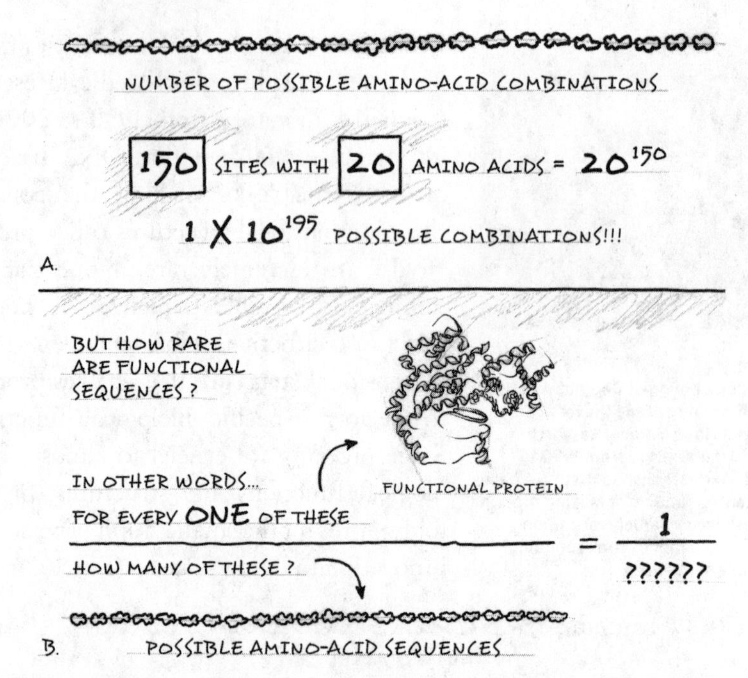

FIGURE 10.9

TOP: This depicts the problem of combinatorial inflation as it applies to proteins. As the number of amino acids necessary to produce a protein or protein fold grows, the corresponding number of possible amino acid combinations grows exponentially.

BOTTOM: This depicts graphically the question of the rarity of proteins in that vast amino acid "sequence space."

How rare are they? Axe set out to answer this question using a sampling technique called site-directed mutagenesis (Fig. 10.9). His experiments revealed that, for every one DNA sequence that generates a short *functional* protein fold of just 150 amino acids in length, there are 10^{77} *non*functional combinations—combinations that will *not* form a stable three-dimensional protein fold capable of performing a specific biological function.[18]

In other words, there are vastly more ways of arranging nucleotide bases that will produce nonfunctional amino-acid chains than there are ways of arranging nucleotide bases that will produce folded and functional proteins. Indeed, for every functional gene capable of coding for a protein fold there is an almost unimaginably large number of corresponding nonfunctional sequences through which the evolutionary process would need to search. To return to our lock illustration, the ratio Axe found implies that the difficulty of a mutational search for a new gene or novel protein fold is equivalent to the difficulty of searching for just one combination on a lock with ten digits on each of seventy-seven dials (Fig. 10.10)!

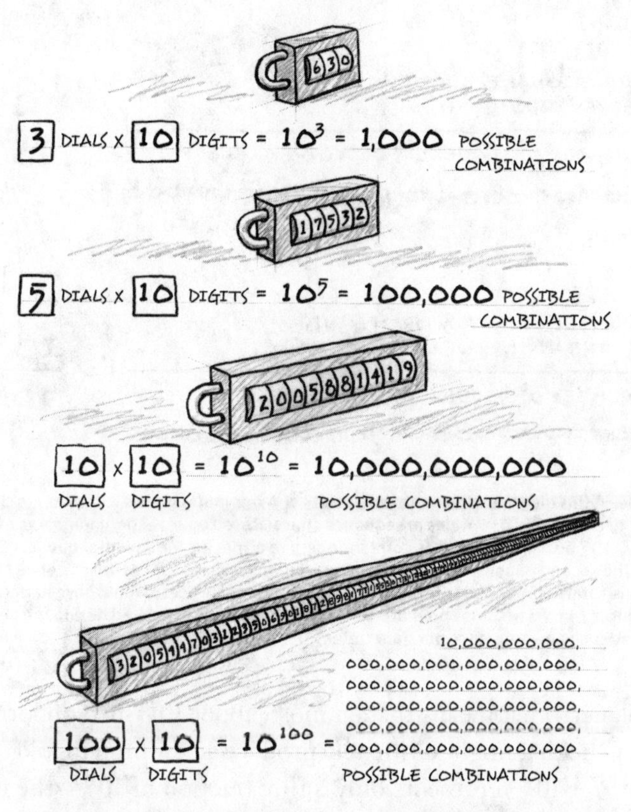

FIGURE 10.10

The problem of combinatorial inflation as illustrated by bike locks of varying sizes. As the number of dials on the bike locks increases, the number of possible combinations rises exponentially.

Clearly, 10^{77} represents a huge number. To put it in context, there are only 10^{65} atoms in our galaxy. But could random genetic mutations effectively search a space of possibilities that large in the time available to the Cambrian explosion or even the entire history of life on earth? To answer that question, we still need to know how many opportunities the evolutionary process would have had to search this huge number of possibilities—as Axe himself has emphasized.

Estimating the Probabilistic Resources

Consider that every time an organism reproduces and generates a new organism, an opportunity occurs to mutate and pass on a new gene sequence. And during the 3.85-billion-year history of life, biologists estimate that about 10^{40} individual organisms—a huge number—have lived

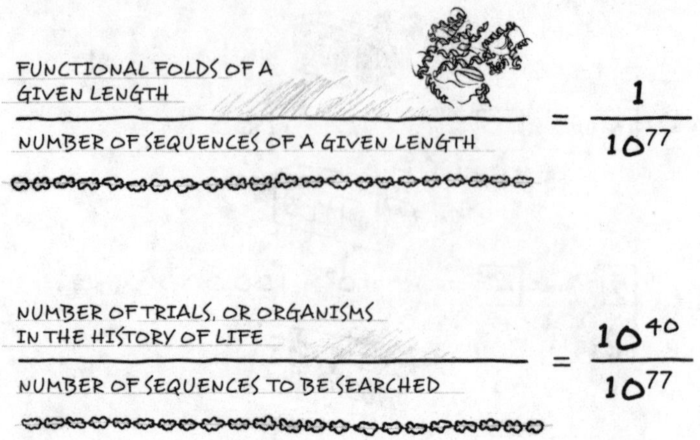

FUNCTIONAL FOLDS OF A GIVEN LENGTH / NUMBER OF SEQUENCES OF A GIVEN LENGTH = $\frac{1}{10^{77}}$

NUMBER OF TRIALS, OR ORGANISMS IN THE HISTORY OF LIFE / NUMBER OF SEQUENCES TO BE SEARCHED = $\frac{10^{40}}{10^{77}}$

FIGURE 10.11

The top panel in this diagram represents the results of Axe's mutagenesis experiments showing the extreme rarity of functional proteins in sequence space. Based on his experiments Axe estimated that there are 10^{77} possible sequences corresponding to a specific functional sequence 150 amino acids long. The second panel shows that functional amino acid sequences are extremely rare even in relation to the total number of opportunities the evolutionary process would have had to generate novel sequences (on the assumption that each organism that has ever lived during the history of life produced one such new sequence per generation).

on our planet. That means that, at most, about 10^{40} such opportunities to mutate a gene that might ultimately produce a new protein fold could have occurred. Yet 10^{40} represents only a tiny fraction of 10^{77}—the number of non-functional sequences corresponding to each protein fold of modest length (Fig. 10.11). Indeed, the fraction 10^{40} divided by 10^{77} equals 1 part in 10^{37}, or 1 part in ten trillion times a trillion times a trillion, to be exact.

This means that for even one relatively modest-size novel protein fold to arise, the mechanism of random mutation and natural selection would have time to search just a tiny fraction of the total number of relevant sequences. In other words, the number of trials available to the evolutionary process turns out to be incredibly small *in relation to* the number of *possible* sequences that need to be searched. Or to put it differently, the size of the relevant spaces that need to be searched by the evolutionary process dwarfs the time available for searching—even taking into account life's 3.85-billion-year history.

It follows that the mechanism of random mutation and natural selection has not had enough time to generate or search anything more than a minuscule fraction (one ten trillion trillion trillionth, to be precise) of the total number of possible nucleotide base or amino-acid sequences corresponding to a single protein fold.

It is therefore overwhelmingly *more likely than not* that a random mutational search would have *failed* to produce even one new functional (information-rich) DNA sequence capable of coding for one new protein fold in the entire history of life on earth. Consequently, the hypothesis that such a random search succeeded is more likely to be *false* than true. And, of course, building new animals would require the creation of *many* new proteins and protein folds, not just one. It follows that the standard neo-Darwinian mechanism does not provide an adequate explanation for the origin of the genetic information necessary to produce the major innovations in biological form that have arisen in the history of life on earth.

Design: Apparent or Intelligent?

The lack of creative power associated with the mechanism of random mutation and natural selection has raised the question of design in biology in a new context. To see why, it's important to remember how Darwinism, and later neo-Darwinism, dispensed with the idea of intelligent design in the first place.

During the nineteenth century, leading up to the publication of Darwin's *Origin of Species*, biologists were struck by how living organisms seemed well adapted to their environments. They attributed this adaptation to the ingenuity of a powerful designing intelligence.

Darwin attempted to show that natural selection acting on random variations could account for this appearance of design. He did so by drawing an analogy to the well-known process of artificial selection, also known as selective breeding. A Scottish sheepherder, for example, might breed for a woollier sheep to enhance its chances of survival in a cold northern climate or to harvest more wool. To do so, she would choose only the woolliest offspring to breed. If, generation after generation, she continued to select and breed only the woolliest sheep, she would eventually produce a woollier breed of sheep—a breed *better adapted* to its environment.

As Darwin pointed out, nature also has a means of sifting: defective creatures are less likely to survive and reproduce, while those offspring with beneficial variations are more likely to pass on their advantages. Darwin argued that this process—natural selection acting on random variations—could alter the features of organisms just as intelligent selection by human breeders can do.

Imagine, for example, that a series of unusually cold winters in the Scottish highlands ensures that all but the woolliest sheep die off. Now, again, only very woolly sheep will remain to breed. If the cold winters continue over several generations, won't the population of sheep eventually become discernibly woollier?

This was Darwin's great insight. Nature could have the same effect on a population of organisms as the intentional decisions of an intelligent agent. Nature would favor the preservation of certain features over others. The resulting change or increase in fitness (or adaptation) would then have been produced not by a breeder choosing a trait or variation—not by "artificial selection"—but instead by a wholly natural process, "natural selection."

In this way Darwin explained the appearance of design without appeal to a designing intelligence. As he put it, "There seems to be no more design in the variability of organic beings and in the action of natural selection, than in the course in which the wind blows."[19] Or as Harvard evolutionary biologist Ernst Mayr, an architect of the modern neo-Darwinian synthesis, explained a century later: "The real core of Darwinism . . . is the theory of natural selection. This theory is so important for the Darwinian because it permits the explanation of adaptation, the 'design' of the natural theologian, by natural means."[20]

But what if minor improvements in the adaptation of organisms to their environment—sheep getting a bit woollier or finch beaks getting a bit longer or shorter—are not the only example of apparent design in the living world? What if other more striking and fundamental features of life—such as the genetic information necessary to build new forms of animal life in the first place—have not been explained by natural selection or any other undirected mechanism? Even such staunch neo-Darwinists as Richard Dawkins recognize that the presence of digitally encoded information in DNA in living organisms represents at least a striking *appearance of design*. But if neither Darwin's original "designer substitute" mechanism nor the updated neo-Darwinian version of that mechanism explains this salient appearance of design, what could?

Information Explosions as Evidence of Intelligent Design

In *Darwin's Doubt*, I acknowledged that evolutionary biologists have recently proposed several new alternative evolutionary mechanisms in an attempt to remedy the "explanatory deficits" of neo-Darwinism. I also noted that biologists today recognize that building new animals requires not only new genetic information, but also what is called "epigenetic" or "ontogenetic" information.[21] Ontogenetic information is not stored in DNA, but instead in higher-level structures within cells and organisms. Even so, I showed that these newer post-neo-Darwinian theories of evolution—self-organization, evolutionary developmental biology, neo-Lamarckian epigenetic inheritance, neutral theory, natural genetic engineering, and others—have failed to account for both the genetic and ontogenetic information necessary for structural innovation in the history of life. Indeed, invariably either these new theories of evolution do not explain the origin of necessary genetic and ontogenetic information or they simply presuppose unexplained, preexisting sources of such information.

And since the Cambrian explosion of animal life and other similar events represent explosions of information as well as of biological form, that raises a question. Is it possible that the dramatic increases of biological information at periodic episodes throughout the history of life not only pose a difficulty for materialistic theories of biological evolution, but also provide positive evidence *for* intelligent design? Could this *unexplained* (from a materialistic point of view) appearance of design point to actual *intelligent* design?

It does.

A Cause Now in Operation

In the last chapter I used the method of multiple competing hypotheses (or inference to the best explanation) to evaluate the "causal adequacy" of proposed explanations for the ultimate origin of biological information. I showed that chemical evolutionary models fail to identify a cause capable of producing the digital information in DNA and RNA necessary to produce the first life. We have seen in this chapter that the main theories of *biological* evolution also fail to account for the origin of the information necessary to build new forms of animal life. Yet, again, we do know of a cause that has demonstrated the power to produce functional or specified information. That cause is intelligent agency.

Intelligent agents, due to their rationality and consciousness, have demonstrated the power to produce functional information in the form of linear sequence-specific arrangements of characters. We know that such agents generate information in the form of software code, ancient inscriptions, meaningful text in books, encrypted military codes, and much else. The generation of functional information is, to quote Henry Quastler again, "habitually associated with conscious activity."[22] Our uniform experience confirms this obvious truth. This suggests that intelligent design meets the key "causal adequacy" requirement of a good historical-scientific explanation as discussed in the previous chapter.

We also know of no materialistic (nonmental) "cause now in operation" that generates large amounts[23] of specified information (especially in a digital or alphabetic form). As I show in more detail in *Darwin's Doubt*, a long and painstaking search for such a cause, by some of the best minds in evolutionary biology, has failed to turn up a cause capable of producing the information necessary for genuine innovation in the history of life. Yet intelligent agents routinely produce vast amounts of specified information in order to communicate and to build a variety of new structures. Thus, *only* intelligent design meets the requirement of causal adequacy. In other words, our uniform experience of cause and effect shows that

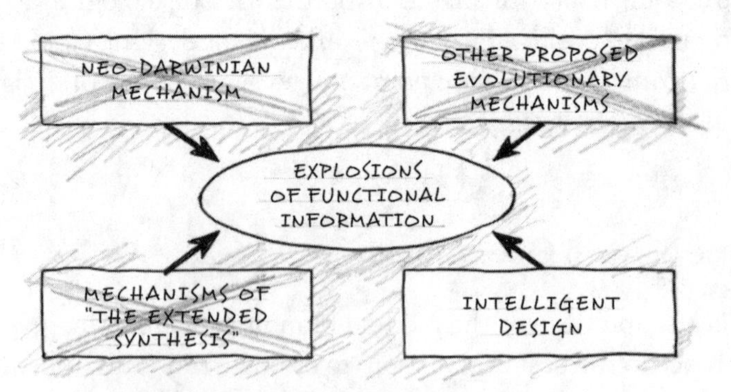

FIGURE 10.12

In this chapter, I show that the neo-Darwinian mechanism of random mutation and natural selection does not provide a plausible (or "causally adequate") explanation for the origin of the functional or specified information in living systems. In *Darwin's Doubt* and other published work, I also show that more recently proposed evolutionary mechanisms associated with the "extended synthesis," as well as those associated with theories of self-organization and punctuated equilibrium, also fail to explain the origin of the information necessary to build novel forms of life. These new evolutionary mechanisms—such as species selection, neutral evolution, natural genetic engineering, neo-Lamarckian epigenetic inheritance, niche construction, and evolutionary developmental processes—invariably either do not address the problem of the origin of genetic and ontogenetic information or they presuppose prior unexplained sources of such specified information. Yet, we know that intelligent agents can and do produce specified information. Consequently, I infer intelligent design as the best, most causally adequate explanation for the explosions of functional or specified information evident in the Cambrian explosion and other similar events in the history of life.

intelligent design is the *only known cause* of the origin of large amounts of functionally specified information. It follows that the great infusion of such information in the Cambrian explosion and comparable events in the history of life is best explained (Fig. 10.12) as the activity of an intelligent cause—what the great nineteenth-century paleontologist Louis Agassiz described as "acts of mind."[24]

Intelligent design provides the best explanation for the origin of genetic information for another reason: purposive agents have just those necessary powers that natural selection lacks as a condition of its causal adequacy. We have seen that natural selection lacks the ability to generate novel information precisely because it can only act *after* new functional information has arisen. Natural selection can favor new proteins and genes, but only after they arise and confer a functional advantage (positively influencing reproductive output). The job of generating new functional genes, proteins, and systems of proteins falls entirely to random mutations. Yet without functional criteria to guide a search through the vast space of possible sequences, random variation is probabilistically doomed. What is needed is not just a source of variation or a mode of selection that can operate after the fact of a successful search, but instead a means of selection that (a) operates during a search—*before* success—and that (b) is guided by information about or knowledge of a functional target.

Demonstration of this requirement has come from an unlikely quarter.

Simulating Intelligent Design

Genetic algorithms are programs that allegedly simulate the creative power of random mutation and natural selection. Richard Dawkins, Bernd-Olaf Küppers, and others have developed computer programs that putatively simulate the production of genetic information by mutation and natural selection.[25] Yet these programs succeed only by the illicit expedient of having an *intelligent programmer* provide the computer with a "target sequence" and then treating proximity to *future* function (i.e., the target sequence), not actual present function, as a selection criterion—thus, actively directing the program to the target sequence. Such foresight has no analog in nature. In biology, survival depends upon maintaining present function. Natural selection, therefore, cannot look forward or devise plans that anticipate future needs or desirable outcomes. The process, as evolutionary theorists Andrei Rodin, Eörs Szathmáry, and Sergei Rodin note, works strictly "'in the present moment,' right here and right now . . . lacking the foresight of potential future advantages."[26]

What unguided evolutionary mechanisms lack, intelligent design—purposive, goal-directed selection—provides. Rational agents can arrange matter and symbols with distant goals in mind. They also routinely solve problems of combinatorial inflation. In using language, the human mind routinely "finds" or generates highly improbable linguistic sequences to convey a *pre*conceived idea. In the process of thought, functional objectives precede and constrain the selection of words, sounds, and symbols to generate functional (and meaningful) sequences from a vast ensemble of meaningless alternatives.[27]

Similarly, the construction of complex technological products, such as bridges, circuit boards, engines, and software, results from goal-directed constraints on the possible arrangements of matter or symbols.[28] Indeed, in all functionally integrated complex systems where the cause is known by experience or observation, designing engineers or other intelligent agents applied constraints to limit possibilities in order to produce improbable but useful forms, sequences, or structures. Rational agents have repeatedly demonstrated the capacity to constrain possible outcomes to actualize improbable but initially unrealized future functions. Repeated experience affirms that intelligent agents (minds) uniquely possess such causal powers.

Analyzing the problem of the origin of biological information, therefore, exposes a deficiency in the causal powers of natural selection and other undirected evolutionary mechanisms that corresponds precisely to powers that agents are known to possess—uniquely so in our experience. Intelligent agents can select functional goals *before* the goals are physically instantiated. They can devise or select material means to accomplish those ends from among an array of possibilities. They can then actualize those goals in accord with a *pre*conceived design plan or set of functional requirements.

These causal powers—ones that natural selection and other undirected evolutionary mechanisms lack—are habitually associated with the attributes of consciousness and rationality—with purposive intelligence. Thus, by invoking intelligent design to explain the origin of the specified information that arises abruptly and episodically in the history of life, contemporary advocates of intelligent design do not posit an arbitrary explanatory element unmotivated by a consideration of the evidence. Instead, we posit an entity possessing just the causal powers that the information explosions in the history of life as well as all other discontinuous increases in functional information require as a condition of their production and explanation.

Multiple Acts of Mind?

Thus, the information problem associated with the origin of the first life does not represent an isolated anomaly, but instead a fundamental challenge to theories of chemical *and* biological evolution. Indeed, repeated abrupt appearances of new biological form and information in the history of life are not at all what we should expect to observe if a purely materialistic evolutionary process lacking "design" and "purpose" was at work.[29] Instead, if a purposive intelligence had acted periodically during the history of life on earth, we might well expect—given our experience of intelligent agents generating information—to find evidence of episodic bursts of new information in the biosphere.

Part III

Inference to the Best Metaphysical Explanation

11

How to Assess a Metaphysical Hypothesis

In April 2000, as a young professor in the philosophy of science, I attended an unusual conference at Baylor University. Titled "The Nature of Nature," it convened philosophers of science and scientists, including several Nobel laureates. The conference included sessions addressing the origin of the universe, the fine tuning of the laws of physics, and the origin of life, among other topics. In addition, the event was designed to explore the overarching question of whether nature as a whole points to a reality beyond itself or whether nature can be better understood as an autonomous, self-existent, and self-organizing system. In other words, the conference addressed the question of whether the worldview of naturalism (or materialism), on the one hand, or something like theism or deism, on the other, better explains key scientific discoveries.

For the first time in such a large academic forum, I presented my case for intelligent design as the best explanation for the origin of the genetic information necessary to produce the first life. I did so with some trepidation, in part because the organizers had slated Christian de Duve (Fig. 11.1), a prominent origin-of-life biochemist and Nobel laureate, to speak directly after me. De Duve had recently published a book, *Vital Dust*, exploring the possibility that life had arisen by self-organizational processes—processes that might make

FIGURE 11.1
Biochemist and Nobel laureate Christian de Duve.

life inevitable once the right conditions on earth had arisen. I planned to critique the self-organizational approach as part of my talk, so I anticipated that de Duve might well challenge my scientific analysis.

To my surprise, de Duve began his talk by saying he had agreed entirely with "the previous speaker" except for my discussion of "that last slide." My first thirty-odd slides had presented a scientific critique of the ability of current chemical evolutionary theories, including self-organizational theories, to explain the origin of genetic information. My last few slides explained the logic of the inference to intelligent design, and my very last slide explained why I rejected methodological naturalism, the principle that scientists must limit themselves to strictly naturalistic explanations.

Over dinner that night de Duve, an elegant Belgian, graciously, and memorably for me, put me at ease by complimenting my knowledge of molecular biology. He also acknowledged, despite his support for a self-organizational approach, that such theories had not yet solved the crucial information problem. Nevertheless, as in his talk earlier that day, he made clear that he favored a strictly naturalistic approach to science as well as a naturalistic answer to the questions motivating the conference.

Unlike de Duve, many scientists don't think about foundational worldview questions such as the nature of nature. Instead, many have long assumed an answer to them. Indeed, since the late nineteenth century, many scientists have agreed with the perspective of Carl Sagan that I quoted in the Prologue: "The cosmos is all that is, or ever was, or ever will be."[1]

FIGURE 11.2
Sean Carroll, Caltech physicist and proponent of scientific naturalism.

More recently, astrophysicist Neil deGrasse Tyson and Caltech cosmologist Sean Carroll have helped to popularize this worldview (Fig. 11.2). Tyson has done so with a rebooted *Cosmos* series, and Carroll in his popular science books and lectures. Carroll defines naturalism not only as the idea that "there's only the natural world"[2] and "no spirits, no deities, or anything else," but also as the idea that "there is a chain of explanations concerning things that happen in the universe, which ultimately reaches to the fundamental laws of nature and stops."[3]

Carroll's approach to these deep questions is refreshing. He also doesn't just *assume* naturalism as the only answer to the question of the nature of nature. He acknowledges the existence of other worldviews and offers naturalism as the best explanation of what science has discovered about reality.

In his bestselling book *The Big Picture*, Carroll takes care to explain what naturalists believe. "The broader ontology typically associated with atheism is naturalism—there is only one world, the natural world, exhibiting patterns we call the 'laws of nature,' and which is discoverable by the methods of science and empirical investigation," he writes. "There is no separate realm of the supernatural, spiritual, or divine; nor is there any cosmic teleology or transcendent purpose inherent in the nature of the universe or in human life."[4]

In previous chapters I described several key scientific discoveries about the origin of the universe and life. But what do they tell us about the nature of nature? By framing the Baylor conference around that question, the organizers assumed, as Sean Carroll has done, that there are many competing metaphysical hypotheses about the nature of reality.

Metaphysics is the discipline of philosophy that addresses the fundamental nature of reality. Ontology, a subdiscipline of metaphysics, is concerned with questions of "being" or ultimate reality. It asks, "What is the thing or the entity or the process from which everything else comes?" Philosophers recognize several main worldviews with different answers to this ultimate, or "prime-reality," question. "Naturalism" (or materialism) views matter and energy and the laws of nature as the prime realities. "Pantheism" asserts an impersonal deity present in matter and energy as the prime reality. "Theism" affirms a personal, intelligent, transcendent God who also acts within the creation. And "deism" affirms a personal, transcendent, intelligent God who does *not* act within the created order after its initial origin (Fig. 11.3, see page 220).

These four worldviews represent four possible ways of answering three basic questions about ultimate reality: Does God exist? If so, is God personal or impersonal? If personal, does God act only at the beginning of the universe or also after it within the created order?

The decision tree in Figure 11.4 (see page 221) shows how the answers given to these three questions allow us to classify major worldviews. Of course, people hold variations of these basic views, but each of those variants typically affirms the core tenets of one of the four fundamental

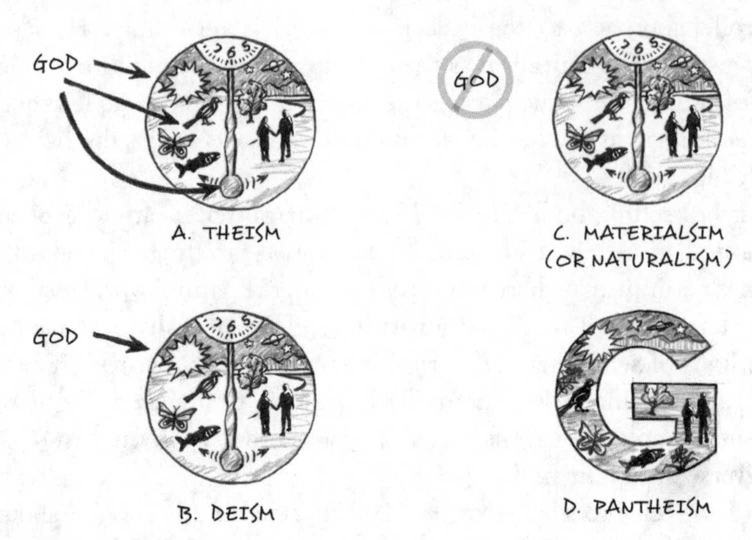

FIGURE 11.3
Philosophers recognize several main worldviews with different answers to the "prime reality" question. Theism affirms a personal, intelligent, transcendent God who also acts within the creation. Deism asserts a personal, transcendent, intelligent God who does *not* act within the created order after its initial origin. Naturalism (or materialism) affirms matter and energy and the laws of nature as the prime realities. Pantheism asserts an *impersonal* deity present in matter and energy as the prime reality. In these diagrams portraying these four great systems of thought, the circles represent the physical universe, the drawings inside the circle depict various living and nonliving entities within the universe, the pendulum represents the laws of nature, and "the big G" represents God. Notice that in Theism, God is depicted as separate from but also active in the universe; in Deism, God is depicted as separate from but not active in the universe; in Naturalism or Materialism, God is portrayed as nonexistent, and in Pantheism, God is shown as present in, or "co-extensive" with, every aspect of the material universe but not existing in any way separate from it.

systems of thought just mentioned. For example, scientific materialism, dialectical materialism, and atheistic existentialism have different views of human nature—that is, whether human beings have free will or not and, if not, what determines human behavior or the development of human history. Nevertheless, all represent different forms of naturalism. All deny the existence of God, and all hold that matter and energy constitute the prime realities. Similarly, the worldview of theism includes Christian, Jewish, Islamic, and nonreligious forms of theism, all of which affirm a personal God as the prime reality, but each of which has a differing view about the nature or attributes of that God.

These competing worldviews also offer differing answers to the question of the nature of nature. For example, theism holds that nature represents an orderly system of cause and effect within an open system—one in which God might act discretely as an agent within the system of natural laws that God otherwise upholds. Naturalism regards nature as

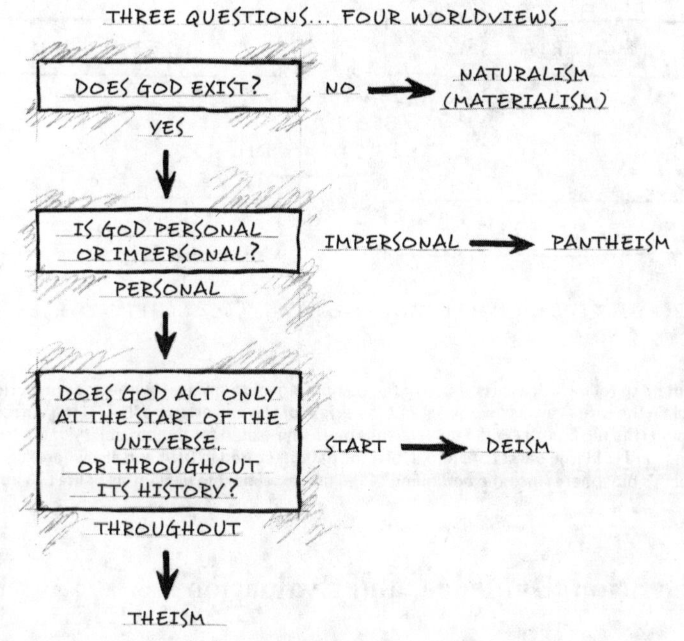

FIGURE 11.4
The four worldviews of theism, deism, pantheism, and materialism represent four possible ways of answering three basic questions about ultimate reality: Does God exist? If so, is God personal or impersonal? If personal, does God act only at the beginning of the universe or also after the beginning within the created order?

an orderly system of cause and effect within a closed system, one where nothing outside of nature could act upon it. Pantheism sees God as an impersonal force or a mystical unity pervading all of nature. It thus regards nature as part of God and God as wholly present within nature. Deism regards nature as a created order that is nevertheless closed to outside influences, because God left the natural world to run on its own after it was created.

Many people acquire their worldview by osmosis from their surrounding culture. They often have unexamined presuppositions about the nature of nature, of human beings, or of ultimate reality. An important question, therefore, is: Do we just have to accept some worldview or another as a presupposition through which we interpret reality? Or can we rationally evaluate different worldviews as competing metaphysical hypotheses and determine which, if any, is most likely to be true? And if so, can scientific evidence, perhaps even evidence about biological and cosmological origins, help us evaluate the likely truth of competing systems of thought (Fig. 11.5)?

FIGURE 11.5
The chapters to follow will assess which of the competing metaphysical hypotheses or worldviews best explain the three key discoveries about the origin of the universe and life: (1) the universe had a beginning (the big bang); (2) the universe has been fine-tuned for the possibility of life from the beginning; and (3) large bursts of biological information (stored in DNA and elsewhere) have arisen in the earth's biosphere since the beginning of the universe making new forms of life possible.

Expectation, Evidence, and Evaluation

In previous chapters, I have cited another prominent contemporary proponent of naturalism and his claim that the universe exhibits "precisely the properties we should expect if there is, at bottom, no design, no purpose . . . nothing but blind, pitiless indifference."[5] I have highlighted this claim not because Richard Dawkins is necessarily the most formidable proponent of scientific naturalism. In fact, I regard Sean Carroll as a more nuanced advocate. I also understand that many scientific naturalists themselves are uncomfortable with Dawkins's frontal attacks on religion, even if they share his conviction that nothing beyond nature—including God—exists. Rather I've focused on his claim because, whatever his reputation among some scientists as a popularizer and a provocateur, Dawkins has a talent for framing issues clearly. His statement raises the critical issue of what "we should expect to observe" in nature if either naturalism or theism were true.

Dawkins, of course, thinks that our observations of nature "are exactly what we would expect" if the worldview of naturalism were true. I appreciate his forthrightness. By invoking the language of *expectation*, Dawkins suggests a way to address the question of the nature of nature and the possible existence of God by making observations of nature itself. Indeed, his statement presupposes that such observations can help to evaluate competing metaphysical hypotheses just as much as competing scientific hypotheses.

His use of the language of *expectation* is also significant, because it dovetails with developments in the philosophy of science. Philosophers of science over the last century and a half have shown how our expectations about evidence enable us to evaluate hypotheses. They have shown how the logical implications of hypotheses allow us to evaluate them by making observations—whether in the laboratory, in the field, or in our ordinary experience—and then comparing those observations to the expectations that flow from the hypotheses in question.

Abduction and the Logic of Confirmation of Hypothesis

In Chapter 9, I briefly introduced a form of inference, or reasoning, known as abduction.[6] Historical scientists of every stripe use it to infer the possible causes of events in the remote past—events such as the origin of life—that they did not have the luxury of witnessing, still less under controlled laboratory conditions. Historical scientists use abduction to infer *past* conditions or causes from *present* clues or evidence. Philosophers of science and logicians have shown that not only historical scientists, but forensic scientists, detectives, theoretical physicists, astronomers, medical diagnosticians, philosophers of religion and science, and anyone making inferences about unobserved (or unobservable) causes will typically use this form of reasoning. Since questions about the origin of life and the universe and the possible existence of God involve questions of ultimate causal origins, investigations of these questions make ample use of abductive reasoning.

FIGURE 11.6
Charles Sanders Peirce, the American philosopher and logician who explicated abductive reasoning.

Abduction was identified as a separate type of logical reasoning by the nineteenth-century logician Charles Sanders Peirce (Fig. 11.6). Peirce described the modes of inference that we use to derive conclusions from facts or data. He noted that in addition to inductive and deductive arguments, we often employ abductive inferences, or what he called "the method of hypothesis."[7] To explain abduction, he contrasted it directly with deduction. To see the difference between these two types of inference, consider the following argument schemata:[8]

Deductive Schema:
Logic: If A is true, then C is a matter of course.
Data: A is given and plainly true.
Conclusion: Hence, C must be true as well.

Abductive Schema:
Logic: If A were true, then C would be as a matter of course.
Data: The surprising fact C is observed.
Conclusion: Hence, there is reason to suspect that A is true.

Notice that in the logic of the deduction, if the premises are true, the conclusion follows with certainty. The logic of abduction, however, does not produce certainty, but instead plausibility or possibility. Unlike deduction, in which the data or minor premise affirms the antecedent variable (A), abductive logic affirms the consequent variable (C). In deductive logic, affirming the consequent variable (with certainty) constitutes a fallacy, one that derives from the failure to acknowledge that more than one antecedent might explain or generate the same evidence.

To see why, consider the following argument:

If it rains, the streets will get wet.
The streets are wet.
Therefore, it rained.

or symbolically:

If R, then W.
W.
Therefore, R.

Obviously, this argument has a problem. It does not follow that because the streets are wet, it necessarily rained. The streets may have gotten wet in some other way. A fire hydrant may have burst or a snowbank may have melted. Nevertheless, that the streets are wet *might* indicate that it has rained. Thus, amending the argument as follows avoids the fallacy:

If it rains, then we would expect the streets to get wet.
The streets are wet.
Therefore, *perhaps* it rained.

or symbolically:

If R, then W.
W.
Perhaps R.

As the above shows, in an abductive argument, even if one may not affirm the conclusion with certainty, one may affirm it as a possibility. And this is precisely what abductive reasoning does. It provides a reason for considering that a hypothesis might be true. Indeed, it gives a reason for believing a hypothesis, even if one cannot affirm the hypothesis (or conclusion) with certainty.

Notice also the role *our expectations* play in this reasoning. The major premises in abductive inferences typically depend upon our expectations of what ought to follow from some previous state of affairs. Thus, Peirce would often articulate the major premise in an abductive inference by describing how, given some antecedent A, some consequent C would follow "*as a matter of course*." Nevertheless, he might just as well have said that, given some antecedent A, we *should expect* C to follow "as a matter of course."

Both the natural and historical sciences employ such logic routinely. In the natural sciences, if we have reason to expect that some state of affairs will ensue given the truth of some hypothesis, and we find that such a state of affairs has occurred, then we say that our hypothesis has been confirmed. This method of "confirmation of hypothesis" functions to provide evidential support for many scientific hypotheses, though, again, obviously not proof. Given Copernicus's heliocentric theory of the solar system, astronomers in the seventeenth century had reason to expect that the planet Venus should exhibit phases. Galileo's discovery that Venus does exhibit phases, therefore, supported the heliocentric view. The discovery did not prove the heliocentric theory, however, since other theories might—and in fact could—explain the same fact.[9]

Peirce acknowledged that abductive inferences on their own may constitute a rather weak form of evidential support. He noted, "As a general rule [abduction] is a weak kind of argument. It often inclines our judgment so slightly toward its conclusion that we cannot say that we believe the latter to be true; we only surmise that it may be so."[10]

Yet as a practical matter Peirce acknowledged that abduction often yields conclusions that are difficult to doubt even if they lack the air-tight certainty that accompanies the logic of deduction. For instance, Peirce argued that skepticism about the existence of Napoleon Bonaparte was unjustified even though Napoleon's existence could be known only by abduction. As Peirce put it, "Numberless documents refer to a conqueror called Napoleon Bonaparte. Though we have not seen the man, yet we cannot explain what we have seen, namely, all these documents and monuments, without supposing that he really existed."[11] Thus, Peirce suggested that by comparing the explanatory power of a hypothesis against other competing hypotheses, historians or scientists can often strengthen particular abductive inferences, rendering them—for all practical purposes—beyond reasonable doubt.

Strengthening Abductive Inferences: Assessing Comparative Explanatory Power

Since Peirce's time, philosophers of science have refined his understanding of how abductive inferences can provide "epistemic" support for hypotheses. Epistemology is the subdiscipline in philosophy concerned with the basis of knowledge and questions about "how we know what we know." Epistemic support refers to any evidence, axiom, or chain of reasoning that provides justification for a given proposition or belief.

The abductive framework often provides a weak form of evidential support, since it can leave open many possible explanations for the same evidence. This limitation typically forces scientists to evaluate the explanatory power, or predictive[12] success, of competing possible hypotheses. As Peirce noted in his discussion of the evidence for the existence of Napoleon, considerations of comparative explanatory power may establish an inference beyond reasonable doubt, even if the abductive logical form of the inference cannot categorically exclude all other possibilities.

As I noted in Chapter 9, this method of comparing the explanatory power of competing hypotheses is sometimes called the "method of multiple competing hypotheses"[13] or "inference to the best explanation."[14] It often reduces the uncertainty, or what philosophers of science call the "underdetermination of theory by data," associated with abductive reasoning. In this method of reasoning, the explanatory power of a potential hypothesis determines which among a competing set of possible explanations is the best. Scientists infer the hypothesis among a competing group that would, if true, provide the best explanation of some set of relevant data.

Consider a homespun illustration.[15] Suppose Ms. Jones falls asleep on the couch on a warm weekend afternoon while watching television. On awakening, she steps outside and sees that (1) the driveway of her house is glistening with water and (2) the car in the driveway is also wet. She decides to investigate.

From those two pieces of evidence, she might conclude using the above abductive syllogism (if it rains, the streets will get wet, etc.) that it rained while she was asleep. But based on the two facts at hand she might just as logically conclude that the automatic sprinklers came on or that someone washed her car. With only the data that the driveway and the car are wet, these explanations are equally plausible.

But suppose our groggy investigator also sees that (3) the lawn and the street are perfectly dry and (4) there isn't a cloud in the sky. Now what might she conclude? Although the sprinkler hypothesis and the rainstorm hypothesis are still possible, these explanations seem much less likely in the light of the additional evidence (facts 3 and 4).

Now, finally, suppose she looks a little harder and sees (5) a bucket with soapy water and a sponge sitting behind the car. With the final piece of data, the best explanation for observations 1 through 5 becomes obvious: someone washed the car.

This everyday scenario provides a good example of making an inference about a past cause based upon present observations using abductive reasoning enhanced with the method of inference to the best explanation. Ms. Jones was asleep when whatever happened in the driveway took place. Those events were unobserved history for her. What remained were clues, signs, or pointers: wet places, dry places, weather conditions, a bucket, a sponge, soapy water, and so on.

Our investigator weighed several possible explanations to see which would make the best sense of the clues. She evaluated the competing explanations using the available evidence and what she knew about cause-and-effect relationships in the world (e.g., people often wash cars in their own driveways, rain doesn't usually fall on single small areas). She then worked backward in time to what probably happened when she was not around to see it. The best explanation was the one that explained more of the evidence more simply than any other.

This example also shows that considerations of *causal adequacy* often determine which among a set of possible explanations will constitute the best. Our sleepy friend wanted to know what had *caused* the evidence she observed. She weighed the competing possible explanations by reference to her knowledge of cause and effect. Of course, it might have rained. But only above her driveway? With no clouds in the sky? And would a rainstorm produce a bucket of soapy water? Not likely. Therefore, the rain hypothesis did not seem causally adequate as an explanation of some of the relevant clues.

Indeed, notice that as our investigator used the method of "inference to the best explanation,"[16] she not only considered several possible hypotheses, but she also compared the known (or theoretically plausible) *causal powers* of the various postulated explanatory entities. She then progressively eliminated causally inadequate explanations.

This example also shows how, based on our background knowledge about how the world works, we typically avoid unnecessarily convoluted explanations—explanations that multiply causal postulations. For example, Ms. Jones didn't seriously consider that (a) it had rained just over her driveway *and* (b) her son had gotten out the bucket and put soapy water in it only to leave it there with no intent of washing the car. Though that was a possible explanation of the facts, it seemed unnecessarily complex and improbable.[17]

This method of inferring to the best explanation has several advantages over either deduction or simple abduction. Deductive inferences produce certainty, but only if the premises are known to be true. Yet major premises in deductive arguments typically affirm universal statements or generalizations about the world (such as "All swans are white") that depend upon some prior inductive inference that may itself be uncertain. Thus, the standard of deductive certainty may be hard to meet.

On the other hand, abductive inferences either provide weak epistemic support for a merely possible conclusion or—if their conclusions are affirmed with certainty—commit the fallacy of affirming the consequent. If Ms. Jones had jumped straight from "wet driveway and car" to "rain," she would have been guilty of affirming the consequent, referred to more colloquially as "jumping to conclusions." Unless abductive inferences are strengthened using a process of elimination showing various alternative hypotheses to be implausible, they will remain inconclusive.[18]

But by systematically evaluating the explanatory power of competing hypotheses and by eliminating those that lack causal adequacy or plausibility given our background knowledge, alternative hypotheses can often be eliminated, sometimes leaving only one plausible explanation. In such cases, the method of inference to the best explanation can help scientists arrive at a definitive, if not absolutely certain, conclusion.

Of course, in some situations several hypotheses may explain the available data equally well, even taking considerations of causal adequacy and simplicity into consideration. Ms. Jones, in our car-washing example, initially encountered a set of facts that could be explained equally well by several different hypotheses. Typically, in such situations investigators will look for more evidence in order to discriminate between the explanatory power of various hypotheses—as Ms. Jones did when she looked around to discover not only a wet driveway and car, but also a dry lawn, dry street, and a bucket of soapy water.[19]

Similarly, to cite an example from science discussed earlier, during the history of twentieth-century cosmology, three different explanations—the big bang, the steady-state, and the oscillating-universe models—initially accounted for the evidence of an expanding universe equally well. Nevertheless, as astronomers and physicists discovered more relevant evidence—such as the cosmic background radiation and the value of the mass density of the universe—they eventually eliminated the steady-state and oscillating models. Only the big bang remained as a causally adequate explanation of the main classes of relevant evidence.

The Causal Powers of Theoretical Entities

The different cosmological models proposed during the last century highlight another issue in assessing competing hypotheses, especially concerning events in the distant past. In Chapter 9, I noted that historical scientists typically assess the causal adequacy of a hypothesis by reference to our experienced-based knowledge of cause and effect. I used the example of how volcanoes have been observed to produce white ash. I did so to show how a past volcanic eruption might best explain some later still-visible white powdery ash.

Nevertheless, scientists must sometimes posit the existence of causes, conditions, or entities whose effects they have not or cannot directly observe. In such cases, their expectations about the observable consequences of a postulated entity will derive from more theoretical considerations about the postulated properties or causal powers of that entity.

As the philosopher of science Michael Scriven explained in his description of inference to the best explanation (or what he called "retrospective causal analysis"), when historical scientists lack "previous direct experience of [a cause's] actual efficacy" in producing an effect of the type in question, "there might be *theoretical* grounds for thinking it a possible cause."[20] Other historians and philosophers of science have explained that extrapolation from the known causal powers of a "relevantly similar"[21] cause might also play a role in justifying such a postulated cause.

Indeed, scientists sometimes invoke theoretical considerations to assess causal adequacy. The big bang theory postulated an initial infinitely hot dense concentration of all matter and energy at the beginning of the universe (Fig. 4.14). Clearly, no one had ever directly observed an infinitely hot dense concentration of matter producing some definite effect. Nevertheless, physicists understood something about the attributes of "blackbodies." Recall that blackbodies are idealized objects that would perfectly absorb electromagnetic radiation and reemit radiation in specific spectral "signatures." Having theoretical reasons for thinking that a near perfect blackbody would have existed soon after the beginning of the universe, scientists then deduced what they would expect to observe if such a hot dense state had existed.

By using their theoretically derived knowledge of the spectral emission signatures of blackbodies and calculating how the wavelengths of light would stretch out as space expanded after the universe cooled from its initial plasma state, physicists predicted that an infinitely hot dense concentration of matter at the beginning of the universe would eventually produce cosmic background radiation at a specific blackbody temperature with a specific spectral signature.

Since the steady-state model did not postulate such an initial dense concentration of mass-energy, steady-state proponents did not expect to observe a pervasive background radiation. Thus, the two differing postulations about the past and theoretical reasoning about near-perfect blackbodies (and their postulated properties) allowed scientists to generate two different sets of expectations about what ought to be observed in the universe today. Those different expectations allowed cosmologists, upon observing the emission signature of the cosmic background radiation, to decide which of the two cosmological models better explained the evidence of observational astronomy.

Darwin used a similar strategy to establish—at least initially—the causal adequacy of his mechanism of random variation and natural selection. By drawing an analogy between artificial and natural selection, he suggested that natural selection could produce morphological change in organisms just as artificial selection could. Darwin then invoked the theoretical consideration that natural selection would have had more time to operate. Next, he extrapolated from the observed causal powers of artificial selection operating over a relatively short time to justify the claim that natural selection operating over a much longer stretch could produce much greater morphological change. Though biologists cannot directly observe natural selection producing the amount of change Darwin postulated, his extrapolation provided a theoretical justification for concluding that natural selection could cause significant morphological innovation.

Recent discoveries and other new considerations—for example, the extreme rarity of functional genes and proteins—have since cast doubt on the merits of this particular extrapolation (see my discussion in Chapter 10). Yet this example illustrates how scientists often extrapolate from the powers of a known entity or process to establish the causal adequacy of a relevantly similar entity possessing greater causal powers. Historical scientists commonly posit a cause of the same type as, but of a different magnitude from, a known cause as a way of demonstrating the causal adequacy of an explanation. Philosophers and scientists have long accepted this as a valid method for establishing the adequacy of a postulated cause when direct observations of the cause-and-effect relationship under consideration are impossible.

The use of such extrapolation and theoretical reasoning can make the method of inference to the best explanation uniquely useful in evaluating the explanatory power of competing worldviews or metaphysical hypotheses. Indeed, though metaphysical hypotheses about the prime or ultimate reality often do not allow direct observation of the entities they postulate producing specific effects, such hypotheses typically do posit (albeit unobservable) entities or past states with specific properties that should, if real, give us reason *to expect* specific observable effects—as we shall see.

The Bayesian Turn

Philosophers of science often use a mathematical formalism known as the probability calculus to help them assess the plausibility of a hypothesis or to compare the plausibility of competing hypotheses. A formalism is simply a procedure expressed in mathematical or logical symbols. The probability calculus for assessing hypotheses is based upon a theorem established by the eighteenth-century English clergyman and mathematician Thomas Bayes. It provides a quantitative method of estimating the strength of a hypothesis or the relative probability of competing hypotheses given some body of evidence. The probability estimates used in this formalism are typically based on what we would, given our knowledge of how the world works, *expect to observe* in the world if one or another of the competing explanations were true. Thus, these probability estimates complement or enhance the abductive reasoning used in the method of inference to the best explanation.

Typically, in a Bayesian analysis,[22] philosophers use what are known as conditional probabilities. A conditional probability is the probability of one thing given or "conditioned on" the observation of another. Two different kinds of conditional probabilities figure in assessing the strength of a hypothesis or in comparing the strengths of competing hypotheses.

First, Bayesian probability analysis requires estimating the probability of some evidence E given a specific hypothesis H—written $P(E \mid H)$. This conditional probability (of E given H) estimates *how much we ought to expect* a given piece of evidence if a specific hypothesis were true. Thus, these conditional probabilities correspond to the major premise in an abductive syllogism—the premise that has the form "If H were true, then the surprising fact E would be a matter of course" (or "If H were true, then E *would be expected*"). Since conditional probabilities measuring degrees of expectation also measure the *probability* of a given piece of evidence occurring (given some hypothesis), Bayesians call these probabilities "likelihoods."

Bayesian analysis often involves comparing "likelihoods"—that is, assessing whether we ought to have *greater* expectation of observing some evidence given one hypothesis as opposed to another. A scientist who judges the probability of observing some piece of evidence to be greater given hypothesis A than hypothesis B can express that judgment symbolically by writing $P(E \mid A) > P(E \mid B)$.

We usually base such judgments about relative probabilities on our wealth of prior experience (what philosophers call "background knowledge") and especially on our prior experience of cause and effect. Indeed, judgments of likelihoods in Bayesian analysis often turn on assessments of causal adequacy, just as evaluations of comparative explanatory power turn on such assessments in the more qualitative approach to inference to the best explanation already presented.

But there is another conditional probability in Bayesian analysis. That probability is the one we ultimately want to know: the probability that a specific hypothesis is true. Called a "posterior probability," this expresses the probability of the truth of a hypothesis H after the fact of actually observing some evidence E—written "the probability of H given E," or P(H | E). Knowing this conditional probability helps us decide how confident we should be in a hypothesis given the presence of some evidence or how much our confidence should change upon observing some new or unexpected evidence. Posterior conditional probabilities also allow us to express the *relative* probability of one hypothesis compared to another. In the case that some evidence E gives hypothesis A a much higher probability than hypothesis B, Bayesians express that fact symbolically as P(A | E) >> P(B | E).

In Bayesian probability, formulas allow philosophers and scientists to calculate how likely a hypothesis is to be true (i.e., the probability of a *hypothesis given the evidence*) *if* they can estimate how much *we ought to expect the evidence* in question (i.e., the probability of the evidence) *given the hypothesis.* The formulas in probability theory that make calculating posterior probabilities possible also sometimes require making an initial assessment of the probability of the hypothesis based upon our background knowledge before any new evidence has come along, written P(H). These probabilities are called "prior probabilities."

Though it can add a quantitative dimension to hypothesis assessment, the logic of Bayesian analysis resembles that of the abductive reasoning employed in inference to the best explanation. If we observe some piece of evidence that *would be expected* given a particular hypothesis but would *not* be so strongly expected if that hypothesis were false, the observation of that evidence will confer support on the hypothesis.[23] Moreover, the greater the expectation of the evidence given that hypothesis (all other things being equal), *the more* the observation of that evidence will increase our confidence in the hypothesis in question. If we have much greater reason to expect a particular piece of evidence given one hypothesis (say, hypothesis A) as opposed to a competing hypothesis (say, hypothesis B), the observation of the expected evidence will make us relatively more confident in the superiority of hypothesis A over B—thus suggesting hypothesis A as the better explanation.[24]

The reasoning of physicists about competing cosmological models can be explicated in Bayesian terms. Physicists thought that we had much more reason to expect the presence of the cosmic microwave background radiation (CMBR) given the big bang than we did given the steady state. A statement to that effect, "The probability of observing evidence of the CMBR given the big bang is much, much greater than the probability of that evidence given the steady state," would express that judgment as a comparison of likelihoods, or symbolically as $P(E_{CMBR} \mid H_{bb}) >> P(E_{CMBR} \mid H_{ss})$. As a consequence of these differing expectations, the discovery of actual evidence of the CMBR led cosmologists to conclude that the big bang hypothesis was much *more likely* to be true than it was before—so much so that it essentially replaced the steady-state model. Bayesian logicians would express that claim symbolically as $P(H_{bb} \mid E_{CMBR}) >> P(H_{ss} \mid E_{CMBR})$.

Here is a more homespun illustration of Bayesian analysis. As I'm hiking through the forest, I come upon a rundown cabin in a clearing ahead.[25] Based upon its appearance (and my background knowledge), I assume tentatively that the cabin is abandoned. In Bayesian terms, I reflexively assign a prior probability of the house's being inhabited at a bit below 50 percent. To my surprise, though, upon entering the cabin, I find a fresh cup of tea steeping on the kitchen table. I assume that observing a steeping cup of tea (*the evidence*) is far more expected or probable given the hypothesis of an inhabited cabin than given the hypothesis of an abandoned one. Consequently, I begin to change my mind. I now conclude that the *hypothesis* of an inhabited cabin (H_i) is more probable than the hypothesis of an abandoned cabin (H_a). We could write this as $P(H_i \mid T_s) > P(H_a \mid T_s)$ where T_s represents tea steeping.

I then observe other evidence that would be expected in a house with people living in it—food in the refrigerator, dishes in the sink, and the sound of running water in the bathroom. I quickly realize that I'd better leave lest I get arrested for trespassing (or shot)! Clearly, the inhabited-cabin hypothesis is now much more probable than the alternative and provides a better explanation for the evidence.

Notice that in making this judgment, I do not actually need to know the exact probability of the evidence given the inhabited-cabin hypothesis, but only that the evidence observed is more strongly expected given that hypothesis than the alternative. This illustrates how Bayesian analysis can often yield decisive judgments about the *relative* strength of competing hypotheses even without assigning exact numeric probabilities. Notice too the role that causal-adequacy considerations played in my judgment. Since an abandoned cabin would have meant the *absence* of a personal agent who could have made the tea, purchased the food, cleared the dishes, or turned on the shower, it clearly lacked causal adequacy.

In Bayesian terminology, given the abandoned-cabin hypothesis, none of those observations would have been expected. Instead, we would have *more* reason to expect such observations—because of the known causal powers of people versus empty cabins—given the inhabited-cabin hypothesis. Indeed, however much it might have been difficult to quantify the exact probability of observing a steeping cup of tea or running shower on the inhabited-cabin hypothesis, we certainly have good reason for thinking that the probability of observing those things is *higher* on that hypothesis than on the alternative.

Notice, finally, that though a Bayesian analysis does a good job of describing how we assess hypotheses, we typically don't need to make explicit use of it. I did not need to make estimates of prior probability, likelihoods, posterior probabilities, Bayes's theorem, or the formulas derived from it to judge whether the cabin was more likely to be inhabited than not. Often, our background knowledge of cause and effect (or our theoretical understanding of the causal powers of a postulated entity) will enable us to come to sound assessments of the merits of competing hypotheses without making explicit use of Bayesian probability calculus—even if our reasoning can also be explicated in Bayesian terms.

Even so, the Bayesian probability calculus can often clarify our thinking and add a helpful quantitative dimension to our assessments. Consequently, the next three chapters will at times employ Bayesian concepts and analysis to enhance and complement the use of abductive reasoning and the method of inference to the best explanation.

Abduction, Explanatory Power, and Metaphysical Hypotheses

As a PhD student at Cambridge, I discovered that Charles Darwin and other historical scientists made abductive inferences and then attempted to strengthen them using the method of inference to the best explanation. My study of Charles Sanders Peirce made clear that scientists, philosophers, and people in ordinary life use this reasoning all the time, often in a way that Bayesian analysis can illuminate or complement. I became intrigued with the possibility that these forms of reasoning could be used to address not just day-to-day questions about rainy streets or possibly abandoned cabins or even significant scientific questions about the causes of different phenomena—but also really big questions about the nature of nature and the possible existence of a deity.

What would the use of abductive reasoning and inference to the best explanation tell us about the status of the God hypothesis if we used them to analyze recent discoveries in cosmology, physics, and biology? Is it possible that with respect to the evidence, theism might have greater explanatory power than competing metaphysical hypotheses? In 1999, writing in an interdisciplinary journal, I published a preliminary essay exploring this possibility.[26] Since then I've become increasingly convinced that this type of reasoning can help address deep worldview questions.

Ironically, I've found support for that conviction from leading defenders of scientific naturalism. Certainly, Richard Dawkins's assertion that the universe has "precisely the properties we should expect" if "blind, pitiless" materialism were true implies that observations of nature can provide support for a metaphysical hypothesis. His provocative claim assumes what many philosophers of science have argued, namely, that a metaphysical hypothesis, just as much as a scientific one, can be evaluated by evaluating whether the evidence we observe matches what we would logically expect if the hypothesis were true.

Dawkins's famous statement can, in fact, be fairly reformulated as an abductive inference. Consider:

Logic: If "blind, pitiless" matter and energy rather than a Mind is the prime reality from which all else originated, then we would expect no evidence of intelligent design in life and the universe, rather only evidence of apparent design.

Data: Life and the universe do not exhibit evidence of actual design, only apparent design.

Conclusion: Therefore, we have reason to believe that life and the universe are the product of blind materialistic forces rather than a preexisting Mind.

Dawkins's way of using his observations to attempt to confirm the hypothesis of naturalism or materialism and to disconfirm theism turns out to be helpful for another reason. He doesn't claim that he can absolutely prove God does *not* exist. Rather he says that, given our observations of the natural world—and given *what we would expect* to see if God did exist or had acted to bring nature into being—the God hypothesis seems "incredibly improbable."

He also claims, invoking Ockham's razor, that the God hypothesis is superfluous—that key evidence concerning biological and cosmological origins can be explained *better and more simply* without any reference to a transcendent creator or intelligent designer. Dawkins—like those nineteenth-century scientists who established scientific materialism as the dominant worldview—argues that he has "no need of that hypothesis."

By using a method of reasoning that seeks to confirm, rather than prove, a hypothesis, Dawkins recognizes that we can have good reason for believing something even if we can't establish the hypothesis with certainty. We can even have much *better* reason for believing one hypothesis than another, even if we can't absolutely prove the better one.

Dawkins and other contemporary proponents of scientific materialism, of course, claim that scientific evidence provides good reason for affirming that nature is all that exists and for denying evidence of a purposive or designing intelligence behind the universe. Indeed, Dawkins argues that we have much *better* reason for believing that God does not exist than we do for believing that God does exist—a belief he characterizes as a "delusion," because "no evidence for God's existence has yet appeared."[27]

What applies in support of his argument *against* the existence of God might, however, apply—in light of other evidence—to an argument *for* the existence of God. Indeed, if it's possible that one pattern of evidence might provide reason for affirming naturalism over theism, then it's also logically possible that a different pattern of evidence might give us *better* reason to affirm theism over naturalism. To say otherwise would treat naturalism as an untestable axiom or dogma rather than a genuine metaphysical hypothesis that could be true or false depending on the evidence—precisely what Sean Carroll, for instance, is loath to do.

The evidence presented in the last several chapters might have left readers expecting a formal proof of God's existence. Though I don't deny that some proofs for God's existence using standard deductive logic may have persuasive force, I'm not going to attempt such a proof in this book.

My doing so would assume an unnecessary and unrealistic burden of proof. Arguments based on empirical observations of nature, as opposed to mathematical axioms, rarely if ever provide proof in the sense of deductive certainty.

Instead, in the spirit of the inquiry that Dawkins has advanced, I will evaluate whether the discoveries in cosmology, physics, and biology discussed in the preceding chapters might provide a *good reason* for believing in God, or even a *better* reason for believing in God than in naturalism or materialism, for example. As I've already suggested, recent scientific discoveries concerning biological and cosmological origins might be "just what we should expect" if a transcendent and intelligent designer acted to produce life and the universe. Since these same observations of nature may not be what we would expect assuming scientific materialism (or other nontheistic worldviews), the God hypothesis could in principle provide the best metaphysical explanation of the relevant scientific evidence. In other words, even if we can't prove God's existence with absolute certainty, we may have better reasons for affirming a theistic view of the "nature of nature" and the "prime reality" than for affirming other metaphysical systems of thought.

The next three chapters explore this possibility in light of the evidence already presented.

12

The God Hypothesis and the Beginning of the Universe

Over the years, I've engaged in several public debates with a noted scientific naturalist and religious agnostic, Michael Shermer (Fig. 12.1). As editor in chief of *Skeptic* magazine and former columnist for *Scientific American*, Shermer has made a living debunking spoon benders, UFO sightings, parapsychology, astrology, and the like. He lumps belief in God and the theory of intelligent design in with these other, more dubious enterprises.

FIGURE 12.1
Michael Shermer, historian of science and editor in chief of *Skeptic* magazine.

One encounter between us proved memorable because of a conversation afterward. We had just debated at Westminster College in Fulton, Missouri, where Winston Churchill made his famed Iron Curtain speech in 1947. As Shermer often does, he began by telling his "deconversion" story. He explained how he lost the religious faith of his youth because of his growing scientific knowledge. Apparently for him, scientific discoveries seemed to provide good reasons for *not* believing in God—or at least *better* reasons for accepting a naturalistic or materialistic worldview than a theistic one.

Though Dr. Shermer and I have profoundly different worldviews and scientific perspectives, he has always presented his position with good humor and in a congenial manner. Consequently, we've developed a good rapport. After our debate, we shared a limousine ride back to the St. Louis airport. This gave us the chance to talk more candidly away from the audience.

I took the opportunity to ask him about the reasons for his loss of religious belief and how science contributed to it. He told me that the general success of science seemed to him to eliminate any need for belief in God. He cited various scientific discoveries, including the discovery of the big bang and the structure of DNA by Watson and Crick. I pointed out that theists such as myself also celebrate such successes. And, I contended, I didn't see how the structure of DNA in any way undermined faith in God, especially since chemical evolutionary theorists had failed to explain the *origin of the information* in DNA—a point he had acknowledged in our first debate. Shermer again agreed that no one knew how the first life might have evolved, but he characterized that as an isolated problem.

"But is it?" I asked. I pointed out that science as he defined it, as a strictly materialistic enterprise, also had no explanation for the origin of the universe or the fine tuning of its physical laws and constants.

He conceded the first point, but then reminded me that physicists had proposed a "multiverse" (the existence of many other universes) to explain the fine tuning.

"But do you really believe that?" I asked.

He smiled and simply said, "Nah."

I asked him about human consciousness. He acknowledged that too, from a materialistic point of view, was a deep mystery—as was the origin of the universe itself. No one knew what had *caused* the big bang, he readily admitted.

As we talked, it became apparent that the science Shermer admired had done a great job of explaining how the universe and life *operate*, but that it had not offered adequate *materialistic* explanations for the *origin* of life, mind, or the universe. But that suggested to me that materialism as a worldview lacked significant explanatory power.

Is it possible, instead, that a theistic worldview—a God hypothesis—might help explain what materialism or naturalism has not—or perhaps cannot? If so, would the explanatory power of such a God hypothesis provide a reason for favoring theism over other competing metaphysical hypotheses? At the time, I remember reflecting on this possibility. This chapter now begins to consider it.

Deductive Proofs, Theism, and Epistemic Support

Given the dominance of scientific materialism during the last century, it is not entirely surprising that contemporary science popularizers such as Dr. Shermer portray science and theistic belief as standing in conflict. What is surprising is that even many theologians and theistic philosophers deny that scientific evidence can *support* theistic belief. Theologians who deny this possibility typically do so because they assume that only deductive arguments can provide "epistemic support" for belief in God and because they know the history of failed attempts to prove God's existence using such arguments during the Middle Ages and the Enlightenment.

These theologians seem to assume that scientific evidence (A) can provide epistemic support for a theological proposition (B) only if that proposition (B) follows from scientific evidence (A) with deductive certainty. That is, they assume that any argument supporting theism must exhibit a logical form such as:

If A, then B.
<u>A.</u>
Therefore, B.

Many arguments for God's existence have been framed in precisely such a form.[1] Logicians call deductive arguments or syllogisms exhibiting the above logical form *modus ponens* arguments (from the Latin for "method of affirming"). If the premises of these (and other) deductive arguments are true and can be known to be true, then the conclusion follows with certainty. In such arguments, logicians say the premises "entail" the conclusion.

Identifying premises known to be true with certainty can be very difficult, however. Many deductive arguments for God's existence failed for exactly this reason. For example, in the seventeenth century Descartes used a famous argument in an attempt to prove the existence of God. He argued (first premise) that many people have a clear and distinct idea of a perfect being. He then asserted (second premise) that nothing short of the existence of such a being could cause that idea. From these two premises, he *deduced* that God must exist. Philosophers found this unconvincing, because they judged both premises in the argument (especially the second) to be far from certain.[2]

Nevertheless, deductive entailment from true premises does constitute a perfectly legitimate, if infrequently attained, form of epistemic support.[3] Even so, scientists rarely prove theories (or laws) with absolute certainty from empirical evidence. Consequently, deductive entailment involves a far stronger standard of epistemic support than empirical science can attain. And if the natural sciences can't attain that standard, then natural theology (based as it is upon observations of the natural world) can't either.

All this may help explain why many theologians today rightly deny that scientific evidence can "prove" God's existence with deductive certainty. And since many theologians and philosophers have assumed deductive entailment as the only possible form of epistemic support for theistic belief, they tend to deny that scientific evidence can provide epistemic support for theism at all.[4]

Consider Ernan McMullin (Fig. 12.2), a prominent philosopher of science and theologian at the University of Notre Dame until his death in 2011. McMullin explicitly denied that the big bang theory provides any evidential support for theistic belief, though, curiously, he admitted that if one assumed the Judeo-Christian doctrine of creation, one might expect to find evidence for a beginning to the universe. As he put it, "What one could say . . . is that if the universe began in time through the act of a Creator, from our vantage point it would look something like the big bang that cosmologists are talking about. What one cannot say is . . . that the big bang model 'supports' the Christian doctrine of Creation."[5]

But what if deductive proof is not the only way to provide epistemic support for such a proposition? If superior explanatory power, rather than deductive entailment, can confer support upon a hypothesis, is it possible that recent developments in cosmology, physics, and biology actually do support, even strongly support, the God hypothesis?

With that possibility in mind, this chapter considers the metaphysical implications of one of the three key discoveries of modern science discussed in the preceding chapters—specifically, (1) the discovery, or the evidential and theoretical indications, that the material universe *had a beginning*. The next two chapters consider the implications of discoveries about the *design* of the universe and life, including the discoveries of (2) *the fine tuning* of the universe for life and complex chemistry from the very beginning of the universe, and (3) the large discontinuous *increases in the functional information* of the biosphere since the beginning. This chapter and the next two will thus make a preliminary argument for

FIGURE 12.2
The late Notre Dame philosopher and theologian Ernan McMullin.

the God hypothesis based upon the discoveries and evidence described up until now.

In the next section of the book (Chapters 15–19), I strengthen this argument by entertaining objections to the *prima facie* case for the God hypothesis presented in this section, including objections based on alternative models of cosmology, physics, and biology as well as on new discoveries in these fields. For example, in Chapters 17–19 I address newer alternative cosmological models that either deny the universe had a beginning or attempt to explain it using something called quantum cosmology. For now, however, let's consider the worldview implications of the discovery, taken at face value, that the universe did have a beginning.

Theism, Confirmation of Hypothesis, and the Beginning of the Universe

In Chapter 5, I cited many astrophysicists who have perceived clear theistic implications in the evidence for a beginning. Nevertheless, Ernan McMullin, among other theologians, has denied that such evidence supports theistic belief. Curiously, though, in the very passage in which McMullin denies this, he acknowledges that the evidence for a beginning actually confirms theistic expectations about the universe having a beginning. As he notes, "If the universe began in time through the act of a Creator . . . it would look something like the big bang that cosmologists are talking about."[6]

To put it another way, if someone posits the existence of God—and, say, the Judeo-Christian doctrine of creation—as a metaphysical hypothesis, then he or she would expect evidence of a finite universe. Arno Penzias, the physicist who won the Nobel Prize for his discovery of the cosmic background radiation, has said as much. As he notes concerning the big bang: "The best data we have are exactly what I would have predicted, had I nothing to go on but the first five books of Moses, the Psalms, the Bible as a whole."[7] In making this connection, Penzias evidently had in mind the famous first words of the Bible, "In the beginning, God created . . ."[8] He might have been thinking about passages in the Old Testament that seem to affirm an expanding universe (in which they discuss God "stretching out the heavens") or other passages that refer to "the beginning of time."[9]

But then doesn't Penzias's statement, like McMullin's, suggest that the big bang theory provides a confirmation of the Judeo-Christian understanding of the origin of the universe (treated as a metaphysical hypothesis) and, with it, an affirmation of the existence of a divine creator?

The discussion in Chapter 11 of abductive inferences and confirmation of hypothesis shows how it could. Consider how Judeo-Christian theism might express its expectations about the origin of the universe and how the evidence for a finite universe might supply a confirmation of such a theistic hypothesis:

Major Premise: If a Judeo-Christian view of the origin of the universe and its affirmation of a divine creator are true, then we have reason to expect the universe had a beginning.

Minor Premise: We have otherwise surprising evidence that the universe had a beginning.

Conclusion: We have reason to think that the Judeo-Christian view of the origin of the universe and its affirmation of a divine creator may be true.

This syllogism suggests that the big bang theory confirms the metaphysical hypothesis of Judeo-Christian theism, providing epistemic support, though not deductive proof, in much the same way that empirical observations confirm scientific theories.

Theism, Naturalism, and a Finite Universe

The Bible aside, there are also philosophical reasons that theists might expect the universe to have a beginning. Theism holds that God is a personal agent with causal powers and free will—that is, it holds that God has the ability, uncompelled by other factors or conditions, to actualize potential states of affairs from among many possibilities. Other agents (human beings) with such powers are known to cause new things to come into existence that did not exist before. Inventors, novelists, musical composers, and others exercise similar powers routinely. Thus, by extrapolating from our experience of other "relevantly similar" causal actors, we can reasonably expect that if a being such as God existed, that being could be expected to have caused new things to come into existence, including the universe itself.

Theism also holds that God is the creator of *all* things. Or in philosophical language, theists conceive of God as the prime reality, the ontological basis of reality, from which everything else comes, including matter, energy, space, and time. Given this conception, theists might reasonably expect to find evidence that the universe itself, including time and space, began to exist.

Further, scientifically informed theists know that modern physics teaches that matter and energy exist in space and time. They also know that (1) matter and energy are linked (by Einstein's equation $E = mc^2$); (2) matter and space are linked (as John Archibald Wheeler put it, "Space tells matter how to move, and matter tells space how to curve"[10]); and (3) space and time are linked, as the concept of spacetime in general relativity implies. Consequently, these theists think of time and space as much as matter and energy as created entities. But that means that any act of creation that brings matter and energy into existence would also bring space and time into existence. Thus, both the universe and the time in it should have a beginning.[11] For these and other reasons,[12] theists who conceive of God as a personal agent might reasonably expect to find evidence of a temporally finite universe—evidence that the universe had a beginning. This expectation can be expressed as part of a syllogism to show how the discovery of a finite universe can underwrite an abductive inference in support of the God hypothesis:

Major Premise: If theism is true, then we would have reason to expect evidence showing the universe had a beginning.
Minor Premise: We have otherwise surprising evidence that the universe had a beginning.
Conclusion: We have reason to think that theism may be true.

Thus, the evidence that the universe had a beginning provides abductive confirmation of the God hypothesis, at least, again, where God is conceived as a creator in the sense just described. Yet the evidence for the beginning of the universe can provide an even stronger form of epistemic support. Specifically, that evidence provides support for theism as an inference to the best (or at least a better) explanation among the other competing metaphysical hypotheses.[13]

To see why, let's first compare the explanatory power of theism to that of basic scientific materialism or naturalism. By *basic*, rather than *exotic*, I mean the version of naturalism that does not posit other unobservable universes or mathematical realities or laws of physics beyond our known universe. (I'll address such versions of naturalism in later chapters of the book.)

As leading naturalists such as Sean Carroll have explained, naturalism typically denies the existence of anything beyond nature itself, where nature consists of matter and energy within space and time. Scientific naturalists think of elementary particles (the smallest material components of the universe) and quantum fields as the entities from which everything else came. Since in their view nothing else besides the natural world exists, naturalists usually regard the universe as an eternal, self-organizing, and self-existent system rather than one created a finite time ago by some external agency. Consequently, it has long seemed to follow from naturalism that the material universe must have existed for an infinitely long time. For naturalists, an infinitely old universe would, as the physicist Robert Dicke put it, "relieve us of the necessity of understanding the origin of matter at any finite time in the past."[14]

Thus, from a naturalistic point of view, the evidence supporting a finite universe has seemed quite unexpected, even "surprising" in Charles Sanders Peirce's sense of the term in his description of abductive reasoning. Many leading scientific naturalists have themselves acknowledged this "unexpectedness." Some have done so explicitly. Others have done so implicitly, often by seeking to eliminate the dissonance between their worldview and a finite universe by proposing alternatives to a beginning.

Einstein, when he was still a strict philosophical materialist, tacitly acknowledged this dissonance when he chose the value of his cosmological constant to depict the universe as static and temporally infinite. Fred Hoyle admitted the challenge when, for explicitly philosophical reasons, he proposed his steady-state theory to retain the concept of an infinite universe—despite its flagrant violation of the law of conservation of energy. Arguably, proponents of an oscillating-universe and eternally chaotic inflation models have tacitly acknowledged the "unexpectedness" of a finite universe, given a naturalistic worldview, by the elaborate, and arguably contrived, cosmological models they have formulated to avoid the conclusion of an ultimate beginning. Sir Arthur Eddington acknowledged the dissonance when he confessed that he found the big bang theory and the idea of a cosmic beginning philosophically "repugnant."[15]

Causal-Adequacy Considerations

So why does a cosmic beginning seem unexpected from a naturalistic point of view? It does so primarily because naturalism or materialism can offer no ready causal explanation for such a beginning. If the "cosmos is all that is," per Carl Sagan, then nothing else exists beyond or separate from it that could act as its cause. Naturalists before the discovery of the beginning of the universe felt confident in positing an eternally and necessarily existing universe that did not require a causal explanation. But several classes of observational astronomical evidence (see Chapters 4 and 5), the Borde-Guth-Vilenkin (BGV) theorem, and even the Hawking-Penrose-Ellis singularity theorems (see Chapter 6) all indicate that the physical universe had a beginning.

Consequently, the origin of the universe would seem to require—by the principle of causality or sufficient reason—a cause. But since, according to naturalism, nothing exists except the natural world (i.e., the universe of matter, energy, space, and time), then nothing else could have functioned as the cause of its coming into existence. The beginning of the universe thus raises a question that naturalists, almost by definition, cannot answer, namely, "What *caused* the whole of nature or the physical universe itself to come into existence?" For this reason, naturalism, in its basic form at least, does not qualify as a *causally adequate* explanation for the presumed fact, variously attested, of the beginning of the universe.

The Hawking-Penrose-Ellis singularity theorems amplify this conclusion. If sometime in the finite past, either the curvature of space reached an infinite and/or the radius and spatial volume of the universe collapsed to zero units, then at that point there would be no space and no place for matter and energy to reside. Consequently, the possibility of a materialistic explanation would also evaporate, since at that point neither material particles nor energy fields would exist. Indeed, since matter and energy cannot exist until space (and probably time) begins to exist, a materialistic explanation involving either material particles or energy fields—before space and time existed—makes no sense. As I used to tell my students, "If you extrapolate back all the way to a singularity, you eventually reach a point where there is no matter left to do the causing."

As we saw in Chapter 6, physicists now question whether the Hawking-Penrose-Ellis result can be extended all the way back to the very beginning. Many have instead adopted eternal chaotic inflationary models of the origin of the universe. But, as we saw, the Borde-Guth-Vilenkin theorem applies just as much to these inflationary models as it does to the standard big bang model. Consequently, this theorem leaves "no escape," as Vilenkin has put it, from the conclusion that the universe ultimately did have a beginning.

Moreover, as we saw in Chapter 6, the Borde-Guth-Vilenkin theorem implies that time had a beginning. And since time and space are linked (not only in general relativity, but in newer theories of quantum gravity), affirming a beginning to time would seem to imply a beginning to space as well, even if space began with a finite (nonzero) volume. In any case, if the universe began to exist, as BGV affirms, then that would imply that whatever properties we associate with the universe—space, matter, and energy as well as time—also began to exist. Indeed, it makes no sense metaphysically to say that the universe of space and time, matter and energy began to exist a finite time ago and also to affirm that "before that" (i.e., in time) time or space or matter or energy already existed as well. Since space, matter, and energy are fundamental features of the universe, the proposition "The universe began to exist a finite time ago" implies that those features of the universe came into existence as well.

For this reason, the BGV theorem reinforces the same metaphysical implications as the Hawking-Penrose-Ellis singularity theorems, whatever physicists think of the applicability of the singularity theorems to the universe in its very earliest and smallest phase. Indeed, the BGV theorem implies the causal *in*adequacy of all materialistic explanations for the origin of the universe, since, again, before the ultimate beginning of the material universe, neither matter nor energy would have yet existed.[16]

A Bayesian Take

Bayesian analysis can clarify and amplify this conclusion. In the first place, since basic naturalism denies the existence of any entity beyond the universe that could act to cause the origin of the universe, proponents of basic naturalism expect an eternal self-existent universe and would not expect evidence of the origin of the universe a finite time ago. Yet since theism does affirm the existence of a transcendent entity beyond the space, time, matter, and energy of our universe, theists might well expect to find evidence of the universe having a temporally finite beginning. Thus, using Bayesian analysis, we can affirm that the likelihood of evidence of a temporal beginning is greater given theism than given basic naturalism. Or stated symbolically, $P(E \mid T) \gg P(E \mid N_b)$.

Many philosophers think that there is no compelling *a priori* reason to regard either a naturalistic or a theistic worldview as much more probable than the other.[17] If so, we can also employ Bayesian reasoning to affirm that the probability of the theistic hypothesis given the evidence of a cosmic beginning is greater than the probability of basic naturalism given that same evidence. Or stated symbolically, $P(T \mid E) \gg P(N_b \mid E)$. Thus, theism provides a better explanation of the temporal beginning of the universe than does naturalism.

Bayesian analysis can also help resolve a possible objection to the argument presented here. Up to this point, I've argued that theists have various reasons to expect that the universe *might have* had a definite beginning in time. Nevertheless, given that theists conceive of God as a personal agent with free will—indeed, one with a unique (and debatable) relationship to time—some could reasonably object to this claim. For example, some could argue that since God is an agent with free will, we have no way of knowing whether God would have created time a finite time ago or whether God might have chosen to create by maintaining all moments of time into existence from eternity past (as proponents of the cosmological argument from contingency presuppose as a possibility). Consequently, some might argue that it is impossible, given a theistic conception of God, to establish that God *would necessarily* have created a temporally finite universe. Therefore, it could be argued that theists do not have definitive grounds for *expecting* a finite universe.

Of course, theists do not have *absolute* grounds for expecting a finite universe. Yet it does not follow that they do not have *greater* grounds for expecting such a universe than philosophical naturalists do. Indeed, even though theists cannot establish that God would necessarily have created a temporally finite universe, theism does offer reasons (as explained above) for suspecting that God might well have done so. Indeed, since theism posits the existence of a being with relevant causal powers beyond space and time, the God of theism could have acted to bring time and space into existence, thus leaving evidence of a universe with a beginning. On the other hand, given basic naturalism, no entity outside space and time exists that could have acted to bring the universe, space, and time into existence a finite time ago. Consequently, the tenets of basic naturalism do not lead us to expect evidence of a temporally finite

FIGURE 12.3
This chapter evaluates which of the competing metaphysical hypotheses or worldviews (theism, deism, pantheism, or materialism) best explains the evidence suggesting the universe had a beginning.

beginning of the universe, whereas theists might—or *might well*—expect such evidence. Thus, theism offers a greater expectation of such evidence than naturalism. Using Bayesian terms, the likelihood of evidence for a cosmic beginning is greater given theism than basic naturalism, or $P(E \mid T) \gg P(E \mid N_b)$.

The assessment of Bayesian likelihoods, based upon causal-adequacy considerations, helps us to characterize the relative strength of our expectations of observing specific evidence given different hypotheses. That in turn allows us to make judgments about which of the competing hypotheses is more likely to be true, even if we cannot put exact numbers on the probabilities (Fig. 12.3).

Recall the analogous, if homespun, example from the previous chapter. Though I, while hiking through the forest, could not say in advance how likely it would be that a person would leave a steeping cup of tea on the table, I recognized immediately that the cup of tea was better explained by the hypothesis of an inhabited cabin than by the reverse.

Why? Because I knew that a person could have made a cup of tea, whereas nothing in an uninhabited house could have done so. Consequently, had I been inclined to use Bayesian analysis, I would have also understood that the *probability* of observing a steeping cup of tea was much higher if the house was inhabited than the reverse—even, again, if I could not say definitely in advance that a person would necessarily leave such evidence and even if I could not have quantified precisely the probability of finding such evidence in that case.

In a similar way, even though theists cannot say in advance that God definitely would have to create a universe exhibiting evidence of a temporal beginning, we recognize that the evidence of such a beginning is more likely given theism than naturalism. If there is, further, no overriding reason *a priori* to prefer naturalism over theism (see n. 17), it follows, given Bayesian probabilistic equations, that the probability of the hypothesis of theism is much greater than the probability of naturalism given the evidence of a beginning of the universe. Thus, theism provides a better explanation of that evidence than does naturalism.

An Uncaused Universe?

Leading proponents of philosophical naturalism effectively, if tacitly, acknowledge this obvious conclusion and try to work around it. For example, Sean Carroll, one of the most prominent proponents of naturalism, has acknowledged that naturalism has not explained the origin of the universe, precisely because it can offer no cause capable of producing it. He suggests, however, that the origin of the universe does not necessarily require a causal explanation; it might "just be." Nevertheless, because the evidence indicates that the universe has not existed infinitely, but instead *began to exist*, it would seem to require—by the principles of causality and sufficient reason—a cause. Saying otherwise undermines one of the basic presuppositions of scientific investigation and indeed of reason itself, namely, that "whatever begins to exist must have a cause."

Philosopher William Lane Craig (Fig. 12.4) brought this point home forcefully in an interview I conducted with him several years ago. He pointed out that saying the universe might have popped into existence uncaused for no reason at all is no different from saying that a freight train or a Bengal tiger might have done so. Though, he pointed out, there was no way to disprove such possibilities, reason and the scientific investigation of the world depend upon the opposite assumption—that all such material events do have causes. He said those who deny this principle and who yet also decry the reasonableness of theism "should, therefore, forever be silenced."[18]

FIGURE 12.4
William Lane Craig, the philosopher of science and proponent of the Kalām cosmological argument.

The evidence supporting the big bang theory, the Hawking-Penrose-Ellis solutions to the field equations of general relativity, and the Borde-Guth-Vilenkin theorem all point to a beginning to the universe.[19] It follows that any entity capable of explaining the origin of the universe, to which these indicators attest, must transcend the space and time, matter and energy of the universe. Naturalism fails to explain the origin of the universe because it denies the existence of any entity external to nature, but theism postulates the existence of precisely such a transcendent entity as a cause. Thus, insofar as God, as conceived by theists, transcends space and time, matter and energy and insofar as the causal explanation of the universe itself requires the existence of some entity separate from the universe to "do the causing," the God hypothesis provides a better, more causally adequate explanation than naturalism for the evidence of a beginning to the universe.

Personal Agency, Libertarian Freedom, and the Beginning

The God hypothesis provides a more causally adequate explanation for another reason. Theists conceive of God as a personal agent possessing free will. For this reason, their conception of God as the first cause helps resolve an otherwise knotty conceptual problem for materialists or naturalists.

Consider, again, that naturalism in its basic form denies the existence of any entity beyond nature. Some naturalists posit a prior material state or event before the big bang, as they do with eternal chaotic inflation or as they might do with the idea of an eternal "primeval atom" (see below). If naturalists take this tack, then that state or event must have possessed the necessary and sufficient conditions for the production of our universe. But then that state or event in turn must have been produced by some earlier state or event possessing the necessary and sufficient conditions for producing it, and so on. Consequently, at some point in the past either *an uncaused* first material state or event must have occurred, *or* the necessary and sufficient conditions for the origin of the universe must have always existed. In other words, either naturalists must posit an uncaused first event or they must posit that, as the philosopher J. P. Moreland notes, "the necessary and sufficient conditions for the first event existed from all eternity" in a timeless, changeless state.[20]

Both options pose difficulties. An uncaused first event violates the principles of causality and sufficient reason with all the destructive consequences for rationality discussed above. If, however, naturalists posit that the necessary and sufficient conditions of the origin of the universe existed from all eternity, then we would expect to observe evidence of

FIGURE 12.5
J. P. Moreland, philosopher of the mind and proponent of mind-body dualism.

an infinitely old universe. But we do not. Indeed, as soon as the necessary and sufficient conditions for the production of a given event occur, that event will occur. If the necessary and sufficient conditions for the production of the universe always existed back into the infinite past, then the universe itself should have come into existence an infinitely long time ago (when those conditions "first" occurred) and we should have evidence of that. But, again, we do not. Instead, we have evidence of a finite, not an infinite, universe.

As J. P. Moreland (Fig. 12.5) and William Lane Craig have shown, positing the action of a personal agent with free will resolves this dilemma. The concept of free will, also called libertarian agency, entails the idea that an agent with such freedom of will can initiate a new chain of cause and effect without being compelled by any prior material conditions. Since minds with free agency can initiate new chains of cause and effect without being compelled, the action of a free agent eliminates the need for an infinite regress of prior material states—and thus an infinite universe at odds with empirical observations.

Free agency also eliminates the need to posit an *uncaused material* first cause, which would violate the principles of causality and sufficient reason. It does so because having free will—familiar to us all because of our own introspective awareness of the powers of our own minds[21]—means that our decisions or acts of mind can alter material states of affairs without being wholly determined by a prior set of necessary and *sufficient* material conditions.

Moreover, it is at least reasonable to consider positing the action of a free agent as the explanation for the beginning of the universe. Most people already accept the reality of their own free will and think that their choices can cause new material states of affairs to occur. Those who don't accept this possibility typically deny the existence of their free will only because philosophical arguments have convinced them that their perception of free will is an illusion. But that implies that people at least have an intuitive understanding of the concept of free will. Thus, the concept of a freely chosen decision does not represent an exotic, *ad hoc*, or arbitrary explanatory postulation, but rather one that we routinely employ to explain other changes of state or states of affairs.

Indeed, positing the action of a free agent gives a perfectly cogent account of how the universe could have begun to exist consistent with our own experience of possessing free will. After all, free agents cause things to exist that did not exist before. At the same time, positing a prior material state to explain the beginning of the material universe generates an explanatory conundrum for naturalism. Therefore, positing the choice of a free agent—a mind—provides a better explanation for the beginning of the universe than naturalism.[22]

There is another aspect to this. Those who acknowledge the free will of personal agents hold that the *freely chosen decisions* of personal agents represent an exception to the principle of causality—that is, the rule that "everything that begins to exist must have a cause." Nevertheless, allowing such an exception underwrites, rather than undermines, human rationality. The idea of reason itself *requires* that human thought is not wholly determined by impersonal material forces (e.g., by external physical stimuli or chemical reactions in the brain). If our thoughts were wholly determined by impersonal material forces or chemical reactions, we would have no reason to trust the reliability of our thoughts, since such forces and chemical reactions have no obvious relationship to the object of our thinking. As the British biologist and philosopher J. B. S. Haldane once said, "If mental processes are determined wholly by the motions of atoms in my brain, I [would] have no reason to suppose my beliefs are true . . . and hence no reason for supposing my brain to be made of atoms."[23]

Thus, theism posits the one kind of entity—a free personal agent—that can initiate new sequences of cause and effect without itself being caused to do so and without, at the same time, undermining confidence in either human rationality or the intelligibility of the physical world. In so doing, it resolves the explanatory puzzle that confronts naturalism as the result of the evidence that supports the universe having a beginning in time.[24]

An Eternally Existing Primeval Atom?

This line of reasoning also applies to those who might posit an eternally existing "primeval atom." Recall that as astronomers extrapolate back in time, they ultimately reach a point in their mind's eye in which all the matter and energy of the universe would have been compressed into a nearly infinitely hot, dense point. Knowing that the Hawking-Penrose-Ellis singularity theorems do not allow extrapolating back to an absolute beginning, some naturalists could claim that this nearly infinitely dense concentration of matter might have existed eternally as a primeval atom, "waiting" for just the right moment to begin to disperse as space began to expand. Yet this hypothesis raises a question about what might have changed to cause the sudden beginning of that expansion.

In so doing, it generates another version of the same dilemma just discussed. Perhaps the necessary and sufficient conditions for the expansion of the universe always existed, in which case we should have evidence that the expansion of the universe has been going on for an infinitely long time. But our best evidence indicates that the expansion of the universe began a finite time ago.

Alternatively, perhaps the necessary and sufficient conditions for the beginning of the expansion of space first occurred a finite time ago, but only after the primeval atom had existed for an infinitely long time before that. In that case, naturalists would need to explain what caused that change of state (i.e., the expansion) by positing some material change in the primeval atom or the space in which it resided. They would also need to explain why that change had not already happened an infinitely long time ago.

Yet it's difficult to see how any material change could have caused the sudden expansion of the universe from an eternally existing primeval atom. Presumably all the matter and energy of the universe was already present in such an atom and had been for an infinitely long time before the expansion began. If so, what possibly could have been added from a materialistic or physical point of view to a primeval atom if that "atom" contained all the matter and energy of the universe? Where could this additional something—to cause the sudden expansion of space—have come from? What could have induced a sudden change in the state or configuration of the matter and energy of the primeval atom, since again, by definition, nothing other than the primeval atom existed before the universe began to expand?

And even if some other material entity was available to cause such a change, why had it not done so an infinite time ago? An infinitely existing primeval atom would have afforded an infinite number of opportunities for such a change to occur, and any one of these opportunities could have occurred an infinitely long time ago. Why then would a sudden change occur only a finite time ago if there had been an infinite number of opportunities for such a change of state to occur over an infinite time?

As physicists Anthony Aguirre and John Kehayias have noted, "It is very difficult to devise a system—especially a quantum one—that does nothing 'forever,' then evolves. A truly stationary or periodic quantum state, which would last forever, would never evolve, whereas one with any instability will not endure for an indefinite time."[25] Invoking personal agent causation again resolves these puzzles. A volitional act can account for a sudden change of state without having to postulate the addition of new matter or energy where none was available and without having to invoke an uncaused material change. Positing a personal creator can also explain why an abrupt change of state occurred without leaving unanswered the question of why the expansion hadn't already occurred infinitely long before.

In any case, since theism posits the existence of an entity separate from the universe capable of causing it to begin to exist (and expand), and since naturalism denies the existence of such a transcendent causal entity, theists have greater reason to expect evidence of a beginning than do naturalists, even if theism does not *necessarily* entail a finite universe. Thus, the probability of a finite universe assuming theism is greater than the probability of a finite universe assuming naturalism. It follows that the evidence of a finite universe confers greater evidential support on theism than it does on basic naturalism and that theism provides a better explanation of the evidence for a finite universe than does naturalism.

Theism, Pantheism, and the Origin of the Universe

But what about pantheism, the worldview implicit in many Eastern religions and in the writing of Western philosophers such as Spinoza?[26] Can it explain the origin of the material universe a finite time ago?

Many diverse religious traditions, especially forms of Hinduism, embrace different facets of pantheism, making the characterization of Eastern philosophy or Eastern religions as a unified perspective impossible. Yet pantheism as a philosophy or worldview does represent a coherent system of thought that permits a general characterization. In general, pantheism affirms the existence of an *impersonal* god (*brahman*) as the ultimate reality and basis for a mystical unity pervading all of reality and all of the physical world (*prakriti*).[27] Pantheism also treats god and nature as well as god and the sentient self (*atman*) as ultimately part of the same oneness or unity. Pantheists equate god (*brahman*) with the whole of nature (*prakriti*) as well as with the spirit within each person (*atman*), each living thing, and even each inanimate object. Thus, Eastern pantheists assert "*atman* is *brahman*," meaning "the soul of the self" or "the soul of the world," *is* "the soul of the One," where the unified but impersonal oneness of all things represents the ground of all being—all that is ultimately real.[28]

Though a pantheistic worldview affirms the existence of a god, it fails to explain the origin of the universe for much the same reason that naturalism does. The god of pantheism exists within, and is coextensive with, the physical universe. Thus, god as conceived by pantheists cannot act to bring the physical universe into being from nothing physical, since such a god does not exist independently of the physical universe. If at some finite point in the past the physical universe did not exist, then a pantheistic god would not have existed either. If the pantheistic god did not exist before the universe began, it could not cause the universe to begin to exist. Thus, pantheism does not meet the test of causal adequacy.

Moreover, since the god of pantheism is not a personal agent, let alone one possessing libertarian freedom, a pantheistic notion of god does not help to resolve the explanatory dilemma posed by the abrupt change of state at the beginning of the universe.

For both reasons just stated, we have more reason to expect a finite universe given theism than we do given pantheism. Or put differently, the probability of a finite universe assuming theism is greater than the probability of a finite universe assuming pantheism. It follows that the evidence of a finite universe confers greater evidential support on the hypothesis of theism than it does on that of pantheism.

On the other hand, since deism does posit a transcendent, personal, and free agent, deism, like theism, does provide a causally adequate explanation for the origin of the universe from a discrete beginning in time. Theism and deism, thus, seem rationally preferable to pantheism or naturalism (Fig. 12.6, see page 258).

God and Ultimate Origins

Since the discovery of the expanding universe and the formulation of the big bang theory, cosmologists, astronomers, and physicists have sensed that an ultimate beginning had profound theological implications. The

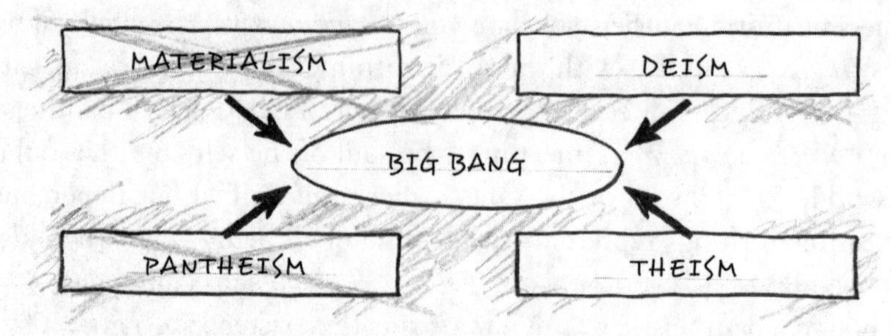

FIGURE 12.6
This chapter argues that theism and deism provide better, more causally adequate explanations for the origin of the universe—and the evidence that the universe had a beginning—than either materialism or pantheism.

titles of some popular books by noted physicists—for example, *God and the Astronomers*, *Cosmos and Creator*, and *The Science of God*—reflect this. As alternative cosmological theories (such as the steady-state or the oscillating-universe) have successively failed to account for relevant astronomical observations, scientists and philosophers have increasingly faced those implications. Yet many theologians, made defensive by the long intellectual dominance of scientific materialism and the history of failed proofs for the existence of God, have been reluctant to consider the possibility that scientific evidence could have implications supportive of theistic belief. To avoid an untenable and excessive rationalism, many have adopted fideism—a religious epistemology that affirms the sufficiency of faith without either reason or evidence—or "faith in faith alone."

But with developments in the philosophy of science showing how evidence can provide strong support for a hypothesis without having to prove it absolutely, and with mounting evidence indicating the universe had a beginning, this defensive intellectual stance now seems unnecessary. The discovery that the universe had a beginning not only allows us to harmonize scientific and theological beliefs about ultimate origins; it provides strong *epistemic* support for theism. The scientific evidence and theoretical developments pointing to a beginning of the universe have helped to revive the God hypothesis.

13

The God Hypothesis and the Design of the Universe

In the United States, the currently reigning understanding of the separation of church and state has come to mean that the Constitution forbids any discussion of God in the public square. Some contemporary advocates of strict separation go further. They react vehemently against any public discussion of ideas that might express a theistic perspective or, even more modestly, that might have implications supporting one. For this reason, our academic and media culture regards the possible theistic implications of the theory of intelligent design as a reason to reject it outright—and even to censor and stigmatize its proponents.

This reality came home to me with particular irony while I was participating in an interview and roundtable discussion on an MSNBC program several years ago. The host, Dan Abrams, was interviewing me about a case in a federal court in Pennsylvania involving an ill-considered local attempt to introduce public-school students to intelligent design. In the on-air discussion both the host and another participant in the interview, Eugenie Scott, of the National Center for Science Education, attempted to discredit the theory of intelligent design by insisting it necessarily implied the existence of God. In their view, if the theory could be shown to support belief in God, it would qualify as religion, thereby negating any scientific basis for it.

In my discussion with Abrams I sought, in the face of multiple interruptions, to explain the difference between (1) the scientific and evidential basis of intelligent design as a theory that makes a limited claim about a creative intelligence of some kind best explaining the origin of biological information and (2) the possible theistic implications of the theory. I acknowledged that I personally thought that the designing intelligence responsible for life was God, but that the evidence from biology alone could not definitively establish that and, in any case, other scientists had proposed other possible designers as candidates. Though my reluctance as a theist to claim any more reflected a concern to avoid overstating my case, my opponents characterized it as a disingenuous attempt to smuggle religion into the public schools, something Dr. Scott in particular had made a career of seeking vigilantly to prevent.

In the discussion that followed, Scott seemed to assume, as Abrams did, that if she could show that intelligent design had theistic implications, then she would have proved it to be illegitimate. After accusing me of dishonesty, she argued that the only kind of designing intelligence that could account for the evidence I had presented was one with godlike attributes. As she put it, "Now, to say, 'Well, we're not claiming who the designer is,' is just a sham. Either the designer is God or somebody with the same skill set."[1]

Actually, Scott was onto something, even if her phrasing lacked nuance. As it happens, I do think explaining the full range of scientific evidence presented in this book—from astronomy and cosmology to physics *and* biology—points to a transcendent designer with the attributes—"the skill set"—that theists ascribe to God. I was pleased to see that Scott intuitively understood that evidence of intelligent design might well lead one logically to identify God as the designer, even if her understanding of the separation of church and state would not allow her to consider such a possibility. Even so, I don't think the evidence from biology alone can establish that the intelligence responsible for life necessarily has all the attributes that theists ascribe to God.

As I explained in Chapters 9 and 10, the information necessary to produce the first living organism as well as other fundamentally new forms of life arose long after the beginning of the universe. Consequently, if intelligent design best explains the origin of biological information, then either a transcendent or a preexisting immanent intelligence (one within the cosmos) could, at least in principle, explain that evidence of design. So the evidence of design in life, taken by itself, does not *necessarily* point to a transcendent intelligence (or God).

But what if we consider the whole ensemble of evidence discussed in this book? Clearly, the activity of an immanent designer within the cosmos (a "space alien," for example) cannot explain the origin of the cosmos itself. But could a designing intelligence of the kind posited by advocates of "panspermia," an idea I described in Chapter 9, account for the fine tuning of the laws and constants of physics or the initial arrangement of matter and energy at the beginning of the universe?

If not, could a wholly naturalistic explanation account for the fine tuning? Is the fine tuning, as Richard Dawkins has framed the issue, "just what we would expect" if "at bottom there was no purpose, no design . . . nothing but blind, pitiless indifference?"[2] Similarly, do we have reason to think "there is a chain of explanations" concerning the fine tuning of the laws and constants of physics and the initial conditions of the universe "which ultimately reaches to the fundamental laws of nature and stops,"[3] as Sean Carroll says naturalism requires?

Or might, again, an explanation involving a purposive agent and intelligent designer provide a better explanation? If so, what kind of intelligent agent would the evidence suggest: an immanent intelligence within the cosmos, a deistic God beyond the cosmos, or a transcendent and active intelligence of the kind envisioned by classical theism? Which constitutes the best explanation of the cosmic fine tuning?

Fine Tuning, Intelligent Design, and Confirmation of Hypothesis

In Chapter 8, I argued that various naturalistic hypotheses did not provide causally adequate explanations for the fine tuning of the universe for life, but yet the fine tuning was just the kind of evidence that we might expect if an intelligent agent had acted to design the universe (Fig. 13.1). Indeed, in our experience, intelligent agents often produce finely tuned machines, systems, or strategies in order to achieve discernible functional outcomes. A finely tuned digital computer, for example, has a number of improbably arranged (finely tuned) parts that perform the discernible function of processing information. A finely tuned musical instrument also has a number of improbably arranged parts that perform a discernible function, and those parts may also have been precisely set or adjusted ("tuned") to allow the instrument to be played "in tune." And a finely tuned recipe will specify a series of precise steps, ingredients, durations, and measures to produce a culinary masterpiece—another kind of functional outcome.

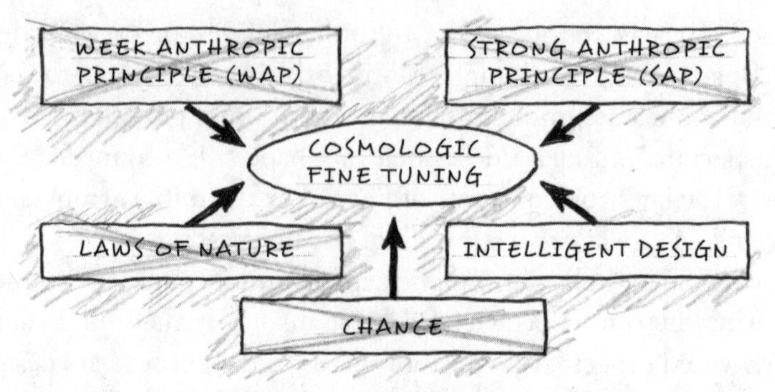

FIGURE 13.1
This chapter builds on the discussion in Chapters 7 and 8 in which I argued that intelligent design best explains the evidence of the fine tuning of the laws and constants of physics and the initial conditions of the universe. This chapter will also suggest that since this evidence of design is present from the beginning of the universe, it points to the need for a transcendent, rather than an immanent, intelligent agent as the best explanation. It does not yet consider the "exotic" naturalistic hypothesis of "the multiverse."

Recall that I showed, drawing on the work of mathematician William Dembski, that physical systems or structures that (1) manifest an extremely improbable combination of factors, conditions, or arrangements of matter and (2) exemplify a significant or "recognizable pattern" or a "set of functional requirements" (what Dembski calls a "specification") invariably arise from intelligent design rather than undirected material processes.

Further, to achieve various goals or objectives, intelligent agents typically must choose among (or constrain) a range of possibilities to actualize an otherwise improbable outcome. The act of choosing among options is what is meant by "fine tuning," and it is precisely what intelligent agents with free will do. (The Latin roots, *inter lego*, of the English word "intelligence" mean "to choose between.") Consequently, we have good reason to expect that, if an intelligent agent acted in the past to design the universe to be a life-sustaining place, we ought to expect to observe something like the cosmological fine tuning.[4]

Not surprisingly, then, some of the physicists who first discovered the fine tuning of the universe thought intelligent design provided "a common-sense interpretation" (in Fred Hoyle's phrase) of the fine-tuning evidence. Those scientists might have, therefore, expressed this interpretation as a consequence of their expectation of how intelligent agents manifest purposive activity—an expectation arising from their experienced-based knowledge of how intelligent agents often choose among a range of possibilities to "fine-tune" various parameters in service of a discernible goal or significant outcome. Scientists supportive of the design hypothesis might articulate this interpretation as an abductive argument:

Major Premise: Given what we know about the features of intelligently designed objects, if an intelligent agent acted to design the universe, we might well expect the universe to exhibit (a) discernible functional outcomes (such as living organisms) that (b) depend upon finely tuned or highly improbable conditions, parameters, or configurations of matter.[5]

Minor Premise: We observe (b) highly improbable conditions, parameters, and configurations of matter in the fine tuning of the laws and constants of physics and the initial conditions of the universe, and these finely tuned parameters (a) make life (a discernible functional outcome) possible.

Conclusion: We have reason to think that an intelligent agent acted to design the universe.

Theism, Panspermia, and Fine Tuning

Yet the cosmological fine-tuning evidence does not just support a generic intelligent design hypothesis. It also provides support for either a theistic or deistic design hypothesis—in other words, a God hypothesis. Here's why.

Recall from Chapters 7 and 8 that physicists think that the fine tuning of the laws and constants of physics, and (obviously) the *initial* conditions of the universe, would have been set from the beginning of the universe. Since both theism and deism conceive of God as existing outside of our time and as having acted to create and design the universe at the beginning of time in this universe, both theism and deism would expect to find evidence of design present from the beginning. Since the cosmological fine tuning provides such evidence, it provides abductive confirmation for a God hypothesis and not just for a generic intelligent designer. The following argument shows why:

Major Premise: If God acted to design the universe, we would expect evidence of fine tuning from the beginning of the universe.

Minor Premise: We have evidence of fine tuning from the beginning of the universe.

Conclusion: We have reason to think that an intelligent agent that transcends the universe—also known as God—acted to design the universe in a way that makes it conducive to life.

In this way, the evidence for cosmological fine tuning provides abductive support for theism and deism as *possible* metaphysical explanations for the origin of the cosmological fine tuning. But what about other possible hypotheses? What about, for example, the possibility of an immanent intelligence?

As I've noted, some scientists such as Francis Crick,[6] Chandra Wickramasinghe, Fred Hoyle,[7] and even Richard Dawkins[8] have postulated that an intelligence elsewhere but within the cosmos might explain the origin of the first life on earth (Fig. 13.2). Crick proposed this idea in 1981 after candidly acknowledging the prohibitively long odds against life arising spontaneously here.[9] He consequently proposed that life first arose by some undirected process of chemical evolution somewhere else in the universe and then continued to evolve, eventually producing an intelligent form of alien life. This immanent intelligence—an extraterrestrial agent rather than a transcendent God—designed and then "seeded" a simpler form of life on earth. Hence, the term "panspermia" (from the Greek *pan*, "all," and *sperma*, "seed").

PANSPERMIA
(SPACE ALIEN DESIGNER HYPOTHESIS)

FIGURE 13.2
Some prominent biologists have proposed that life was seeded on earth by an extraterrestrial intelligence. While this hypothesis known as "panspermia" might in theory explain the evidence of design in living systems on earth, it cannot explain the origin or fine tuning of the universe, since both those events would have preceded all forms of life in the universes, including any putative intelligent aliens.

As I mentioned, Richard Dawkins himself has advanced this idea in a popular documentary, *Expelled: No Intelligence Allowed*, featuring the actor, lawyer, and economist Ben Stein. Dawkins's comments are worth reviewing in a bit more detail. The film documents the suppression of the academic freedom of scientists who support the theory of intelligent design and who work in mainstream universities and research institutes. In an interview with Stein, Dawkins acknowledged that neither he nor anyone else knows how life first evolved on earth. Stein then asked, "What do you think of the possibility that intelligent design might turn out to be the answer to some issues in genetics or in evolution?" To Stein's surprise, Dawkins proceeded to speculate that life on earth might have originated as the result of a higher alien intelligence elsewhere in the universe, which then seeded life here. As Dawkins explained:

> It could be that at some earlier time somewhere in the universe a civilization evolved by probably some type of Darwinian means to a very, very high level of technology and designed a form of life that they seeded onto, perhaps, this planet. Now that is a possibility, and an intriguing possibility, and I suppose it's possible that you might find evidence for that; if you looked at the details of biochemistry and molecular biology you might find a signature of some sort of a designer.[10]

As it happened, I was also interviewed in the *Expelled* documentary and so attended a premier on opening night. At the time, I was finishing my book about the origin of life, *Signature in the Cell.* Consequently, Dawkins's acknowledging that molecular biology might reveal a "signature of some sort of a designer" got my attention. Of course, he quickly clarified his admission by explaining that this higher intelligence must have resided "elsewhere in the universe" and "would itself have had to have come about by some explicable or ultimately explicable process"—by which he meant a fully naturalistic evolutionary one.

What should we make of this proposal? I've previously conceded that, as an explanatory possibility, the information in living systems might have arisen from an immanent intelligence in the cosmos. Nevertheless, I've never found that explanation satisfying, even as an explanation for the origin of the first life. For one thing, any theory of the origin of life, whether purporting to explain the first life on earth or elsewhere in the cosmos, must account for the origin of the functional or specified information necessary to configure matter into a self-replicating system—something that most biologists (including Dawkins) take as a *sine qua non* of a genuinely living organism. Yet those who propose panspermia do not explain the problem of the ultimate origin of that functional biological information.[11]

Simply asserting that life arose somewhere else out in the cosmos does not explain how the information necessary to build the first life, let alone the first intelligent life, could have arisen. It merely pushes the explanatory challenge farther back in time and out into space. Indeed, positing another form of preexisting life only presupposes the existence of the very thing that all theories of the origin of life must explain and have yet to explain—the origin of functional biological information.

Beyond that, panspermia certainly does not explain the origin of the cosmological fine tuning. Since the fine tuning of the laws and constants of physics and the initial conditions of the universe date from the very origin of the universe itself, if intelligent design best explains the fine tuning, then the designing intelligence responsible for the fine tuning must have had the capability of setting the fine-tuning parameters and initial conditions from the moment of creation. Yet, clearly, no intelligent being *within* the cosmos that arose after the beginning of the cosmos could be responsible for the fine tuning of the laws and constants of physics that made its existence and evolution possible. Such an intelligent agent "inside" the universe might reconfigure or move matter and energy around in accord with the laws of nature. Nevertheless, no such being subject to those laws could possibly change the constants of physics simply by changing the material *state* of the universe. Similarly, no intelligent being arising after the beginning of the universe could have set the initial conditions of the universe upon which its later evolution and existence would depend. It follows that an immanent intelligence (an extraterrestrial alien, for instance) fails to qualify as a causally adequate explanation for the origin of the cosmic fine tuning.[12]

Thus, even if we concede as a logical possibility that an immanent intelligence *might* explain the origin of life on earth (since such an entity could possibly precede it), the panspermia hypothesis does not explain either the ultimate origin of life in the universe or the fine tuning of the universe—to say nothing of the origin of the universe itself. Instead, if intelligent design best explains the fine tuning of the universe, then the kind of intelligence necessary to explain the fine tuning of the universe must in some way preexist or exist independently of the material universe. Indeed, any designing intelligence responsible for the cosmological fine tuning must have had the capability of setting the parameters and initial conditions from the beginning.

Since both theism and deism conceive of God as having an existence independent of the material universe—either in a timeless eternal realm or in another realm of time independent of the time in our universe—both can account for (a) the origin of the universe in time (i.e., at a beginning) and (b) the fine tuning of the universe from the beginning of time. In other words, since both theism and deism posit the prior (either ontological or temporal) existence of a transcendent intelligent agent, the creative and causal act of such an agent in choosing to design the universe with a specific suite of life-permitting parameters would explain the origin of the fine tuning from the beginning of the universe. Thus, theism or deism can provide a causally adequate explanation for the origin of the fine tuning, whereas an immanent intelligence within the cosmos cannot. In other words, both the God of theism and deism would have the "right skill set" to explain the origin of the universe and its fine tuning.

The God Hypothesis: A Better Explanation than Naturalism

Even so, often there is more than one possible or adequate explanation for the same evidence. So what about naturalism? We've seen that specific naturalistic hypotheses such as the weak and strong anthropic principles fail to explain the fine tuning. But do naturalists generally (or other naturalistic approaches) expect evidence of fine tuning?

Recall that Sean Carroll defined naturalism not only as denying the existence of anything outside of nature, but also as affirming that the "chain of explanations concerning things that happen in the universe . . . ultimately reaches to the fundamental laws of nature and stops."[13] In other words, naturalism is committed to the belief that the fundamental laws of physics can ultimately explain all phenomena.

Unfortunately for proponents of naturalism, the laws of physics do not, and cannot, explain either the fine tuning of the constants of proportionality within the laws of physics or the fine tuning of the initial conditions of the universe. Indeed, the fundamental laws of physics cannot, in principle, explain why the constants of proportionality have the values that they do. As I explained in Chapters 7 and 8, (1) the structure of the laws allows them to have other values and (2) the specific values of the constants represent features of the laws themselves, not aspects of nature that the laws could conceivably explain. Similarly, the laws of physics do not explain why the universe had the precise set of initial conditions it did. The laws *apply* to those material conditions, and the laws must *presuppose* them to describe the universe accurately or to make definite predictions about what would have happened after the beginning of the universe. But the laws do not *explain* their origin.

To see why, consider the following example of a law of physics that describes the behavior of a simple physical system. Figure 13.3 shows a harmonic oscillator—a wire fixed between two pegs at opposite ends of a board. Many musical instruments such as guitars and pianos use taut wire stretched between two pegs to produce sounds at different frequencies of vibration. Let's say a piano tuner has come to your home to tune your piano. Hooke's law, established in the seventeenth century by physicist Robert Hooke, and the equations derived from it describe harmonic motion.[14] They state that the wire, when plucked, will vibrate or oscillate in the form of a sine wave at some regular frequency. To know the resulting wave's precise height from trough to peak (its amplitude), physicists must know how hard the piano tuner plucked the wire—in other words the initial condition of the wire during the period of time the physicist chooses to study its motion. Yet Hooke's law and its derived equations do not tell physicists how strongly the wire was plucked. In order to solve the equation that describes harmonic motion, that information must be provided by measurement.

Similarly, the law does not specify the allowed frequencies of vibration of the wire that will result once the piano tuner plucks it. Those values will depend upon what physicists call boundary conditions. Boundary conditions define the constraints upon, or the extent of, the physical system under study. These boundary conditions make it possible to limit

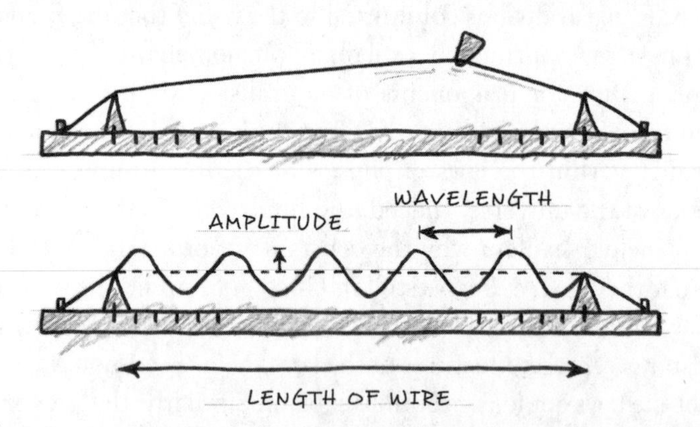

FIGURE 13.3

The top diagram depicts a wire between two pegs being plucked. The plucking results in the wire forming oscillating sine waves. The bottom diagram depicts the resulting wave form of a specific wavelength. The force of the plucking determines the amplitude (height) of the wave. The length between the pegs determines the possible wavelengths.

the number of solutions to the equations describing the behavior of that system. In this case, knowing (among other factors) the distance between the two pegs to which the wire is attached allows physicists to determine the allowable frequencies.

Here's the point. Neither the initial conditions (how hard the wire is plucked) nor the boundary conditions (the distance between the pegs) are determined by Hooke's law itself. Instead, Hooke's law of harmonic motion allows a vast array of different values for initial conditions. The law also applies to a similarly large ensemble of systems with different boundaries (in this case, the different distances between the fixed ends of the wire). Consequently, the law does not explain why the system has the boundaries it does or why the initial conditions of the string were what they were. That information is logically extrinsic to the structure and form of the law.

What does this have to do with naturalism and the fine tuning of the universe? Well, just as Hooke's law does not determine how hard a piano string is plucked, the fundamental laws of physics do not determine the finely tuned initial configurations of mass and energy at the beginning of the universe. Those are additional contingent factors not determined by the fundamental laws of physics that the physicist must take into account from time zero moving forward. In a similar way, the laws of physics do not determine or explain the contingent and finely tuned values of the constants of proportionality. Instead, physicists must determine those contingent values by measurement, independent of their knowledge of the general mathematical form of the relevant laws.

Nevertheless, some have proposed that physicists will eventually discover a new, more general law, sometimes called a "theory of everything," that will explain the origin of the fine tuning of the universe. They hope that such a theory will show that there are "no free parameters"—that is, that there are no parameters or values for the physical constants left undetermined by an underlying all-encompassing general law. Though this proposal may sound plausible, it betrays a significant confusion about the logical structure of scientific laws and how scientists use them. When scientists describe the behavior of physical systems, they not only must know the relevant laws of nature. They must also know specific facts about the objects to which the laws apply—in other words, the conditions or states in which the relevant objects under study find themselves.

These facts may include the tension in a string, the angle of an inclined plane, or the force applied by a cue to a pool ball, for example. Such specific facts allow physicists to describe nature accurately, using various laws of physics (Hooke's law, laws of mechanics, momentum exchange, etc.). But without these extrinsic inputs of information—what physicists call initial and boundary conditions—the laws themselves will not yield precise descriptions of any material system. As Michael Polanyi (Fig. 13.4) noted, "Physics is dumb without the gift of boundary conditions forming its frame";[15] "a boundary condition is always extraneous to the process which it delimits. In Galileo's experiments on balls rolling down a slope, the angle of the slope was not derived from the laws of mechanics but was chosen by Galileo."[16] Indeed, all known laws of physics require such extrinsic inputs of information about the features of the systems (i.e., their initial and boundary conditions) or about the objects to which they apply.

FIGURE 13.4
The Hungarian physical chemist and philosopher of science Michael Polanyi.

All physical laws also require such extrinsic inputs of information about the values of their own constants. Recall from Chapter 7 that the values of the physical constants represent information about, as I put it there, the "net effect of all the other factors (not specifically represented in the relevant equations) that affect the magnitude of the forces in question." Consequently, these constants also provide information about the universe itself or the part of it described by a physical law. There is, therefore, no reason to think that any explanatory principle with the features of a law of physics—that is, a principle stating some general relationship between physical variables—could explain the fine tuning of the universe without also including specific information about contingent features of the universe.

Moreover, even if some future "superlaw" were discovered that generated all the values of the present constants of physics, such a law would not eliminate the need for initial and boundary conditions as well as other constants with specific values. All known laws require such "free parameters" in order to describe the universe in all its specificity. For a new, more fundamental law to describe a universe with any specificity at all, it would also need to have its own constants and incorporate independent information about the initial and boundary conditions of the universe to describe it accurately or to make any specific predictions about it. Thus, to say that there is some as yet undiscovered general law of nature that will ultimately explain all the specific values of the fine tuning, without itself having to have finely tuned initial and boundary conditions and constants, again ignores the "nature" of natural laws and what they need in order to describe nature accurately. It follows that naturalism, with its commitment to explaining everything by reference to the fundamental laws of nature alone, cannot explain the fine tuning of the universe—at least not without presupposing exquisite prior unexplained fine tuning of other contingent parameters.

Appealing to some as yet *undiscovered* law to explain the fine tuning of the physical constants seems implausible for another related reason. Natural laws by definition describe phenomena that conform to regular or repetitive patterns. Yet the idiosyncratic values of the different physical constants and initial conditions constitute a highly irregular and nonrepetitive ensemble. It seems unlikely, therefore, that any more fundamental laws could explain why all the fundamental constants have exactly the values they do—why, for example, the permittivity constant in Coulomb's law should have the value 8.85×10^{-12} Coulombs^2 per Newton-meter, *and* the electron mass should have exactly the value of 9.11×10^{-31} kilograms, *and* the electron charge to mass ratio is exactly 1.76×10^{11} Coulombs per kilogram, *and* the fine-structure constant should have the value .00729, *and* the Higgs field vacuum expectation has the value of 246 GeV.[17] These constants specify a highly irregular array of values that describe either ratios between completely different types of quantities or completely different quantities (e.g., mass, charge, and energy) or quantities without units at all. As a group, they do not exhibit the kind of regular pattern that is at all likely to be subsumed under a single physical law.

Considerations of causal adequacy reinforce the above arguments. In our experience, we have often observed intelligent agents producing finely tuned systems, whether in a digital computer, an internal combustion engine, an arrangement of flowers conveying a message, or a recipe for an exquisite French dish. The very idea of fine tuning implies that some conditions or parameters were precisely and improbably set to achieve a purpose—one that points, based upon our uniform and repeated experience, to the action of a *purposive* or intelligent agent.

On the other hand, we lack experience of undirected material processes producing obviously finely tuned systems—ones exhibiting both extreme improbability and functional specificity. Computers, engines, meaningful sequences of letters, and recipes arise from purposeful "fine-tuners," not undirected processes. For this reason, scientific naturalists have typically attempted to explain the fine tuning not by positing a known law or process, but instead by positing alternate explanations that attempt to reduce the *surprise* associated with the discovery of the fine tuning. Both the weak anthropic principle and the strong anthropic principle provide good examples of this strategy. Nevertheless, as we saw in Chapter 8, both of these explanations failed to meet the test of causal adequacy and for compelling reasons of basic logic.

Of course, in an attempt to explain the extreme improbability of the fine tuning, some contemporary physicists have postulated the existence of other universes—that is, the "multiverse."[18] This hypothesis provides an example of what I call "exotic naturalism," by which I mean naturalistic hypotheses that posit other unknown realms of nature beyond this observable universe to explain naturalistically otherwise inexplicable phenomena of this universe, such as its beginning and fine tuning.

I evaluate the multiverse hypothesis in Chapter 16 in the next section of this book. For now, suffice to say that *basic* naturalism, or materialism, clearly fails to explain the fine tuning, because none of the popular proposals for doing so—neither brute chance, nor the laws of physics, nor the weak and strong anthropic principles—meets the test of causal adequacy. Indeed, since our experience affirms that finely tuned systems arise from intelligent activity, and since naturalism denies the existence of an intelligent agent before the beginning of the universe, basic naturalism lacks recourse to an entity with the causal powers to produce the effect in question. By contrast, theism and deism affirm the existence of a transcendent intelligent agent prior (either ontologically or temporally) to the beginning of the universe. Thus, theism and deism posit an agent with causal powers—the "right skill set"—to produce the fine tuning of the universe from its beginning, whereas naturalism does not.

An Objection Answered

Sean Carroll defines naturalism as the idea that "the things that happen in the universe" are ultimately explicable by "the fundamental laws of nature." Consequently, scientific naturalists could object by insisting that the laws simply *include* the finely tuned values for the physical constants, thus making a life-permitting universe expected given the laws of nature (and naturalism). In other words, naturalists might simply treat the values of the constants as givens. Yet even if one defines the laws as including the constants, the laws *certainly* do not contain information about the *initial conditions* of the universe. Yet those conditions also exhibit extreme fine tuning that the laws don't explain. Moreover, simply defining constants as part of the laws of physics still does not explain why the physical constants themselves have the exact values they do. And since nothing in fundamental physical theory explains why the constants have these precise values, the values still require explanation—an explanation that the laws of physics can't provide, since those laws can't explain themselves (or their own contingent features).

What Naturalism Expects: A Bayesian Turn

Bayesian reasoning, in comparing the *expectations* of competing metaphysical hypotheses, reinforces this conclusion. Recall again Richard Dawkins's claim that the universe has just the properties that we should expect if "there was no purpose, no design . . . nothing but blind, pitiless indifference."[19] In fact, however, the hypothesis of materialism does not lead naturally to the expectation of a finely tuned universe capable of sustaining life. Since in our experience fine tuning results from intelligent agency, and since naturalism denies the existence of any intelligent agent preexisting the universe, philosophical naturalists should not expect to observe a universe in which life depends upon exquisite fine tuning. Instead, they should expect a universe in which all phenomena can be explained by reference to the fundamental laws of physics. But, as we have seen, those laws themselves do not explain either the fine tuning of the initial conditions of the universe or the contingent features of the physical laws (the fine tuning of their constants) necessary to maintain a universe capable of sustaining life.

A naturalistic worldview should, given the discovery of the extreme fine tuning necessary for life, actually lead us to expect a sterile universe incapable of hosting life anywhere. Here's why. The fundamental laws of physics are consistent with a *vast array of other possible universes*: universes that evolved from different initial conditions and universes with different physical constants in their laws, or both. Since the laws of nature do not determine the values of the physical constants or the initial conditions of the universe, they do not render any one of these possible universes—including our own—any more or less probable than another. Yet the overwhelming majority of these potential universes preclude the existence of life.[20]

The physicist Luke Barnes (Fig. 13.5) has argued that the improbability of the fine tuning thus provides an objective *quantitative* measure of our expectation—the Bayesian *likelihood*—of observing a life-permitting universe given naturalism. That's because the range of possible values and initial conditions compatible with the basic laws of physics allows for vastly and quantifiably more sterile universes than life-permitting universes. Since naturalism affirms the laws of physics as the ultimate realities and explanatory principles, the probability of a sterile universe given naturalism is vastly and quantifiably greater than that of a life-conducive universe. Thus, we have a quantifiably greater reason to *expect* a sterile universe given naturalism.

FIGURE 13.5
The Australian physicist Luke Barnes, coauthor of *The Fortunate Universe.*

Indeed, if we just consider the fine tuning of the initial entropy, we can calculate that, given the fundamental laws of physics—that is, given naturalism—we should expect vastly fewer life-conducive universes than sterile universes by a factor of 1 to $10^{10^{98}}$. (See n. 11 in Chapter 8.) Barnes argues philosophical naturalists should also strongly expect a universe in which the constants of the physical laws would render life *im*possible—for, indeed, the overwhelming majority of those values would also produce a lifeless universe. Thus, again, our present observation of a life-friendly universe with its improbable fine tuning contradicts what we should—*in all probability*—expect if the hypothesis of naturalism were true.

Moreover, as I've argued, the observation of extreme fine tuning confirms precisely what we might well expect if a purposive intelligence—indeed, a theistic or deistic creator—had acted to design the universe and life. We certainly have *more* reason to expect a universe fine-tuned for life (or a life-permitting universe that depends upon fine tuning)[21] assuming theism or deism than we do assuming naturalism. We can express that idea using Bayesian symbolism (where T represents theism, D represents deism, N_b represents basic naturalism, and E_{ft} represents evidence of fine tuning) as follows: $P(E_{ft} \mid T) >> P(E_{ft} \mid N_b)$ and $P(E_{ft} \mid D) >> P(E_{ft} \mid N_b)$.

Since, again, we don't have decisive reasons *a priori* to regard either a naturalistic or deistic or theistic worldview as more probable than the others, we can also affirm that the probability of a God hypothesis given the fine-tuning evidence is greater than the probability of basic naturalism given that same evidence. Or stated symbolically, $P(T \mid E_{ft}) >> P(N_b \mid E_{ft})$ and $P(D \mid E_{ft}) >> P(N_b \mid E_{ft})$.

To convey this conclusion, Barnes uses an illustration reminiscent of ones I have used in a different context (see Chapter 10). In a lecture I had the pleasure of hearing, Dr. Barnes asked his listeners to consider several possible scenarios involving a burglar and a bank vault. First, imagine that you are a bank clerk tipped off that a burglar has slipped into your bank and has tried to open the safe through pure luck. Call this the "random-fiddling" hypothesis. You hurry to the vault to see what happened. Barnes points out that, given the "random-fiddling" hypothesis, you should *not* expect to find an open safe. That's because the overwhelming majority of the possible turns of the dials on the lock will *not* open the safe, and before the alarms sound the burglar has an extremely limited amount of time to search randomly for the correct combination.

But what if you were tipped off that a burglar broke into the bank vault with inside information about the combination of the safe and how to elude the security guards and cameras. Then as you rush to the vault, you might well expect to find the safe wide open, since the burglar's knowledge could enable him to gain quick access to the contents and make an expeditious escape. You would certainly have greater reason to expect to find an open safe given the "inside-job" hypothesis than you would given the "random-fiddling" hypothesis.

Now imagine a third scenario involving two conflicting tips. Say that one of your neighbors told you that a burglar was headed to the bank to try to slip in and open the safe on pure dumb luck. And another neighbor told you, no, the burglar is heading to the bank and has inside help—that is, an intelligently designed bank heist is about to occur. You rush to the bank and find that the safe indeed has been breached and the contents stolen.

What would you conclude? Which of the tips given to you by your two neighbors is more likely to have been true? Clearly, the observation of the open safe lends greater support for the "inside-job" hypothesis. In the same way, Barnes argues, since we have a greater reason to expect a life-permitting universe given theism than naturalism, the observation of a life-permitting universe that depends upon extreme fine tuning provides greater support for theism than for naturalism. Or, as I frame the argument with a slightly different emphasis,[22] since we have a greater reason to expect *an exquisitely finely tuned universe capable of hosting life* given theism than naturalism, the observation of a finely tuned life-permitting universe provides greater support for theism. We can again state that conclusion using Bayesian symbolism: $P(T \mid E_{ft}) >> P(N_b \mid E_{ft})$. In other words, theism (or deism) provides a better explanation for the cosmic fine tuning than does basic naturalism.

Theism, Pantheism, and Fine Tuning

But what about pantheism? Does this worldview provide a causally adequate explanation for the origin of the cosmological fine tuning?

Recall that the god of pantheism is not a conscious being or person with whom one can, for example, communicate by prayer. Remember also that though the physical world reveals apparent distinctions between different objects (sentient and nonsentient), pantheists regard all of nature and god as ultimately One. Thus, many pantheists regard the observable differences in the physical world of nature (*prakriti*) as an illusion (*maya*). Although there are theistic versions of Hinduism that affirm the rationality and consciousness of God and also affirm distinctions between God, the self, and the world, the pantheistic versions of Hinduism do not. To strict pantheists, ultimate reality is an impersonal unity.[23]

Pantheism, thus conceived, offers no explanation for the origin of the fine tuning of the laws and constants of physics or the initial conditions of the universe. As I've argued, based upon our experience of cause and effect, explaining systems exhibiting fine tuning requires an intelligent cause. Pantheism, with its *impersonal* god, denies the existence of such a foundational intelligence or conscious mind.[24]

Moreover, since the fine tuning of the universe was established at the beginning of the universe, explaining it also seems to require a *preexistent* intelligence. Yet in pantheism nothing preexists the natural world, since the natural world (*prakriti*) and the god of pantheism (*brahman*) possess self-existence coextensively. Nature and god didn't come into existence.

They simply are. Moreover, god is in nature and the whole of nature is god. Thus, pantheism, like naturalism, lacks reference to any preexistent or transcendent entity, still less an intelligent or conscious one, that could account for the fine tuning *from the beginning* of the universe itself. As such, it lacks causal adequacy as an explanation.

Thus, a theistic or deistic God hypothesis provides a more causally adequate explanation of the origin of the fine tuning of the universe (and the origin of the universe itself) than either pantheism or naturalism.

The Right Skill Set

Testing hypotheses using the method of inference to the best explanation involves assessing whether a postulated cause has the attributes necessary to explain the effects in question. Scientists and philosophers can do this by observing the effects that a given cause produces. When cause-and-effect relationships cannot be directly observed, they can do so by analyzing the likely effects of a postulated cause based upon theoretical considerations (of the properties of that postulated cause) or by extrapolating from knowledge of what "relevantly similar" causes produce.

The discovery of the fine tuning of the universe, like the discovery of the beginning of the universe itself, represents an effect that requires a cause with specific attributes, including both transcendence and intelligence. Both theism and deism posit an entity, namely, God, with those attributes as well as other relevant attributes such as creativity and power. Consequently, God, as conceived by both theists and deists, possesses the "right skill set" or "postulated properties" or "causal powers" to create and finely tune the universe for life (Fig. 13.6).

Moreover, positing such an unobserved superintelligence to explain the fine tuning of the universe does not violate any principles of sound reasoning. Instead, it represents a natural extrapolation from our knowledge of the kind of finely tuned, information-rich systems that relevantly similar human intelligences produce. Similarly, positing a specifically *transcendent* intelligence to explain the fine tuning is warranted by the nature and the timing of the appearance of the effect itself—and by the attributes of the posited cause, namely, God.

Effects result from causes distinct from themselves.[25] Since the effects in question include the beginning of the universe as a whole and the fine tuning of the whole universe from the beginning, the nature of the ef-

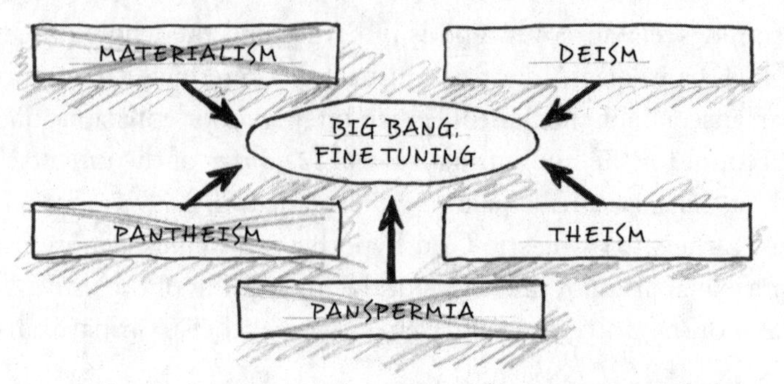

FIGURE 13.6

This chapter has evaluated which of the competing metaphysical hypotheses (theism, deism, pantheism, materialism, or panspermia) best explain the fine tuning of the universe. It has argued that theism and deism provide causally adequate explanations for this evidence whereas neither pantheism, materialism, or panspermia do. Similarly, neither pantheism, materialism, nor panspermia explain the evidence for the beginning of the universe as well as theism or deism do. Thus, given these two classes of evidence (i.e., the big bang and fine tuning), theism and deism remain as possibly the best explanations.

fects requires a cause beyond the universe—a transcendent cause. Since God, as conceived by both theists and deists, possesses this attribute (as well as intelligence), God could plausibly function as an explanation for the beginning of the material universe and the origin of its fine tuning.

Naturalism, materialism, and pantheism, on the other hand, deny the reality of any intelligent agent existing before or independently of the universe. Consequently, these systems do not provide causally adequate explanations of either the fine tuning or the origin of the universe. Panspermia, for its part, not only pushes the mystery of life's ultimate origin out of view without explaining it; it does not offer an alternate explanation for the fine tuning of the laws of physics or the initial conditions of the universe. For all these reasons, cosmic fine-tuning evidence reinforces the conclusion of the last chapter by showing that theism and deism again provide better explanations than their competitors.

But that raises another question: Is there any evidence that might distinguish the explanatory power of theism from that of deism? The next chapter explores this question.

14

The God Hypothesis and the Design of Life

In the debate over biological origins, proponents of a position known as "theistic evolution" have often opposed the case for intelligent design as strenuously as any atheist. The sociological reasons for such opposition among people otherwise committed to a belief in a divine creator are complex. But the scientific viewpoint of theistic evolutionists is fairly easy to understand, as it closely mirrors that of mainstream evolutionary theorists.

Most theistic evolutionists affirm the adequacy of various evolutionary mechanisms to account for the origin of new forms of life. Like many evolutionary biologists, theistic evolutionists deny any evidence of intelligent design in living systems. Unlike typical evolutionary biologists, however, many equate the alleged creative power of mutation and natural selection and other evolutionary mechanisms with the creative power of God. They thus affirm that God, albeit in a completely undetectable way, used, but did not actively or discernibly guide, the evolutionary process to generate new forms of life. They also typically accept that God created and fine-tuned the universe in the beginning. But they hold that God then used fully naturalistic processes to generate life after that, without intervening in, or directly guiding, those processes.

I've already critiqued the scientific basis of this particular version of theistic evolution. I did so—in effect—in Chapters 9 and 10, and more fully in my previous books, by showing that neither mutation and natural selection nor other evolutionary mechanisms possess the creative power *to generate* the specified or functional information needed to produce the first life or subsequent major morphological innovations in the history of life. In contrast, our uniform and repeated experience has shown that intelligent agents can and do produce such information, including information in a digital form. Thus, I argued that intelligent design provides a better, more causally adequate explanation for the origin of the information necessary for the origin and development of life than either biological or chemical evolutionary processes.

Some theistic evolutionists acknowledge that known evolutionary mechanisms lack the creative power to produce large amounts of new biological form and information. But they propose that the information necessary to build at least the first living cell might have been present in the finely tuned initial conditions of the universe and in the laws and constants of physics from the beginning of the universe itself. Thus, they affirm a different version of theistic evolution, what might be called a "front-loaded" version of the design hypothesis.

Some religious critics have characterized theistic evolution, especially this latter version, as tantamount to a deistic perspective in which God creates the universe at the beginning but then takes a "hands-off" approach after that. Strictly speaking, however, theistic evolution does not actually qualify as deism. Deism denies *any* divine activity in the natural world after the first moment of creation. Yet most theistic evolutionists think that God *actively sustains* the orderly concourse of nature—the laws of nature—on a moment-by-moment basis.

Since the late eighteenth century, the popularity of deism has waned among philosophers and scientists. Few leading figures debating the origin of life and the universe today represent a deistic worldview. Nevertheless, some versions of theistic evolution—the versions that affirm front-loaded design—share a key tenet with deism, even if they don't qualify as fully deistic. In particular, like deism these versions of theistic evolution advance the idea that God set up the universe in the beginning so as to make life's later origin inevitable without any need of further *creative* acts (or "interventions"). Thus, evaluating the front-loaded design hypothesis (as advanced by some contemporary theistic evolutionists) will allow us to assess whether deism could provide an adequate explanation of the origin of life and the information needed to produce it. That in turn will help us address the question raised at the end of the last chapter about whether theism or deism has greater overall ex-

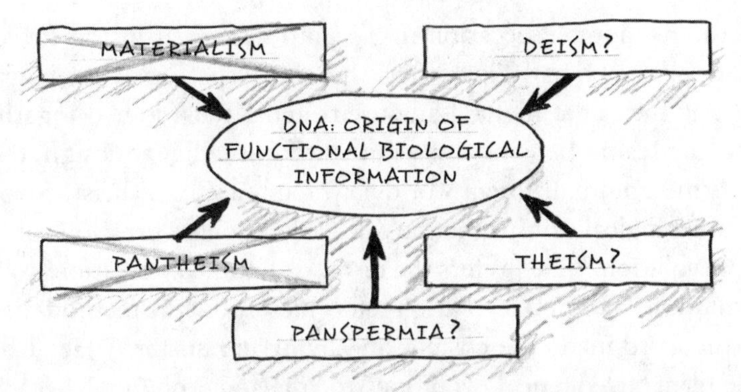

FIGURE 14.1
This chapter will evaluate whether theism or deism (or possibly panspermia) provides a better, more causally adequate explanation for the explosions of the functional biological information that have occurred in the history of life on earth.

planatory power—i.e., the question of which of these two worldviews better explains the whole ensemble of evidences about cosmological and biological origins that we have been considering (Fig. 14.1).

Understanding Deism and Front-Loaded Design

If biological information arose well after the beginning of the universe and did so by intelligent design, as I've argued in Chapters 9 and 10, then that would seem to suggest a designing intelligence acting well after the beginning of time. Since deism conceives of God as acting only at the beginning of the universe, discrete increases in functional biological information throughout the history of life would seem, *prima facie*, to preclude deism as a causally adequate explanation for such information. But, as noted, some theistic evolutionists have suggested, *as deists also would have to do*, that the information necessary to produce the first living cell and other novel forms of life could have been present in the fine tuning of the initial conditions of the universe.

I first encountered the front-loaded view of design in the work of a theistic evolutionist named Denis Lamoureux, a professor of science and religion at St. Joseph's College, University of Alberta. Though Professor Lamoureux is not the only proponent of this view, he has articulated it with perhaps the most clarity and prominence.

Lamoureux and I have debated both in print and in person. That includes the memorable occasion in Toronto described in the Prologue, where Lamoureux participated as the third speaker with Lawrence Krauss and me. That night, Lamoureux and Krauss joined together as a sort of tag team opposing my argument for intelligent design, though each affirms a quite different worldview. Lamoureux, a theist, advocates a view of biological origins that he calls "evolutionary creation" or "teleological evolution."[1] He prefers the term "evolutionary creation" to "theistic evolution." For him, evolution refers merely to "the method through which the Lord made the cosmos and living organisms."[2] He also prefers the term "teleological evolution" to "theistic evolution," because he wants to affirm, against most other neo-Darwinists, that evolution is "a planned and purpose-driven natural process."

But what exactly is meant by this idea? And does it provide an adequate scientific explanation for the origin and development of life?

According to Lamoureux, the theory of evolutionary creation affirms that "the Creator established and maintains the laws of nature, including the mechanisms of a purpose-driven teleological evolution." Lamoureux provides several illustrations to convey what he has in mind. For example, he suggests that "God organized the big bang, so that the deck was stacked"[3] to produce life. He also likens God to an expert billiards player who can sink all the balls on the table in one shot. He compares the precise arrangement of the balls and the billiards player's single shot to God's initial act of creativity in bringing the universe into being with a precise arrangement of matter. Just as the billiards player can clear the table with one shot, God can create everything (the universe and life) with an initial act of creativity in which he arranges matter just right at the beginning and then lets it unfold deterministically in accord with the laws that God also established then. Thus, God needs no additional shots, no further acts of creativity, to bring life into existence. Lamoureux also compares the process of biological evolution to embryological development, in which an organism develops deterministically from a fertilized egg in accord with the laws of nature and the information-rich initial conditions of the fertilized egg.[4]

Thus, Lamoureux affirms a form of design that could fit nicely within a deistic framework, even as he himself disavows such a worldview. He sometimes calls his position "evolutionary intelligent design,"[5] holding that God arranged the initial conditions of the universe with no need for any additional intelligent input. As he explains, "Design is evident in the finely tuned physical laws and initial conditions necessary for the evolution of the cosmos through the big bang, and design is also apparent in the biological processes necessary for life to evolve."[6]

Although he uses the term "intelligent design," he objects to the contemporary theory that goes by that name. According to Lamoureux, to invoke a specific instance of design after the initial creation would imply a violation of natural law by invoking the activity of a "God of the gaps." (See Chapter 20 for my discussion of the "God-of-the-gaps" objection.) As he explains, "Intelligent Design Theory . . . is a narrow view of design and claims that design is connected to miraculous interventions (i.e., God-of-the-gaps miracles that introduce creatures and/or missing parts) in the origin of living organisms." Thus, he continues, "ID Theory should be termed *Interventionistic* Design Theory."[7]

Lamoureux prefers a view that, like deism, confines God's *creative* activity to the very beginning of the universe. As such, evaluating his view will also allow us to evaluate the plausibility of a strictly deistic, front-loaded design hypothesis, as an explanation for the origin of biological information.

So is it possible that the laws of physics and chemistry and the initial configurations of mass-energy could have had imbedded within them all the information necessary to produce the first living cell and other more complex forms of life? Did they?

Scientific Problems with a Front-Loaded Design Hypothesis

The view articulated by Lamoureux and others is scientifically problematic. To see why, first, recall my discussion in Chapter 9 of the information-bearing properties of DNA. DNA is composed of strings of four different chemicals, called nucleotides, that carry in their linear sequence the specified information necessary to build proteins.[8] Neo-Darwinian biologists, computer software engineers, and leading biotechnologists all acknowledge that the information in DNA and RNA resembles digital computer code.[9] That raises the question of how that information—information necessary to produce the first life—arose.

Denis Lamoureux does not directly address the problem of the origin of the first life. He doesn't say which specific naturalistic theory of life's origin—if any—he favors. Nevertheless, the metaphors he employs

(God stacked the deck at the beginning; God as cosmic billiards player; teleological evolution as embryological development, etc.) imply that he thinks deterministic *laws* caused life to self-organize or self-assemble from some highly configured, and therefore information-rich, set of initial cosmic conditions. Indeed, he emphasizes deterministic laws and specific initial conditions over the mechanism of random mutation and natural selection I critiqued in Chapter 10. Nevertheless, he does not say whether he thinks (a) that *all* the information necessary to produce the first and subsequent forms of life was entirely present in the initial conditions of the universe, or (b) that the laws of nature added new information during the subsequent "self-assembly" process.[10] Let's consider the two possibilities, both problematic, starting with the second.

Are Laws Creative?

Like other theistic evolutionists, Lamoureux sometimes speaks as though he thinks the physical laws of nature might be *generating* the information necessary to produce new forms of life. He refers to evolution as "a planned and purpose-driven natural process" and affirms "that humans evolved from pre-human ancestors, and over a period of time the Image of God and human sin were gradually and mysteriously manifested."[11] Since Lamoureux disavows specific acts of divine creation as illicit appeals to a "God of the gaps," and since he affirms that humans, at least, acquire new characteristics during the evolutionary process, it might be that he thinks that the laws of nature are *generating* the new information necessary to produce novel living systems. It might also be that he thinks that the mechanism of random mutation and natural selection generates new information, but, as noted in Chapters 9 and 10 the extreme rarity of functional genes and protein folds in "sequence space" underscores the implausibility of that view.

So what about the laws of nature? Do they generate information?

There are good reasons to doubt this. To see why, imagine that a group of small radio-controlled helicopters hovers in tight formation over the Rose Bowl in Pasadena, California. From below, the helicopters appear to be spelling a message: "Go USC." At halftime, with the field cleared, each helicopter releases either a maroon or gold paint ball, one of the two University of Southern California colors. Gravity takes over and the paint balls fall to the earth, splattering paint on the field after they hit the turf. Now on the field below, a somewhat messier but still legible message appears. It also spells "Go USC."

Did the law of gravity, or the force described by the law, produce this information? Clearly, it did not. The information that appeared on the field already existed in the arrangement of the helicopters above the stadium in "the initial conditions." Gravitational forces played no role in causing the information on the field to self-organize. Gravity merely transmitted preexisting information from the helicopter formation to the field below.

There is a deeper reason that laws can transmit, but not generate, information. Scientific laws describe highly regular phenomena or structures, possessing what information theorists refer to as *redundant order*. On the other hand, the arrangements of symbols (or chemical subunits functioning in the same way) in information-rich text, including in DNA, possess a high degree of specified complexity, not redundant order.

To illustrate the difference, compare the sequences:

ababababababababababababababababababababababababab
That's one small step for man, one giant leap for mankind.

The first sequence is repetitive and ordered, but not complex or informative. The second sequence is not ordered, in the sense of being repetitious, but it is complex and also informative. The second sequence is complex, because its characters do not follow a rigidly repeating, law-bound pattern, and thus it is not what information theorists call "compressible." That means it cannot be generated by a few instructions or simple rules such as "repeat ab 25 times." It is also informative because, unlike a merely complex sequence such as "sretfdhu&*jsa&90te," the particular arrangement of characters is specified[12] so as to perform a communication function. In any case, informative sequences have the feature of complexity or aperiodicity and thus are qualitatively distinguishable from systems characterized by the kind of periodic order that natural laws describe or generate.

Indeed, to say that the processes that natural laws describe can generate functionally specified informational sequences betrays a confusion of categories. Laws are the wrong kind of entity to generate the informational features of life. To look to the laws of nature to generate information is to search for the improbable and specific where it is least likely to be found: in the domain of the recurring and the general.

And yet some scientists say we must await the discovery of new natural laws to explain the origin of biological information. German chemist Manfred Eigen has argued that "our task is to find an algorithm, a natural law, that leads to the origin of information."[13] Yet clearly this statement betrays a category error. Physical laws do not generate or describe complex sequences, whether functionally specified or otherwise; they *describe* highly regular, repetitive, and periodic patterns of events. This is not to malign the laws of physics and chemistry. It's simply to accurately state what they do.

But there is another reason that we will not discover such a law. According to classical information theory, the amount of information present in a sequence is inversely proportional to the probability of the sequence occurring. Yet the regularities we refer to as laws describe highly deterministic or predictable relationships between antecedent conditions and subsequent events. Indeed, laws describe patterns in which the probability of each successive event (given the previous event) approaches 1 (i.e., 100 percent, the highest probability possible).

Yet potential information content mounts as *im*probabilities multiply. A fair coin has a 50 percent probability of turning up heads when flipped. When it does turn up heads, it conveys one bit (one binary digit) of information. If a coin is flipped twice, it has a 25 percent chance of turning up heads both times. If that occurs, the coin transmits more information, two bits.

But what if we all know it's a trick coin, one that will always land on heads? In that case, because we know it will always turn up heads by physical necessity, the coin will eliminate no uncertainty each time it lands on heads, because we already know what's coming. In that case, each new flip will convey *no* new information.

The moral of this story? Information is conveyed whenever one event among an ensemble *of possibilities* (as opposed to a sole necessity) occurs. The greater the number of possibilities and the greater the *im*probability of any one possibility occurring, the more information is transmitted when a particular possibility is fixed, specified, or elected. On the other hand, as philosopher of science Fred Dretske notes, "No information is associated with, or generated by, the occurrence . . . of events for which there are no possible alternatives."[14]

Since natural laws describe situations in which specific outcomes follow specific conditions by necessity, they do not generate, or describe the generation of, new information. Indeed, to the extent a sequence of symbols or a series of events results from a predictable law-bound process, the information content of the sequence will be limited or effaced by redundancy. Thus, natural laws cannot in principle generate or explain the *origin* of information, whether specified or otherwise.

What about Initial Conditions?

But what about the other possibility that Denis Lamoureux might have in mind? Could the *initial conditions* of the universe have contained the information necessary to produce the first living systems?

To suggest they might, Lamoureux draws an analogy between the origin and evolution of life, on the one hand, and the embryological development of a particular animal, on the other hand. He describes how an adult organism unfolds from preexisting genetic information in an embryo as an analogy to the evolution of life starting from information-rich initial conditions. He also likens a billiards player clearing the table with a single shot to life arising from the initial conditions of the universe. In that analogy he emphasizes how the precise initial arrangement of the balls and the perfect placement of the shot in accord with the deterministic laws of physics ensure a desired outcome. By these analogies, he suggests that the precise arrangements of mass and energy, at or just after, the beginning of the universe contained the information necessary to produce the first living organism, and possibly all subsequent ones.

For Lamoureux's scenario to be plausible, the arrangements of matter and energy at the beginning of the universe must have had, first, a high degree of *biologically relevant* specificity. Indeed, those initial arrangements of matter and energy must have stored or encoded or been able to specify, at the very least, the information necessary to produce the first living cell. After all, building living cells requires a vast amount of genetic information contained in DNA or RNA.

This fact raises two precise and analytically tractable questions by which Lamoureux's proposal can be evaluated: Was the information necessary to produce information-rich DNA (or RNA) molecules present in the arrangement of mass and energy just after the beginning of the universe? If so, could some physical law or mechanism then faithfully transmit that specified information over billions of years and across vast distances without additional informational inputs to make the origin of life on earth inevitable long after the origin of the universe?

In both cases, the answer is: "almost certainly not." Here's why.

First, the biologically relevant chemical subunits of DNA themselves do not contain the information necessary for producing the specified information DNA contains. And if they do not contain such information, then the simpler and less biologically relevant arrangements of elementary particles (or distributions of mass-energy) present at the beginning of the universe almost certainly did not contain such information either.

Our examination of the DNA molecule in Chapter 9 revealed that lawlike forces of chemical attraction do not account for the *information* in DNA. Instead, DNA's ability to carry information depends on the *absence* of lawlike forces of chemical attraction dictating the sequence of nucleotide bases along the double helix.

You may remember (see Fig. 9.7) that DNA depends upon several chemical bonds: (1) the bonds between the sugar and phosphate molecules forming the two twisting backbones of the DNA molecule, (2) the bonds attaching individual nucleotide bases to the sugar-phosphate backbones, and (3) hydrogen bonds stretching horizontally across the molecule between nucleotide bases that hold the two complementary copies of the DNA message text together, making replication of the genetic instructions possible.

Nevertheless, recall that there are no chemical bonds between the information-carrying bases along the axis of the DNA molecule where the genetic instructions are encoded. Consequently, forces of chemical attraction do not determine the arrangement of the bases any more than laws of physics determine the arrangement of letters in a line of Shakespearean verse.[15]

When I was a professor, I used to illustrate this fact with a simple analogy. I would place magnetic letters on a metallic surface and then show how the letters could be arranged and rearranged in many possible ways. I did this to show how the forces of magnetic attraction between the metal surface and the magnetic letters did not determine any specific arrangement of letters. After arranging the letters in a short message such as "Students Rock," I'd ask students if the magnetic forces of attraction had determined the message on the board.

Of course, the students recognized that the magnetic forces were responsible for the attachment of the letters to the metallic board, but not their arrangement. I then compared the metallic backboard and the meaningful arrangement of magnetic letters on it to the sugar-phosphate backbone of DNA and the information-rich arrangement of nucleotide bases affixed to it. In both cases the relevant forces of attraction between the information-carrying characters and the medium (the sugar-phosphate backbone or magnetic backboard) did not determine the information-rich arrangements.

This fact about DNA was first noticed in 1967 by the physical chemist Michael Polanyi. In a seminal article, "Life Transcending Physics and Chemistry,"[16] Polanyi showed that the physical-chemical laws governing the assembly of the chemical subunits in the DNA molecule allow (or "leave indeterminate") a vast ensemble of possible arrangements of nucleotide bases. But that meant the chemical properties of the constituent parts of DNA (and the laws governing their attractions and interactions) do not determine the specific sequencing of the bases in the genetic molecule.

Yet the specific sequencing of the nucleotide bases in DNA and RNA constitutes precisely the feature of the cell that origin-of-life biologists need to explain.[17] If lawlike processes of chemical attraction do not determine the specific sequencing of nucleotide bases, then biochemists cannot reasonably invoke such "self-organizational" processes as the *explanation* for the *origin* of that information contained in the nucleotide base sequences. (It turns out that the information stored in RNA and proteins also defies explanation by self-organizing forces of chemical attraction.)[18]

What does this have to do with the front-loaded concept of design? Lamoureux's model of front-loaded design envisions some law-governed process acting on a specific information-rich set of initial conditions at the beginning of the universe to ensure that life will eventually arise. In Lamoureux's billiard-ball illustration, for example, a law of nature (the conservation of momentum) ensures a transition from an initial arrangement of matter (in the specifically arranged balls) to a desired end point (all the balls resting in the pockets). He uses such examples to support a kind of self-organizational theory of the origin of life and extols the "self-assembling character of the natural world."[19]

Yet the bonding affinities between the subunits of DNA do not ensure that they will self-organize into functional information-rich sequences. And that's the kicker: if the complex and biologically relevant chemical subunits of DNA—the nucleotide bases, sugars, and phosphates—lack the self-organizational capabilities necessary to produce information-rich DNA sequencing, then the far less complex, less specifically configured, less biologically relevant elementary particles or energy fields at the beginning of the universe certainly lacked such biologically relevant self-organizational capability.

Recent work in mathematical and molecular biology only reinforces this conclusion. The Dutch structural biologist Peter Tompa and the American biophysicist George Rose have shown that assembling a cell requires confronting a kind of "Humpty Dumpty" problem.[20] The classic nursery rhyme speaks of a broken Humpty, whom not even "all the king's horses and all the king's men" can reassemble into a properly organized egg. Similarly, Tompa and Rose have demonstrated mathematically that even having all the requisite parts for a cell will not ensure that those ingredients will self-organize into a living system.

Instead, the cell, like individual genes or proteins, faces an extreme combinatorial problem. Tompa and Rose calculate, building on the work of protein scientist Cyrus Levinthal, that there are a whopping $10^{79,000,000,000}$ different ways of combining just the proteins in a relatively simple unicellular yeast. That number only grows exponentially larger when biologists attempt to calculate the number of possible ways of combining all the proteins and all the other large molecular components necessary for that one-celled organism, including the DNA and RNA molecules, ribosomes, lipids and glycolipid molecules, and others. The number of possible combinations of these cellular components (called the "interactome") vastly exceeds the number of elementary particles in the universe (10^{80}) and even the number of events since the big bang (10^{139}).

Thus, even if the evolutionary process started not with elementary particles or even with sugars, phosphates, nucleotide bases, and amino acids, but instead with all the functional proteins (or even with all other macromolecular components) necessary to sustain a one-celled organism, a trial-and-error process could not plausibly "search" the correspondingly vast space of possible combinations and have a realistic chance of finding one of the few special combinations consistent with the living state. Indeed, given the number of different possible ways to put together just the necessary proteins, in all likelihood (to vastly understate the case) a cell would not self-organize *even from such a biologically relevant set of components.* As Tompa and Rose conclude: "The functional state is selected from a staggering number of useless or potentially deleterious alternatives. In particular, a simplified calculation is sufficient to show that the number of distinguishable states of the interactome exceeds comprehension. Consequently, the cell cannot self-organize by random assembly of its components."[21]

But if that is so, then the front-loaded hypothesis borders on the absurd. If even starting with a biologically relevant soup of information-rich *macro*molecules (not just their constituent parts) does not ensure that life would self-organize, then it follows *a fortiori* that the much less biologically relevant configurations of elementary particles or matter and energy fields present at the beginning of the universe would not ensure such a self-organizational origin of life either.

If so, then the deck was *not* stacked from the beginning to *ensure* the production of life. Instead, the improbable fine tuning of the laws and constants of physics and the initial conditions of the universe—important as they are—turn out to be merely necessary, but not sufficient, conditions for the production of a living cell. So additional biologically relevant information must have arisen after the beginning of the universe in order to produce life on earth. And yet, as we have seen, self organizational laws as well as all other chemical evolutionary scenarios (see Chapter 9) have failed to account for the origin of the genetic information necessary to produce the first cell.[22]

Could intelligent design have played a role? In Chapter 9, I argued as much. I did so not just because chemical evolutionary models fail to account for the origin of the information necessary to produce the first cell, but also because we know—based upon our uniform and repeated experience—that intelligent agents, and no other kind of cause, can and do generate specified digital information—the kind of information present in the biomacromolecules necessary for life.

So could a deistic evolutionist, or a theistic evolutionist of Lamoureux's stripe, posit the kind of intelligent design needed to explain the origin of biological information? Clearly not. We've seen that biologically relevant information must have arisen *after* the beginning of the universe for life to arise on earth. Yet, Lamoureux insists, and committed deists necessarily must insist, that God's direct, discrete, or special creative activity played no role in the history of the universe after the big bang. Consequently, Lamoureux committed himself, as deists also must do, to a front-loaded design hypothesis. But if front-loaded design fails to explain the origin of the information needed to build the first cell, as indeed it does, then not only Lamoureux's hypothesis, but any deistic design hypothesis will also fail to explain the origin of such information. Indeed, a deistic design hypothesis certainly fails to explain any new bursts of biological information in the earth's biosphere as well as does a theistic design hypothesis, the latter of which posits the activity of an intelligent designer both at and after the beginning of the universe.

Limits on Transmitting Information with Fidelity

Two other considerations support such a theistic design hypothesis. *Even if* the information necessary to produce a living cell had somehow been front-loaded at the beginning of the universe, the laws of physics and chemistry by themselves could not have transmitted that information faithfully in a form that would have made the origin of life possible without additional infusions of information. Both thermodynamic and information-theoretic considerations undergird this conclusion. The second law of thermodynamics implies that any biologically relevant information-rich configurations of mass and energy present at the beginning of the universe would dissipate over time. In addition, Claude Shannon's tenth theorem of information theory teaches that only the application of an external source of information (or control) could prevent or constrain that tendency.

This latter point occurred to me recently in a conversation with a fellow philosopher of science at a small private conference. There I had the opportunity to preview the argument of this book with several colleagues. Before dinner one night, I explained why the facts of molecular biology render a front-loaded view of design implausible. The philosopher of science, although himself sympathetic to theism, wanted to probe the strength of this claim. After I explained the relevant facts about the chemical structure and composition of the DNA molecule, he quickly conceded my point that the biologically relevant subunits of DNA (the sugars, phosphates, and bases) lack the ability to self-organize into a functional information-rich genetic sequence. He also conceded that if this were true, then it stands to reason that the less complex and less biologically relevant configurations of mass and energy at the beginning of the universe would also lack this capability.

Nevertheless, he asked, "Isn't it still possible that God *could have* set up the universe with enough information at the beginning to ensure that life would arise in accord with strictly deterministic laws of nature?" I replied that I wasn't trying to prove there was no possible way that God *could have* provided enough information at the beginning to make life arising later inevitable. My claim was rather that the *evidence* we have about this universe and about DNA suggests that the information necessary to produce the first cell did not reside in the elementary particles or energy fields at the beginning of the universe. Consequently, additional information must have arisen after the big bang.

He acknowledged this but, as philosophers are prone to do, kept probing by reframing the same objection. He asked, "Couldn't God have somehow arranged the initial mass-energy *so perfectly* and given the universe the exactly right push, so that all those elementary particles would have eventually congealed into more complex atoms, and those atoms assembled into bio-specific molecules, and those molecules arranged themselves into cells, and so on?" In other words, he envisioned the same kind of scenario that Lamoureux conveyed with his billiards table analogy.

Like Lamoureux, my philosopher colleague envisioned God arranging matter and energy at the beginning so that he would not have needed to take any further action (or provide any further informational inputs) to create life. I again acknowledged this as a *logical* possibility, but not one consistent with the facts of biology and physics as we observe them. Nevertheless, it occurred to me later that the front-loaded concept of design was scientifically implausible for another reason.

Building the living cells that we find on earth requires, as I have noted, a vast amount of genetic information stored in essentially digital form. For this reason, the basic laws and theorems of information theory that apply to the storage, generation, and transmission of information also apply to the analysis of the genetic information in living systems.

One of those laws is known as Shannon's tenth theorem,[23] named for Claude Shannon, the MIT scientist who developed information theory in the late 1940s. This theorem recognizes that as an information-rich signal or message travels across a communication channel (or through a medium of transmission), it will typically experience what information theorists describe as degradation, corruption, or loss of fidelity, a process described by the second law of thermodynamics. Shannon's tenth theorem states that in order for the information-rich message to arrive from a sender to a receiver without loss, information from outside the system must be added to compensate for the inevitable loss of information that will occur as the result of entropy-producing buffeting from the surrounding environment. In other words, to transmit information with a high degree of fidelity, information from a "correction channel" outside the main channel must be applied to constrain the inevitable noise.

Shannon's theorem states that compensating for such an informational loss requires a correction channel with a channel capacity (measured in bits per second of data) that equals or exceeds the information loss.[24] Moreover, it presupposes that such a correction channel exists independently of the transmission channel and adds compensatory information into the transmission channel *after* the original transmission of information begins.

Applying Shannon's tenth theorem to the hypothetical case in which the information present in the initial conditions of the universe is somehow sufficient to generate the first life yields an interesting conclusion. Applying the theorem to such a scenario implies the need for an extrinsic source of information that adds information into the process putatively responsible for the origin of life but *after* the universe's origin—and *during* the transmission—of the information present in the initial conditions of the universe. In short, Shannon's tenth theorem and the thermodynamic realities that underlie it imply that additional information beyond the amount present at the beginning of the universe would have to be added after the big bang in order to build a living system.

But What If . . . ?

Those persistent advocates of the front-loaded design hypothesis might still object, however. "Couldn't a deistic designer," they might ask, "have compensated for this inevitable informational loss by providing *even more* information in the initial conditions of the universe?" Perhaps, as an imaginary or strictly logical possibility, but again practical, empirical considerations render such purely hypothetical speculations implausible in the extreme.

To see why, assume that the initial conditions of the universe did contain enough information to produce a living organism as well as some extra entropy-compensating information to boot. As noted, ordinary entropy-producing physical processes would quickly begin to degrade any biologically relevant information in the initial conditions of the universe by dispersing and disrupting those initial arrangements of matter and energy. Any extra entropy-compensating information front-loaded in the initial conditions of the universe would begin to experience degradation as well, especially given the intense energy driving the initial expansion of the universe. Moreover, according to cosmologists, the elementary particles present in the early universe would not have even cooled enough to form stable atoms until about 380,000 years after the big bang. Is it even remotely plausible that biologically relevant information in any physical medium or configuration could have survived the intense heat of the early universe and then have been transmitted with adequate fidelity across vast expanses of space to direct the construction of a living cell on earth 13 billion years later?

In any case, unpredictable quantum fluctuations in the location and energies associated with subatomic particles would have further and irreversibly exacerbated information loss. Such fluctuations—constantly at work in the subatomic realm—underlie what physicists call "quantum indeterminacy." These unpredictable fluctuations make it effectively impossible to forecast the evolution of subatomic material states, because the fluctuations destroy information about the location, energy levels, and trajectories of subatomic particles.[25] And since a deistic God would not be involved in the universe after the beginning, such a deity would have no control or influence over these otherwise inherently random fluctuations and thus no basis for knowing or forecasting their effects. That would have made it effectively impossible to know at the beginning of the universe *just what kind of information* or configurations of mass-energy would be needed later to overcome or compensate for such unpredictable entropy-producing quantum effects.

Theism: A Better Explanation than Deism

So clearly a front-loaded (deistic) design hypothesis is extremely *im*plausible given our current understanding of physics. Admittedly, a truly omniscient God could know in advance every quantum fluctuation that would ever take place. Thus, someone could conceive of a creator who sets up the initial conditions of the universe in such a way as to anticipate its own *future* influence over quantum fluctuations. But a God so intimately involved in the detailed workings of nature after its beginning and through all time would not qualify as a deistic creator. Instead, such a creator comports more closely with a theistic, rather than a deistic, conception of God.[26]

Of course, *theistic* evolutionists could still argue that since the God they envision does control the universe after the beginning, front-loading is not logically *impossible*. I would agree, though the version of "front-loading" they envision would seem to require a great deal of divine action after the beginning to ensure that quantum fluctuations come out just right—and that, again, suggests a theistic God involved with the creation, not a deistic one.

Even so, my case for theism over deism as a better explanation of the origin of biological information does not require proving that all versions of front-loading are *logically impossible*, but instead only that a fully deistic view of front-loading is *scientifically implausible*. And, in fact, it is.

The universe that Lamoureux and my philosopher friend at the conference were envisioning is a universe that would, like balls on a billiards table, unfold in a perfectly predictable and deterministic way.[27] That depiction matches the early nineteenth-century understanding of physics championed by Pierre Laplace, who thought that if scientists knew the initial conditions of the universe and the laws of physics, they could calculate every subsequent state with precision. But it does not match the universe as described by twenty-first-century physics or information theory.[28]

Indeed, given the facts of molecular biology, the axioms of information theory, the laws of thermodynamics, the high-energy state of the early universe, the reality of unpredictable quantum fluctuations, and what we know about the time that elapsed between the origin of the universe and the first life on earth, explanations of the origin of life that deny the need for new information after the beginning of the universe clearly lack scientific plausibility.

And since deism denies that God could have or would have acted to add any such necessary information after an original act of creation, deistic and other truly front-loaded design hypotheses cannot account for the origin of the first life. Since, on the other hand, theism does posit an intelligent agent who acts in a creative way (in addition to sustaining the laws of nature) after the beginning of the universe, theism provides a better explanation for the origin of the information necessary to produce the first cell as well as subsequent innovations in the history of life.

Thus, of these two worldview hypotheses, theism provides a better overall explanation than deism of the three key facts about biological and cosmological origins under examination: (1) the material universe had a beginning; (2) the material universe has been finely tuned for life from the beginning; and (3) large discontinuous increases in functionally specified information have entered the biosphere since the beginning. Deism can explain the first two of those facts; theism can explain all three.

A More Respectable Hypothesis

In their books, television interviews, *YouTube* videos, and lectures, the New Atheists and others have assured millions that scientific evidence, especially as it concerns the origin of life and the universe, supports a materialistic or atheistic outlook. They have claimed or assumed, as Sean Carroll and Michael Shermer have done, that the fundamental laws of nature alone will prove sufficient to explain the most salient features of life and the universe. They have argued, as Richard Dawkins has done, that "the universe we observe has precisely the properties we should expect if there is, at bottom, no design, no purpose . . . nothing but blind, pitiless indifference."

But the evidence examined so far suggests the need to reassess such claims. That the universe *had a beginning*, that it was finely tuned *from the beginning*, and that our planet has experienced dramatic discontinuous increases in biological form and information *since the beginning* are not at all what proponents of a naturalistic worldview would most "naturally" expect. Yet theists might well expect evidence of such discontinuity and design. Theism does, in any case, offer *causally adequate* explanations for the origin and fine tuning of the universe and the origin of biological information. Consequently, many scientists and philosophers have begun to question a default commitment to scientific materialism and to consider what physicist and theologian John Polkinghorne has called "a new natural theology." As the historian of science Frederic Burnham observed, the God hypothesis "is now a more respectable hypothesis than at any time in the last one hundred years."[29] Or as astronomer Allan Sandage commented in 1985, "If God did not exist, science would have to . . . invent the concept to explain what it is discovering."[30]

Burnham's comment, offered in 1992, came in response to the discovery of the fluctuations in the cosmic microwave background radiation

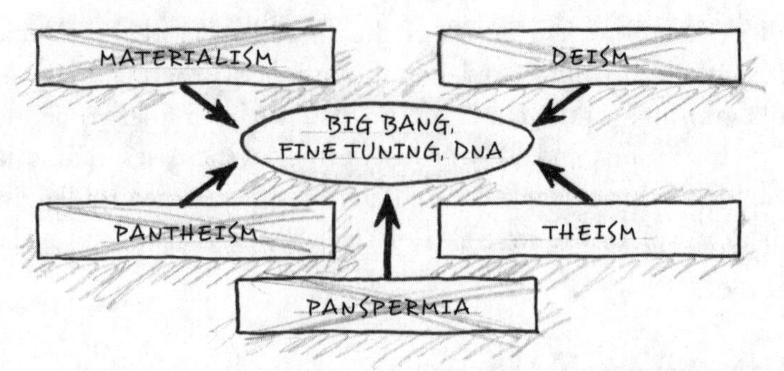

MULTIPLE COMPETING METAPHYSICAL HYPOTHESES

FIGURE 14.2
This chapter and the previous two have shown that only theism provides a causally adequate explanation for the whole ensemble of evidence about biological and cosmological origins under consideration. Deism can explain the origin of the universe and its fine tuning, but not subsequent infusions of functional biological information into the earth's biosphere. Panspermia might *in theory* explain the origin of biological information on earth, but it does not explain the ultimate origin of biological information. Nor can it explain the origin of the universe or its fine tuning. Materialism and pantheism fail to account for all three key classes of evidence since they deny a preexistent transcendent intelligence.

(CMBR) by the COBE satellite, providing another dramatic confirmation of the big bang model and its implication of a beginning. Yet it is not only cosmology that has rendered the "God hypothesis" newly respectable. As one surveys several classes of evidence from the natural sciences—cosmology, astronomy, physics, biochemistry, molecular biology, and paleontology—the God hypothesis emerges as an explanation with unique scope and power. Theism explains an ensemble of metaphysically significant events in the history of the universe and life more simply, more adequately, and more comprehensively than major competing metaphysical systems, including not only materialism and naturalism, but also pantheism and deism[31] (Fig. 14.2).

Again, this does not *prove* God's existence, since superior explanatory power does not constitute deductive certainty. It does show, however, that the natural sciences now provide strong *epistemic support* for the existence of God as conceived by Judeo-Christian and other traditional theists.[32]

Part IV

Conjectures and Refutations

15

The Information Shell Game

Authors are sometimes startled by responses to their work. I have had that experience on a couple of personally memorable occasions, each introduced by an excited phone call from my office. After my book *Signature in the Cell* was published, a colleague phoned to tell me about a review in a prominent publication. Good reviews are always gratifying, but this one was notable because of the altogether unlikely identity of the reviewer. It was December 2009 and, as my co-worker informed me, *Signature* had just been selected by the (London) *Times Literary Supplement* (*TLS*) as a book of the year[1] by the philosopher Thomas Nagel of New York University. Nagel is a well-known atheist and recognized eminence in the American academy, and his commendatory review of a book arguing for intelligent design surprised not just me, but a lot of other people (Fig. 15.1). For many, unlike for me, it was not a pleasant surprise.

FIGURE 15.1
Thomas Nagel, the eminent New York University philosopher of science and author of *Mind & Cosmos: Why the Materialist Neo-Darwinian Conception of Nature Is Almost Certainly False.*

The review provoked many denunciations of Nagel. The protests began with a series of letters to the *TLS*, starting with one from a British chemist at Loughborough University. The chemist, Stephen Fletcher, mocked Nagel for entertaining the "malicious and absurd" argument for intelligent design, which he characterized as prescientific superstition, comparable with belief in "gods, devils, pixies, fairies, you name it."[2] Fletcher cited the "RNA-world hypothesis" (see Chapter 9 of this book) to suggest that scientists had effectively solved the origin-of-life problem, rendering the thesis of *Signature in the Cell* obsolete.

Subsequently *TLS* allowed both Nagel and me to respond. Several years later, Nagel wrote a book, *Mind & Cosmos: Why the Materialist Neo-Darwinian Conception of Nature Is Almost Certainly False*, noting again his assessment of intelligent design, which ignited still greater fury. An American magazine, the *Weekly Standard*, pictured him on its cover as if about to be burned at the stake; the headline read "The Heretic." (The *Standard* was not calling for an auto-da-fé, but protesting the rough treatment given to Nagel by his fellow intellectuals.) The protracted controversy around Nagel's critique of Darwinism and his sympathetic consideration (though not acceptance) of intelligent design undoubtedly attracted more attention than his original review. It certainly offered readers an occasion to consider the strength of my case in the face of spirited opposition.

In September 2013, I received another alert from my office. It was about another review, this time of my second book, *Darwin's Doubt*. The call told me that the most prestigious American scientific journal, *Science*, had just published a review by Charles Marshall, a paleontologist and evolutionary biologist at the University of California–Berkeley.[3] I had long admired Marshall's work on the Cambrian explosion. The review was the first to address the main information-based argument of the book and, despite a respectful tone and a few complimentary comments, it offered a decidedly negative assessment of my thesis. Sympathetic colleagues were nevertheless pleased because the review was respectful and the argument of the book had gained the attention of *Science*. I was delighted by what Marshall said in critique.

Everyone likes praise and respect, but Nagel's and Marshall's reviews and the discussions they provoked were important for another reason. When I was PhD student at Cambridge, a supervisor once told me to "beware the sound of one hand clapping." By that he meant that it is impossible to assess an argument without assessing the counterarguments. Later, as a college professor, I developed a corollary to this principle. I used to tell my own students that the *best* way to weigh an argument was to see how well it withstood critical scrutiny and how well its proponents could respond to the strongest objections to their case.

Medieval philosophers presupposed this principle in their disputational method. They would first make an affirmative argument for a proposition and then respond one by one to possible objections to it. In his book *The Discourses of Science*, Italian philosopher of science Marcello Pera argues that science necessarily has a similar rhetorical dimension. Science advances as scientists *argue* about how to interpret evidence.[4] Karl Popper, one of the great philosophers of science of the twentieth century, characterized the scientific method as a process of making affirmative "conjectures" and seeking (or evaluating) possible "refutations." The method of inference to the best explanation, used throughout the investigation in this book, embodies that approach. It recognizes that, in science and philosophy, testing a hypothesis requires comparing its strength against competing hypotheses and assessing the arguments for—and against—those competitors.

Up to this point, I've made a case for theism as an inference to the best metaphysical explanation for the scientific evidence that we have concerning the origin of (1) the universe, (2) the fine tuning of the universe, and (3) the information necessary to produce the first life as well as subsequent novel forms of life. In so doing, I've compared the explanatory power of theism to competing metaphysical hypotheses as well as competing design hypotheses (such as directed panspermia). To strengthen this argument, it remains to demonstrate that it can withstand critical scrutiny. Consequently, this penultimate section of five chapters will respond to objections to the case developed so far.

When presenting an argument as an inference to the best explanation, objections come in two closely related forms. Critics may object by challenging the evidential basis of the argument, or they may challenge the interpretation of the evidence. I've addressed a number of both types of objections already, but other salient, and more current, challenges remain.

Let's look first at the scientific objections to my case for the intelligent design of life, and especially those objections posed since my books *Signature in the Cell* and *Darwin's Doubt* were published.

Challenge from Chemical Evolutionary Theorists

Stephen Fletcher's letters denouncing Thomas Nagel's review of *Signature in the Cell* in the *TLS* affected an air of superiority typical of those who think they are defending "science" against irrational superstition and malevolent religiosity. "The belief that we share this planet with supernatural beings is an old one," said Fletcher in his opening attack. "While the scullery maid sleeps, they are busy in the kitchen making the milk go sour. For a society with no concept of bacteria, this is, perhaps, a forgivable conceit. But for a modern university professor [i.e., Nagel] to take this idea seriously is, I think, mind-blowing."[5]

Of course, thinking that a designing intelligence might have had something to do with the origin of life—no better than positing pixie "magic" in Fletcher's view—does not deny that ordinary natural processes, such as bacteria causing milk to sour, occur with predictable regularity. As we saw in Chapter 2, belief in the intelligent design of the universe and life *and* in the orderly concourse of nature went hand in hand during the scientific revolution and gave us the idea of the laws of nature.

Clearly, Fletcher's opening salvo constituted a straw-man attack, since Nagel would not only doubtless disavow belief in fairies, pixies, or devils, but also because he explicitly disavows belief in God. As Nagel explained, "[Since] I am not tempted to believe in God, I do not draw Meyer's conclusions." Instead, as he explained, he endorsed the book for its "careful treatment of a fiendishly difficult problem" and for its "challenge to the dogma that everything in the world must be ultimately explainable by chemistry and physics."[6]

Even so, Fletcher's critique was not entirely lacking in scientific substance. Instead, he also presented spirited scientific challenges to the thesis of *Signature in the Cell*. Recounting these challenges will offer readers an excellent opportunity to evaluate the strength of my argument for intelligent design as the best explanation for the origin of the information necessary to produce the first life.

The RNA World Revisited

In his first letter, Fletcher claimed that my critique of standard chemical evolutionary theories had failed to recognize that "natural selection is in fact a chemical process as well as a biological process, and it was operating for about half a billion years before the earliest cellular life forms appear in the fossil record." He then faulted Nagel and me for not recognizing that "natural selection does not require DNA" and that "before DNA there was another hereditary system at work, less biologically fit than DNA, most likely RNA (ribonucleic acid)." Fletcher concluded by recommending with some lively invective that "readers who wish to know more about this topic are strongly advised to keep their hard-earned cash in their pockets, forgo Meyer's book, and simply read 'RNA World' on Wikipedia."[7]

Nagel responded a week later by pointing out that I had extensively reviewed and critiqued the RNA-world proposal in *Signature in the Cell*, something Fletcher presumably should have known, since his statement, "It is hard to imagine a worse book," implied that he read the book. "If he has," Nagel declared, "he knows that it includes a chapter on 'The RNA World,' which describes that hypothesis for the origin of DNA at least as fully as the Wikipedia article that Fletcher recommends." He then noted that the tone of Fletcher's letter exemplified "the widespread intolerance" among scientists who think, as Nagel had said before, that everything "must be ultimately explainable by chemistry and physics."[8]

Another British chemist, John Walton, from St. Andrews University, also wrote to the *TLS* challenging Fletcher's specific scientific claims. Walton noted (1) that "intense laboratory research has failed to produce even one nucleotide (RNA component) under geologically plausible conditions"; and (2) that scientists had encountered "insuperable problems" in explaining the origin of the information that would need to be present in "the chains of nucleotides required for the RNA world."[9]

Fletcher responded again, reiterating his assertions about the sufficiency of the RNA world and by comparing both Nagel and Walton to "someone who has attended a Uri Geller spoon-bending demonstration, and has decided that the laws of metallurgy are in urgent need of repair."[10]

Over a month after Fletcher's original broadside, the *TLS* published my letter in response. Since in *Signature* I had detailed numerous problems with the RNA-first origin-of-life hypothesis, I focused on the crucial problem: the problem of the origin of specified genetic information. The RNA-world hypothesis posits that life first arose from a process of chemical evolution that gained traction after self-copying RNA molecules putatively first made prebiotic natural selection possible. That's why Fletcher insisted that "natural selection is in fact a chemical process" and that "natural selection does not require DNA."

In reply, I noted that the RNA-world hypothesis *presupposed, but did not explain*, the origin of genetic information:

> Everything we know about RNA catalysts, including those with partial self-copying capacity, shows that the function of these molecules depends on the precise arrangement of their information-carrying constituents (i.e., their nucleotide bases). Functional RNA catalysts arise only once RNA bases are specifically arranged into information-rich sequences—that is, function arises after, not before, the information problem has been solved. For this reason, invoking prebiotic natural selection in an RNA world does not solve the problem of the origin of genetic information; it merely presupposes a solution in the form of a hypothetical, information-rich RNA molecule capable of copying itself.[11]

For good measure, I quoted the Nobel laureate Christian de Duve, whom we met in Chapter 9 and who noted that postulations of prebiotic natural selection fail because they "need information which implies they have to presuppose what is to be explained in the first place."[12]

New Evidence for the RNA World?

Fletcher replied vigorously with yet a third letter in the *TLS*. Here he cited several recent scientific papers that he claimed demonstrated the plausibility of the RNA-world hypothesis. For example, Fletcher cited a study by chemist John Sutherland (Fig. 15.2) of the University of Manchester.[13] Sutherland, with colleagues Matthew Powner and Béatrice Gerland, partially addressed one of the many difficulties associated with the RNA-world scenario. Starting with several simple chemical compounds, Sutherland and colleagues successfully synthesized a pyrimidine ribonucleotide, one of the two types of the four bases of the RNA molecule. (Of the four information-carrying nucleotide bases in DNA and RNA, chemists classify two as "pyrimidines" and two as "purines" due to differences in chemical structure.)

FIGURE 15.2
The University of Manchester origin-of-life biochemist John Sutherland.

Fletcher also cited work by Tracey Lincoln and Gerald Joyce[14] that ostensibly established the capacity of RNA to self-replicate. He concluded: "It is Stephen Meyer's bad luck to have published his book in 2009, the very year that the RNA-world scenario became eminently plausible. . . . I am afraid that reality has overtaken Meyer's book and its flawed reasoning."[15]

Fletcher's third challenge provided a critical test of the strength of the case for intelligent design as an explanation for the origin of the first life. Others have noted as much, including the anonymous editors of the *Wikipedia* entries. The online encyclopedia, the source that Fletcher initially recommended, is not always reliable, especially on controversial subjects. Interestingly, though, *Wikipedia* now cites Fletcher's letters to the *TLS* as a key factual refutation of the case for intelligent design in *Signature in the Cell*, thus forming a neat circle: in challenging the case for intelligent design, Stephen Fletcher appealed to the authority of *Wikipedia*, while *Wikipedia* appealed to the authority of Stephen Fletcher. Meanwhile, since 2010, to demonstrate the plausibility of the RNA world and the general claim that life could have arisen from simple chemicals without intelligent guidance, proponents of chemical evolution have repeatedly cited the same studies that Fletcher did.

So have prebiotic simulation experiments rendered the RNA-world hypothesis "eminently plausible" and in so doing made the design hypothesis unnecessary as an explanation for the origin of the information necessary to produce the first life?

Not exactly.[16] Actually, not at all. Let's take a closer look at the two experiments that Fletcher cited.

Too Clever by Half

Fletcher cited, first, the experiments, mentioned above, in which chemists Sutherland, Powner, and Gerland synthesized one of the two types of the four nucleotide bases of the RNA molecule. Nevertheless, this work did nothing to address the much more acute problem of explaining how the nucleotide bases in DNA or RNA acquired their specific information-rich arrangements. In effect, the Powner study helped explain the origin of two of the "letters" in the genetic text, but not the specific arrangements of the four different "letters" into functional genetic "words" or "sentences" (Fig. 15.3).

Moreover, Powner and his colleagues only partially addressed the problem of generating the constituent building blocks of RNA under plausible prebiotic conditions. The weakness in their demonstration, ironically, was their own skillful intervention. To ensure a biologically rel-

FIGURE 15.3

The RNA-world scenario in seven steps. *Step 1:* The building blocks of RNA arise on the early earth. *Step 2:* RNA building blocks link up to form RNA oligonucleotide chains. *Step 3:* An RNA replicase arises by chance and selective pressures ensue favoring more complex forms of molecular organization. *Step 4:* RNA enzymes begin to synthesize proteins from RNA templates. *Step 5:* Protein-based protein synthesis replaces RNA-based protein synthesis. *Step 6:* Reverse transcriptase transfers genetic information from RNA molecules into DNA molecules. *Step 7:* The modern gene expression system arises within a proto-membrane. Each of the steps in this scenario is biochemically implausible, particularly steps 3 and 4, which presuppose significant sources of unexplained genetic information.

evant outcome, they had to intervene—repeatedly and intelligently—in their experiment: first, by selecting only the "right-handed" versions of sugar that life requires (sugars, like amino acids, come in two mirror-image chemical structures called isomers); second, by purifying their reaction products at each step to prevent interfering cross-reactions; and third, by following a precise procedure in which they carefully selected chemically purified reagents and then choreographed the order in which those reagents were introduced into the reaction series. As my colleague David Berlinski pointed out, "They began with what they needed and purified what they got until they got what they wanted."[17]

Thus, not only did this study fail to address the problem of getting nucleotide bases to arrange themselves into functionally specified sequences, but to the extent that it succeeded in producing biologically relevant chemical constituents of RNA, the study illustrated the indispensable role of *intelligence* in generating such chemistry.

The Lincoln and Joyce study that Fletcher also cited, the one that purportedly established the capacity of RNA to self-replicate, illustrates this problem even more acutely. The "self-replicating" RNA molecules in this experiment did not copy a template of genetic information from free-standing nucleotides as protein machines (called polymerases) do in actual cells. Instead, in the experiment, a presynthesized *specifically sequenced* RNA molecule merely catalyzed a single chemical bond, fusing together two other presynthesized partial RNA chains. Their version of "self-replication," therefore, amounted to nothing more than joining two sequence-specific premade halves together.

More significantly, Lincoln and Joyce themselves *intelligently arranged* the base sequences in these RNA chains. They generated the sequence-specific functional information that made even this limited form of replication possible. Thus, the experiment not only demonstrated that even a limited capacity for RNA self-replication depends upon information-rich RNA molecules; it also lent additional support to the hypothesis that intelligent design is the only known means by which functional information arises.

What Are We Simulating?

Since 2010, origin-of-life researchers have conducted other experiments that putatively establish the plausibility of key steps along the way to life from simpler prebiotic chemistry. But these studies illustrate the problem of "investigator interference" as well, since their success also depends crucially on the choreography of the *intelligent* chemists doing the experiments.

Professor James Tour (Fig. 15.4), of Rice University, one of the leading synthetic organic chemists in the world, has written two seminal re-

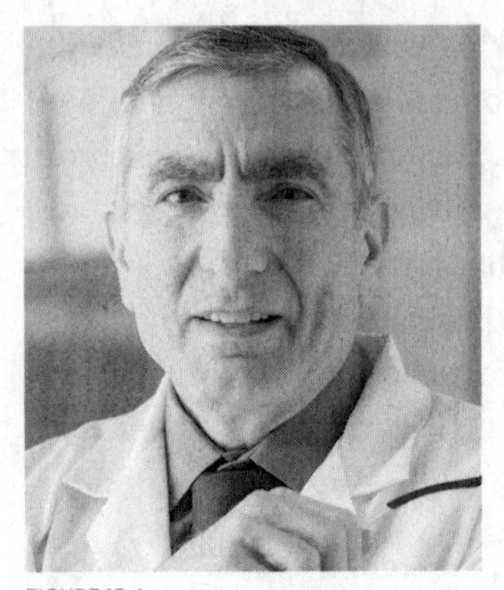

FIGURE 15.4
James Tour, Rice University, organic chemist.

view articles since 2016 assessing the status of chemical evolutionary theory.[18] He has concluded that all current chemical evolutionary scenarios lack plausibility precisely because they depend upon intelligently purified chemical reagents, intelligently designed chemical recipes, and intelligently guided experimental protocols.

Moreover, as I showed in *Signature in the Cell*, whenever chemists set up or interfere in a reaction sequence—or whenever they otherwise apply constraints to a chemical system—to ensure one outcome and preclude others, they effectively input information into that system. In so doing, they inadvertently simulate, if anything, the need for intelligent design to generate biologically relevant chemistry and information.

Challenge from Evolutionary Biologists

In my second book, *Darwin's Doubt*, I extended the information-based argument of *Signature in the Cell* by showing that intelligent design best explains the origin of the information necessary to build fundamentally new forms of animal life and their novel body plans. Whereas *Signature* presented intelligent design as an alternative to materialistic theories of chemical evolution, *Darwin's Doubt* presented intelligent design as an alternative to materialistic theories of biological evolution. Chapter 10 of this book summarized core parts of that argument. Subsequent scientific criticisms have afforded excellent opportunities to test its strength as well.

The review of *Darwin's Doubt* that most directly challenged the main thesis of the book was written by Berkeley paleontologist Charles Marshall and published in *Science*. Marshall, a thoroughly mainstream evolutionary biologist, has taken little interest in the kind of antireligious polemics advanced by Richard Dawkins and other New Atheists. At the same time, he acknowledges holding a materialistic worldview and thinks that some contemporary form of evolutionary theory can (or will ultimately) explain the major innovations in biological form in the history of life. A leading expert on the Cambrian era, Marshall has posited various explanations for the abrupt appearance of new forms of life in that period.

Unlike many other reviewers, Marshall grappled directly with my main arguments about the problem of the origin of biological information and morphological novelty. Yet his review demonstrated—if inadvertently—that evolutionary biologists have not solved that problem and do not have a better explanation than intelligent design.

To rebut my claim that evolutionary mechanisms lack the creative power to generate the information necessary to produce new forms of animal life, Marshall did not defend the sufficiency of mutation and natural selection (or any other materialistic evolutionary mechanism). Instead, he disputed that significant amounts of new genetic information *would have been necessary* to build new animals and their distinctive body plans.

Marshall claimed that "rewiring" of what are called developmental gene regulatory networks (dGRNs) could produce new animals from preexisting genes. Developmental gene regulatory networks comprise networks of genes and gene products (DNA-binding proteins and regulatory RNAs) that control the timing and expression of genetic information during animal development. The components in these networks transmit *signals* (known as transcriptional regulators or transcription factors) that influence the way individual cells develop and differentiate. For example, exactly *when* a signaling molecule gets transmitted often depends upon when a signal from another molecule is received, which in turn affects the transmission of still others—all beautifully coordinated to perform specific time-critical functions. These networks of genes and gene products function much like integrated circuits and ensure that the developing organism produces the right proteins at the right times to service the right types of cells during embryological development (Fig. 15.5).

This "rewiring" hypothesis formed the basis of Marshall's critique. As he argued:

> [Meyer's] case . . . rests on the claim that the origin of new animal body plans requires vast amounts of novel genetic information coupled with the unsubstantiated assertion that this new genetic information must include many new protein folds. In fact, our present understanding of morphogenesis indicates that new phyla were not made by new genes but largely emerged through the rewiring of the gene regulatory networks (GRNs) of already existing genes.[19]

Superficially, Marshall's proposal sounded plausible. Nevertheless, it too presupposed significant and unexplained sources of biological information.

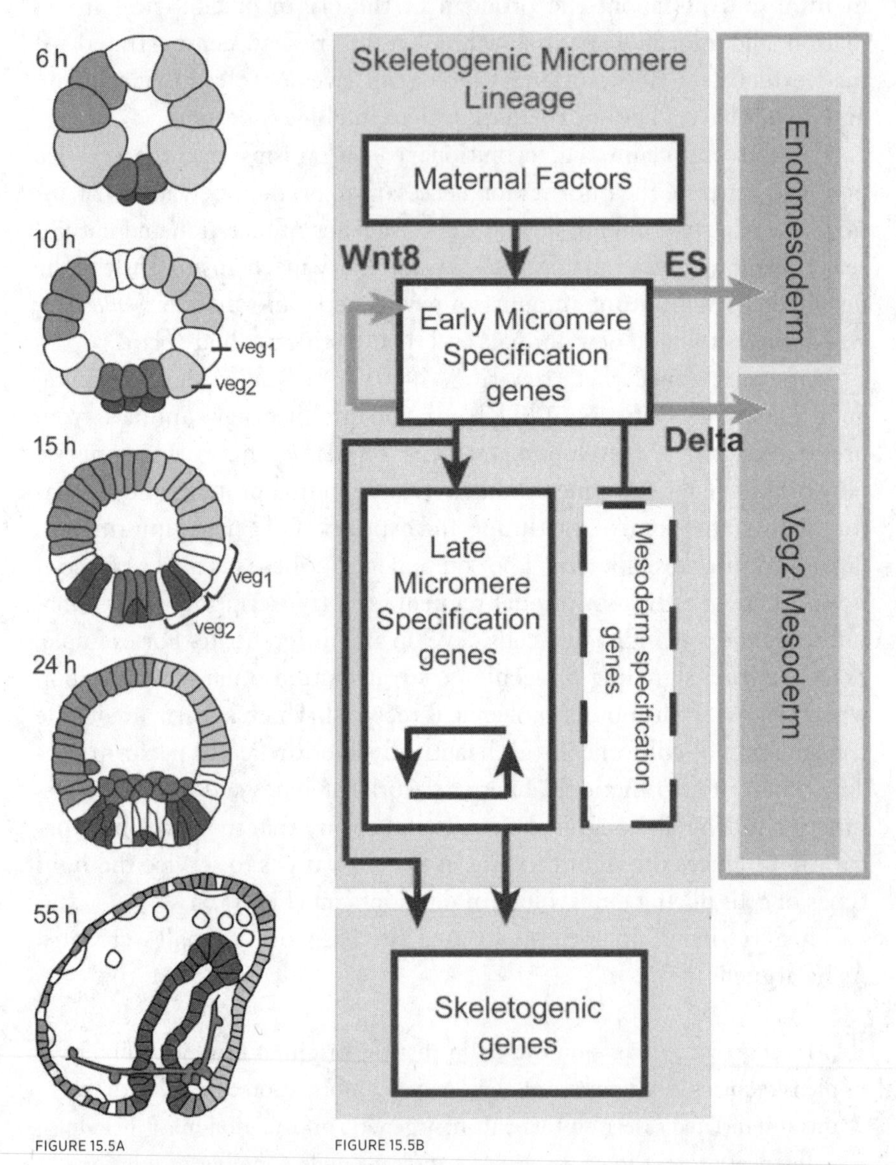

FIGURE 15.5A FIGURE 15.5B

Developmental gene regulatory networks (dGRNs) coordinate the timing and expression of genetic information during animal development from embryo to fully developed adult form. When developmental biologists map the functional relationships in these coordinated networks of genes and gene products (including proteins or regulatory RNAs) the resulting schematics look strikingly similar to integrated circuits. Figure 15.5a (left, above) shows the development of a purple sea urchin, *Strongylocentrotus purpuratus*, starting at six hours after fertilization and progressing through cell division to fifty-five hours when the larval skeleton appears. Figure 15.5b (right, above) depicts the major classes of genes involved in specifying the larval skeleton. Figure 15.5c (facing page) shows the detailed genetic circuitry implicated in the overall "gene regulatory network" controlling the construction of the larval skeleton.

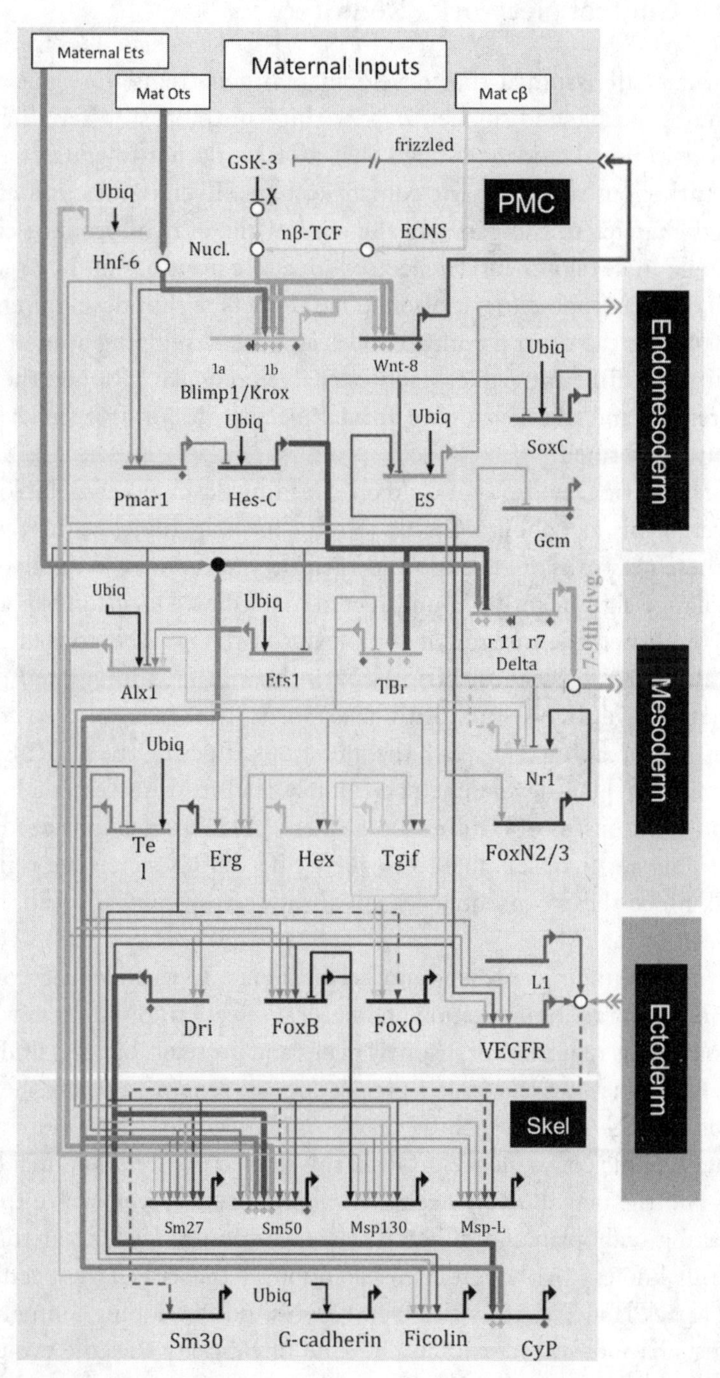
Maternal Ets
Maternal Inputs
Mat Ots
Mat cβ
frizzled
GSK-3
PMC
Ubiq
nβ-TCF
ECNS
Nucl.
Hnf-6
Wnt-8
1a
1b
Blimp1/Krox
Ubiq
SoxC
Pmar1
Hes-C
ES
Gcm
Endomesoderm
7-9th clvg.
Alx1
Ets1
TBr
r 11 r7
Delta
Mesoderm
Nr1
Te
l
Erg
Hex
Tgif
FoxN2/3
L1
Dri
FoxB
FoxO
VEGFR
Ectoderm
Skel
Sm27
Sm50
Msp130
Msp-L
Sm30
G-cadherin
Ficolin
CyP

FIGURE 15.5C

Elastic Control Networks Required

First, Marshall assumed that developmental gene regulatory networks were more flexible and subject to "rewiring" in the past.[20] Yet all available observational evidence shows that dGRNs do not tolerate changes or perturbations to their basic control systems. Even modest mutation-induced changes to the genes in the core of the dGRN produce either no change in developmental trajectory (due to a preprogrammed redundancy) or catastrophic (most often lethal) effects within developing animals. Disrupt the central control nodes and the developing animal does not shift to a different viable, stably heritable body plan. Rather, the system crashes, and the developing animal usually dies or, if it survives, is severely malformed.[21]

The late Eric Davidson, of Caltech, a leading developmental biologist, discovered this fact about dGRNs.[22] In his investigations, he discovered what these networks of genes do and what they never do; what they never do is change significantly via undirected mutations. Davidson explained why. He likened the integrated complexity of the dGRNs to that of an integrated circuit on an electrical circuit board. This integrated complexity makes dGRNs stubbornly resistant to fundamental restructuring without breaking.[23] Instead, the mutations affecting the dGRNs that regulate body-plan development inevitably lead to "catastrophic loss of the body part or loss of viability altogether."[24] Davidson emphasized that "there is always an observable consequence if a dGRN subcircuit is interrupted. Since these consequences are always catastrophically bad, flexibility is minimal."[25]

Davidson's findings present another challenge to the adequacy of the mechanism of random mutation and natural selection. Building new animal body plans requires not just new genes and proteins, but *new* dGRNs. But to build a new dGRN from a preexisting one requires altering the preexisting dGRN—the very thing Davidson showed does not occur without catastrophic consequence.[26] Given this, how could a new animal body plan—and the new dGRNs necessary to produce it—ever evolve from a preexisting body plan and dGRN? Davidson himself made clear that no one really knows.[27] Although many evolutionary theorists have speculated about early "labile" (highly flexible) dGRNs, no developing animal that biologists have observed exhibits the kind of elasticity that the evolution of new body plans requires. Davidson, when discussing these hypothetical labile dGRNs, thus acknowledged that evolutionary biologists are speculating "where no modern dGRN provides a model."[28]

But there is a more fundamental and obvious problem. Marshall claimed that building new forms of animal life does not require new sources of genetic information. But his account of body-plan building (morphogenesis) *presupposes* many unexplained sources of such information. Indeed, he presupposes at least three. Let's examine each in turn.

The Genetic Information in dGRNs

Marshall presupposed unexplained genetic information, first, by invoking preexistent developmental gene regulatory networks. The many genes that code for signaling proteins and RNAs in developmental gene regulatory networks contain a vast amount of genetic information—the origin of which Marshall does not explain.

In his scientific papers and in his discussion of how "rewiring" gene regulatory networks could generate new body plans, Marshall clearly recognizes that preexisting genes would be necessary to produce new animals. He emphasizes that *Hox* genes, in particular, must have played a significant causal role in producing the origin of the first animals during the Cambrian explosion.[29] *Hox* genes are information-rich regulatory genes that coordinate the expression of other genes and thus play important roles in many dGRNs. Nevertheless, he does not explain the origin of these or any other information-rich genes in dGRNs. Thus, his proposal begs the question as to the origin of at least one additional, significant, and necessary source of genetic information.

A "Genetic Toolkit" for Anatomical Novelties

When Marshall wrote that new animals "emerged through the rewiring of the gene regulatory networks (GRNs) of already existing genes," he did not specify whether he meant already existing genes in genetic regulatory networks or other preexisting genes such as those necessary for building the specific anatomical structures that characterize the Cambrian animals (the expression of which dGRNs regulate). Yet when writing elsewhere, Marshall has emphasized that building new animal body plans would require many other preexisting genes, indeed, a preexisting, preadapted "genetic toolkit" for building specific anatomical parts and structures.[30]

In a 2006 paper, "Explaining the Cambrian 'Explosion' of Animals," Marshall noted: "Animals cannot evolve if the genes for making them are not yet in place. So clearly, developmental/genetic innovation must have played a central role in the [Cambrian] radiation."[31] Indeed, Marshall emphasized, in addition to *Hox* genes, the need for "gene novelties" for building the anatomical structures and other novel features of the various animals that arose in the Cambrian period.[32]

Of course, he's right about this. Building new animals would have required a whole range of different proteins to build and service specific forms of animal life. Different forms of complex animal life exhibit unique cell types, and typically each cell type depends upon other specialized or dedicated proteins—which in turn require genetic information.[33] New forms of animal life would have needed various specialized proteins: for facilitating adhesion, for regulating development, for building specialized tissues or structural parts of specialized organs, for producing eggs and sperm, and many other distinctive functions and structures. These proteins must have arisen sometime in the history of life, but Marshall does not explain how the information for building them originated.

Rewiring Networks and Informational Inputs

Finally, "rewiring" genetic circuitry as Marshall envisions it would itself require new information. To see why, consider what would be needed to rewire the circuitry of the 1950s vintage electric guitar shown in Figure 15.6. Notice that the material components of the three different designs of the circuitry in the figure are the same in all three guitars, though the musical tones produced by the rewired guitars will differ perceptibly in accord with the designer's intent. Rewiring requires the deliberate selection of a specific configuration of parts out of a much larger range of possible options. Thus, it requires an infusion of specified information to transform the original system into new and different arrangements of parts. Notice too that such an informational input will be required whether the individual parts of the circuit remain largely the same or whether new parts must be introduced.

In a similar but greater way, given the complexity of an animal compared to a guitar, rewiring the circuitry of a gene regulatory network would also require new inputs of information. It would require multi-

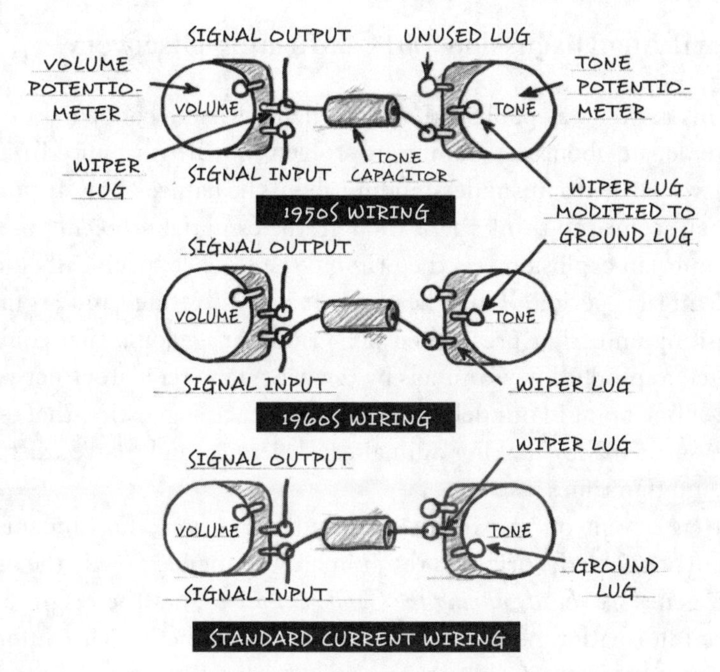

FIGURE 15.6
Three different circuitry designs for different electric guitars. Notice that though the material components of the three designs are the same in all three guitars, converting one design to another would require a reconfiguration of the parts and, thus, an input of information.

ple coordinated changes in the sequences of bases within the individual genes and/or changes to the arrangement or timing of expression of whole genes within the developmental gene regulatory network. Such reconfiguring would entail fixing certain material states and excluding a vast ensemble of others. Thus, it would constitute a substantial infusion of new functional information into the dGRN.[34] Thus, even if it were possible to rewire genetic regulatory networks without destroying a developing animal, Marshall's "rewiring" proposal itself presupposes, but does not explain, the need for an additional source of information.

Note, finally, the inescapably teleological (or purposeful) language of Marshall's "rewiring" proposal. Any electrician or electrical engineer—indeed, anyone who works with actual circuitry and a power supply with current passing through the circuit—knows that successful rewiring requires well-informed decisions, that is, both information *and* intelligent design. What rewiring manifestly does *not* allow is *random* changes. That's a great way to burn down your house or blow out the motherboard on your computer.

A Clarifying Discussion and Confirming Discovery

After his review was published, Marshall and I had a congenial ninety-minute debate about the Cambrian problem on British radio. In it, we may have clarified a misunderstanding about the nature of my argument. Marshall seemed to think that I thought the essential problem posed by the Cambrian explosion was the origin of new genes specifically during the Cambrian period. Thus, he thought he had refuted my argument by positing an earlier preadapted pre-Cambrian genome that could be activated to produce new animals by rewiring gene regulatory networks. Thus, at one point in our debate he cited evidence suggesting that one of the key proteins for making animal exoskeletons might have existed in pre-Cambrian times.

But, as I went on to explain in that conversation, the fundamental problem posed by the origin of the animals was not necessarily the origin of new genes *specifically during the Cambrian period*, but the origin of the genetic information necessary to build animals, whether that information first arose in the Cambrian period or earlier. Pushing the origin of genetic information back into the pre-Cambrian period left unanswered the question of its ultimate origin.

Recall also from Chapter 10 that mutagenesis experiments have established the *extreme* rarity of functional genes and proteins among the many possible ways of arranging nucleotide bases or amino acids within their corresponding "sequence spaces."[35] This rarity makes it overwhelmingly more likely than not that a series of random mutational searches will *fail* to generate even a single new gene or protein fold within available evolutionary time. Marshall did not explain *how* a random mutational search could have located the extremely rare functional sequences of nucleotide bases capable of building protein folds within an exponentially large sequence space of possible arrangements. In other words, he does not explain how any evolutionary mechanism could have solved the search problem described in Chapter 10. Instead, he simply assumes that the necessary genes for building new forms of animal life arose earlier in the history of life, without explaining *how* they did.

I did and still do suspect that much of the genetic information necessary to account for the abrupt appearance of the Cambrian animals arose in the Cambrian period. (Recent genetic analyses have confirmed my view.) But I acknowledged that the genes necessary to build the Cambrian animals might have arisen earlier without in any way solving the fundamental problem. I noted that positing *preexisting* genetic information (e.g., for building animal exoskeleton proteins) left unanswered the question of the earlier origin of that genetic information.[36] To that, Marshall replied, "Fair enough." In so doing, in my view, he effectively acknowledged the reality of the problem of the ultimate origin of genetic information.[37]

Challenge from Theistic Evolutionists

Since Marshall's review, two prominent theistic evolutionists have also challenged my critique of the efficacy of mutation and natural selection as a mechanism for generating new genetic information. Deborah Haarsma, of the BioLogos Foundation, a prominent group dedicated to winning Christians over to evolutionary thinking, has claimed that new studies show functional genes and proteins are *not* extremely rare. This is despite what the experiments conducted by Douglas Axe and others (see Chapter 10) have indicated.[38] Dennis Venema, a biologist at Trinity Western University, likewise associated with BioLogos, has argued that the evolution of an enzyme capable of digesting synthetic nylon shows that functional genes and proteins cannot be extremely rare. Consequently, they argue, new information capable of building a new protein fold can arise by mutation and selection in the available evolutionary time.

Haarsma's claim is false. Venema's claims are inaccurate and misleading.

First, at least four other studies using different methods of estimating the rarity of functional proteins[39] have confirmed Axe's multiyear experimental study.[40] Moreover, recent work at the Weizmann Institute in Israel buttresses Axe's original conclusions. Protein scientist Dan Tawfik has shown that protein folds lose their structural and thermodynamic stability as more and more mutations accumulate. Experimentally, Tawfik found that he could completely destroy the stability of numerous different protein folds with between three and fifteen random mutational changes.[41] Yet to turn one protein fold into another requires many more than just fifteen mutations.

So just as a series of random changes to computer code will destroy the function of the software before a new program could arise, a small handful (typically between 3 and 15) of random changes to the amino acid sequence in a protein will destroy the stability of the protein fold well before enough mutations could accumulate to generate a novel fold.

In fact, function-ready protein folds will degrade more quickly than English sentences. Moreover, Tawfik showed that this same thing occurs in large classes of what are called globular proteins, not just the beta-lactamase enzyme that Douglas Axe studied, suggesting that Axe's results can be generalized.

In addition, as the German paleontologist Günter Bechly and two other Discovery Institute colleagues, complex systems physicist Brian Miller and mathematician David Berlinski, note: "Calculations based on Tawfik's work confirm and extend the applicability of Axe's original measure of the rarity of protein folds. These calculations confirm that the measure of rarity that Axe determined for the protein he studied is actually representative of the rarity for large classes of other globular proteins."[42] They also note that Tawfik (no friend of intelligent design) has reluctantly described the origination of a novel protein fold as "something like close to a miracle."

Nevertheless, Haarsma cites work by an Italian research group that allegedly contradicts Axe's findings.[43] That study sought to evaluate how often randomly generated amino-acid chains (polypeptides) organize themselves into stable three-dimensional structures. Yet the test the Italian group used to identify stable structures couldn't distinguish folded functional proteins from nonfunctional aggregations of amino acids. The group did report two folded structures, but they discovered that, except in strongly acidic environments, these structures formed insoluble aggregates, not protein folds. This means these amino-acid chains would not fold in actual living cells. Thus, nothing in the Italian study refutes either Axe's or Tawfik's results.

Venema's claims are also deeply problematic.[44] Venema points to the discovery in the 1970s of an enzyme (a protein) called "nylonase" to refute my claim that a random mutation–driven search for a novel protein fold is overwhelmingly more likely to fail than to succeed even in the multibillion-year history of life on earth. The nylonase enzyme can break down synthetic nylon. Venema claims the rapid origin, in just forty years, of the nylonase enzyme demonstrates the power of evolutionary processes to generate a "brand-new protein." As he argues, "Since nylon is a synthetic chemical invented in the 1930s, this indicated that these bacteria had adapted to use it as a food source in a mere forty years—less than a blink of the eye, in evolutionary time scales. . . . Rather than being a modified version of another enzyme, this functional sequence of amino acids had popped into existence in a moment, through a single mutation."[45]

He further claims that this enzyme appeared "de novo," basically "brand-new," via a single but major "frame-shift" mutation. A frame-shift mutation occurs when a single nucleotide letter is randomly inserted into the genetic sequence, causing the protein machine that transcribes the genetic message to shift its starting point by one nucleotide—one genetic "letter"—as it transcribes, or "reads," the sequence. Venema thinks that the origin of nylonase via such a mutation demonstrates that functional protein folds must be much more common in sequence space than Axe has argued. As Venema put it, "If only one in 10 to the 77th proteins are functional, there should be no way that this sort of thing could happen in billions and billions of years, let alone 40."[46]

Nevertheless, contrary to what Venema has claimed, nylonase did not arise *de novo* via a single frame-shift mutation; it is not a "a brand-new protein"; and it certainly does not represent a new protein fold.

First, the Japanese researchers Venema cites who have most extensively studied nylonase postulated that it arose by two minor point mutations, not a dramatic frame-shift mutation. These mutations produced just two amino-acid changes or substitutions[47] to a preexisting 392 amino-acid protein—hardly a dramatic *de novo* origin event.

Second, based on their study, the researchers also inferred that the original gene from which the nylonase gene arose coded for a protein with limited nylonase function even before nylon was invented. This seems likely because a naturally occurring "cousin" of nylonase—an enzyme with a high degree of sequence similarity to it—has measurable (if weak) nylonase activity and can be converted to greater nylonase activity with just two minor mutations. This suggests that such mutations optimized the function of a preexisting protein with modest nylonase activity.[48]

Most important, the evidence indicates that nylonase *does not represent a new protein fold*, but instead displays the same stable, complex three-dimensional fold (specifically, a beta-lactamase fold) as both its cousin and likely ancestral protein. The very researchers that Venema cites as his source for the story of the origin of nylonase make this clear. As the Japanese researchers note, "We propose that amino-acid replacements in the catalytic cleft of a *preexisting* esterase with the *beta-lactamase fold* resulted in the evolution of the nylon oligomer hydrolase."[49] Note the terms "preexisting" and "beta-lactamase fold." These words indicate that the mutations responsible for the origin of nylonase did not produce a gene capable of coding for a *new* protein fold, but instead a gene that coded for the same beta-lactamase fold as its predecessor.

Thus, the nylonase story confirms what Axe and I have argued, namely, that the mechanism of random mutation and natural selection can *optimize* (or even shift) the function of a protein, provided it does not have to generate a new fold. Given the extreme rarity of protein folds in sequence space, however, the number of mutational changes necessary to produce a novel *fold* (to *innovate* rather than *optimize*) exceeds what can be reasonably expected to occur in available evolutionary time. The nylonase story confirms, rather than refutes, that claim. It suggests that the mechanism of random mutation and natural selection fails to explain the origin of the *amount* of new information necessary to generate a new protein fold and thus any significant structural innovations in the history of life.[50]

Challenge from Atheistic Evolutionists

Finally, prominent atheistic evolutionists have attempted to refute my critique of the mechanism of random mutation and natural selection as an explanation for the origin of genetic information.

I began this book by describing my March 2016 debate with Lawrence Krauss at the University of Toronto. Krauss, and later Richard Dawkins in defense of Krauss, claimed that my critique misrepresented the evolutionary mechanism as a purely random process. Instead, both Krauss and Dawkins insisted, in Dawkins's words, that "natural selection is a *nonrandom* process," implying that it presumably could succeed in finding the extremely rare functional arrangements of nucleotide bases and amino acids within the space of possible arrangements in available evolutionary time.

After the debate, Dawkins defended Krauss on the blog site *Why Evolution Is True*, operated by University of Chicago biologist Jerry Coyne, another New Atheist. "Meyer was terrible," Dawkins wrote. "When will these people understand that calculating how many gazillions of ways you can permute things at random is irrelevant. It's irrelevant, as Lawrence said, because natural selection is a *nonrandom* process."[51]

I was pleased that Dawkins had decided to weigh in. He had previously declined an invitation from the president of our institute to debate me, stating, "Your people haven't earned it." So his direct engagement not only represented a concession of sorts; it also yielded an opportunity to test the strength of my case against an objection from a prominent critic of intelligent design.

As it turned out, in their attempts to circumvent the information problem, both Dawkins and Krauss had to misrepresent how the neo-Darwinian mechanism works. Natural selection itself is arguably a "non-random process," as Dawkins insisted. Rates of reproductive success do correlate with the traits that organisms possess. Those with fitness advantages will, all else being equal, out-reproduce those lacking such advantages.

Yet clearly there is more to the neo-Darwinian mechanism than just natural selection. The standard neo-Darwinian evolutionary mechanism comprises (1) natural selection and/or (2) genetic drift acting on (3) adaptively random genetic variations and mutations. As conceived from Darwin to the present, natural selection "selects," or acts to preserve, those random variations that confer a fitness or functional advantage upon the organisms that possess them. But it "selects" only *after* such advantageous variations or mutations have arisen. Thus, selection does not *cause* novel variations; rather, it sifts what is delivered to it by the random changes (i.e., mutations) that *do* cause variations. This has been neo-Darwinian orthodoxy for many decades.

All this means that natural selection does nothing to help *generate* functional DNA base (or amino-acid) sequences, that is, new genetic information. It can only *preserve* such sequences (if they confer a functional advantage) *once they have originated*. Adaptive advantage accrues only *after* the generation of new functional genes and proteins—after the fact, that is, of some presumably successful random mutational search. Thus, the evolutionary mechanism *as a whole* depends upon an ineliminable element of randomness—a point that even other evolutionary biologists acknowledged after the debate in Toronto. Laurence Moran and P. Z. Myers, both outspoken atheists, criticized Krauss and Dawkins for mischaracterizing the neo-Darwinian mechanism as wholly nonrandom, and Moran specifically blamed Krauss's uncritical reliance upon Dawkins as the source of his misinformation.[52]

In any case, the need for random mutations to generate novel base or amino-acid sequences before natural selection can play a role means that precise quantitative measures of the rarity of genes and proteins within the sequence space of possibilities are, contrary to what Dawkins claimed, highly relevant to assessing the alleged creative power of the mechanism of mutation and selection. Moreover, the empirically based estimates of the rarity of protein folds (set conservatively by Axe at 1 in 10^{77}) do pose a formidable challenge to those who claim that the evolutionary mechanism provides an adequate means of producing novel genetic information—at least in amounts sufficient to generate such folds.[53]

Why a formidable challenge? Again, because random mutations *alone* must produce *exceedingly rare* functional sequences among a vast combinatorial sea before natural selection can play any significant role. As I discussed in Chapter 10, even in the entire multibillion-year history of life on earth, there have not been enough replication events to generate or "search" anything more than a minuscule fraction (one ten trillion trillion trillionth, to be exact) of the total number of possible nucleotide base or amino-acid sequences corresponding to a single functional gene or protein fold. As with my hypothetical bike thief confronted with many more combinations than he has time to explore, the mechanism of random mutation and natural selection turns out to be much *more likely* to *fail* than to succeed in generating even a single new gene capable of producing a new protein fold in the history of life on earth.

Information All the Way Down . . . to Mind

My interactions with staunch critics reveal that even the best and most prominent scientists defending evolutionary theory have failed to identify a materialistic process that can generate enough information to produce a new protein fold, let alone fundamentally new forms of life. Dawkins and Krauss misrepresented the mechanism of random mutation and natural selection and thus failed to answer the information challenge at the heart of my argument. Venema and Haarsma falsely portrayed the studies they cited and claimed more creative power for the evolutionary mechanism than those studies, or any others, can demonstrate. Fletcher and Marshall did address my information-based argument for intelligent design. But in order to refute it, they presupposed, rather than explained, the origin of the relevant genetic information, as other proponents of evolutionary theory have frequently done.[54]

The simulation experiments that Fletcher cited and a wealth of ordinary experience confirm that intelligent agents do have the causal power to generate new specified or functional information, especially in a digital form. Absent any other known cause, intelligent design stands as the best, most causally adequate explanation for the origin of functional or specified information necessary to produce fundamentally new forms of life. The failure of critics to refute this claim only reinforces the strength of the biological argument for intelligent design.

16

One God or Many Universes?

In making a case for the God hypothesis as the best explanation of biological and cosmological origins, I've mainly compared theism as a metaphysical explanation to the standard formulation of the main competing metaphysical hypothesis on offer in the West, namely, scientific naturalism, or materialism. I have not yet compared theism (or a theistic design hypothesis) to what I call *exotic* naturalism—that is, more elaborate versions of naturalism involving various auxiliary hypotheses to explain evidences that seem otherwise unexpected on more standard formulations of naturalism.

Standard naturalism posits only the matter and energy in this universe as the entities from which all else comes. Proponents of exotic forms of naturalism seek to enhance the explanatory power of naturalism by positing other material universes or dimensions of reality. Some physicists have proposed the existence of other universes or "a multiverse" in part to explain[1] the extreme fine tuning of our universe. Another example of exotic naturalism is called "quantum cosmology." It attempts to explain the origin of our universe as the outcome of a set of possibilities described by the mathematics of quantum mechanics (more on this heady concept in Chapters 17–19).

So to rebut objections to my case based upon these popular exotic forms of naturalism,[2] this chapter and the next three address, first, the multiverse hypothesis and, next, quantum cosmology as possible materialistic alternatives to the God hypothesis. The dictionary defines "exotic," by the way, as "originating in or characteristic of a distant foreign country,"[3] which seems an appropriate description of attempts to explain the origin of our universe by positing other unobservable universes!

Fine Tuning Revisited

Recall from Chapters 7, 8, and 13 that we live in a kind of "Goldilocks universe" where both the laws and constants of physics and the initial arrangement of matter and energy at the beginning of the universe appear finely tuned for life. Some physicists have suggested intelligent design as an explanation for the precise combination of these fine-tuning parameters. The conjunction of the incredible improbability of the fine tuning *and* the way in which the actual values of the finely tuned parameters and conditions match the requirements of a life-friendly universe has suggested design as a "commonsense" interpretation. In Chapter 13, I also argued that since this evidence of design was present from the beginning of the universe, it also suggests the activity of a *transcendent* rather an immanent intelligence.

In making this case, I critiqued several nontheistic interpretations of the fine tuning: (1) the weak anthropic principle, (2) the strong anthropic principle,[4] (3) explanations based upon natural law, and (4) explanations based upon chance. In the last case, I critiqued only explanations that draw on the probabilistic resources of *this* universe. As it happens, most scientific materialists themselves now find these four interpretations of the fine tuning intellectually unsatisfying. Consequently, many have proposed an even more imaginative alternative. This explanation, known as the multiverse, cleverly revives a kind of chance hypothesis, one that attempts to render the cosmological fine tuning probable after all.

The Multiverse

To explain cosmic fine tuning, some physicists have postulated not a "fine-tuner" or intelligent designer, but the existence of a vast number of other universes. This "multiverse" concept not only posits many other universes, but also various mechanisms for producing these universes. Having a mechanism for generating new universes would, according to proponents of this idea, increase the number of opportunities for generating a life-friendly universe. Thus, they portray our universe as something like the lucky winner of a cosmic lottery and the universe-generating mechanism as something like a roulette wheel or a slot machine turning out either life-conducive winners or life-unfriendly losers with each spin or pull on the handle.[5]

It's important to understand why proponents of the multiverse need a universe-generating mechanism. Most proponents think of the different universes that they postulate as causally isolated or disconnected from each other. Thus, for the most part, they do not expect to have any direct observational evidence of universes other than our own.[6] Consequently, nothing that happens in one universe should have any effect on things that happen in another universe. Nor would events in one universe affect the *probability* of events in another universe, including the probabilities of whatever events were responsible for setting the values of the fine-tuning parameters in another universe—or ours.

Yet if all the different universes were produced by the same underlying causal mechanism, then it would be possible to conceive of our universe as the winner of a cosmic lottery, where some winner with just the right laws, constants, and/or initial conditions would eventually emerge. Postulating a "universe-generating machine" could conceivably render the probability of getting a universe with life-friendly conditions quite high and in the process explain the fine tuning as the result of a randomizing element—like the action of a giant slot machine.

Physicists have proposed two leading cosmological models with different mechanisms to explain where new universes might have come from. One model, proposed by Andrei Linde, Alan Guth, and Paul Steinhardt, is called inflationary cosmology.[7] The other model is based on string theory. Both approaches were advanced initially to address specific puzzles in physics and then were later appropriated as multiverse explanations for the fine tuning of our universe.

The Inflationary Multiverse

Let's first review the inflationary cosmological model we examined in Chapter 6 (Fig. 16.1). Proponents of inflationary cosmology posit that just after the big bang the universe expanded at an extremely rapid rate. Then after a tiny fraction of a second, the rate of expansion settled down to a more sedate pace. Physicists first proposed inflationary cosmology to explain several puzzling features of the universe from the perspective of standard big bang cosmology—its relative homogeneity, especially in

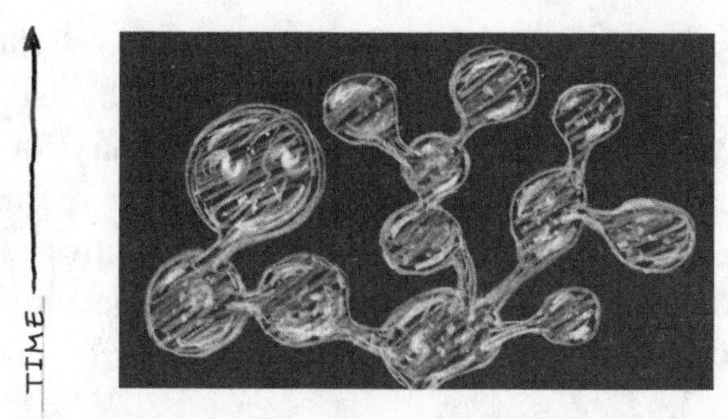

FIGURE 16.1
The inflationary multiverse envisions new universes emerging from older universes. To explain the rapid expansion of space (in all universes), it posits the existence of an inflaton field. To explain the origin of these new universes, it further posits that when the energy of the inflaton field shuts off in precise ways in local areas of individual universes, new "bubble" universes will emerge. Though these new universes would not have different laws and constants of physics, they could, according to proponents of this model, have quite different configurations of mass and energy, making the events and structures that exist in these new universes different from our own.

the temperature of the cosmic background radiation, the flatness of the universe, and the absence of magnetic monopoles.

Proponents of inflationary cosmology posit a specific universe-generating mechanism. According to the currently dominant "eternal chaotic inflationary model," an outward-pushing field with vacuum energy, dubbed "an inflaton field," causes the expansion of a wider space in which our universe and others arose. As the inflaton field expands, the energy of the field sporadically decays in isolated locations. When that happens, the inflaton field spawns other, lower-energy "bubble" universes as the original inflaton field continues to expand into the infinite future. Since new bubble universes expand more slowly than the bubbles that contain them, collisions rarely, if ever, happen. Consequently, a multiverse of causally isolated nested bubble universes results.[8]

Some physicists have appropriated this model to explain fine tuning, though only the fine tuning of the initial conditions, not the laws and constants of physics, since the laws of physics would be the same in all the bubbles within the larger universe.[9] Even so, proponents of the inflationary multiverse argue that, since the inflaton field can produce an infinite number of other universes, every *event* that has occurred in our universe is bound to occur somewhere endlessly many times. It follows that events or conditions that *appear* extremely improbable, considering only our universe, are actually highly probable—or even inevitable. Sooner or later some universe had to acquire the finely tuned conditions necessary to sustain life. Our universe just happened to be the lucky one. According to this theory, since we only observe the bubble universe in which we live, we *falsely* think the conditions necessary for life are extremely improbable, when in fact, given the action of the inflaton field as a universe-generating mechanism, a life-friendly universe must inevitably arise in some universe somewhere—we just happen to be living in that lucky universe.[10]

The String-Theory Multiverse

To explain the fine tuning of the laws and constants of physics, physicists have also appropriated what is known as string theory, itself a pretty heady idea involving many highly abstract entities and physical concepts. I confess ahead of time that string theory can seem a bit jargon-ladened and even arcane. Even so, much of the rest of this chapter discusses the theory, for which I beg your indulgence. I've worked hard to explain the theory as clearly and accessibly as possible, so I think even nonphysicist readers will find understanding it worth a try. Doing so will make it possible to understand a key part of my argument. Indeed, it's important to understand at least the basics of the theory in order to understand currently popular multiverse theories—theories that now stand as the leading alternatives to theistic design as an explanation for cosmic fine tuning.

According to string theory,[11] the fundamental units of matter are tiny one-dimensional strings or filaments of energy rather than elementary particles such as photons, quarks, and electrons. These filaments of energy form different vibrational patterns including both "open" strings and "closed" strings (Fig. 16.2). In fact, string theory teaches that all elementary particles are just manifestations of the underlying behavior of differently vibrating strings.[12]

The earliest version of string theory, which focused on elementary particles known as bosons, thought to carry the strong nuclear force, required a twenty-six dimensional spacetime (including twenty-two unobservable dimensions of space). In expanding string theory to account for matter particles as well as force particles, string theorists have since found that just six or seven extra spatial dimensions will suffice.

That, of course, raises the question of where these unobserved dimensions of space reside. String theorists currently postulate that these other

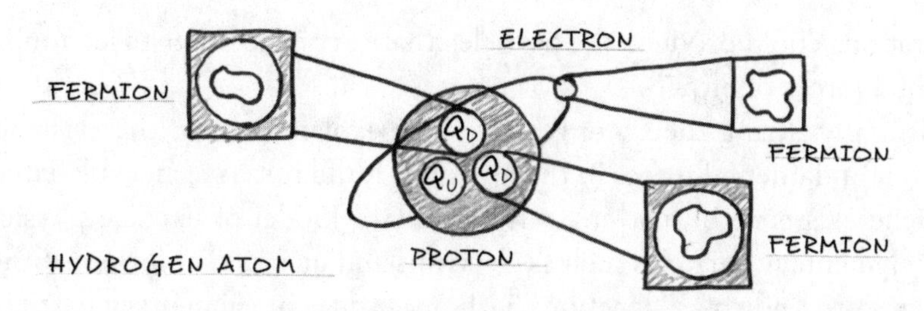

FIGURE 16.2
According to string theory, the fundamental units of matter are composed of vibrating filaments of energy called "strings." Elementary particles or "fermions" are made of "closed" strings. The particles called "bosons" that transmit the fundamental forces of physics are made of open strings. This figure shows the relationship between closed strings and the different elementary particles or fermions that constitute the hydrogen atom. It shows how, according to string theorists, different closed strings make up electrons as well as the "up quarks" and "down quarks" that in turn make up protons.

dimensions of space are curled up into various "topological" or spatial structures on a miniaturized and unobservable scale. They call these various structures "compactifications" of space, or "vacua." These other dimensions of space are postulated to exist inside 10^{-35} of a meter, the spatial radius of what physicists call the Planck length, the distance scale in which quantum gravitational phenomena should occur.

String theorists envision the strings of energy vibrating in various ways inside these tiny structures (compactifications) containing the six extra spatial dimensions. The different kinds of vibrations in those six extra dimensions produce the particle-like phenomena we observe in our three macroscopic dimensions of space. Moreover, just as strings vibrate *within* these compactified dimensions, so-called lines of flux wrap around the outside of them to hold the spaces together in specific shapes. You might think of these lines of flux as functioning roughly like the lines of force in a magnetic field, which determine the orientation of iron filings in the area around the magnet.

String theory is fundamentally a particle physics–based theory of gravity on a vanishingly small quantum scale. A consequence of string theory is the existence of "gravitons." String theorists understand gravitons as massless, closed strings that transmit gravitational forces over long distances at the speed of light. During the 1980s string theory seemed to hold out the possibility of reducing all other fundamental physical forces to gravity, thus suggesting that it might provide a "theory of everything." String theorists posited that specific vibrations of gravitons are responsible for gravitational attraction, while other graviton vibrations could produce the particles that carry the other three fundamental forces of physics.

Initially, string theory only offered an explanation for the existence of the fundamental forces in the universe. It did not have an explanation for the existence of matter. In the standard model of particle physics, the elementary particles such as electrons and quarks that form material objects are known as "fermions," whereas types of elementary particles known as "bosons" transmit forces. Initially, string theorists proposed that vibrating gravitons accounted for the existence of bosons as well as the forces they transmit. Nevertheless, the original version of string theory did not offer an explanation for the existence of fermions (and thus matter).[13]

To extend string theory to describe matter particles (fermions), string theorists invoked a principle known as "supersymmetry." Supersymmetry specifies that for every bosonic elementary particle that exists, a complementary fermionic partner particle must also exist, and vice versa. This supposition reduced the number of required spacetime dimensions in string theory from the originally required twenty-six, to ten. It also made string theory applicable to our universe, since clearly our universe has both forces *and* matter. Thus, string theorists posited not only the existence of gravitons, but also their complementary supersymmetric fermionic partner strings called "gravitinos."[14] They postulated that different vibrations of gravitinos give rise to the different matter particles just as the different vibrations of gravitons produce the bosons. By postulating supersymmetry and multidimensional compactifications of space, string theorists hoped that they would succeed in describing matter and the four fundamental forces of physics in a single, unique, and comprehensive mathematical structure.

Nevertheless, string theory encountered a theoretical difficulty early on. The mathematical structure that allegedly described the four fundamental forces of physics did not have a unique solution corresponding to the physics of our universe. Instead, the mathematical structure of string theory had infinitely many solutions (possible compactifications of the extra dimensions of space), each of which described a different physics. The hope for a unique solution describing the physics of our world quickly receded. Indeed, just for solutions to the equations of the string theory that have a positive cosmological constant (as our universe does), there are anywhere between 10^{500} and $10^{1,000}$ different vacua.

Physicists now call this vast collection of possible solutions or vacua or compactifications of space the "string landscape."[15]

The String Landscape and Fine Tuning

At first, physicists regarded the existence of so many solutions to the mathematical equations as an embarrassment, since they were using string theory to model fundamental physical reality *in our universe*. But some string theorists attempted to turn this vice into a virtue by using the multiplicity of possible solutions as a way of addressing the fine-tuning problem. They proposed that each solution to the equations of string theory corresponds to *a different universe with different physical laws and constants of physics*. They further suggested that the shape of the folded spaces associated with each different solution determines the *laws* of physics manifested in the three observable spatial dimensions of these different universes. They also theorized that the number of lines of flux around the folded spaces (as well as the contours of those spaces) determines the different *constants* of physics.[16]

These string theorists also proposed a mechanism for generating the 10^{500} to $10^{1,000}$ possible universes corresponding to these solutions,[17] thus rendering probable, they supposed, the fine tuning of the laws and constants in *our* universe. They envisioned an initial high-energy compactification of space representing a universe containing one quantum gravitational field (corresponding to one solution to the equations of string theory). They then postulated a process that could cause that compactification of space to change shape and size. More specifically, they explored the idea that as the lines of flux holding the compactification in place began to decay or lose energy, new compactifications of space (containing lower-energy quantum gravitational fields) would result. These new compactifications of space (vacua) would constitute new universes with different physical laws and constants. Recall that the *shape* of the vacua determines how strings vibrate and the resulting forces, and thus the *laws* of physics, whereas the different *sizes* of the components of the vacua (such as its contours and lines of flux) determine the values of the *constants* of physics.

As this process of energy decay continued, one universe would morph into another as each successive universe would "cascade down the landscape" of possible compactifications. Since this process would "explore" many of the possible compactifications[18] with different possible combinations of laws and constants of physics, eventually our life-friendly universe

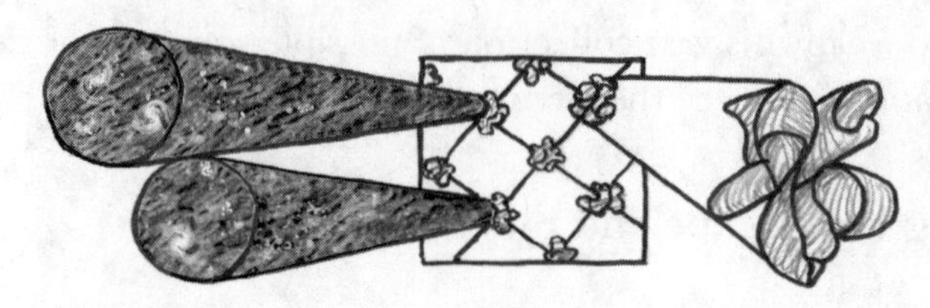

FIGURE 16.3
According to proponents of the string theoretic landscape, each solution to the string theoretic equations corresponds to a multidimensional compactification of space containing different strings of energy. Proponents of the string landscape theorize that each of these compactifications (or vacua) could also correspond to a different universe with different laws and constants of physics. This diagram shows (on the right) a possible compactification of space, (in the middle) an ensemble of such compactifications and (on the left) two universes with presumably different sets of laws and constants of physics corresponding to two possible compactifications. The whole ensemble of possible compactifications or universes is known as the "string theoretic landscape."

would arise. Thus, the process of "exploring the landscape" would render the apparently improbable combination of fine-tuning parameters in our universe an inevitable outcome of a random search process (Fig. 16.3).

Assessing the Multiverse

So do either inflationary cosmology or string theory give an adequate account of the fine tuning of the laws and constants of physics and/or the initial conditions of the universe? Do either of these cosmological models provide a better explanation of cosmic fine tuning than theistic design?

Many physicists today regard the argument over fine tuning as "a wash." Some leading physicists have told me—in all candor—that they regard the multiverse hypothesis as a speculative metaphysical hypothesis, not a scientific one.[19] For them, since neither other universes nor God can be observed or measured, the choice between the two theories comes down to subjective preference.[20] They would deny any evidential or theoretical reasons for preferring one hypothesis over the other.

I've come to a different conclusion. Because both scientific *and* metaphysical hypotheses can be evaluated by comparing their explanatory power to that of their competitors (see Chapter 11), we can assess the relative merits of the theistic design and multiverse hypotheses. And there are many reasons for preferring theistic design over the multiverse.

Reasons to Prefer Theistic Design over the Multiverse

First, as the Oxford philosopher Richard Swinburne has argued, the theistic design hypothesis constitutes a simpler and less *ad hoc* explanation for cosmic fine tuning.[21] Swinburne affirms here the principle of Ockham's razor, which states that when attempting to explain phenomena we should, as much as possible, avoid "multiplying (theoretical) entities." In other words, when evaluating competing hypotheses, we should prefer, all other things being equal, the simpler hypothesis with fewer such theoretical entities.

Swinburne notes that the God hypothesis requires the postulation of only one explanatory entity, an intelligent and powerful transcendent agent, rather than multiple entities, including an infinite number of causally separate universes and the various universe-generating mechanisms posited by multiverse advocates. As he argues, "It is the height of irrationality to postulate an infinite number of universes never causally connected with each other, merely to avoid the hypothesis of theism. Given that . . . a theory is simpler the fewer entities it postulates, it is far simpler to postulate one God than an infinite number of universes, each differing from each other."[22]

Philosopher of physics Bruce Gordon (Fig. 16.4) has amplified this argument by pointing out that accepting the multiverse hypothesis requires accepting two distinct types of universe-generating mechanisms to explain two distinct types of fine tuning. He notes that inflationary cosmology could conceivably explain the fine tuning of the initial conditions of the universe, but it does not explain the origin of the fine tuning of the laws and constants of physics. That's because the inflaton field operates in accord with the same laws of physics across its entire expanding space. As it generates new bubble universes, those universes have the same laws and constants as the inflaton field from which they were birthed. As new bubble universes break off, only new initial configurations of mass-energy could arise.

FIGURE 16.4
Bruce Gordon, the Canadian philosopher of physics and staunch critic of the multiverse hypothesis.

In contrast, string theory might be used to explain the fine tuning of the laws and constants of physics, but in most models it does not generate multiple sets of initial conditions for each choice of physical laws. (I address an exception in n. 26). This means that to formulate a multiverse theory capable of explaining both types of fine tuning, physicists must postulate *two* types of universe-generating mechanisms operating in combination, one based on string theory and one based on inflationary cosmology.

That need has led many theoretical physicists to embrace a synthetic multiverse model dubbed the "inflationary string landscape model." Several theoretical physicists, most prominently Raphael Bousso, Joseph Polchinski, and Leonard Susskind (Fig. 16.5),[23] have advanced this model. They envision the decay of lines of flux around an initial high-energy compactification of space producing different vacua or compactifications (and separate universes) with different laws and constants of physics, just as advocates of the standard string-theory multiverse envision (see above). They also envision that within each string vacuum inflation will begin to occur.

As these new vacua (universes) begin to inflate as the result of the action of their own particular inflaton fields, local fluctuations in energy cause these vacua to decay in a specific area of the inflating space, leading to new bubble universes, each with different initial conditions. If such a proliferation of bubble universes happens to occur in a compactification of space with the right set of laws and constants, and if one of the bubble universes has the right initial conditions and shutoff energies, a life-friendly universe with the right laws and constants *and* the right initial conditions will arise.

Believing Six (or More) Impossible Things Before Breakfast

In theory, at least, the inflationary string landscape model can explain the whole range of fine-tuning phenomena, but only at the cost of what philosophers of science call "a bloated ontology" (recall that ontology is the study of what really exists). That is, it does so by positing an extraordinary number of purely hypothetical and abstract entities for which we have no direct evidence.[24]

Indeed, to explain the fine tuning of both the initial conditions and the laws and constants of physics, the combination of inflationary cosmology and string theory needs to affirm numerous purely hypothetical entities, abstract postulates, and unobservable processes. In particular for an inflationary string landscape model to explain both types of fine tuning, it must make the following postulations:

1. An inflaton field exists.
2. The decay of inflaton fields will produce new bubble universes with different initial conditions.
3. The process of inflation will continue eternally into the future.
4. An infinite number of bubble universes exists (or will eventually exist).
5. Unimaginably small vibrating strings of energy exist.
6. Six or seven extra hidden spatial dimensions exist.
7. The vibrating strings of energy within string vacua create the physical phenomena we observe.
8. Lines of flux around the compactifications of space exist, making them quasi-stable with a positive cosmological constant.
9. Supersymmetry applies to fundamental strings, so that both gravitons and gravitinos exist and their different vibrational modes correspond to all forms of radiation, matter, and the fundamental forces of physics.
10. Every mathematical solution to the equations of string theory corresponds to an actually existing universe with different laws and constants of physics (i.e., the string landscape exists).

In addition, multiverse advocates must affirm that an inflaton field plus some string-theory universe-generating mechanism can together produce *enough* different universes to render probable the origin of the finely tuned initial conditions, laws, and constants of *our* universe.[25]

Bruce Gordon likens accepting all these postulations to believing "six impossible things before breakfast," as in the Alice in Wonderland story—with, that is, a few more impossible (or at least implausible) things thrown in for good measure. Kidding aside, he argues that the theistic design hypothesis—if adjudicated by Ockham's criterion—provides a much simpler explanation of cosmological fine tuning than the multiverse, because theistic design affirms one clear simple postulate (the activity of a transcendent fine-tuner) and avoids the un-

FIGURE 16.5
Stanford physicist Leonard Susskind, a key architect of the "inflationary string multiverse."

necessary and profligate multiplication of abstract theoretical entities entailed by the inflationary string multiverse.[26]

A Dinosaur Bone–Producing Field?

Philosopher of physics Robin Collins (Fig. 16.6) makes a related argument. He argues, all things being equal, we should prefer hypotheses "that are natural extrapolations from what we already know about the causal powers of various kinds of entities."[27] When it comes to explaining the fine tuning of our universe, the multiverse hypothesis fails this test. The theistic design hypothesis does not.

To illustrate, Collins asks his readers to imagine a paleontologist who posits the existence of an electromagnetic "dinosaur bone–producing field," as opposed to actual dinosaurs, as the cause of large fossilized bones.[28] Although certainly such a field qualifies as a *possible* explanation for the origin of the fossil bones, we have no experience of such fields or of their producing fossilized bones. Yet we have observed animal remains in various phases of decay preserved in sediments and sedimentary rock. Thus, most scientists rightly prefer the actual dinosaur hypothesis over the apparent dinosaur hypothesis (i.e., the "dinosaur bone–producing field") as an explanation for the origin of fossils, since it's based on a natural extrapolation of a causal process we have observed.

FIGURE 16.6
The Messiah College philosopher of physics Robin Collins, a prominent proponent of the fine-tuning design argument.

In the same way, we have no experience of anything like inflaton fields with precisely calibrated shutoff energies or string landscapes of compactified extradimensional spaces or anything else (that is not itself designed) producing finely tuned systems. Yet we do have extensive experience of intelligent agents producing finely tuned systems such as Swiss watches, fine recipes, integrated circuits, written texts, and computer programs. Thus, according to Collins, postulating "a supermind" to explain the fine tuning of the universe is a natural extrapolation from our experience-based knowledge of the causal powers of human intelligent agents, whereas the universe-generating mechanisms of the various multiverse proposals lack a similar experiential basis.

Prior Unexplained Fine Tuning

There is yet another reason to prefer theistic design as an explanation of fine tuning over exotic versions of naturalism that postulate a multiverse. In order to explain the origin of the fine tuning in our universe, both inflationary cosmology and string theory (and versions of the multiverse that combine them) posit universe-generating mechanisms that themselves *require prior unexplained fine tuning.*

Let's look at this problem first as it arises within inflationary cosmology. Recall from Chapter 6 that the architects of inflationary cosmology proposed it to explain the absence of certain features of the universe that were expected on the basis of standard big bang cosmology, including the homogeneity and flatness of the universe. Yet for inflationary cosmology to explain the homogeneity of the cosmic background radiation and the flatness of the universe, the postulated inflaton field would need a certain minimum initial energy to drive the exponential expansion of the original volume of space. The inflaton field would also need to decay in just the right amount to produce a habitable bubble universe.

But these conditions imply that the "shutoff" energy of the inflaton field requires precise fine tuning. In fact, depending on the inflationary model, to produce a life-compatible universe the shutoff energy of the inflaton field requires fine tuning of between 1 part in 10^{53} and 1 in 10^{123}. In addition, the shutoff *interval* of the inflaton field must also be precisely finely tuned. In current models, inflation begins at around 10^{-37} seconds after the big bang and ends about 10^{-35} seconds after the big bang, during which the radius of space itself expands by a factor of at least 10^{26}.[29]

In addition, two theoretical physicists, Sean Carroll and Heywood Tam, have shown that an overwhelming majority of universes that a hypothetical inflaton field would produce will not inflate and develop in a life-conducive way.[30] They note that the inflaton field is, in theory, subject to random quantum fluctuations, the vast majority of which would generate universes incompatible with life. In particular, Carroll and Tam have shown that the fraction of realistic cosmologies—cosmologies generating life-friendly universes—resulting from inflation is exceedingly small, *approximately* 1 in $10^{66,000,000}$, an unimaginably small fraction. This vanishingly small ratio, and corresponding degree of *im*probability, implies the need for another source of *extreme* fine tuning.

Making Matters Worse

Thus, not only does the universe-generating mechanism in inflationary cosmology require prior unexplained fine tuning. It actually requires more fine tuning than it was proposed to explain, making the fine-tuning problem it was designed to solve significantly worse. Consider the fine tuning associated with the flatness problem (the ratio of the actual mass density of the universe to the critical mass density). In standard big bang cosmology it is about 1 part in 10^{59}. Consider also that the fine tuning associated with the homogeneity of the cosmic microwave background radiation is a more modest 1 part in 10^5.[31] Yet inflationary cosmology requires several sources of extreme fine tuning that collectively dwarf that left unexplained by the standard big bang model.

In the first place, as we've seen, the fine tuning of the inflation shutoff energies necessary to produce new habitable bubble universes ranges from between 1 part in 10^{53} and 1 part in 10^{123}. Second, the fine tuning associated with the choice of inflationary models (and the various parameters specified in these models) is 1 part in $10^{66,000,000}$ as estimated by Carroll and Tam.

Third, inflationary cosmology makes the already acute fine tuning problem associated with the initial "low-entropy state" of our universe exponentially worse. Recall that low entropy corresponds to a highly ordered state, and high entropy to a more disordered state. In a cosmological context, the "initial low-entropy state" of the universe refers to an initial highly ordered, homogeneous distribution of mass-energy (and a "smooth spacetime"). Indeed, as we saw in Chapter 8, for highly ordered, highly condensed matter such as stable galaxies and planetary systems to have developed, an even more highly ordered arrangement of matter and energy (and a "smoother spacetime") must have existed from the beginning of the universe. That's because an undirected energy flowing through a system will generate more entropy (more disorder and a lumpier spacetime). Think, again, of a tornado going through a junkyard. So if our universe manifests highly ordered arrangements of matter now, the initial arrangement of matter and energy must have been even more highly ordered at the beginning of the universe.

Inflationary cosmology not only does nothing to explain this initial-entropy fine tuning—estimated by Roger Penrose at 1 part in $10^{10^{123}}$—it actually exacerbates it.[32] It does so because the massive energy of expansion during inflation would increase the entropy/disorder of the universe more than the expansion envisioned by standard big bang cosmology. Thus, inflation would imply the need for a greater initial homogeneity (order) in the configuration of mass-energy (and initial smoothness of spacetime) to account for the high degree of order we see today. Since inflationary cosmology posits an exponentially larger surge of energy—an even bigger tornado!—than standard big bang cosmology, it would generate *exponentially* more entropy. It therefore requires an *exponentially* larger corresponding order (lower entropy state) at the beginning of the universe.

Cumulatively, the unexplained fine tuning necessary to produce life-compatible conditions given an inflationary cosmological model dwarfs that of standard big bang cosmology. Bruce Gordon quips that the use of inflationary cosmology to solve the fine-tuning problem associated with the standard big bang model is like "digging the Grand Canyon to fill a pothole."[33]

Preexisting Unexplained Fine Tuning in String Theory

String theory also requires fine tuning of its universe-generating mechanism. Recall that string theorists assume that a particular compactification of space (i.e., a separate universe) containing an extremely high-energy quantum gravitational field begins the process of "exploring the landscape." As the result of the decay of lines of flux wrapped around this compactification, the compactification changes shape and adopts a lower-energy state (or quantum gravitational field). As this process continues and each successive compactification/universe "cascades down the landscape," new universes with new laws and physical constants arise.[34]

Though this process may, in theory, explore many of the possible universes with different sets of laws and constants, it too requires exquisite prior fine tuning. Recall that in string theory just the solutions that produce universes with a positive cosmological constant represent 10^{500} (or possibly $10^{1,000}$) possible solutions.[35] Many more such solutions exist that do not meet this criterion. Thus, in order to ensure that a specific universe cascades to as many of the lower-energy universes as possible (thus increasing the odds of finding one with life-compatible laws and constants), the initial universe must start at an extremely high-energy (presumably something close to the high*est*) level. Moreover, since the Borde-Guth-Vilenkin theorem also applies to standard versions of the inflationary string landscape model, exploring the landscape fully would require not only a beginning, but a beginning with just such a specific high-energy universe or compactification. That, in turn, implies the need for exquisite *initial-condition* fine tuning as measured by the rarity of the highest-energy solution(s) (roughly 1 part in 10^{500} or more) within the array of possible solutions or compactifications or universes.

Even so, however, there is no guarantee that all the possible compactifications will get explored this way. String theorists have no way of knowing that this mechanism would explore "every path down the mountain." That requirement implies the need for some unknown finely tuned mechanism or *directed process* to ensure that "exploring the landscape" lands on one of the few, extremely improbable, life-compatible compactifications.[36]

In addition, specific string-theoretic models invariably manifest other forms of fine tuning. For example, the cyclic ekpyrotic model that posits the creation of new universes as the result of two colliding "branes" (containing many universes each) requires extensive fine tuning in the positioning of the branes. Physicists Renata Kallosh, Lev Kofman, and Andrei Linde have shown the two postulated branes must maintain a parallel positioning in order to prevent large inhomogeneities in the resulting universes.[37] Further, the two universes must remain parallel in the multidimensional space that contains the branes (called "the bulk") to better than 1 part in 10^{60} across a distance 10^{30} times greater than the distance between them in order to produce a life-friendly universe. (Think of two large pieces of paper stacked on top of each other maintaining equidistant spacing across the whole of their widths and lengths.) Kallosh, Kofman, and Linde have also shown that the energy potential associated with the multidimensional colliding branes has to be fine-tuned to 1 part in 10^{50}.[38]

Robin Collins has a clever way of characterizing this whole situation. He likens physicists who attempt to explain fine tuning solely by reference to universe-creating mechanisms, without intelligent design, to a hapless soul who denies any human ingenuity in the making of a freshly baked loaf of bread simply because the baker used a breadmaking machine. Clearly, argues Collins, such a benighted fellow has overlooked an obvious fact: the breadmaking machine itself required prior ingenuity and design, as did the recipe for and the preparation of the dough that went into it. Similarly, even if a multiverse hypothesis is true, it would support, rather than undermine, the intelligent design hypothesis, since the multiverse hypothesis depends upon the specific features of universe-generating mechanisms that invariably require prior and otherwise unexplained fine tuning.[39]

Deflating the Inflationary String Bubble?

In addition to all these problems, there are empirical reasons to doubt the separate cosmological models that jointly form the currently favored inflationary string landscape hypothesis.[40] In the first place, inflationary cosmology provides neither a necessary nor the necessarily *best* explanation of the homogeneity and flatness of the universe or of the absence of magnetic monopoles—the main evidence it was designed to explain. Instead, positing initial-condition fine tuning in conjunction with standard big bang cosmology explains both the homogeneity[41] and the flatness of the universe just as well as various elaborate inflationary cosmological models. Indeed, both the homogeneity and the flatness problems are only considered problems by those who regard the existence of fine tuning as a problem.

Yet, as noted, many kinds of fine tuning remain unexplained, whatever cosmologists may propose as an explanation for these features of the universe. And, as noted, inflationary cosmology actually makes the fine-tuning problem exponentially worse. Thus, on the assumption that good explanations necessarily minimize postulated fine tuning, inflationary cosmology would fail to provide a *better* explanation of the evidence than standard big bang models (at least those that also posit some initial-condition fine tuning). Similarly, the absence of magnetic monopoles can be explained without invoking inflation by assuming that monopoles don't exist and that the grand unified theories that predict them—theories known to be inadequate on other grounds—are false.[42]

In addition, several key predictions of inflationary cosmology have failed to materialize. The simplest inflationary multiverse models predict: (1) much larger variations than physicists have observed in the temperature of the cosmic background radiation, (2) detectable gravity waves as a consequence of random local fluctuations in the gravitational field, and (3) something called "near-scale invariance" in the imaging of the variations of the cosmic background radiation.[43] Physicists have yet to observe evidence of the second of these phenomena and their observations directly contradict the first and third.[44]

As a result of these failed predictions, several leading physicists, including Roger Penrose, Abraham Loeb (head of the Department of Astronomy at Harvard University), and Paul Steinhardt (one of the architects of the original inflationary model), have rejected inflationary cosmology.[45] Steinhardt notes that the failed predictions of inflationary cosmology have forced its advocates to formulate increasingly contrived models of the function and contours of the inflaton field energy. As he observes, "For the first time in more than 30 years, the simplest inflationary models, including those described in standard textbooks, are strongly disfavored by observations. Of course, theorists rapidly rushed to patch the inflationary picture but at the cost of making arcane models of inflationary energy and revealing yet further problems."[46]

The Failed Predictions of String Theory

String theory, for its part, requires "supersymmetry" as a necessary consequence of its attempt to unify the fundamental forces of physics. Recall, for example, that string theory postulates differently vibrating gravitons as the entities responsible for the fundamental forces of physics and that it requires gravitinos to account for the existence of matter. Physicists did not expect to detect either individual gravitons or their supersymmetric partner particles gravitinos (at least directly) in high-energy supercollider experiments. That's because neither supersymmetric particle was thought to have enough energy, even when coupled to other fields, to allow detection.

Nevertheless, higher-energy supersymmetric particles that string theory also requires should be detectable in supercollider experiments such as those performed at the Large Hadron Collider (LHC) near Geneva, Switzerland. Indeed, string theory predicts the detection of supersymmetric elementary particles under specific conditions in such high-energy supercollider experiments.[47] Yet experiments conducted at the LHC have repeatedly failed to detect such supersymmetric particles.[48] This failure provides a straightforward refutation of a predictive test of string theory using the standard logic of scientific falsification.

These failed predictions as well as the embarrassment of an infinite number of string-theory solutions have engendered a growing skepticism about string theory among many leading physicists. As the Nobel Prize–winning theoretical physicist Gerard 't Hooft has explained:

> I would not even be prepared to call string theory a "theory," rather a "model," or not even that: just a hunch. After all, a theory should come with instructions on how . . . to identify the things one wishes to describe, in our case, the elementary particles, and one should, at least in principle, be able to formulate the rules for calculating the properties of these particles, and how to make new predictions for them. Imagine that I give you a chair, while explaining that the legs are missing, and that the seat, back, and armrests will be delivered soon. Whatever I gave you, can I still call it a chair?[49]

A Divine Foot in the Door?

So why, despite these many liabilities, does the multiverse remain the go-to explanation of cosmological fine tuning for so many physicists? In fact, as I learned in my candid conversation with Michael Shermer (see Chapter 12), many scientific materialists do not actually believe that a quasi-infinite array of other universes really exists.

Yet, there are genuine proponents of the multiverse and they have indicated why they affirm the idea in spite of what others may regard as its transparently obvious implausibility. Consider what Stanford physicist Leonard Susskind, one of the architects of the string-theory multiverse, has himself said about the underlying impulse behind the construction of this immense theoretical superstructure. "If, for some unforeseen reason," he says, "the [string] landscape turns out to be inconsistent—maybe for mathematical reasons, or because it disagrees with observation," [then] "as things stand now we will be in a very awkward position. Without any explanation of nature's fine tunings we will be hard pressed to answer the ID [intelligent design] critics."[50]

Other leading physicists familiar with the research program have observed such a strong metaphysical predilection among their colleagues. As University of London theoretical physicist Bernard Carr has observed, "To the hard-line physicist, the multiverse may not be entirely respectable, but it is at least preferable to invoking a Creator. Indeed, anthropically inclined physicists like Susskind and Weinberg are attracted to the multiverse precisely because it seems to dispense with God as the explanation of cosmic design."[51]

None of this will surprise anyone with any acquaintance with the ethos or sociology of contemporary science. Since the rise of scientific materialism during the end of the nineteenth century, many scientists have regarded it as their duty to explain all events and phenomena, even singular events such as the origin of the universe, life, or consciousness, without reference to a designing intelligence. Some have come to see their vocation as scientists as part of a long struggle against what they regard as the irrationality of religion. Thus they have vigilantly resisted considering any discovery or explanation with implications favorable to theism, whatever the cost to the coherence of the scientific world picture. As retired Harvard evolutionary biologist Richard Lewontin wrote in the *New York Review of Books* in 1997:

> Our willingness to accept scientific claims that are against common sense is the key to an understanding of the real struggle between science and the supernatural. We take the side of science in spite of the patent absurdity of some of its constructs . . . in spite of the tolerance of the scientific community for unsubstantiated just-so stories, because we have a prior commitment, a commitment to materialism. . . . Moreover, that materialism is absolute, for we cannot allow a Divine Foot in the door.[52]

Of course, noting atheistic or materialistic motives in those promoting the multiverse hypothesis does not refute it, any more than identifying theistic motives in proponents of the God hypothesis refutes theism. What matters in assessing any worldview or hypothesis is its coherence, parsimony, and explanatory power with respect to the relevant evidence. But as we have seen, the multiverse hypothesis lacks precisely these attributes. Just its inability to explain fine tuning without invoking prior unexplained fine tuning should give us pause, to say nothing of its lack of parsimony.

Remember our discussion in the last chapter of the evidence for intelligent design in biology. Leading evolutionary biologists sought to solve the problem of the origin of functional genetic information by invoking prior unexplained sources of such information. Something similar is going on here. Physicists have offered many materialistic explanations for the origin of the fine tuning that attempt to explain the fine tuning without invoking a "fine-tuner" or mind. Yet these proposals either do not explain the fine tuning (as the weak and strong anthropic principles do not) or they invariably attempt to explain it only by tacitly invoking other sources of prior unexplained fine tuning.

Yet we saw in Chapter 8 that what we call fine tuning exhibits precisely those features—extreme improbability and functional specification—that rightly and invariably trigger an inference to intelligent design based upon our uniform and repeated experience. It follows that if the multiverse explanation cannot account for fine tuning without begging the question as to the origin of prior fine tuning and without endlessly multiplying theoretical entities, then intelligent design still stands as the best explanation for the fine tuning of the laws and constants of physics and the initial conditions of the universe.

17

Stephen Hawking and Quantum Cosmology

When I rose to the podium to debate Lawrence Krauss at the University of Toronto, I wasn't thinking about just the theory of intelligent design, though I would soon give a short presentation on the biological evidence for it. Instead, I was eagerly anticipating the opportunity to discuss Krauss's view of the origin of the universe. Krauss had recently published a book titled *A Universe from Nothing: Why There Is Something Rather Than Nothing*, which presented a theory of the origin of the universe known as quantum cosmology. In preparing for the discussion, I had immersed myself not only in Krauss's work but in the technical papers of another physicist, Alexander Vilenkin, whose work Krauss had popularized. My predebate study reawakened in me an intense, but partially dormant, interest in cosmology.

I first encountered quantum cosmology when I attended a lecture series in Cambridge by Stephen Hawking in the late 1980s just a couple years after I heard Allan Sandage's talk about the big bang. In the lecture series, Hawking presented a version of quantum cosmology that he was introducing to the general public about this time in a bestselling book, *A Brief History of Time: From the Big Bang to Black Holes* (1988).

Though Hawking had helped to prove the singularity theorems with Roger Penrose[1] in 1970 and George Ellis in 1973,[2] he found their implication of an absolute beginning of time and space philosophically disturbing and scientifically unsatisfying. Consequently, he began to formulate a cosmological model that he hoped would eliminate the implications of a beginning. He did so by applying the physics of the very small, known as "quantum mechanics," to analyze the universe when it *was* very small. In so doing, he not only challenged the idea that the universe had a definite beginning; he also posed an objection to the evidential basis for the cosmological argument. This chapter and the next two will address that challenge and objection.

The Physics of the Early, Tiny Universe

Quantum mechanics describes the interactions and motions of subatomic particles that manifest both wavelike and particle-like behavior. Whereas the expansion of the universe has produced a vast spatial volume in the present, at some point in the finite past the universe would have been small enough that physicists would need to consider how quantum mechanical effects would influence gravity. Physicists think that, in that small space, Einstein's theory of gravity—general relativity—would break down. Indeed, Hawking and Ellis acknowledged this possibility in their original work on the singularity theorems in 1973.[3]

Instead, many physicists have suggested that gravitational attraction would have worked differently, since the early universe would have been subject to quantum mechanical principles and unpredictable quantum fluctuations. As one science writer has explained: "Einstein's general theory of relativity fails to take into account the quantum fluctuations which must be present in any physical process involving gravity; therefore general relativity cannot be extrapolated in an unmodified form to predict what will happen at or below the Planck length."[4]

To date, physicists have not formulated an adequate theory of "quantum gravity," one that coherently synthesizes general relativity with quantum mechanics. Nevertheless, Hawking thought that physicists could at least anticipate the contours of such a theory.[5] Based on those expectations, he and his American coauthor, physicist James Hartle, of the University of California–Santa Barbara, first applied quantum mechanical ideas about how gravity might work on a subatomic scale to describe the universe in its earliest state.[6]

Hawking found, however, that in order to make precise mathematical calculations about the probable state of affairs in the early universe, he needed to use a calculating device that required introducing the concept of imaginary time. You may recall from math classes that *i*, the letter mathematicians use to represent imaginary numbers, equals the square root of –1, so that $i^2 = -1$. Hawking introduced imaginary time into one of Einstein's mathematical expressions (called the spacetime metric) that describes the geometry (or curvature) of spacetime.[7] Since this expression includes a time variable, Hawking simply equated time with imaginary time ($t = i\tau$) to make it possible to calculate the probabilities associated with different possible early states of the universe.[8] Mathematicians call this transformation a "Wick rotation."

When Hawking performed the Wick rotation, the resulting mathematical construct depicted a universe with spatial dimensions but no preferred temporal direction—and no temporal beginning. Indeed, the resulting mathematical depiction of the universe treated time as essentially another dimension of space, thus "spatializing time," as physicists describe it. Consequently, what Hawking called his "mathematical device" eliminated the *temporal* singularity—at least as long as he continued to describe the geometry of space using imaginary rather than real time.[9] This description of spacetime depicted a universe that is "finite but unbounded" by a specific temporal point of origin.[10]

In *A Brief History of Time*, Hawking presented this result as a challenge to the idea that the universe had a definite beginning in time. He argued that this mathematical model implied the universe would not, therefore, need a transcendent creator to explain its origin. After he explained how this mathematical manipulation eliminated the singularity, he famously observed, "So long as the universe had a beginning, we would suppose it had a creator. But if the universe is really completely self-contained, having no boundary or edge, it would have neither beginning nor end; it would simply be. What place, then, for a creator?"[11]

Hawking's proposal attained vast popular exposure. His book eventually sold over ten million copies, increasing his already immense international celebrity. By providing a singularity-free description of the early universe, the beginning of which he had previously helped to prove, Hawking's proposal also created the widespread impression that he had refuted the Kalām (first-cause) cosmological argument for the existence of God. Certainly, Hawking presented his model as a challenge to the second premise of that argument, the statement that "the universe began to exist."[12]

During the 1990s and early 2000s when I was a college professor, I frequently discussed Hawking's version of quantum cosmology with my philosophy of science students. By then, many had heard of Hawking's new cosmological model, his rejection of the temporal singularity, and his alleged refutation of the cosmological argument. Indeed, by the 1990s, Hawking singlehandedly began to shift perceptions about the implications of the big bang theory. During the early 1980s many astronomers, astrophysicists, philosophers, and cosmologists (Penzias, Jastrow, Ross, Gingerich, Sandage, Schroeder, and Jaki) were publishing books or giving public lectures exploring the theistic implications of the big bang. But the publicity surrounding Hawking's popular presentation of quantum cosmology and his presumed authority undermined this interpretation. His celebrity and charisma, related in part to his heroic resolve in the face of his physical disability, contributed to his influence.

Nevertheless, his key claim to have eliminated the need for a temporal beginning in the depiction of the universe proved vulnerable to an obvious critique. Physicists and philosophers pointed out that his decision to equate time with imaginary time has no physical justification apart from its expediency in making the calculations that Hawking wanted to make.[13] Indeed, Hawking himself acknowledged this expediency as his reason for introducing imaginary numbers to represent time.[14]

Consequently, Hawking's mathematical description of the geometry of spacetime lacks applicability and intelligibility as a physical description of our universe. Physicists familiar with his mathematical manipulations have explained why. When Hawking substituted $i\tau$ (imaginary time) for the variable t (real time) in Einstein's mathematical expression, the resulting depiction of the geometry of spacetime *did not correspond to anything in the real universe*.

The problem was twofold. First, the use of imaginary time functions in mathematics merely as a calculating device—one that allows mathematicians to solve certain otherwise intractable problems. It does so by, first, allowing physicists to transform mathematical expressions (in this case a functional integral)[15] into the domain of complex numbers (with imaginary time) and, then, after solving the resulting transformed (complex) equation, converting those mathematical expressions back into the real domain with real-time variables. All this is fair enough. Hawking used a mathematical method that mathematicians and physicists have long used to solve mathematical problems.

The problem with Hawking's presentation was not the mathematical techniques he used, but instead the metaphysical interpretation that he assigned to the intermediate steps in his mathematical manipulations. Indeed, the intermediate steps in his mathematical procedure produced an expression that does not describe the spacetime geometry of the *real* universe. Instead, time, when confined to the imaginary axis of the complex plane (as it is in the mathematics of complex analysis), *has no physical meaning*. Hawking himself acknowledged as much. As he explained, imaginary numbers "are a mathematical construct; they don't need a physical realisation; one can't have an imaginary number of oranges or an imaginary credit card bill."[16] Thus, in his *Brief History* he candidly described his use of imaginary time as a "mathematical device (or trick)."[17]

Second, the specific mathematical transformation that Hawking performed allowed him to treat time as a dimension of space for the purposes of his calculations. But collapsing time into space in this way, again, does not result in a mathematical expression with physical meaning, still less one that changes *over time* as our universe does. Though in general relativity, time and space are linked or fused, they are treated differently. Time is not the same thing as space. Events occur in space, but also *in temporal sequence*. Collapsing time into space (or "spatializing time")[18] eliminates the possibility of describing this reality of our universe and renders Hawking's intermediate mathematical description of the geometry of "spacetime" inapplicable as a description of our universe.

In his technical papers with James Hartle, Hawking did not draw any metaphysical implications from his mathematical expression describing the universe with imaginary time. Instead, he and Hartle merely used the Wick rotation as a calculating device to develop the mathematical implications of his larger cosmological model (more on this to come). In *A Brief History of Time*, however, he seemed to speak out of both sides of his mouth. In some places, he conceded the lack of realism associated with his intermediate mathematical construct; in others, he drew significant scientific and metaphysical conclusions from it, most notably in making the claim that his construct using imaginary time showed the universe did not have a beginning and, thus, need a creator.

Yet Hawking also acknowledged that once his mathematical depiction of the geometry of space is transformed back into the real domain with a real-time variable—the domain of mathematics that *does* apply to our universe—the singularity reappears. As he noted, "When one goes back to the real time in which we live, however, there will still appear to be singularities. . . . Only if [we] lived in imaginary time would [we] encounter no singularities. . . . In real time, the universe has a beginning and an end at singularities that form a boundary to space-time and at which the laws of science break down."[19]

All this brings me back to that night in Toronto in 2016. By then, I had thought about Hawking's quantum cosmology for many years. I knew well the criticisms that philosophers of science and physicists had lodged against his model over the previous three decades.[20] I often explained these criticisms to my students and had colleagues who had written extensively about them.[21] Yet I also knew that, partially in response to these criticisms, theoretical physicists had formulated other versions of quantum cosmology.

In particular, five years after the publication of Hawking and Hartle's first technical paper on quantum cosmology, Alexander Vilenkin formulated another version of quantum cosmology.[22] His theory began to attract widespread notice after he explained some of his ideas about it in his 2007 book *Many Worlds in One*. Vilenkin's theory did not attempt to eliminate the singularity. Instead, it presupposed the singularity and sought to explain how the origin of the universe came from it—indeed, from nothing physical at all. This was the version of quantum cosmology that Krauss had popularized and that I had come to Toronto hoping to discuss.

"Ordinary" Quantum Mechanics and the Discovery of Wave-Particle Duality

Vilenkin's formulation of quantum cosmology, like Hawking's, depends upon an application of *quantum mechanics* to the physics of the early universe. Quantum mechanics was developed to describe the nature of light (as well as electrons and other subatomic particles) once physicists in the early twentieth century determined that light behaves like both a wave and a particle. Understanding quantum cosmology, therefore, requires understanding the intellectual origins of "ordinary" quantum mechanics—though, of course, there is nothing ordinary about quantum mechanics at all.[23]

During the seventeenth century, Newton showed that light travels in a straight line. He did so in a series of experiments using prisms in which he separated light into distinct colors.[24] To Newton this suggested that light acted like a particle, or "corpuscle." He argued that conceiving of light as a corpuscle better explained the laws of optical reflection than did the rival

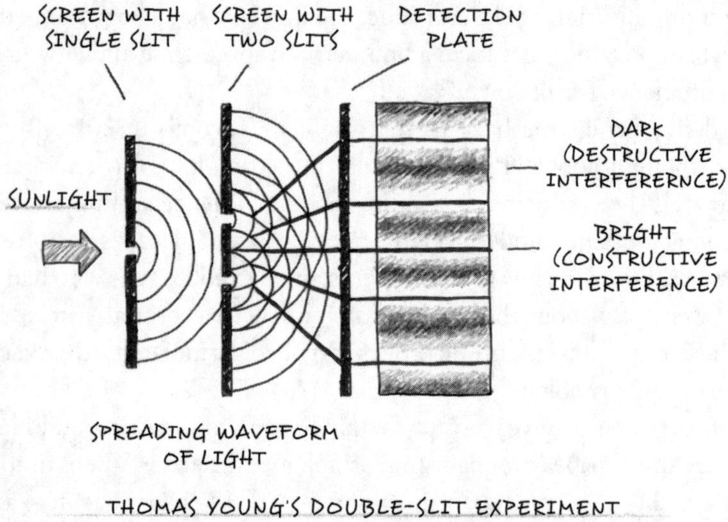

SCREEN WITH SINGLE SLIT
SCREEN WITH TWO SLITS
DETECTION PLATE
DARK (DESTRUCTIVE INTERFERERNCE)
BRIGHT (CONSTRUCTIVE INTERFERENCE)
SUNLIGHT
SPREADING WAVEFORM OF LIGHT

THOMAS YOUNG'S DOUBLE-SLIT EXPERIMENT

FIGURE 17.1
Interference pattern. When electrons or light with a specific frequency pass through two different slits, they produce an interference pattern on a terminal screen placed at a specific distance behind the slits. This pattern is the result of the waves from each slit either adding together or canceling each other out to form the corresponding light and dark lines. If a series of individual electrons or photons are emitted over time, the interference pattern will gradually appear. The individual electrons or photons passing through one of the slits act as though a wave has passed through both slits.

wave theory of light advanced by his contemporary, the Dutch physicist Christiaan Huygens. Huygens, for his part, argued that the wave theory better explained light refraction—that is, the tendency of light to bend as it passes from one medium (such as air) to another (such as water).[25]

In 1801, the English physicist Thomas Young performed a clever experiment that seemed to clarify the issue. In his famed "double-slit" experiment, Young passed a single wavelength of sunlight through a slit-like opening in a screen.[26] He then passed the remaining light through another screen with two slits in it (Fig. 17.1). On the other side of that second screen he placed a third detecting screen. Young saw that after passing through the double slits the light reaching the far screen exhibited a classic interference pattern characteristic of intersecting waves.

Interference patterns result when two propagating waves collide. For example, if two pebbles are dropped near each other in a pond, two waves will result and eventually intersect. When the waves collide, the height (or amplitude) of the resulting waves will increase in places where the peaks of the two waves converge and will cancel out in places where the peaks and troughs match up. Physicists call the former interaction "constructive interference" and the latter "destructive interference." Young's experiment detected a pattern of alternating constructive and destructive interference in the light as it reached the terminal screen. This result confirmed the wavelike properties of light.

Yet near the end of the nineteenth century the German physicist Heinrich Hertz observed a phenomenon known as the photoelectric effect, which again underscored the particle-like properties of light.[27] The photoelectric effect refers to the emission of electrons from a metallic surface when the metal is bombarded with light. Proponents of the classical wave theory of light predicted that the amplitude of such light should determine the kinetic energy of the resulting electron emissions, just as the height of a water wave passing through a harbor in a storm would determine how many small boats washed up on shore.[28] Instead, experiments showed that the frequency, not the amplitude, of the incident light determined the kinetic energy of the emissions.

This and other unexpected observations suggested to Einstein that light energy propagated in more concentrated, less wavelike packets called "photons." Since experiments also showed that these photons would not cause emissions of electrons unless they reached a certain minimum threshold energy, Einstein proposed that light energy propagated in discrete bundles or packets of energy called "quanta."[29] Thus, Einstein explained the photoelectric effect by postulating that light consisted of particle-like photons rather than spatially extended waves.[30] Moreover, he argued that when these photons contained discrete packets of sufficient energy, they could bump electrons loose from an irradiated metal surface. Thus, Einstein concluded, in addition to its demonstrable wavelike properties, light also acted like discrete particles or packets of radiant energy.

Another double-slit experiment, performed in 1909 by the physicist Geoffrey Taylor, confirmed this dual nature of light.[31] In 1907, following Einstein's explanation of the photoelectric effect, the British physicist J. J. Thomson proposed that the interference patterns in previous double-slit experiments resulted from individual photons somehow interfering with each other. Thomson thought that lowering the intensity of light would separate the individual photons from each other as they passed through the slits in the apparatus. If on average, for example, only one photon per second passed through a slit, then the leading photon ("Photon A") would not—in theory—interfere with the trailing photon ("Photon B") and no interference pattern should emerge.

Taylor tested Thomson's hypothesis by lowering the intensity of light to the point that individual photons passing through the two slits in the barrier would produce single-grain impressions on photographic film on the far barrier in his apparatus.[32] Even so, over time, as more and more photons hit the film, an interference pattern still emerged (Figs. 17.2a and 17.2b), just as had occurred in Young's experiment in 1801. In other words, the photons made impressions across the width of the photographic film in a way that suggested the constructive and destructive interference of two spreading but colliding wave forms. But this implied,

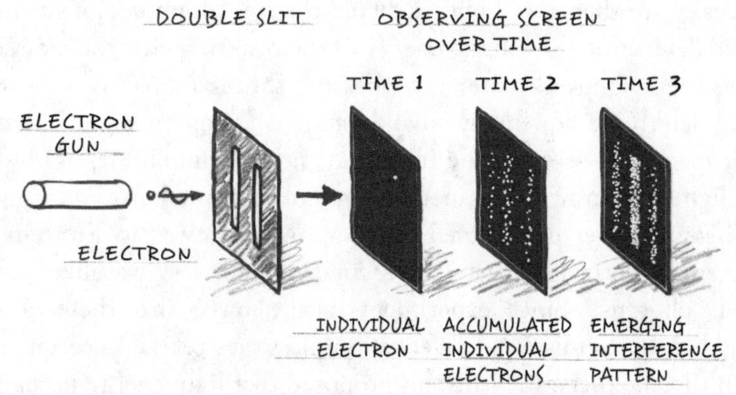

FIGURE 17.2A
Double-slit experiment (twentieth century). One version of the double-slit experiment uses an electron gun that emits individual electrons over an extended period of time. The electrons can pass through one of two slits and then hit a horizontal detection screen a specific distance beyond the slits. Over time, the distribution of detected electrons forms an interference pattern on the vertical detection screen demonstrating that electrons behave as *waves* with a characteristic wavelength.

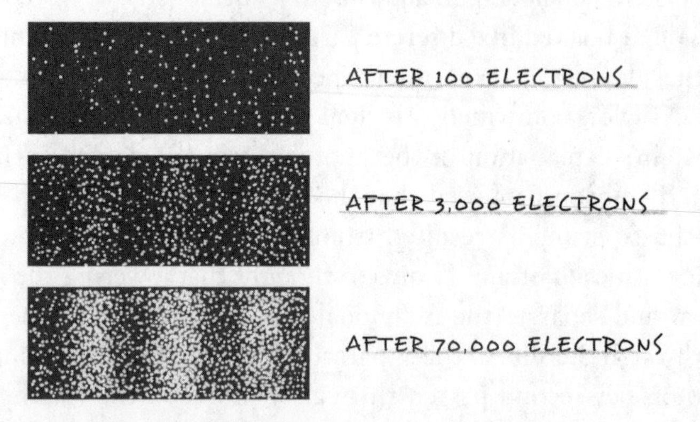

FIGURE 17.2B
Detection plate from a double-slit experiment. As electrons hit the vertical detection screen they initially create a fairly random scatter effect. But over time an interference pattern of light and dark bands emerges. The electrons hit with greater probability in the whiter regions than in the darker ones.

strangely, that the single photons making the impressions on the film had first passed through both slits as a spreading wave and then generated two smaller derivative waves on the other side of the slits that, in turn, interfered with each other before hitting the detection film. Then, most strangely, upon hitting the detection film the waves "collapsed" to manifest a particular position for each photon. After several hours or days, depending on the intensity of the light, the interference pattern always emerged, manifesting the wavelike character of the propagating photons.

Interestingly, no interference pattern emerged when Taylor passed light through just a single slit, suggesting that, in this case, a single wave continued to propagate on the other side of the barrier without another wave to interfere with it. That result reinforced the interpretation of the double-slit experiment as an interference pattern produced by individual photons passing through both slits as a single wave and then forming two interfering waves on the other side. Indeed, the double slit in the barrier allowed two waves to emerge on the other side, making interference possible.

In any case, since Taylor's use of low-intensity filtered light (through smoked-glass screens) ensured that photons passed through the slits and arrived at the detection plate essentially *one at a time* as discrete particles, the presence of the interference pattern confirmed the dual wave- and particle-like nature of the photons all in one experiment. Experiments performed in the 1920s (and later) confirmed that electrons, atoms, and other subatomic particles exhibit this same dual nature.[33]

The Formation of Quantum Mechanics

FIGURE 17.3
Physicist Erwin Schrödinger, architect of quantum mechanics and the famed Schrödinger equation.

The task of explaining, or at least accurately describing, these bizarre results fell to physicists in the 1920s and 1930s. They proposed a mathematical apparatus to characterize the phenomenon of wave-particle duality. The equation devised to do so—known as the Schrödinger equation,[34] after Austrian physicist Erwin Schrödinger (Fig. 17.3), who derived it—allowed physicists to calculate the probability that a photon (or any subatomic particle) would manifest itself at any given location, once

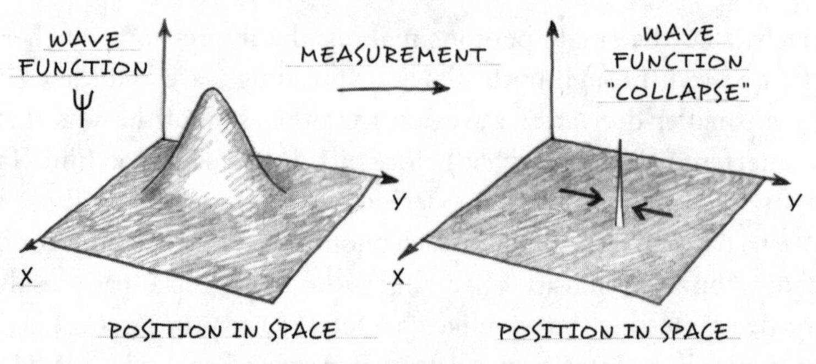

FIGURE 17.4

The Copenhagen interpretation of quantum mechanics. According to the Copenhagen interpretation of quantum mechanics, a proton or electron traveling through space exists in a "superposition" of possible positions (or possible values for its momentum) at the same time. What physicists call the "wave function" for such a quantum mechanical system represents the ensemble of possible states that the photon or electron might exhibit, and its magnitude squared equals the probability distribution for the position, momentum, or other variables taking on particular values when measured. When physicists take a measurement, the wave function "collapses" into a specific state corresponding to a specific measured value. For instance, if the position of a particle is measured, the wave function will collapse into a state corresponding to the specific measured position. The diagram illustrates how the wave function $\psi(x,y)$ initially has a broad peak representing many possible positions. After an apparatus measures the particle at a given position (x, y), the function becomes narrowly peaked at the measured position.

the spreading wave form hit a detection plate or other device. Since the photon evidently behaved like a spreading wave with spatial extension, physicists could not determine before making such an observation where exactly the photon (in its particulate character) was located—or how the wave packet would "collapse." Once observed, the light could manifest its particulate nature at any place along that spreading wave form (Fig. 17.4).

As the German physicist Max Born soon showed,[35] quantum mechanics provided a way of calculating the *probability* of the photon emerging at various possible locations at any given time. It did not, however, make it possible to calculate its actual location with certainty before detection, since the photon would not exist in a particular location until an observation was made. Weirder still, since explaining the interference pattern on the detection plate implied that a photon behaved like a wave until detection, most physicists *denied that it had a specific location* until such an observation occurred.

The Schrödinger equation that physicists use to describe the dual nature of light is a good example of what mathematicians call a "differential" equation. Differential equations differ from other equations in that their solutions typically do not represent specific numbers or values, but rather whole functions. By contrast, algebraic equations are functions or relationships between different variables (such as the function $y = 3x^2 + 2$) that, when solved, allow mathematicians to determine the values of the variables in the function directly (in the example, if $x = 1$, then $y = 5$). The solutions to differential equations, however, are *functions* that have *unspecified constants* (e.g., $y = ae^{kx}$, where x and y are variables, a and k are unknown constants, and e is a known constant[36]). Mathematicians can solve these equations only after they fix the values for those constants by providing what are called boundary and/or initial conditions (see my discussion in Chapter 13). These boundary conditions reflect features of the physical system that the equation in question describes, such as, for example, the distance between two pegs to which a vibrating string is attached.

When solved, Erwin Schrödinger's equation generates what is known as a "wave function," which physicists represent with the Greek letter ψ (psi). Like other differential equations, the Schrödinger equation can only be solved once physicists fix boundary conditions in accord with, for example, the features of the experimental apparatus used in the double-slit experiment.

The wave function ψ allows physicists to calculate the probability of the photon (or electron) having a particular location or momentum upon detection.[37] Prior to observation the photon does not have a specific momentum or location. Indeed, the double-slit experiment implied that the photon exists as a spatially extended wave, not a spatially discrete particle. Moreover, the wave *function* itself does not exist as an entity in space and time.[38] Rather, the wave function is a mathematical concept describing possibilities that might exist in space and time once the photon as a wave encounters an observer or detector and the "probability wave" collapses. The wave function also depicts what physicists call "superposition," the idea that prior to observation subatomic particles "exist" as mathematical possibilities in multiple indeterminate states at once, occupying an actual place (or acquiring a specific momentum) in space and time only after detection.

The idea that a photon or other subatomic particle would have no definite character until observed by a conscious agent (or a detection device placed in position by such an agent) rocked physics as well as common sense. Leading physicists during the twentieth century repeatedly commented on the strangeness of this apparent "observer-dependent" aspect of fundamental subatomic reality. The Caltech physicist Richard Feynman mused that he could "safely say that no one understands quantum mechanics."[39] The Danish physicist Niels Bohr, for whom the observer-dependent "Copenhagen interpretation of quantum mechanics" is named,[40] commented that those "not shocked by quantum mechanics when they first come across it cannot possibly have understood it."[41] Indeed, the physics of the very small turned out to be the physics of the very weird—so weird that the double-slit experiment and the mathematics of quantum mechanics have spawned numerous competing philosophical interpretations of wave-particle duality to this day.

Quantum Cosmology

What does this physics of the tiny realm of subatomic particles have to do with the origin of the largest object we know—the universe? As noted in previous chapters, an expanding universe in the forward direction of time implies a much smaller universe in the remote past. By extrapolating backward, astrophysicists envision a time in the first

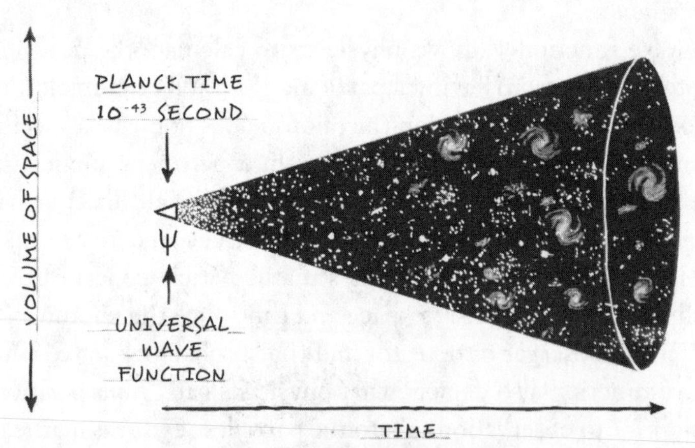

FIGURE 17.5

Quantum cosmology. Cosmologists have attempted to synthesize general relativity with quantum mechanics to generate "quantum cosmological" models for the earliest stage of the universe. Quantum cosmologists seek to determine a wave function for the universe, which they represent with the same Greek letter ψ as used in standard quantum mechanics. The ψ function describes different universes with different possible gravitational fields in "superposition." Knowing ψ allows physicists to calculate the probability that a specific universe with a specific gravitational field will appear, that is, a universe with a specific spatial geometry and matter field (and a resulting mass-energy configuration). Constructing a universal wave function allows physicists to calculate the probability that different possible universes with different gravitational fields existed "inside of" or will "emerge from" the universe as it existed inside "Planck time"—that is, in the first 10^{-43} seconds after the beginning of the universe when the universe would have been small enough to be subject to quantum effects.

FIGURE 17.6A

FIGURE 17.6B

Physicists John Wheeler and Bryce DeWitt, architects of the Wheeler-DeWitt equation.

fractions of a second after the big bang—up until the first 10^{-43} of a second to be exact—when the universe would have been small enough that quantum mechanics would have been relevant for understanding how gravity works. In that subatomic realm, physicists think that general relativity—Einstein's theory of gravity that applies to macroscopic objects—would break down. That insight has led physicists to seek to formulate a quantum theory of gravity, though as yet no such theory has attained anything like widespread assent.

To describe how gravity would have worked in such a tiny space during the very earliest period of the universe, quantum cosmologists have developed an equation that synthesizes mathematical concepts from quantum mechanics and general relativity (Fig. 17.5). That equation is called the Wheeler-DeWitt equation, named for the physicists John Wheeler (Fig. 17.6a) and Bryce DeWitt (Fig. 17.6b) who developed it.[42] Some physicists regard the equation as at least a first step in the development of a quantum theory of gravity.[43]

In any case, physicists have developed quantum cosmology as a kind of mathematical analogue to "ordinary" quantum mechanics. In ordinary quantum mechanics, the different solutions to the Schrödinger equation allow physicists to construct a mathematical expression called a wave function. The wave function, in turn, allows physicists to calculate the probability of finding a particle at a given position and time or to determine the probability of that particle having a specific momentum (or to specify the probability of a host of other relevant properties).

In quantum cosmology, solving the Wheeler-DeWitt equation allows physicists to construct a wave function *for the universe.* That wave function then describes different possible universes with different possible gravitational fields, that is, different curvatures of space and different mass-energy configurations (or matter fields). In other words, the universal wave function, the solution to the Wheeler-DeWitt equation, describes the different possible spatial geometries and configurations of matter (in "superposition") that a universe *could* adopt.

Gravitational fields are determined by the shape (or curvature) of space and by the configurations of mass-energy (or matter fields) within those spaces. Thus, the universal wave function describes different possible pairings of spatial geometries and mass-energy configurations.[44] Quantum cosmologists represent these different possible geometries and configurations as ordered pairs in an abstract space of possibilities that they call "superspace."

In mathematical parlance, superspace is the domain of the universal wave function ψ represented as a set of ordered pairs in the following form: ψ (*curv*, *matter*). Moreover, in the analogy that quantum cosmologists have drawn between quantum cosmology and ordinary quantum mechanics, different possible gravitational fields in the universal wave function correspond to the different possible positions of particles (or values of their momentum) described in the wave function of ordinary quantum mechanics.

Further, just as solving the Schrödinger equation allows physicists to determine the wave function for an electron or photon and then to calculate the probability of the electron or photon having different locations or momentum values at different times, solving the Wheeler-DeWitt equation allows physicists to determine the wave function of the entire universe and then to calculate the probability that a given universe exhibiting a specific gravitational field with a specific curvature mass-energy pairing will emerge (or be observed).[45]

Quantum Cosmology and the Origin of the Universe

What does any of this have to do with the question of the origin of the universe? As the British theoretical physicist Christopher Isham has explained, quantum cosmology functions as a theory of the origin of the universe when the universal wave function ψ *includes* the spatial geometry of our universe as a possible or reasonably probable outcome (or technically, "observation") among the ensemble of possible gravitational fields (geometries and configurations of mass-energy) that ψ includes.[46]

Yet ψ acquires specificity as a function only as a solution to the Wheeler-DeWitt equation. It follows that only solutions to this equation that include our universe as a possible or reasonably probable outcome (or observation) will provide what quantum cosmologists regard as genuine explanations for the origin of our universe.

For nontechnical readers and scientists alike, the world of quantum cosmology presents a dizzying array of abstract concepts, analogies to known but mysterious physical processes, and of course complex mathematics. Nevertheless, understanding how physicists use quantum cosmology as an origins theory requires keeping just three main elements in view: first, the origin of our universe with its specific attributes, *the thing to be explained*; second, the universal wave function ψ, *the mathematical entity that does the explaining*; and, third, the Wheeler-DeWitt equation and the mathematical procedure for solving it, the alleged *justifications* for treating the universal wave function ψ as an explanation for the origin of the universe.

Philosophers of science use the Latin term *explanandum* to describe a particular event in need of explanation and the term *explanans* as a synonym for an explanation. They also sometimes use the word "warrant" as a synonym for justification or support. In quantum cosmology: (1) the origin of the universe is the *explanandum*; (2) a specific universal wave function provides the *explanans*; and (3) the mathematical procedure by which physicists solve the Wheeler-DeWitt equation provides the "warrant" for presenting any specific universal wave function as the explanation for the origin of the universe.

Keeping these three elements in focus and understanding their interrelationship will help readers track the ensuing discussion, even without detailed knowledge of the mathematics that quantum cosmologists use to model the origin of the universe. Though the following discussion is predicated upon a careful analysis of the mathematical procedures involved in solving the Wheeler-DeWitt equation—that is, in producing a universal wave function—it won't recapitulate the detailed mathematical steps required to solve that equation.

Nevertheless, it will describe the mathematical procedure of the quantum cosmologists in *general logical terms*, so as to make clear how the universal wave function acts as an explanation and why quantum cosmologists think that constructing a specific universal wave function as a solution to the Wheeler-DeWitt equation justifies treating such a solution as an explanation for the origin of the universe. Both technical and nontechnical readers may find this level of description helpful in evaluating the logical basis of the claimed explanations for the origin of the universe that quantum cosmologists offer.

Hawking-Hartle Revisited

The preceding discussion may also help put in context Stephen Hawking's famous claims to have eliminated the need to posit a creator to explain the origin of the universe. With James Hartle, Hawking developed a quantum cosmological model based on the Wheeler-DeWitt equation. With their model, they were attempting not so much to eliminate the singularity at the beginning of the universe, but instead to describe (or even explain) the origin of the universe as the consequence of a fundamental physical theory—a theory of quantum cosmology (or quantum gravity).

Thus, Hawking and Hartle mainly wanted to determine the wave function of the entire universe by solving the Wheeler-DeWitt equation. If they could do that, they could then calculate the probability that a universe such as ours with its specific gravitational field would emerge.

To solve the Wheeler-DeWitt equation they used what is known as a "sum-over-histories" or "path-integral" method. In ordinary quantum mechanics, physicists use the sum-over-histories method to sum the mathematical expressions (called functional integrals) that describe the possible paths of photons or electrons in some physical setting (e.g., as they pass through double slits on their way to detectors).

This summing procedure may seem opaque. When most of us think of summing things up, we think of adding whole numbers and getting an answer as another whole number, as in 3 + 9 + 7 = 19. Nevertheless, in some types of mathematics, mathematicians sum whole *collections of functions* or other complicated mathematical expressions.

Hawking and Hartle wanted to apply the sum-over-histories technique to sum up the mathematical expressions that describe the "paths" from the presupposed cosmological singularity to the different possible universes (with different possible gravitational fields) that might emerge from the singularity. More specifically, Hawking and Hartle conceived of these different universes as emerging from a singularity, call it "Point A," in the trajectory to a possible universe, "Point B" (or rather points

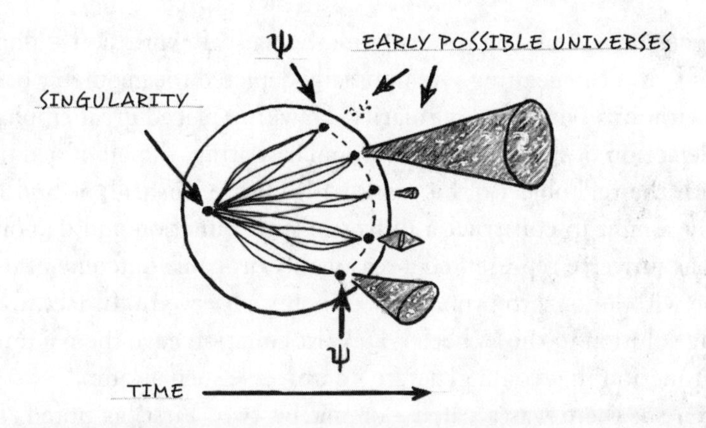

FIGURE 17.7
To solve the Wheeler-DeWitt equation physicists use a path-integral method that requires them to sum up the different mathematical expressions describing different paths from the singularity at the beginning of the universe to different possible universes with different gravitational fields. This diagram shows roughly what physicists envision their mathematical procedure representing. It shows the presumed singularity at the beginning of the universe, some of the different paths (through "superspace"), and an ensemble of possible universes (represented by the resulting universal wave function (Ψ).

B_a–B_z). Each end point in this path from the singularity through superspace represents a possible universe with some possible gravitational field (Fig. 17.7). Summing all the paths to the different possible outcomes, or universes, would allow Hawking and Hartle to construct a universal wave function.

The resulting wave function would then yield, as wave functions do in ordinary quantum mechanics, a probability distribution. (A probability distribution is a function that describes the probability of different possible events at different times or places.) From that Hawking and Hartle could then determine the probability of any particular universe emerging from the singularity using the wave function. If that wave function included a universe like ours as a possible (or reasonably probable) outcome, then Hawking and Hartle would consider the origin of the universe explained by reference to a fundamental theory of physics.

Nevertheless, they realized that solving the Wheeler-DeWitt equation would prove intractable in the domain of real numbers. That's where the mathematical calculating device involving imaginary time (the Wick rotation discussed above) came in. Hawking and Hartle realized that if they were substituting $i\tau$ for t in Einstein's mathematical expression describing the geometry of space (his "spacetime metric"), they could solve the Wheeler-DeWitt equation. When they performed this transformation, the resulting mathematical expression—albeit one without physical meaning—temporarily depicted the geometry of spacetime without a temporal singularity. Hawking placed great emphasis on this depiction of spacetime in his popular writing, though it had little to do with the real object of his work with Hartle. Instead, he and Hartle mainly sought to construct a universal wave function and demonstrate that our universe represented a reasonably probable outcome of it—that is, one with a nonzero probability of being observed.[47] In fact, their resulting solution to the Wheeler-DeWitt equation gave them a universal wave function that could generate[48] a universe such as ours.[49]

Even so, there was a catch—or maybe two. First, as noted, Hawking and Hartle's new quantum cosmological model did not eliminate the singularity at the beginning of the universe. Indeed, their use of the path-integral method actually *presupposed* a spacetime singularity out of which numerous possible universes could emerge.[50] Hawking only eliminated the depiction of a temporal beginning in one of the initial steps of a multistep calculating procedure, and only then by interpreting a mathematical expression with no physical meaning as if it had physical and metaphysical significance.

Second, to solve the Wheeler-DeWitt equation and construct a universal wave function, Hawking and Hartle needed to limit the number of possible universes (with different curvature-matter pairings) under consideration or, in the jargon of quantum cosmology, the number of "paths through superspace." Hawking and Hartle constructed a wave function using the path-integral method. But they chose to do so only using certain "paths." In particular, they only included paths to possible universes that met certain criteria—criteria that they knew would enable their mathematical formalism to produce a viable wave function that included universes such as our own. They chose, for example, only isotropic, closed, and spatially homogeneous universes and only those with a positive cosmological constant.[51] These restrictions generated a "minisuperspace" that provided a much smaller number of allowed curvature-matter pairings (or gravitational fields) in superspace.

Even so, when Hawking and Hartle went to solve the resulting path integral, they found that they could not solve it in a perfectly general way with existing mathematical techniques. Instead, they had to use approximations that further restricted the degrees of mathematical freedom associated with the problem. Nevertheless, after a few more mathematical steps, Hawking and Hartle did succeed in producing a wave function that included a universe like ours as a possible (and reasonably probable) outcome.[52]

Even so, only a few of the many possible matter-curvature pairings or corresponding "paths through superspace" would—when summed—produce such a wave function. That's because only an extremely small number of solutions to the Wheeler-DeWitt equation will generate wave functions that include universes with spatial geometries and mass-energy configurations like ours. Thus, all quantum cosmological models must constrain the number of possible solutions to the Wheeler-DeWitt equation in order to generate a wave function that includes universes like our own. Nevertheless, to generate realistic quantum cosmological models that in some sense explain the origin of our universe, physicists can't just choose those constraints arbitrarily. Instead, to explain our universe as a reasonably probable outcome of a *natural physical* process, they must provide some nonquestion-begging physical rationale for the constraints that they choose.

We'll see in the next chapter that this requirement has proven difficult to meet for both Hawking and Hartle's model and other quantum cosmological models, including those that attempt to explain how the universe emerged *from nothing* or nothing but the laws of physics.

18

The Cosmological Information Problem

As the foregoing chapter likely conveyed to patient readers, quantum cosmology can seem puzzling and paradoxical, to put it mildly. A few years ago, I was on the radio with host Dennis Prager, who in the course of our conversation described an earlier interview he had conducted with Lawrence Krauss. The subject of their conversation was Krauss's book *A Universe from Nothing*. Dr. Krauss explained how he believes the universe could emerge from nothing, and Prager questioned him about the notion of getting something from nothing. Krauss told him, "The word 'nothing' means a lot of different things to people."[1]

"That's when I gave up,"[2] Prager told me when he later recounted the exchange in our interview.

As I prepared for my debate with Krauss, I considered the version of quantum cosmology that he described in his book.

Indeed, Krauss hadn't popularized the Hawking-Hartle model of quantum cosmology, but a parallel proposal based upon the work of Alexander Vilenkin,[3] the same Russian physicist who helped to prove the Borde-Guth-Vilenkin theorem.[4] Vilenkin presupposed that the universe began from a spatial singularity of zero volume and then "quantum-tunneled" into a space of a specific finite volume where it could then expand in an inflationary way. (More on "quantum tunneling" to come.) In Vilenkin's model, the probability of this "tunneling" transition is determined by the universal wave function.[5] In *A Universe from Nothing*, Krauss adopted Vilenkin's idea. There he argued that the "laws of physics" explained how the universe could have emerged from nothing and that those laws, which he seemed to define as part of nothing, also show that "nothing is unstable."[6]

Interestingly, in 2014, Hawking echoed this claim in a book titled *The Grand Design*, coauthored with Leonard Mlodinow. They wrote: "Because there is a law such as gravity, the universe can and will create itself from nothing. Spontaneous creation is the reason there is something rather than nothing, why the universe exists, why we exist." Lest anyone miss the metaphysical implications of this view of cosmology, Hawking made them explicit: "It is not necessary to invoke God to light the blue touch paper and set the universe going."[7] In his book, Krauss developed this same perspective.

When I first read Krauss's book, I discovered that the lion's share of his discussion attempted to describe how material particles in the universe emerged from preexisting energy-rich fields in a preexisting space. This space and energy presumably arose from the singularity at the big bang. Near the end of this discussion, Krauss, clearly sensitive to the objection that neither space nor energy qualified as genuine "nothing," acknowledged that he had not yet established his main claim. Then in a short chapter near the end of the book, he attempted to prove the thesis of his book—that the laws of physics could explain how the universe itself arose from literally nothing.[8] He did so with a cursory description of Alexander Vilenkin's work in quantum cosmology.[9] I found Krauss's discussion of quantum cosmology intellectually unsatisfying. Nevertheless, his discussion of Vilenkin's work spurred me to track down Vilenkin's technical papers and to review his philosophically sensitive book *Many Worlds in One*.

What I found in Vilenkin's work surprised me. In his use of quantum cosmology, Vilenkin was clearly attempting to model the origin of the universe as a consequence of a deeper physical law or theory. But he exhibited a much more profound sense of the difficulty of this endeavor than either Krauss or Hawking. He also showed a keen sense of the paradoxical or even contradictory aspects of invoking a mathematical equation developed in the human mind as the cause of an actual universe. My reading of Vilenkin led me to conclude that quantum cosmology had not explained the origin of the universe in purely physical or materialistic terms. Instead, though he likely would not agree, it seemed clear to me that, to the extent quantum cosmology did accurately describe the origin and early state of the universe, it had several unexpected theistic implications.

The Laws of Physics and the Origin of the Universe

Proponents of quantum cosmology frequently claim the laws of physics explain the origin of the universe. For example, Hawking asserts, "Because there is a law such as gravity, the universe can and will create itself from nothing." Krauss echoes this claim: "The laws themselves require our universe to come into existence, to develop and evolve."[10]

When Krauss and Hawking say the laws of nature or "a law such as gravity" explains the origin of the universe, they refer to the whole mathematical superstructure of quantum cosmology, the universal wave function, the Wheeler-DeWitt equation, and current ideas about quantum gravity.[11] They also assume that the laws of physics *cause* or *explain* specific events, including the origin of the universe.

The claim that the laws of physics *cause* events to occur sounds obviously true to many scientists, because we hear it so often and are trained to think of the laws of nature as the ultimate explanatory principles in nature. Unfortunately, this idea conceals an imprecision in thought and makes what philosophers and logicians call a "category mistake."

To see why, consider the following illustration. If one billiard ball of some given mass bashes into another billiard ball, the law of conservation of momentum accurately describes the interaction. It will even allow us to make predictions about, for example, the change in velocity of the second ball, if we know the masses of the two balls and the velocity of the first ball as required by the equation describing momentum exchange. Physicists write the law of momentum conservation as follows: $m_1v_1 + m_2v_2 = m_1v_1' + m_2v_2'$

Nevertheless, the equation describing that interaction—the law of conservation of momentum—does not *cause* the second ball to move. The cause of the movement of the second ball is the movement of the first ball. The cause of the second ball's movement is an antecedent event—the prior movement of the first ball coming into contact with the second. The law simply describes that interaction.[12]

A similar, but even deeper, confusion attends Krauss's and Hawking's claims about the law of gravity—expressed as a mathematical equation. The law of gravity does not *cause* material objects or space and energy to come into existence; instead, it describes how material objects interact with each other (and with space) once they already *exist*. The law does not *cause* gravitational motion, nor does the law have the causal power to create a gravitational field, or matter or energy, or time or space. The laws of physics *describe* the interactions of things (matter and energy) that already exist within space and time.

This confusion, running all through Krauss's work, brought me back to an idea that I had critiqued years before in my PhD thesis. There, I showed that causes and scientific laws are not the same things. Causes are typically particular events (or sequences of events) that precede other events and meet specific logical and contextual criteria. Laws, by contrast, *describe* general *relationships* between different types of events or variables. Sometimes laws describe antecedent events that do cause other events. Other times they describe noncausal relationships between different events or variables (relationships where one event is a necessary condition but not a cause of some outcome, for example, or relationships involving correlations, not causation). The laws of physics represent only our descriptions of nature. Descriptions in themselves do not cause things to happen.

Admittedly, however, the antecedent material conditions described by some laws of physics *do cause* other events—as in the case of the first billiard ball hitting the second one. So, one might ask, couldn't quantum cosmology include a law that specifies some material antecedent event as the cause of the origin of the universe? Couldn't the universal wave function and/or the Wheeler-DeWitt equation, conceived as a proto-law of quantum gravity, specify a material antecedent *cause* for the origin of matter, space, time, and energy?

The answer is no. Instead, this potential objection to the above argument actually underscores why quantum cosmology does not provide a causal explanation for the biggest event in the history of the universe, namely, its origin.

Recall that the universal wave function merely describes possible universes with different possible gravitational fields. These possible universes represent outcomes or potential observations or effects—universes that *could come into existence*. The universal wave function just describes the "superposition" of all the universes with different spatial geometries and configurations of mass-energy *that could exist* without specifying any antecedent that might cause one of those universes, as opposed to all the others, to come into existence. How could it? Before matter, space, time, and energy first arose, no such entities existed. Moreover, in both main models of quantum cosmology, the outcomes described by ψ, the universal wave function,[13] arise from an initial temporal singularity of zero spatial volume. Quantum cosmology presupposes this singularity but does not provide a physical cause or explanation for the origin of ψ or the possible universes it describes that may emerge out of the singularity.

In my billiard-ball example, the law describing momentum exchange did not cause the second billiard ball to move upon contact with the first. Instead, the movement of the *first* billiard ball did that. In the case of quantum cosmology, prior to the "law" or mathematical function ψ, which putatively "explains" the origin of the physical universe, there are no antecedent "billiard balls," no physical particles, no energy fields, not even time. And ψ, the universal wave function, merely *describes* the possible universes with different possible gravitational fields that could arise from the singularity.[14] Before ψ, no physical universe, and thus no possible physical causal antecedent, would have yet existed (even if, as some physicists interpret the universal wave function, ψ describes already existing universes in superposition).

Since the Wheeler-DeWitt equation has to be solved[15] to generate a universal wave function, one might argue that the equation itself represents a causal antecedent to the different possible outcomes described by the universal wave function. Nevertheless, both the Wheeler-DeWitt equation and the curvature-matter pairings in superspace represent purely mathematical realities or physical possibilities. Indeed, "superspace" itself constitutes an immaterial, timeless, spaceless, and infinite realm of mathematical possibilities. Yet these mathematically possible universes (as well as the presupposed singularity, which also exists as a point in superspace) have no physical, or at least no necessary physical, existence.

And even if they did exist, they would not preexist our universe (as potential causal antecedents), since both our universe and these other possible universes "reside" as possibilities in the same *timeless* mathematical space of possibilities, namely, superspace. Thus, the purely mathematical character of quantum cosmology—even if conceived as a proto-law of quantum gravity—renders it incapable of specifying any material antecedent as a physical cause of the origin of the universe.

Of Math and Minds

How, then, do Krauss and others maintain that purely mathematical entities bring a material universe into being in time and space? In other words, how can a mathematical equation create an actual physical universe?

This question has troubled the leading physicists promoting quantum cosmology—at least in their more reflective moments. In *A Brief History of Time*, Stephen Hawking famously asked, "What is it that breathes fire into the equations and makes a universe for them to describe?"[16] Though Hawking posed this question—perhaps somewhat rhetorically—he never returned to answer it.

Alexander Vilenkin has raised the same question. He notes that, in his version of quantum cosmology, the process of "quantum tunneling" from superspace into a real universe produces space and time, matter and energy. But he acknowledges that even the process of tunneling must be governed by laws that "should be 'there' even prior to the universe itself." He goes on: "Does this mean that the laws are not mere descriptions of reality and can have an independent existence of their own? In the absence of space, time, and matter, what tablets could they be written upon? The laws are expressed in the form of mathematical equations. If the medium of mathematics is the mind, does this mean that *mind should predate the universe*?"[17]

After raising this suggestive possibility in the last chapter of his book, Vilenkin concluded his discussion without answering his own question. Nor did Krauss, in his short popularization of Vilenkin's work, grapple with it.

But Vilenkin's reflective question suggests two basic options, neither of which supports Krauss's atheistic or materialistic viewpoint. Either the laws that he and Vilenkin invoke to explain the origin of space (and

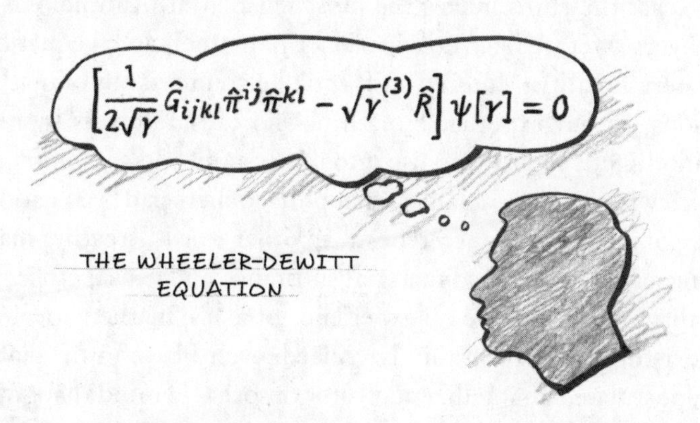

FIGURE 18.1
Matter out of math? Mathematical concepts, expressions, and equations exist in minds. That raises a profound question for quantum cosmologists. How do the mathematical expressions that they use to describe possible universes (or the early universe) cause an actual material universe to come into existence?

energy) are mathematical descriptions that exist only in the minds of physicists—in which case they have no power to generate anything in the natural world external to our minds, let alone the whole universe. Or the mathematical ideas and expressions, including those describing possible universes, exist independently of the human mind. In other words, quantum cosmology suggests either a kind of magic where human math creates a universe (clearly, not a satisfactory explanation) (Fig. 18.1) or mathematical Platonism.[18]

The Greek philosopher Plato argued that material objects such as chairs or houses or horses exemplify immaterial "forms" or ideas in a transcendent, changeless, abstract (immaterial) realm outside our universe. Similarly, *mathematical* Platonism asserts that mathematical concepts or ideas exist independently of the human mind. But this view in turn suggests two possibilities: mathematical ideas exist in an abstract transcendent realm of pure ideas, as Platonic philosophy suggests about the forms, or mathematical ideas reside in and issue from a transcendent intelligent mind.

That then gives us a total of three distinct ways of thinking about the relationship between the mathematics of quantum cosmology and the material universe: (1) these mathematical expressions exist solely in the human mind and somehow produce a material universe; or (2) these equations represent pure mathematical ideas that exist independently of the human mind in a transcendent, immaterial realm of pure ideas; or (3) these equations exist in and issue from a preexisting transcendent mind.

Of those three options, I would argue, based on our uniform experience, that the third makes the most sense. Math can help us describe the universe, yet we have no experience of mathematical equations creating material reality. Material stuff can't be conjured out of mathematical equations. In our experience math has no causal powers by itself apart from intelligent agents who use it to understand and act upon nature. To say otherwise commits a fallacy that philosophers call "reification" or the "fallacy of misplaced concreteness," in other words, treating mathematical concepts as if they had material substance and causal efficacy.

Similarly, we also have no experience of ideas, mathematical or otherwise, existing apart from minds. Indeed, even Plato, in his dialogue *Timaeus*, postulated an intelligent creator of sorts—a mind that gives reality to the forms and ideas that otherwise exist in a purely abstract realm.

We do, however, have a wealth of experience of ideas that start in the mental realm and by acts of volition and intelligent design produce entities that embody those ideas—what the thirteenth-century theologian Thomas Aquinas called "exemplar causation."[19] Therefore, it seems a reasonable extrapolation from our uniform and repeated experience of "relevantly similar entities"[20] (human minds) and their causal powers to think that, *if* a realm of mathematical ideas and objects must preexist the universe, as quantum cosmology implies, then those ideas must have a transcendent mental source—they must reflect the contents of a preexisting mind. When Vilenkin himself tumbled to this realization, however briefly, he raised the possibility of a decidedly theistic interpretation of quantum cosmology.

Quantum Tunneling

But what about the process of quantum tunneling to which quantum cosmologists refer? Does that provide a physical mechanism, rather than just a mathematical equation, for explaining the origin of the universe?

In fact, it doesn't. Recall that quantum cosmology is based upon an analogy with ordinary quantum mechanics. The idea of quantum tunneling extends this analogy. Quantum tunneling in ordinary quantum mechanics refers to a process by which a physically bounded subatomic particle can overcome a potential energy barrier even though the particle in question, according to classical mechanics, lacks sufficient kinetic energy to do so. In the subatomic realm of quantum mechanics, however, the wave function that allows physicists to determine the probability of finding a given subatomic particle in various places also admits the possibility of finding that particle on the other side of a potential energy barrier—a barrier that the subatomic particle could not overcome based solely on its kinetic energy (if only classical mechanics applied).

To illustrate, imagine a car running out of gas as it attempts to climb a hill. As it slows down and loses kinetic energy, it will (absent braking action) gradually roll back to the bottom of the hill. But in the weird realm of the quantum, there is actually a finite probability that the equivalent of the car in the world of subatomic particles could suddenly find itself (or could be observed) on the other side of the hill, even though it didn't have enough kinetic energy to crest the hill. Quantum cosmologists have appropriated this idea by drawing an analogy between energy barriers to enclosed subatomic particles and energy barriers to the development of an expanding universe.

In his quantum cosmology model, Vilenkin proposes "quantum tunneling" to explain how the universe developed from an initial singularity to an expanding universe with our gravitational field. He first posits the existence of a universe beginning in a singularity. As soon as this universe begins to expand (by what mechanism, Vilenkin does not say), its continued expansion would, according to Einstein's field equations, be opposed by an increasing gravitational energy barrier resulting from a matter field that Vilenkin assumes would be present in that expanding space. But that energy barrier would make further expansion impossible and push the universe back to a singularity. In ordinary quantum mechanics there is a greater-than-zero probability of a particle escaping a high-walled enclosure functioning as a potential energy barrier. In a similar way, Vilenkin's "tunneling wave function" suggests the possibility that the initial universe that emerges out of the singularity could overcome the gravitational energy barrier, thus allowing it to grow large enough to continue to expand.

In any case, tunneling occurs in Vilenkin's model *after* a universe—indeed, an expanding universe with a preexisting matter field—has already arisen by some other unspecified means. Thus, his quantum-tunneling scenario does not explain the origin of the universe; it presupposes one.

Hawking and Hartle similarly propose a process of quantum tunneling to account for the transition from an initially "closed" universe to an expanding one. (Closed universes have spherelike positive curvature that will eventually close back in on itself, rather than allowing space to expand indefinitely.) Since expanding universes have more energy than closed ones, Hawking and Hartle envision the tunneling "mechanism" as, again, overcoming an energy barrier.

Even so, Hawking and Hartle's quantum-tunneling scenario also attempts to account for the evolution, not the origin, of the universe. Recall that quantum cosmologists regard their models as explanations for the origin of a specific universe if the universal wave function they construct includes that universe (with reasonable probability). Hawking and Hartle's model generated a solution to the Wheeler-DeWitt equation that included—and thus "explained" the origin of—an initially closed universe destined for recollapse. It then envisioned that universe tunneling into a new state of continual expansion. Insofar as that initially closed universe represents an actually existing material state in their scenario, tunneling from it into an expanding universe might qualify as a material process or mechanism. Nevertheless, the tunneling event that Hawking and Hartle invoke occurs *after the origin* of that universe.

Thus, tunneling in their scenario, as in Vilenkin's, does not explain the ultimate origin of the universe, but at best only its subsequent evolution. Indeed, though physicists sometimes give the impression that quantum tunneling provides a quasi-mechanistic explanation for the *origin* of the universe, it clearly does not.

Pulling a Universe Out of a Hat

This last point merits elaboration. Vilenkin constructs his "tunneling wave function" by assuming a singularity (i.e., a universe with beginning) that can tunnel into an expanding universe. Vilenkin's "tunneling wave function" determines the probability that the nascent universe he presupposes will tunnel into a particular state. Hawking and Hartle envision tunneling as an event that converts a preexisting closed universe (describable by a wave function) into a continually expanding universe. In their case, they construct a wave function that describes possible universes that could exist before the postulated tunneling event would occur.[21]

Yet, in both cases, quantum cosmologists must presuppose the existence of a universe. But that presupposes the very thing, the origin of which, they are attempting to explain. As philosopher of physics Willem Drees notes: "Hawking and Hartle interpreted their wave function of the universe as giving the probability for the universe to appear from nothing. However, this is not a correct interpretation, since the normalization presupposes a universe, not nothing."[22]

To see why, consider this. In ordinary quantum mechanics, an experimental apparatus has to exist before physicists can determine the wave function that describes the probable behavior of the photon within that apparatus. It follows, from the same analogy that justifies quantum cosmology in the first place, that a universe must first exist with possible properties before quantum cosmologists can construct the universal wave function that describes those properties in superposition. Indeed, the mathematics of quantum cosmology begins by describing a universe (or universes) already presupposed to exist.[23]

It has to. Recall that in the double-slit experiment a photon in superposition within that experimental apparatus logically precedes the solution to the Schrödinger equation—a wave function that enables physicists to describe the probable behavior or properties of the photon. In the same way, a universe with certain possible properties logically precedes the mathematical procedures that produce a solution to the Wheeler-DeWitt equation—a universal wave function that allows assigning definite probabilities to the different possible properties or attributes that the universe could possess. Thus, the mathematical procedures that quantum cosmologists use to produce a wave function to explain the origin of the universe tacitly presuppose the existence of a universe.[24] The quantum cosmologists are thus like bakers who allege they have baked a cake before they had the ingredients to do so, but in describing the amazing trick discuss how they used those very ingredients to bake the cake.[25]

Limiting Degrees of Mathematical Freedom

There is another crucial problem with the quantum cosmological models of Vilenkin and Hawking-Hartle. It is the problem I referred to at the end of the previous chapter. Their models not only presuppose a universe in the act of explaining its origin; they also smuggle information into the mathematical calculations they make as they seek to explain it. For this reason, if quantum cosmology provides a correct description of the world, it again inadvertently models the need for a transcendent intelligence.

Here's why. Vilenkin notes that the Wheeler-DeWitt equation, like all differential equations, allows for an infinite number of solutions. To determine a unique solution—a unique universal wave function—theoretical physicists must carefully choose boundary conditions and impose them on the equation at the outset. Yet unlike the boundary conditions imposed on a vibrating string (recall the discussion in Chapter 13), no physical system yet exists that can determine the appropriate constraints on the Wheeler-DeWitt equation.[26]

There is a good reason for this. In physics, differential equations typically describe the behavior of physical systems. The physical parameters of the system determine the initial and boundary conditions that delimit the range of relevant solutions to the equation in question. For example, in the case of a guitar string vibrating in accord with the equation for oscillating motion (Hooke's law), the distance over which the string is stretched determines the relevant boundary conditions. How hard the guitar player plucked the string determines the initial condition.

Yet since the Wheeler-DeWitt equation logically precedes its solution, and since only out of its solution—the universal wave function—will (presumably) an actual universe with physical parameters emerge, no physical system *can* provide information about how to constrain the equation with appropriate boundary conditions. The Wheeler-DeWitt equation, however, like other differential equations, has an infinite number of possible solutions. Consequently, physicists need information about boundary conditions to solve it, as Vilenkin himself has noted.[27]

In a revealing passage in his technical work, Vilenkin describes the need for boundary conditions to restrict degrees of mathematical freedom on possible solutions to the Wheeler-DeWitt equation. He remarks: "In ordinary quantum mechanics, the boundary conditions for the wave function are determined by the physical setup external to the system under consideration. In quantum cosmology, there is nothing external to the universe, and a boundary condition should be added to eq. (9) [the Wheeler-DeWitt equation]."[28]

This passage is revealing because it shows that physicists themselves must arbitrarily restrict the infinite degrees of mathematical freedom inherent in the Wheeler-DeWitt equation in order to solve it. Vilenkin did so by choosing specific boundary conditions to restrict the values of superspace (creating what theorists call a "mini-superspace"). He also made arbitrary assumptions about the nature of the universes that could emerge out of the singularity. In particular, his mathematical apparatus presupposed that such universes would be homogeneous, isotropic, and closed.[29] Only such restriction of possibilities allows physicists to solve the equation and thus to generate a (more or less unique) universal wave function ψ that includes universes like ours.

Vilenkin has explained his procedure in detail.[30] His method of restricting superspace by imposing *carefully chosen* boundary conditions on the Wheeler-DeWitt equation does result in a more or less unique solution to the equation. The resulting universal wave function ψ does include a universe like ours as a probable observation. Consequently, quantum cosmologists regard such an outcome as an explanation of the origin of the universe, indeed, as an explanation of the universe "from nothing," as Krauss puts it. Nevertheless, the specific universal wave function ψ that

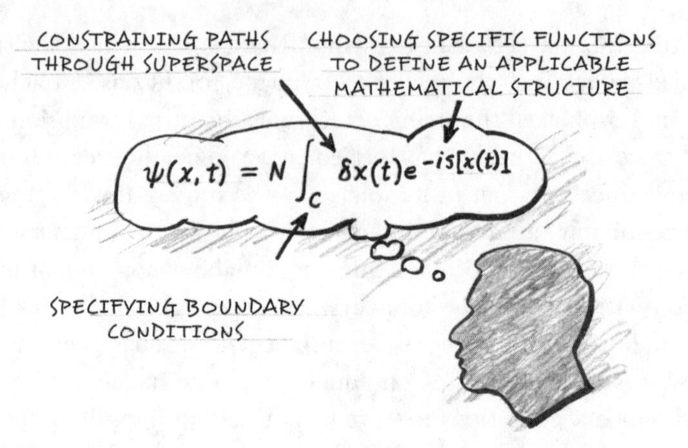

FIGURE 18.2

Solving the Wheeler-DeWitt equation allows quantum cosmologists to construct a universal wave function (Ψ) that describes possible universes with different possible gravitational fields. If our universe is included in the ensemble described by a universal wave function (Ψ), quantum cosmologists will regard (Ψ) as a description or explanation of the origin of the physical universe. This figure shows a mathematical expression called a "path integral" that is used to solve the Wheeler-DeWitt equation and construct the universal wave function (Ψ). The arrows point to variables, functions, and boundary conditions that must be specified to solve the path-integral (and, thus, the Wheeler-DeWitt equation). Because the path-integral, like the Wheeler-DeWitt equation, logically precedes any mathematical expression describing possible universes or the origin of them (as the universal wave function Ψ does), the path-integral does not itself describe a physical system. Consequently, there is no physical system that can determine the boundary conditions (or specify other mathematical parameters) that allow the path-integral to be solved. Instead, physicists themselves must determine these constraints. Quantum cosmologists invariably do this selectively with an end goal in mind, namely, constructing a universal wave function that includes a universe such as ours as a reasonably probable outcome.

"explains" the origin of the universe in this way is entirely an artifact of the restrictions that the theoretical physicists themselves have placed on the possible solutions to the Wheeler-DeWitt equation (Fig. 18.2).

Hawking and Hartle's Constraints on the Wave Function

As noted in the last chapter, Hawking and Hartle's version of quantum cosmology also imposes constraints on the possible solutions to the Wheeler-DeWitt equation, but in a different way. First, and most important, they constrained outcomes by specifying only certain kinds of universes with certain kinds of possible geometries for inclusion in their summing procedure. In particular, they chose to consider only homogeneous, isotropic, and closed universes with a positive cosmological constant. As they explained in their first seminal paper on quantum cosmology, "It is particularly straightforward to construct mini-superspace models using the functional integral approach to quantum gravity. One simply *restricts* the functional integral to the *restricted degrees of freedom* to be quantized. . . . [We use] a particularly simple mini-superspace model. In it we restrict the cosmological constant to be positive and the four-geometries to be spatially homogeneous, isotropic, and closed."[31]

As noted in the last chapter, as a calculational expedient, Hawking and Hartle only summed over those "paths through superspace" (or curvature-matter pairings) that used an imaginary time coordinate in their depiction of possible geometries of space. Wave functions computed from such "paths" do not by definition change over time. Thus, quantum cosmologists refer to those regions of superspace as "nonoscillating."[32] In any case, by choosing to sum over *only those* universes with no real-time variable when computing the universal wave function, Hawking and Hartle imposed further restrictions on the possible universes considered in their construction of ψ.

Hawking and Hartle also restricted degrees of mathematical freedom as a result of the approximation techniques they used in their sum-over-histories method.[33] The use of these techniques further constricted "superspace" in the process of finding an approximate solution to the path integral.[34] Without these techniques, Hawking and Hartle could not solve—even approximately—the Wheeler-DeWitt equation.

Understandable considerations of mathematical expediency dictated the use of these mathematical procedures and approximations. Nevertheless, no physical theory justified the specific constriction of superspace that the use of these techniques entailed. Consequently, from the standpoint of fundamental physics, the use of these methods and certainly the assumptions about the universe that Hawking and Hartle used to restrict superspace constitute *ad hoc* constraints on the process of constructing the universal wave function. In a recent interview, James Hartle acknowledged as much. "I have to tell you in confidence," he explained, "that every time when we do one of those calculations, we have to use very simple models in which lots of degrees of freedom are just eliminated. It's called mini-superspace. . . . It's how we make our daily bread, so to speak."[35]

Daily bread or not, Hawking and Hartle's assumptions about the kind of universes they would consider in the construction of the universal wave function clearly appropriated knowledge of the properties of *our* universe in a question-begging way. They effectively smuggled information into their calculation by winnowing the region of superspace under consideration, so as to virtually ensure that the paths through superspace that they summed would produce a universal wave function that included universes with properties similar to our own.[36]

Indeed, Hawking and Hartle clearly chose the regions of superspace (and possible universes under consideration) that they did with an end in mind—a universe with fundamental physics matching our own with recognizable and relevant spacetime geometries exhibiting the correct relationships between spatial curvature and matter as described by general relativity. In essence, they designed a mathematical procedure and limited inputs into it, so as to give "the right answer."[37] They subtly acknowledged this in the introduction to their 1983 technical paper:

> In any attempt to apply quantum mechanics to the Universe as a whole, the specification of the possible quantum-mechanical states which the Universe can occupy is of central importance. This *specification* determines the possible dynamical behavior of the Universe. Moreover, if the uniqueness of the present Universe is to find any explanation in quantum gravity, it can only come *from a restriction on the possible states* available.[38]

Leading critics of Hawking and Hartle's and Vilenkin's quantum cosmological models have noted the arbitrary nature of the constraints they impose.[39] For example, Christopher Isham has noted that although quantum cosmologists do generate a wave function that includes universes such as ours, they only do so as the result of their use of restrictive mathematical approximations and their own decisions to impose many arbitrary constraints on the possible universes (in "superspace") that they will consider. As he states the situation in discussing Vilenkin's approach, "Various approximate calculations have been performed which do indeed predict a unique state function [i.e., universal wave function]. However, these approximations involve ignoring all but a small number of the infinite possible modes of the universe, and it is by no means clear that the uniqueness will be preserved in the full theory."[40]

Hawking and Hartle tacitly acknowledge that this problem afflicts their technical work, though Hawking does not discuss it in his popular books. They do so by noting that, absent specific restrictions, their method will not generate a unique universal wave function. In other words, *unless* they constrain the space of possible universes under consideration in specific ways, their method will not yield wave functions that include universes such as ours. As the theoretical physicist Jonathan Halliwell notes, "The no-boundary proposal as it stands does not fix the wave function uniquely. There are . . . many no-boundary wave functions, each corresponding to a different choice of contour."[41]

Only by making these arbitrary choices to constrain the space of possible universes under consideration can Hawking-Hartle and Vilenkin construct a universal wave function that will contain universes like ours. This is no small point. The restrictions placed by physicists on possible solutions to the Wheeler-DeWitt equation represent an infusion of information into the mathematical apparatus they use to model the origin of the universe. Indeed, according to basic axioms of Shannon's information theory, anytime someone elects one option and excludes another, he or she inputs one bit of information into a system. Thus, the choice to exclude a nearly infinite number of possible mathematical solutions to the Wheeler-DeWitt equation, whether by (a) directly imposing boundary conditions on the equation, (b) limiting the possible universes under consideration when constructing the universal wave function (limiting "paths through superspace"), or (c) both, represents an enormous input of information into the mathematical equations and procedures that quantum cosmologists use to model the origin and development of the universe. As Halliwell notes of the Hawking-Hartle model, "The wave function is therefore only fixed uniquely after one has *put in some extra information fixing the contour*."[42] Indeed. The source of that "extra information" is precisely what is at issue.

Modeling a Universe by Design

Significantly, the choice of these constraints occurs entirely because of the decision of an intelligent agent—in particular, that of a theoretical physicist. Neither the Wheeler-DeWitt equation itself nor any deeper theory of quantum gravity or other fundamental physical theory determines the choice of these boundary conditions or constraints. They are imposed on the equation by an intelligent choice, a purposeful scientist who selects them with a distant goal in mind. In all modeling of the origin of the universe using quantum cosmology, intelligent agents must restrict degrees of mathematical freedom to generate a desired outcome, a wave function that includes our universe. Philosophers of physics call this the "infinite winnowing problem." Both Hawking-Hartle and Vilenkin solve this winnowing problem by their own intelligent design.[43]

Some physicists have criticized Hawking and Hartle for the ad hoc constraints they impose on their mathematical procedure. I'm not sure that warrants critique as much as simply comment. In my view, both Hawking-Hartle and Vilenkin inadvertently model the need for a designing intelligence to exclude some options and elect others—that is, to impart information—in order to achieve an intended outcome. In different ways, both versions of quantum cosmology illustrate the need for intelligence in much the same way that origin-of-life researchers inadvertently simulate the need for intelligent design when they choreograph and constrain chemical reactions to synthesize life-relevant biochemicals.

The Hawking-Hartle method also reminds me of Richard Dawkins's famed computer simulation of the alleged creative power of natural selection and random mutation. The simulation, described in his book *The Blind Watchmaker*, allegedly generated an information-rich line from Shakespeare's *Hamlet*, "Methinks it is like a weasel," through random changes in a string of text.[44] In both cases, the cosmological and the biological, the modeler had a distant goal in mind. Dawkins had a target sequence in mind; Hawking-Hartle and Vilenkin each had a target wave function in mind. These physicists knew that getting to the target would require making selections among a vast ensemble of possibilities according to certain criteria. Dawkins provided the target sequence to his computer and programmed it with selection criteria that would ensure that it converged on the sequence he wanted. Hawking and Hartle understood the kind of wave function that would have physical and cosmological relevance and then chose possible universes for inclusion in their summing procedure in accord with selection criteria that would ensure *that kind* of wave function. Similarly, Vilenkin knew the boundary conditions he would need to choose, and the assumptions about allowable universes he would need to make, to ensure the construction of a relevant wave function. Both the selection criteria in Dawkins's biological simulation and the physical selection criteria (of the quantum cosmologists) were teleological. Both had ends in mind. And each required intelligent inputs of information to reach those ends.

Thus, these quantum cosmological models inadvertently confirm a major theme of this book: it takes a mind to generate specified or functional information, whether in ordinary experience, computer simulations, origin-of-life simulation experiments, the production of new forms of life, or, as we now see, in modeling the design of the universe. Indeed, even if Professor Krauss could explain the origin of matter, energy, space, and time from nothing, or from nothing but the mathematically expressed laws of physics, he still could not explain the origin of the *information* necessary to express and solve the equations that supposedly explain the origin of the universe. Instead, as we have seen, the quantum cosmological theories subtly depend upon the activity of a mind to model the origin of the universe.

It follows that *if* some version of quantum cosmology provides the correct model for the origin of the universe, then mind, not just matter and energy (or even math), played a causal role in that ultimate origin event. Insofar as quantum cosmology models the origin of the universe, it implies the need for prior intelligent design.

Calling the Bluff

When I met Krauss in Toronto, I was well aware of how the popularization of quantum cosmology had contributed to the perception that science had rendered theistic belief either unnecessary or untenable. Thus, my impending encounter with him occasioned an opportunity to rekindle my interest in Hawking's and Vilenkin's work.

What I discovered in their work reminded me of attempts by mainstream evolutionary biologists to account for the origin of biological information by positing prior unexplained sources of information or by inadvertently simulating its production in their own intelligently designed experimental protocols. It also reminded me of attempts by physicists to explain the origin of the fine tuning by positing universe-generating mechanisms that required prior unexplained fine tuning.

As I studied their papers, I kept stumbling across places where quantum cosmologists, very intelligent physicists all, imposed constraints on their mathematical procedures to ensure outcomes relevant to describing our universe. Of course, they acknowledged this implicitly using technical mathematical terminology, often employing the passive voice, as in "a boundary condition *should be added to* eq. (9) [the Wheeler-DeWitt equation] as an independent physical law."[45]

In the debate, I planned to ask Krauss about what I now call the "cosmological information problem." As I've already related, I did not get to have that conversation about cosmology with him. When I found myself experiencing a migraine, I realized that it was not the best time to initiate a discussion about the universal wave function, superspace, or the application of quantum mechanics to cosmology. For his part, Krauss spent little time that night explaining how the universe could have arisen "from nothing" and still less explaining how quantum cosmology justified that claim.

After my opening talk, a local Canadian physician took me to a dark room to cover my eyes in hopes that I would recover enough to participate in the panel discussion that was to follow the third speaker, Denis Lamoureux. This same doctor had joined me for dinner before the debate and knew that I was looking forward to the discussion with Krauss. He now advised me not to try to raise the subject of quantum cosmology, since Krauss had not discussed his own theory in enough detail to warrant a critique. "The audience," he said, "will be lost." And so the rest of the evening proceeded along more predictable lines: I amplified and defended my opening argument for the intelligent design of life, and Krauss and the other panelist attempted to refute it.

In retrospect, as I mentioned in the Prologue, the unanticipated direction of the discussion was a blessing in disguise. My predebate study left me convinced that quantum cosmologists were gerrymandering their choices of boundary conditions and other assumptions. Nevertheless, the difficulty of the specialized mathematics involved rendered the indispensable contribution of the theorists—their active and intelligent input of information—opaque even to many other scientists. This made it possible for quantum cosmologists to hide behind the mathematical complexity of their models as they made highly dubious, and even absurd, metaphysical claims. I realized after the debate that to make this clear to a wider audience, I would benefit from more time to study the technical details and formulate my critique carefully.

At the debate, Krauss warned me not to challenge him on his cosmological theories. He told the audience, "Now after they lost . . . [the argument about] biology, what I noticed is that cosmology was next. And I'm hoping for his own sake that Stephen doesn't try to do that while I'm here, because it will be a mistake, I promise." Krauss's warning had the exact opposite of its intended effect. He meant to dissuade me from challenging his cosmological claims, lest I find myself embarrassed and out of my depth. That night, for entirely different reasons, no challenge occurred. But his advisory notice seemed to me to telegraph weakness. Krauss is a smart man. I suspected that he knew he couldn't explain how our universe originated from nothing, with or without quantum cosmology. After the debate, I decided to get to the bottom of the matter.

I began a formal research project with two colleagues: Bruce Gordon, a philosopher of physics who had done his PhD at Northwestern University with Arthur Fine, a leading figure in that field, and Brian Miller, a physicist with a PhD in complex systems physics from Duke University.[46]

After three years of focused research and presentations to specialists in private conferences and seminars, I was confident that quantum cosmology does not explain the origin of the universe in purely materialistic terms. Instead, to the limited extent it succeeds, it attributes causal powers to abstract mathematics and depends upon intelligent inputs of information from theoretical physicists as they model the origin of the universe. Thus, it does not dispense at all with intelligent design or with theism as an explanation for the origin of the universe.

Instead, quantum cosmology implies the need for an intelligent agent to breathe, if not "fire into the equations," then certainly specificity and information. Thus, it implies something akin to the biblical idea that "in the beginning was the Word." And that's not nothing—by anyone's definition.

19

Collapsing Waves and Boltzmann Brains

The double-slit experiment that established the wave-particle duality of light (and later of electrons) suggested that the *observation of the light* somehow caused the light wave to manifest a particle-like nature as an individual photon with discrete attributes. Quantum physicists call this transition—from a photon in a "superposition" of many possible states at the same time to its manifestation as a particle with discrete attributes—the "collapse of the wave function" or the "collapse of the wave packet."

Traditionally, quantum physicists have thought that the observation of the photon *caused* the wave packet to collapse. That's because only when it reached the detector did the wave manifest specific particulate attributes (of position and momentum). In another weird version of the double-slit experiment, physicists discovered that observing the wave earlier as it passed through one of the slits caused it to manifest particle-like attributes *at that point.* Even weirder, they discovered that getting an advanced peek at the photons as they passed through the slits by installing detectors there eliminated the interference pattern once the photons reached the detection plate on the far end of the experimental apparatus. As I've noted previously, physicists now call this interpretation of the collapse of the wave function, where the observer somehow causes the collapse, the "Copenhagen interpretation." It is so named in honor of the Danish theoretical physicist Niels Bohr with whom it is associated.[1]

The Collapse of the Universal Wave Function

Up to this point, I've not raised the question of what causes the collapse of the wave function in quantum *cosmology*, as opposed to quantum physics. But since quantum cosmology is based upon an analogy with ordinary quantum physics, and since quantum cosmologists regard "the universe" before Planck time as existing in a superposition of many possible spatial geometries and configurations of mass-energy *simultaneously*, the question inevitably arises: What causes all these different possible universes to collapse suddenly into our universe with its specific spatial geometry and mass-energy configuration?[2] In ordinary quantum mechanics, an observer appears to cause the wave packet to collapse. But does that mean that some "Cosmic Observer" must preexist the collapse that produces our universe out of the universal wave function ψ?[3]

The traditional Copenhagen interpretation of the collapse of the wave packet, when applied to quantum cosmology, would seem to require such a transcendent Cosmic Observer to cause the collapse and thus the emergence of our universe with its specific attributes. Adopting a Copenhagen interpretation of the collapse of the *universal* wave function in a quantum cosmological context appears, therefore, to have clear theistic implications. Indeed, what other than a transcendent godlike being could have been present to observe and thus cause such an event before there was a physical universe[4] (Fig. 19.1)?

Nevertheless, the idea that a Cosmic Observer might have caused our universe to arise out of the universal wave function has held no appeal for the mostly atheistic or religiously agnostic proponents of quantum cosmology. Instead, as we've seen, the chief architects of the theory, such as Stephen Hawking, posed quantum cosmology as a counter to the cosmological argument for God's existence. In Hawking's case, that meant formulating a theory that would undermine the implications of his own proof of the singularity at the beginning of the universe, which is only one of the reasons that Hawking stands as one the most interesting intellectual figures of recent memory.

Nevertheless, other physicists have formulated many other interpretations of the collapse of the wave function. A popular interpretation among many physicists today is known as the modified Copenhagen interpretation. It holds that the wave packet does not collapse because

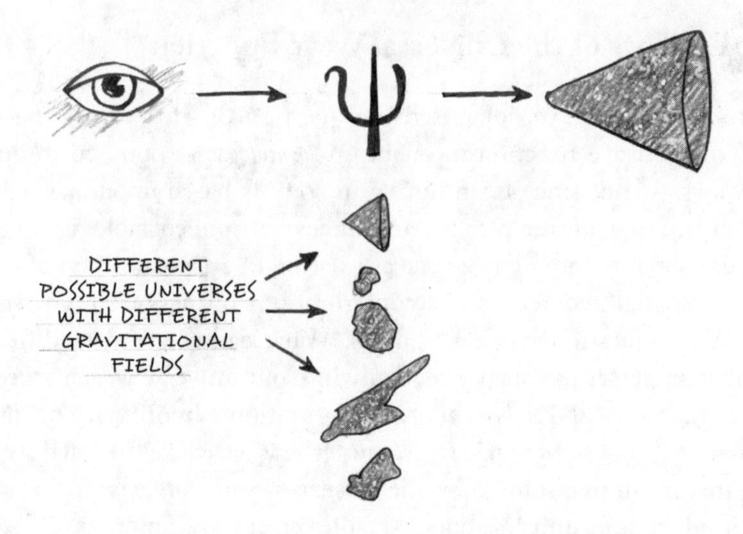

FIGURE 19.1

A Cosmic Observer? The traditional Copenhagen interpretation of the collapse of the wave function—when applied to the universal wave function in quantum cosmology—would seem to require a transcendent "Cosmic Observer" to cause the collapse and, thus, the emergence of a specific universe among the various possible universes described by the universal wave function.

of an observation per se, but instead because the waves of light (or electrons) encounter a large macroscopic object.[5] According to this interpretation, waves of light collapse and acquire definite characteristics whenever they collide with large macroscopic objects of any kind. In the double-slit experiment, the large macroscopic object just happened to have a detector present so that human observers were able to learn of the collapse and perceive the particulate as well as the wave character of light.

This interpretation may circumvent the need to invoke an observer in ordinary quantum mechanics (though there are technical problems with it).[6] Nevertheless, it does not offer any explanation for what might have caused the emergence of a universe out of the indeterminate array of possibilities represented by the universal wave function. The reason for this is fairly obvious. Until a universe emerged from the universal wave function, no large macroscopic objects would have yet existed.

Thus, in a quantum cosmological context, where the Copenhagen interpretation seems to have clear but undesirable theistic implications and the modified Copenhagen interpretation cannot in principle explain the collapse of the universal wave function, most quantum cosmologists[7] have adopted another interpretation known as the many-worlds interpretation (MWI).

According to this interpretation of ordinary quantum mechanics, first posed by the physicist Hugh Everett in his 1957 Princeton University PhD dissertation,[8] the wave function does not actually collapse at all. Instead, photons or electrons exemplifying every possible position or momentum described by the wave function in a quantum mechanical system actually exist. Moreover, they exist simultaneously in separate worlds or universes. In other words, the wave function that describes the different possible positions of a photon, for example, is actually describing different realities in which the photon exists in each of several different locations in detectors in multiple separate universes.

This idea applied to quantum cosmology implies that every possible universe described by the universal wave function also exists as its own separate reality. Thus, this interpretation circumvents the need to explain what *causes* the collapse of the universal wave function to bring a specific universe or, indeed, our universe into existence. That's because, again, the wave function never collapses. All possible universes described by the universal wave function exist, because the universal wave function represents the most fundamental description of reality that we have and every possibility that it describes must exist in some universe.

Vain Imaginings or Fire in the Equations?

The many-worlds interpretation, though speculative in the extreme and problematic on technical grounds, if true, would still reinforce my critique of the use of quantum cosmology to explain the origin of the universe in strictly materialistic terms.

In the first place, by treating all of the merely mathematical possibilities described by the universal wave function as a real universe, the MWI "reifies the math" on a literally unimaginable scale. Yet it still does not answer the question of "what breathes fire" into the relevant equation. The MWI simply asserts that, for some unspecified reason and by some unknown cause, every possibility described by the universal wave function must actually exist—indeed, that every possible *universe* described by the universal wave function must exist as an *actual* universe. Yet this interpretation does not cite a physical cause of the origin of our universe—or any other universe. It simply imputes a specific meaning to the universal wave function by positing the existence of these other universes.

As such, it certainly does not provide a *better* explanation for the origin of the universe than the God hypothesis. Theism posits the action of a powerful preexisting mind—a personal agent—as the cause of the origin of the universe. On the supposition that such a being exists, theism meets the causal adequacy criterion of a good explanation. The MWI, however, doesn't posit anything other than a mathematical expression, the universal wave function, that could conceivably function as a cause of the origin of our universe—or any other. Thus, either the MWI imputes to the universal wave function causal powers that we know purely mathematical descriptions or equations lack (thus, "reifying the math"), or it fails to offer any cause for the origin of the universes it posits.

In the latter case, quantum cosmology would simply fail to meet the causal-adequacy criterion of a good explanation, violating the principle of sufficient reason.[9] In the former case, where cosmologists treat the universal wave function *as* the prime or metaphysical foundation of reality, they necessarily commit themselves to a form of mathematical Platonism in which an equation describing mathematical possibilities somehow determines reality (or multiple realities).

Yet, as Vilenkin has noted, in our experience mathematical ideas reside in minds.[10] Our experience also shows that only minds can use mathematical ideas to influence the structure of matter. Consequently, to the extent that quantum cosmologists do reify the universal wave function as the metaphysical foundation of all existence—and most do so by claiming that the laws of physics explain why there is something rather than nothing, for example—they implicitly affirm a metaphysics that comports better with theism than with scientific atheism or materialism. Whereas theism or deism posits a mind prior to, or independent of, the material universe (or universes) in which mathematical ideas could reside, materialism or atheism, the worldview advanced by most proponents of quantum cosmology, does not. Indeed, it's worth asking: What kind of mind could hold an infinite number of ideas about an infinite number of universes?

Many Worlds and Intelligent Design

There is yet another, more profound problem with using the many-worlds interpretation of quantum cosmology as an alternative to the God hypothesis. Recall that, for quantum cosmologists, identifying a universal wave function ψ that includes our universe as a reasonably probable outcome suffices to explain the origin of the universe. MWI proponents reinforce the use of ψ as a presumably nontheistic explanation by putatively eliminating any need to explain what causes ψ to collapse.

Nevertheless, treating the universal wave function ψ as the ultimate explanatory principle or as the ground of all being fails to account for the dependence of ψ and its specific mathematical features on prior infusions of information into the mathematical operations that generate ψ. Recall that to generate a universal wave function of any kind, quantum cosmologists modeling the origin of the universe first have to solve the Wheeler-DeWitt equation.[11] Recall also that to generate a specific wave function that includes universes such as ours, theoretical physicists must *choose* specific boundary conditions or otherwise arbitrarily limit the degrees of mathematical freedom associated with the Wheeler-DeWitt equation.

It follows that the specific features of any universal wave function (and the universes that it describes) depend upon prior information-rich choices of boundary conditions or restrictions on superspace—those that made the Wheeler-DeWitt equation soluble. Thus, even if physicists interpret the universal wave function as having some reality independent of our minds and that it somehow explains the existence of many universes, the origin of ψ and its specific mathematical features would still require explanation. And based upon the mathematical process by which physicists model the origin of the universes described in ψ (and the origin of ψ itself), any such explanation must reference the prior unexplained information in the form of the specific boundary constraints and/or restrictions that allowed physicists to determine a specific universal wave function in the first place.

In other words, even if the many-worlds interpretation does in some weak sense "explain" the origin of our and other universes, the universal wave function that putatively does that explaining itself reflects prior intelligent design. Physicists don't even get a universal wave function "to reify" until they first solve the Wheeler-DeWitt equation. But the mathematical procedure that quantum cosmology employs to produce ψ implies the need for prior *intelligently chosen* information-rich constraints. As much as any other interpretation of the universal wave function in quantum cosmology, the MWI presupposes a source of information, one that quantum cosmologists themselves provide to solve the Wheeler-DeWitt equation. In that sense, it also presupposes the need for prior intelligent design.

The Mathematical Universe Hypothesis

The Swedish-born MIT physicist Max Tegmark (Fig. 19.2) has proposed an even more radical cosmological theory—one that could be used to address the problem just discussed. Tegmark argues that every possible mathematical structure imaginable has a physical expression in some possible universe. Or as Tegmark puts it, "All structures that exist mathematically exist also physically."[12]

Tegmark proposed this idea to account for quantum mechanics and general relativity as the governing mathematical framework of the universe, realizing that other mathematical frameworks are possible. As other philosophers and physicists have noted, the equations of quantum mechanics and general relativity represent only a small set of the number of possible fundamental mathematical structures that might govern the universe. Metaphorically speaking, quantum mechanics and general relativity represent a tiny region in an abstract space of mathematical possibilities and thus could represent the outcome of an "infinite winnowing" among other possible fundamental mathematical frameworks.

FIGURE 19.2
The Swedish-born MIT physicist and cosmologist Max Tegmark.

Tegmark's theory, known as the mathematical universe hypothesis, denies the need for any such winnowing. Thus, his hypothesis could also be posed as a solution to the problem of the origin of the arbitrary boundary constraints and other restrictions that make any specific universal wave function relevant to explaining our universe. His proposal implies that not only does every possible universe described by the universal wave function exist, but every possible universal wave function—that is, every possible solution to the Wheeler-DeWitt equation—governs reality in some possible universe. His hypothesis also implies that any mathematically expressible physical law or theory exists (or describes reality) in some possible universe somewhere. In Tegmark's cosmology, nothing would need to constrain mathematical degrees of freedom, since all mathematically possible structures, formalisms, equations, or functions, including all possible mathematical solutions to the Wheeler-DeWitt equation, are actualized somewhere in some universe.

If true, Tegmark's cosmology eliminates the need to explain how any particular universal wave function arose. It would also eliminate the need to posit, invoke, model, or simulate prior intelligence to constrain mathematical degrees of freedom. Nevertheless, Tegmark's cosmology does not explain the origin of the universe in materialistic terms. And, if true, it comes at a devastating cost to our ability to trust scientific rationality.

Tegmark's mathematical universe hypothesis constitutes a form of radical mathematical Platonism. It reifies all the possible mathematical structures that humans have conceived or could conceive. Tegmark quite explicitly affirms such a reification, stating that "Physical existence is equivalent to mathematical existence."[13] But such a bold affirmation equating mathematical ideas with physical reality makes Tegmark's hypothesis no less problematic for materialism than quantum cosmological theories that do the same. As noted, we have absolutely no experience of pure math causing anything physical to originate apart from minds that have ideas and act upon physical reality guided by them. Consequently, Tegmark's proposal would also seem to require the existence and activity of a mind—indeed, an infinite mind—as a condition of its plausibility, just as quantum cosmological proposals that reify mathematical entities do.

Epistemological Cost: Undermining Scientific Explanation

Even so, Tegmark's hypothesis also comes at a great cost to scientific rationality. Tegmark posits an infinite multiplicity of universes exemplifying every possible set of mathematical relationships governing every possible configuration of matter. These mathematical relationships could include highly regular laws. They could also express laws that admit exceptions or laws that describe abrupt or irregular deviations from established trends. Or these laws could allow completely random fluctuations of various kinds to occur. As the physical chemist Michael Polanyi pointed out in his classic work of scientific epistemology, *Personal Knowledge*, there are an infinite number of mathematical relationships or functions that describe any finite number of data points, including an infinite number of functions that inscribe highly irregular curves or irregular relationships between variables.

Thus, accepting Tegmark's mathematical universe hypothesis, we may live in a universe in which the laws of physics have *to this point* described highly regular relationships between variables, but also a universe in which those same laws will at some arbitrary point *in the future* begin to describe highly irregular or idiosyncratic relationships.[14] For example, the force of gravity that we currently describe classically with Newton's universal law or Einstein's equations of general relativity could in some universe, maybe our own, have a dynamic element in which the strength of gravitation begins to increase exponentially, but only at some specified time or place after the beginning of that universe. Since some mathematical structure or set of equations could describe such a complex set of relationships between, say, curvature, matter, gravitational attraction, and *time*, such a universe would, given Tegmark's theory, have to exist. We might live in precisely such a universe with an impending radical change in the strength of gravitational force—perhaps occurring the moment you finish reading this sentence—and not yet know it.

If Tegmark's theory aptly describes reality, anything can and will happen somewhere in some universe infinitely many times. Every event or causal antecedent may presage an infinite number of possible consequent events in any one of an innumerable set of possible worlds—worlds governed by an innumerable set of possible mathematically described deterministic laws of physics. In addition, if eternal chaotic inflation aptly models our universe, and Tegmark's theory would include inflationary cosmological models as models expressing a possible mathematical structure, quantum fluctuations can produce multiple bubble universes and unpredictable events[15] all the time.

All this renders scientific explanation fundamentally uncertain, if not impossible. Scientific explanation presupposes the uniformity and regularity of nature, including the uniformity of the fundamental laws of physics and the regularity of patterns of cause and effect. Tegmark's mathematical universe hypothesis implies that such uniformity and regularity may not characterize our universe, however much it might have seemed to do so up until this point. Given Tegmark's infinite-universe cosmology, scientists could attribute any particular event to a cause long known to produce that effect. Or they could, with equal justification, attribute that same event to some alternative cause that had never before produced the effect in question—knowing that in some universe (possibly ours) radically different but mathematically describable cause-effect relationships may apply during brief but specified times.

For example, based upon our uniform and repeated experience up to this point, we would typically explain the sudden condensation of water on a pane of glass by describing how warm moist air of a given humidity and temperature had just interacted with a cold window pane of some lower temperature. But given the way the mathematical universe hypothesis denies any reason to assume the continued uniformity of nature, we might just as well attribute that same event to any number of other events or conditions—none of which had previously produced that effect in our world but which could conceivably do so in accord with some other mathematically-describable relationship between relevant variables. Thus, we could justifiably posit that we lived in one of the infinite universes in which a sudden *rise* in the temperature of the window and *drop* in the temperature and humidity of the air could, for some brief but specified time, also produce condensation. Any scientist taking Tegmark's hypothesis seriously would have to consider that we could live in that sort of a universe.

Moreover, events that we would ordinarily explain by reference to known causes, based upon ordinary experience of cause and effect, can also be just as readily attributed to random quantum fluctuations—fluctuations that would be extremely improbable if our universe was the only universe, but that are inevitable somewhere if an inflaton field, for example, has been busily spitting out universes for an infinite time. In infinite-universe cosmologies, physicists could attribute any event to chance—to a statistically improbable quantum fluctuation that nevertheless had to occur eventually (and an infinite number of times) in any one of an infinite number of possible universes. This means, assuming such a cosmology, that an exquisitely designed machine or an intricately crafted piece of poetry is just as likely to have been produced by random fluctuations in the quantum vacuum as by an intelligent human being. It also means that natural phenomena such as earthquakes are just as likely to have resulted from a series of improbable random fluctuations in a quantum vacuum as they are from a discernible progression of material causes.

Epistemological Cost: Undermining Scientific Prediction

Infinite-universe cosmologies especially undermine confidence in scientific prediction as well as explanation. For example, given Tegmark's theory, every conceivable mathematical relationship must exist in some possible world between a given antecedent and a given ensemble of possible consequents. That means that every antecedent event could generate an infinite number of different possible consequent events. Moreover, those relationships may also exhibit a dynamic element in which the relationships between variables may, unbeknownst to us, change at specific times.

If so, then for the physicist who takes Tegmark's theory seriously, specifying an antecedent will not allow prediction of a specific consequent, no matter how much past experience we may have of a specific antecedent-consequent relationship. We simply cannot know whether we might live in one of those universes in which the relationships are destined or "scheduled" by some higher mathematical relationship to change at some specified time. The assumption that anything can happen and will eventually happen somewhere, and could happen here, utterly destroys our ability to make warranted predictions.

Tegmark himself has recognized how infinite-universe cosmologies undermine scientific prediction, though for a slightly different reason. In a provocative article, "Infinity Is a Beautiful Concept—and It's Ruining Physics," Tegmark describes a problem that physicists call the "measure problem." Imagine, for example, that a physicist who posits an infinite number of universes also wants to know what percentage of those universes will eventually support life. Or put differently, suppose the physicist wants to predict the probability that some universe will eventually produce life. Our physicist suspects that because of the extreme specificity of the initial and boundary conditions needed to ensure a life-permitting universe (or a universal wave function that includes one), the probability of generating such a universe will be extremely low and, therefore, that a huge—indeed, an infinite—number of universes will be incapable of hosting life. Nevertheless, the physicist also knows that in all infinite-universe cosmologies every event will occur an infinite number of times. Therefore, an infinite number of universes will also eventually produce life, even if not all universes do.

To calculate the probability of finding a universe capable of supporting life, our physicist must now divide one infinite quantity by another, yielding a mathematically meaningless ratio. As Tegmark explains:

> Physics is all about predicting the future from the past, but inflation [one of many infinite-universe cosmologies] seems to sabotage this. When we try to predict the probability that something particular will happen, in-

flation always gives the same useless answer: infinity divided by infinity. The problem is that whatever experiment you make, inflation predicts there will be infinitely many copies of you, far away in our infinite space, obtaining each physically possible outcome; and despite years of teeth-grinding in the cosmology community, no consensus has emerged on how to extract sensible answers from these infinities. So, strictly speaking, we physicists can no longer predict anything at all![16]

Thus, infinite-universe cosmologies, including Tegmark's, have an unexpected liability: once they are permitted as a possible explanation for anything, they undermine confidence in practical and scientific reasoning about *everything.*

Absurd Consequences: Boltzmann Brains

To make matters worse, infinite-universe cosmologies entail bizarre absurdities that cast doubt on our ability to trust our senses, our memories, and the reliability of our minds. In particular, infinite-universe cosmologies, including Tegmark's as well as the model of eternal chaotic inflation, imply the existence of an infinite number of "Boltzmann brains." Boltzmann brains are hypothetical observers with brains that could in theory assemble as the result of a chance concatenation of atoms or elementary particles due, for example, to random quantum or thermal fluctuations. Physicists call such brains "Boltzmann brains"[17] for the late nineteenth-century Austrian physicist Ludwig Boltzmann (Fig. 19.3), who first conceived of this seemingly fanciful possibility. (Boltzmann was possibly a man with too much time on his hands.)

FIGURE 19.3
The nineteenth-century Austrian physicist Ludwig Boltzmann, who speculated about the possibility of "Boltzmann brains."

According to quantum mechanics, there is a finite, if extremely tiny, probability of random fluctuations at a subatomic level occasionally generating unexpected macroscopic outcomes such as, for example, the Statue of Liberty waving at you as you fly by it in an airplane. Though such events will in all probability never happen in our solitary universe, any event with a finite probability of occurrence, however small, will inevitably happen in an infinite multiverse. Indeed, it will do so an infinite number of times.

One such incredibly improbable event would be the production of a Boltzmann brain. In an infinite number of universes, an infinite number of such brains would exist, including an infinite number with false memories and perceptions (Fig. 19.4). More troubling, it follows that we ourselves might have such brains rather than so-called natural brains with reliable perceptions and true memories. Consequently, many physicists now worry that positing an infinite number of universes, either to solve the fine-tuning problem or (to the extent that any might be aware of it) the problem of the informational inputs necessary to render quantum cosmology plausible, should lead us to doubt the reliability of our own minds.[18] In other words, positing infinite-universe cosmologies leads inevitably to radical epistemological skepticism.

Physicists and philosophers, perhaps also supplied with an excess of free time, have proposed various solutions to this problem. Initially, they

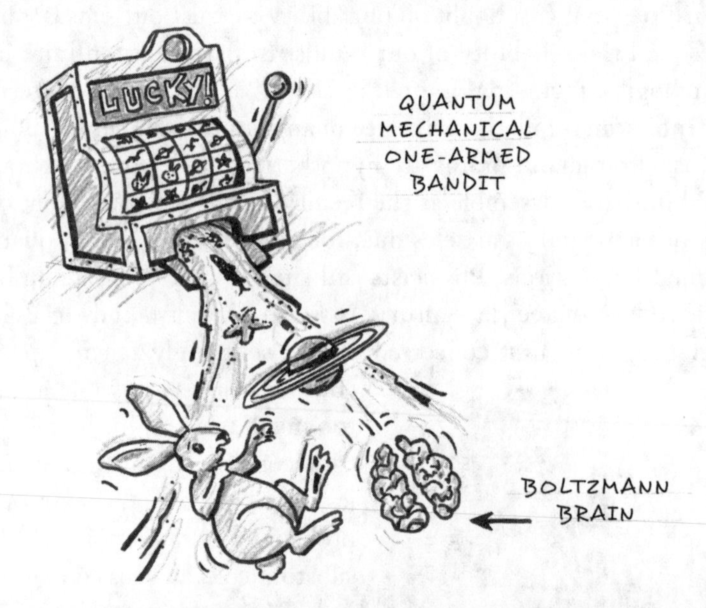

FIGURE 19.4

The Boltzmann brain problem. According to quantum mechanics, there is a finite, if extremely tiny, probability of random quantum fluctuations at a subatomic level occasionally generating unexpected macroscopic outcomes, including the production of fully formed persons with so-called Boltzmann brains containing false memories. Though such events would, in all probability, never happen if our solitary universe were the only universe, any event with a finite probability of occurrence, however small, will inevitably happen in an infinite multiverse. In such a multiverse, an infinite number of brains with false memories and perceptions would inevitably arise. More troubling, there are reasons to think that in such a multiverse we ourselves are more likely to have "Boltzmann brains" than "natural brains" with reliable perceptions and true memories.

attempted to do this by showing that, for inflationary cosmology at least, the ratio or "relative frequency" of Boltzmann brains to natural brains in a given quadrant of the inflaton field would be low. They then sought to show by extrapolation that that relative frequency would only grow smaller as they extended their analysis to more and more of the inflaton field. Thus the probability of any given brain being a Boltzmann brain rather than a natural brain would be extremely low when considering the whole inflaton field (and the infinite number of universes it contained).

Here's a mathematical illustration that shows roughly what these physicists were trying to demonstrate. Imagine that you want to divide the number of prime numbers that exist on a section of the number line by the number of even numbers. Though the total number of primes and evens are both infinite on the number line as a whole, if you were to consider only a finite section of the number line you would find that the ratio of primes to evens would be very low. You would also find that as you considered more and more of the number line, the ratio of the prime numbers to even numbers would get even smaller. As you then extrapolated from the finite portion of the number line to an infinite section of the number line, you would find that the relative frequency of the primes to evens would approach a limit. It would get closer and closer to zero. Thus, even though there are an infinite number of prime numbers sprinkled along the number line, if you attempted to sample numbers at random along any finite stretch of the number line, you would have a much, much higher probability of finding an even number than a prime.

In a similar way, proponents of infinite-universe cosmologies tried to show that the ratio of Boltzmann brains to natural brains in any finite sector of the space produced by the inflaton field would also be low. They tried to show further that the relative frequency of the two types of brains would only become smaller as they extrapolated from one quadrant of space to progressively larger and larger quadrants, with the ratio eventually approaching zero. Thus, as with comparing prime and even numbers, the ratio of Boltzmann brains to natural brains would be incredibly low, making it improbable that any given brain would be a Boltzmann brain even in an infinite-universe scenario.

This proposed solution turned out not to work, however. The physicists proposing it soon realized that, in any given sector of the inflating space, the inflaton field would produce astronomically more extremely young or short-lived universes than extremely old universes such as ours. Why? Because the inflaton field generates new bubble universes extremely frequently, whenever quantum fluctuations occur. Thus, an exponentially vast number of young bubble universes would arise in the time it would take for, say, a single multibillion-year-old universe such as ours to develop.

Why is this a problem? Because many such Boltzmann brains with false memories would arise by spontaneous quantum fluctuations in the young universes in the time that it would take for one or a few conscious intelligent forms of life (i.e., natural brains with real memories and accurate sense perceptions) to evolve in one of the relatively few old universes. Thus, the activity of the inflaton field would ensure that most observers would be Boltzmann brains in universes too young to permit the kind of evolution needed to produce ordinary observers with reliable memories.[19]

Indeed, since so many more young universes would form than old universes, the young universes containing brains with false memories and perceptions would vastly outnumber extremely old universes with people in them like ourselves with (we think!) true memories and perceptions of an ancient universe. It follows that if inflationary multiverse models accurately represent reality, then it is vastly more probable that we ourselves are Boltzmann brains than persons with real memories and accurate perceptions of living in a 13.8-billion-year-old universe. In some models, it's even more probable that a whole universe like ours could have spontaneously fluctuated into existence than it is that our universe with its extraordinarily improbable initial conditions would have evolved in an orderly and lawlike way over billions of years.

Thus, the inflationary multiverse hypothesis generates multiple absurdities. It implies that we are probably not the people we take ourselves to be and that our memories and perceptions are probably not reliable, but quite possibly the result of random quantum fluctuations. Neither is our universe itself what it appears to be under the hypothesis of eternal inflation. The inflationary multiverse would render all scientific reasoning, explanation, and perception unreliable, undermining any basis for accepting the multiverse hypothesis or any scientific hypothesis or conclusion whatsoever. It would be hard to invent a more self-refuting hypothesis than that!

Moreover, Tegmark's theory entails all these same epistemological costs, including the existence of Boltzmann brains. Not only does it include a universe described by inflationary cosmology and its mathematical structure as a possible universe that would have to exist, but Tegmark's theory also allows virtually anything to occur somewhere, provided physicists can construct a mathematical description of the process (or change of state) by which such possible events might occur.[20] But if physicists need not concern themselves with considerations of causal adequacy or empirical plausibility based upon past experience, they can characterize any hypothetical change of state mathematically. They could, for example, describe an initial condition in which no biologically relevant information was present and then a subsequent condition in which an enormous amount of specified biologically relevant information was present—enough to specify the construction of a brain with false memories perhaps. If they don't have to assign an empirically plausible cause to that transition, but can assume that any transition from an initial state to a final state will eventually occur an infinite number of times, as Tegmark's theory implies, then such a transition must have occurred in some possible universe. So why not ours? Thus, Tegmark's theory also implies the existence of Boltzmann brains, with all the epistemological chaos, if not lunacy, that entails.

Tegmark himself has also acknowledged the problems associated with infinite cosmologies by proposing an amendment to his radical theory. Instead of positing that all possible mathematical structures must exist in some possible world, he has more recently suggested that physicists limit themselves to mathematical structures that use only finite numbers and thus exclude all infinities.[21] Ironically, this restriction on the space of possible mathematical structures represents another winnowing of possibilities, the need for which Tegmark set out to eliminate with his mathematical universe hypothesis.[22]

Simplicity and Status of the Dialectic

Up to this point, I've assumed for the sake of argument that quantum cosmology more or less accurately models the early universe and its origin. I've sought to show, moreover, that if it does aptly model the origin of the universe, it nevertheless inadvertently supports a theistic perspective, however much its architects and popularizers might have used it to undermine the cosmological argument for God's existence.

Of course, if these models do not accurately depict the origin of the universe, then they have no force against that argument, in which case the God hypothesis presented as the best explanation for our temporally finite universe still stands. In sum, if quantum cosmology (and/or one of the interpretations of the universal wave function) does accurately model the origin of the universe, then it has idealist and theistic implications. If quantum cosmology (variously interpreted) does not accurately model the origin of the universe, the facts of cosmology have theistic implications for wholly other reasons.

This dynamic holds for the most exotic interpretations of quantum cosmology: the many-worlds interpretation and Tegmark's mathematical universe hypothesis. Since both interpretations ultimately posit nothing but math as possible causes for the origin of the universe, both have idealist and, arguably, theistic implications.

Nevertheless, in both cases there are obvious reasons to question these interpretations of quantum cosmology.[23] The many-worlds interpretation and the mathematical universe hypothesis flagrantly violate the Ockham's razor principle. Indeed, both interpretations multiply theoretical entities infinitely. Consequently, an unadorned God hypothesis clearly qualifies as a theoretically simpler, less convoluted explanation for the origin of the universe and its specific (life-permitting) attributes than either of these interpretations of quantum cosmology. Moreover, if these exotic interpretations are taken seriously, both profoundly undermine confidence in scientific rationality. Both theories result in science-destroying absurdities. They imply that certain explanations we would regard as completely absurd are actually *more likely* to be true than explanations we would ordinarily accept.

Think of it. At the end of this path lie Boltzmann brains, a consequence of the project of defending science from the possibility of intelligent design. We shouldn't leave the notion behind without underlining how ultimately mad, and terrifying, this is. Pulling myself away from thoughts about cosmology and biology, I've considered this when walking the streets near the offices in Seattle where my colleagues and I work.

Here in the city's historic Pioneer Square neighborhood, a combination of debatable public policies has resulted in a heartbreaking spectacle: large numbers of homeless people, most mentally ill, drug-addicted, or both, camping in tents or wandering about. Many seem lost in our world or disconnected from it entirely; it's not unusual for a few of them to be screaming at the sky or at imagined interlocutors.

This is how I think of Boltzmann brains, minds adrift, struggling with false memories and false knowledge of themselves. Yet theories in flight from the God hypothesis—from what materialists call the irrationality of religious belief—imply the reality of innumerable such brains and the unreliability of all of our perceptions, memories, and reason itself.

No God, No Science?

And so Tegmark's hypothesis has brought our discussion full circle to an ironic conclusion. Quantum cosmologists, in particular Stephen Hawking, posed quantum cosmology as an alternative to the manifestly theistic implications of the big bang theory and Hawking's own proof of (or mathematical argument pointing to) a cosmological singularity. Hawking and others argued, especially in popular books and the popular media, that quantum cosmology had undermined the case for the existence, or at least the need to posit the existence, of a transcendent creator. As presented by Hawking and Krauss, quantum cosmology provided the ultimate scientific counterargument to the God hypothesis.

Yet as the dialectic about quantum cosmology unfolded, many unanswered questions have emerged about the cause of the origin of the universe and its contingent features. Most quantum cosmologists adopted the many-worlds interpretation of quantum mechanics in order to eliminate the need to account for how the universal wave function collapsed to produce our *specific* universe. Instead, they posited that all universes described by the universal wave function necessarily existed. But that left further unanswered questions about the origin of the specific information-rich constraints that Hawking and Hartle as well as Vilenkin had to impose on the mathematical procedures used to generate a universal wave function that included a universe like ours (and, on the many-worlds interpretation, a multiplicity of other universes). Max Tegmark's hypothesis could be construed as eliminating the need to account for any such winnowing of mathematical possibilities by proposing that all possible mathematical structures exist, including, by implication, all the possible ways of constraining the Wheeler-DeWitt equation with different boundary conditions and/or restrictions on superspace.

Tegmark's radical theory, with its infinities of universes manifesting every possible mathematical structure, eliminated the need to explain the specificity required to generate our universe. But it would destroy confidence in scientific explanation and prediction and imply the existence of Boltzmann brains, something out of sci-fi horror. Thus, the attempt to refute the case for God based upon the science of cosmology has ultimately resulted in absurd cosmologies that undermine belief in the reliability of science. Unlike the founders of modern science, who understood that their belief in God gave them a reason to trust in the uniformity and intelligibility of the universe and the reliability of the human mind, contemporary physicists averse to theistic belief have proffered ideas that deny, by implication, precisely such uniformity, intelligibility, and reliability.

Throughout this book, I have argued that the scientific evidence we have concerning biological and cosmological origins leads logically to the knowledge of God. Now we see that the attempt to deny the explanatory power of the God hypothesis eventually and necessarily requires positing infinite probabilistic resources and universes—a postulation that denies the possibility of knowledge. Indeed, these resulting cosmologies illustrate a maxim of St. Augustine: *Crede ut intelligas*, that is, "Believe in order to understand." I have argued that we can reasonably believe in the reality of God because of what we know about nature. The absurd implications of infinite-universe cosmologies now raise the possibility that we might well *need* such belief to have confidence in scientific rationality—and thus our ability to know nature at all.

Part V
Conclusion

20

Acts of God or God of the Gaps?

At 2:00 a.m. on a wintery English night in December 1986, I sat outside Isaac Newton's old rooms at the front of Trinity College. As a deep fog moved among the medieval colleges in the center of Cambridge, I reflected on how thinking about science and God had changed since the publication of Newton's great *Principia* in 1687, almost exactly three centuries earlier. In the epilogue to a later edition of that book called "The General Scholium" and in other scientific works, notably the *Opticks*, Newton articulated a profoundly theological perspective. Not only did he extol the order and uniformity of nature as a reflection of God's character and superintending care of creation; he argued for the existence of God based on the design evident in nature—in short, for a God hypothesis.

I had taken a break from writing my first essay about the controversy over universal gravitation. It was well after the undergraduates had returned home for "Christmas vac," and my wife had taken a short trip home to visit family in the States during the break. Without anyone else around, I found myself writing later and later into each night. With my body clock completely off, I walked from our flat near Hobson's Conduit and found myself where Newton once lived and wrote. You may recall that the German philosopher Leibniz, his contemporary, objected to Newton's universal law of gravity, because it failed to specify any material cause for the regular motions of the planetary bodies that the law described. This was the focus of the essay I was working on. I sat for an hour thinking about all that I had been reading: about Newton's reflections on the mystery of action at a distance; about his understanding of divine action in nature; and about the overwhelming impression of design he perceived in the fine tuning of the positions of the planets and in how the integrated complexity of the eye seemed to anticipate the properties of light.

Newton fascinated me in part because his perspective on the relationship between science and theistic belief differed so dramatically from the one I encountered in modern Oxbridge. During the late 1980s, two important books gained enormous prominence there and around the world. In 1986, Oxford University biologist Richard Dawkins published *The Blind Watchmaker: Why the Evidence of Evolution Reveals a Universe Without Design*—a book that eventually sold over three million copies. A line from the first page succinctly captured his thesis: "Biology is the study of complicated things that *give the appearance* of having been designed for a purpose."[1] Since, Dawkins argued, evolutionary theory explains this appearance as the product of the wholly undirected process of mutation and natural selection, it also eliminates the need to posit any role for a designing intelligence in the history of life. And that, he argued, "made it possible to be an intellectually fulfilled atheist."[2]

Then in 1988, Stephen Hawking, at Cambridge, published *A Brief History of Time*. Whereas Dawkins took aim at the design argument in biology, Hawking sought to undermine the cosmological argument for God's existence. Hawking's bestseller eclipsed even Dawkins's, eventually topping ten million copies sold worldwide.

These two books powerfully shaped public opinion. In Cambridge, Hawking's reputation as a physicist and his growing international celebrity hovered over almost all discussions of science and religion. If Stephen Hawking had explained the origin of the universe with a new law of quantum gravity, well, "What place, then, for a creator?" indeed. And if neo-Darwinism had shown that mutation and natural selection could explain away the appearance of design in life, as Dawkins had argued in his brilliantly clear prose, then life might well have resulted from a "blind watchmaker" (or "blind, pitiless indifference," as he later put it) and nothing more.

Dawkins and Hawking both either argued or implied, as Hawking later contended, that "the simplest explanation is that there is no God."[3] Consequently, their books encouraged the perception that science and theistic belief conflict. Many of the scientists I met in Cambridge assumed a similar view, even if many disclaimed Dawkins's needlessly strident rhetoric against religion.

Those few theists that I met in the sciences, including a group in Britain called "Christians in Science," had adopted a defensive posture. Some subscribed to the model I discussed in Chapter 3 called "compartmentalism," or what the late Harvard paleontologist Stephen Jay Gould would later call "nonoverlapping magisteria," or NOMA.[4] Recall that this model holds that science and religion describe completely different realities. Proponents often support this view by quoting an aphorism used by Galileo affirming that the Bible teaches "how one goes to heaven, not how the heavens go."[5] Others subscribed to a closely related idea called "complementarity." Proponents of this view hold that science and religion may sometimes describe the same realities; however, they do so in complementary but ultimately incompatible or "noncommensurable" language.[6]

Proponents of both these views deny that science contradicts religious belief, but they do so by portraying science and religion as such totally distinct enterprises that their claims could not possibly intersect in any significant way. This assumption insulated the God hypothesis from scientific refutation, but it also denied the possibility that science could offer any support for theistic belief.

An Equal and Opposite Reaction

I encountered these ways of thinking just as I had begun to think about an entirely different way of conceiving the relationship between science and theistic belief. My thoughts were driven in part by the ideas of scientists such as Allan Sandage, Dean Kenyon, and Charles Thaxton, whom I encountered at the Dallas and Yale conferences I attended before leaving for Cambridge. My reading of the founders of early modern science also provided an intellectual counterweight to the perspective I was encountering in contemporary Cambridge. Although Newton and Robert Boyle, for example, acknowledged the religious neutrality of many realms of scientific endeavor, both thought that certain aspects of nature pointed to "an intelligent and powerful Being," as Newton put it. Newton also understood that the most fundamental laws of nature either merely describe the observed regularities in nature or they manifest the "constant Spirit action" of a "Divine Sustainer" of the world. He did not think the laws of physics alone explained the origin of the solar system or, still less, the origin of the universe.

My exposure to Newton's understanding of what the laws of physics *do* and *do not explain* left me skeptical of Hawking's bold and popular claims about the origin of the universe. That in turn left me open to the idea that the big bang theory and the evidence for a finite universe might have theistic implications after all. My reading of the Leibniz-Clarke correspondence and the *Principia* acquainted me with Newton's philosophy of nature and natural theology (Fig. 20.1). Newton's advocacy for the design argument, based upon the fine tuning of our planetary system and the integrated complexity of living systems, echoed ideas I had recently encountered in the work of other contemporary scientists. Indeed, one exception to the otherwise defensive intellectual posture among theists in science was the Cambridge physicist John Polkinghorne. Polkinghorne had begun to advocate a robust design argument based upon the anthropic fine tuning of the universe. As for the origin of life, I had my own reasons for thinking that scientific atheists such as Dawkins were overstating their case. I knew already that neither he nor anyone else had explained the origin of the digital code necessary to produce the first living cell.

FIGURE 20.1
Sir Isaac Newton in his later years.

Thus, my study of Newton's natural philosophy—in light of current evidence concerning the origin of the universe and life—suggested to me the possibility of an evidence-based argument for God's existence. As Newton famously observed in his vigorous, but sometimes confusingly punctuated eighteenth-century prose, "And thus much concerning God; to discourse of whom from the appearances of things, does certainly belong to Natural Philosophy."[7] Translation: our observations of nature can tell us a lot about the reality and attributes of God.[8]

The God-of-the-Gaps Objection

A common objection to Newton's view of the relationship between science and theistic belief is known as the God-of-the-gaps objection (hereafter, the GOTG objection). According to those who pose this objection, the GOTG fallacy occurs whenever someone invokes the activity of a creative intelligence or God to explain phenomena or events in the natural world. Such postulations, critics argue, stifle scientific advance by using God (or creative intelligence) to account for phenomena or events that scientists will eventually explain, once they discover new laws of nature or material processes. Allowing God as an explanation only serves to impede or distract scientists from discovering such true explanations.

Some religious scientists also worry that positing God as an explanation for natural phenomena or events in natural history will inevitably bring disrepute to belief in God, once scientists discover new laws or processes that make the God hypothesis unnecessary. Some theists also find it theologically and aesthetically unappealing, even demeaning of the divine person, to think of God as having to "intervene" periodically to "fix" some aspect of the original design of the universe that, presumably, was not properly designed in the first place.

BioLogos, an influential Christian group that advocates theistic evolution and opposes intelligent design,[9] explains the GOTG objection this way:

> God-of-the-gaps arguments use gaps in scientific explanation as indicators, or even proof, of God's action and therefore of God's existence. Such arguments propose divine acts in place of natural, scientific causes for phenomena that science cannot yet explain. The assumption is that if science cannot explain how something happened, then God must be the explanation. But . . . with the continuing advancement of science, God-of-the-gaps explanations often get replaced by natural mechanisms . . . [and] scientific research can unnecessarily be placed at odds with belief in God.[10]

Critics of intelligent design or science-based theistic arguments often cite Newton as the prime example of a scientist[11] who made the GOTG blunder. As the story is often told, after Newton successfully used his universal law of gravity to describe the motion of the planets in the solar system, he discovered that the orbits of the outer planets did not conform precisely to the trajectories he had calculated. Further, he apparently realized that the mutual gravitational attraction between the outer planets would make the solar system unstable. As a result, he allegedly posited episodic interventions of God or—in some versions of the story—angels to put the planets back into correct orbits. Later, when the French physicist Laplace showed how Newton's own laws could account for the observed anomalies, he showed (voila!) Newton's postulation of divine action—his "God of the gaps"—to be unnecessary. In this way, Newton's God went into permanent retreat. Or so the story goes.[12] In the last two sections of this chapter, I'll set the historical record straight.

Opponents of the theory of intelligent design frequently characterize it as a GOTG argument or "an argument from ignorance," often comparing it to Newton's supposed postulation of divine action to fix the solar system. According to this criticism, anyone who infers design from the specified information in living cells uses current *ignorance* or *gaps* in our knowledge of an adequate materialistic cause to justify that inference. Since, the objection goes, design advocates can't imagine a natural process that can produce specified biological information, they resort to invoking the mysterious notion of intelligent design. In this view, intelligent design functions as a placeholder for ignorance, one that its proponents fill by postulating the activity of a creative intelligence. Critics have charged that the argument for intelligent design in the history of life constitutes a fallacious GOTG argument even though it does not mention God. It follows that extending that argument to support an explicit God hypothesis (as this book does) would, in their eyes, commit the GOTG fallacy all the more.

But does it?

An Argument from Ignorance?

Let's look first at whether the argument for intelligent design in biology—which by itself does not attempt to identify the designing intelligence responsible for life—commits the argument from ignorance fallacy. Arguments from ignorance occur when evidence *against* one proposition is offered as the *sole grounds* for accepting an alternative. Thus, they have the following form:

Premise: Cause A cannot produce or explain evidence E.
Conclusion: Therefore, cause B produced or explains E.

Arguments of this form commit an obvious logical error. They do not include a premise providing affirmative support for the preferred conclusion. Instead, they only present evidence *against* the adequacy of some alternate explanation. For example, if someone were to argue that an injury to a particular racehorse (say, "Seattle Slew II") means that another horse (say, "Justify Junior") will necessarily win the Kentucky

Derby, that argument would commit a fallacy. To justify the conclusion, the person needs to provide affirmative evidence demonstrating that the favored horse has shown itself to be faster than other horses in the field.

Does the argument for the intelligent design of life, presented earlier, commit a similar fallacy? It does not. True, that argument does depend *in part* upon critical assessments of the causal adequacy of materialistic explanations for the origin of specified information in DNA. Nevertheless, the "specified complexity" or "specified information" of DNA implicates a prior intelligent cause not only because materialistic origin-of-life scenarios fail to explain it, but also because we *know* that intelligent agents can and do produce information of that kind. We have positive experience-based knowledge of an alternate cause sufficient to produce this effect.

To depict proponents of the theory of intelligent design as committing the GOTG fallacy, critics must misrepresent the case for it. For example, as Michael Shermer claims, "Intelligent design . . . argues that life is too specifically complex (complex structures like DNA) . . . to have evolved by natural forces. Therefore, life must have been created by . . . an intelligent designer."[13] In short, he claims that proponents of intelligent design argue as follows:

Premise: Material causes cannot produce or explain specified information.
Conclusion: Therefore, an intelligent cause produced the specified information in life.

In fact, the case for the intelligent design of life, presented in Chapters 9 and 10 and in my previous books, doesn't rely on such logic. Instead, the argument takes the following significantly different form:

Premise One: Despite a thorough search, no materialistic causes have been discovered with the power to produce large amounts of specified information necessary to produce the first cell.
Premise Two: Intelligent causes have demonstrated the power to produce large amounts of specified information.
Conclusion: Intelligent design constitutes the best, most causally adequate explanation for the origin of the specified information in the cell.

Clearly, in addition to a premise about how materialistic causes lack causal adequacy, this argument affirms the demonstrated causal adequacy of an alternate cause, namely, intelligent agency. The argument as stated does not fail to provide a premise affirming positive evidence for an alternate cause. The argument specifically *includes* such a premise. Therefore, it does not appeal to ignorance or commit a "gaps" fallacy (Fig. 20.2).

Instead, as we saw in Chapter 9, contemporary proponents of intelligent design employ the standard uniformitarian method of reasoning used in the historical sciences. That such arguments for intelligent design necessarily include critical evaluations of the causal adequacy of competing hypotheses is entirely appropriate. Historical scientists must compare the adequacy of competing hypotheses to judge which of several hypotheses qualifies as best. Yet we would not say, for example, that an archaeologist had committed a "scribe-of-the-gaps" fallacy simply

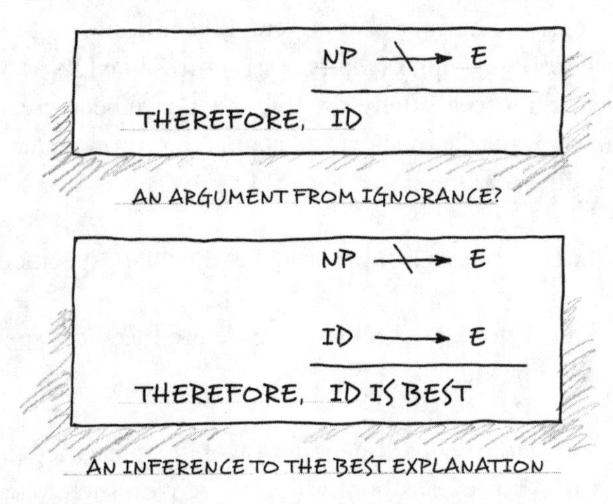

FIGURE 20.2
Some critics of intelligent design portray the case for intelligent design as a fallacious argument from ignorance. They claim proponents of the argument affirm intelligent design only because of the implausibility of various naturalistic processes (NP) as causal explanations for the origin of the specified information, the key effect (E) that needs to be explained in living systems. Nevertheless, the specified information of DNA implicates a prior intelligent cause, not only because various naturalistic or materialistic origin-of-life scenarios fail to explain it, but also because we *know* that intelligent agents can and do produce information of this kind. Thus, in addition to a premise about how natural processes lack causal adequacy, the argument for intelligent design (ID) presented here also cites evidence of the power of intelligent agents to produce functional or specified information. The argument as stated, thus, does not fail to provide a premise affirming positive evidence for the adequacy of a preferred cause. The argument specifically *includes* such a premise. Therefore, it does not commit a fallacious argument from ignorance. The fallacious form of the ID argument as portrayed by ID critics is depicted in the top half of this figure. The valid form of the argument presented in this book as an inference to the best explanation is depicted in the bottom half of the figure.

because—after rejecting the hypothesis that an ancient hieroglyphic inscription was caused by a sandstorm—she went on to conclude that the inscription had been produced by an intelligent scribe. Instead, the archaeologist made an inference based upon her experience-based *knowledge* that information-rich inscriptions invariably arise from intelligent causes. She did not base her inference *solely* on her judgment that no natural cause could explain the inscription.

Moreover, what we know from our uniform and repeated experience about the cause of specified information (especially when we find it in a digital form) allows us to treat such information as a distinctive hallmark of intelligence. In all cases where we know the origin of specified information, intelligent design played a causal role. Thus, when we encounter such information in the biomacromolecules necessary to life, we may infer or retrodict—based upon our *knowledge* of established cause-and-effect relationships—that an intelligent cause operated in the past to produce the information necessary for life's origin.

A "God-of-the-Gaps" Argument?

But what about the argument that this book presents not just for an intelligent designer of unspecified identity, but specifically for a theistic designer and creator—a God hypothesis—as the best explanation for biological and cosmological origins? Is it a GOTG argument?

Again, it is not. Though the argument presented here does concern events that confront *materialistic* accounts of the origin of the universe and life with causal discontinuities or explanatory gaps, it does not affirm the existence or activity of God solely on the basis of those gaps. Instead, it uses straightforward considerations of causal adequacy along with parsimony and other theoretical virtues[14] to assess the explanatory power of competing metaphysical hypotheses and to present theism as an inference to the best explanation, not an argument from ignorance.

The Causal Inadequacy of Materialistic Explanations

In assessing explanations for the origin of the universe, I have, of course, critiqued many materialistic theories, including the steady-state and oscillating-universe theories. I also showed that quantum cosmology—advanced by scientific materialists as an alternative to the God hypothesis—failed to provide a causally adequate *materialistic* explanation for the origin of the universe (and even that it has unintended theistic implications).

In addition, I argued that *all* materialistic theories of the origin of the material universe face a fundamental problem given the evidence we have of a cosmic beginning. Before matter and energy exist, they cannot cause, or be invoked to explain, the origin of the *material* universe. Instead, positing a materialistic process to explain the origin of matter and energy assumes the existence of the very entities—matter and energy—the origin of which materialists need to explain. No truly materialistic explanation can close this particular causal discontinuity or gap—the gap between either nothing or a preexisting *im*material or mathematical reality, on the one hand, and a material universe, on the other.

Nor do appeals to the laws of nature solve this problem. Stephen Hawking's claim that the "laws of science" or "the law of gravity" can explain "why there is something rather than nothing" betrayed a deep philosophical confusion about what the laws of physics can do. Laws of nature describe how nature operates and how different parts of nature interact with one another; they don't cause the natural world to come into existence in the first place. This suggests the futility of waiting for the discovery of some new law of nature or a "theory of everything." No law of nature can close the causal discontinuity between nothing and the origin of nature itself.

Similarly, in Chapters 13 and 16, I argued against the causal adequacy of proposed naturalistic explanations—such as the weak and strong anthropic principles and the multiverse hypothesis—for the fine tuning of the universe. I showed, more fundamentally, that the laws of physics cannot explain their own fine tuning or the fine tuning of the initial configurations of mass-energy at the beginning of the universe. We saw that, in order to describe the behavior of a physical system, all known laws of physics require *extrinsic* inputs of information about the initial and boundary conditions of the system in question and about the values of their own constants of proportionality. By "extrinsic," I mean the information about such conditions that comes from beyond the laws of physics themselves. The values of the physical constants and certainly the initial conditions of the universe constitute such extrinsic information. And yet the initial conditions of the universe and the constants of physics are exquisitely finely tuned. It follows that the laws of physics themselves do not—and cannot—explain the origin of the fine tuning.

This suggests yet another gap that the laws of nature cannot in principle close.[15] Even BioLogos, the group that has strenuously marshaled the GOTG objection to design arguments in biology, acknowledges that the objection does not apply to the fine tuning (and perhaps the origin) of the universe itself. As they explain, "Some critics charge that invoking God as the fine-tuner is a return to the God of the gaps. But there does not seem to be any way to explain the detailed properties of the laws of nature from within science."[16]

The Causal Adequacy of the God Hypothesis

Even so, the argument presented here does not leap straight from critiquing materialistic explanations to concluding that a transcendent and intelligent agent must have caused the universe and life to arise. First, as noted, I provided positive evidence of the causal adequacy of intelligent agency as a cause of the kind of specified digital information needed to produce life. In addition, I offered a positive rationale for affirming the unique causal adequacy of a *transcendent* and intelligent agent as the best explanation for the *ensemble* of evidence considered in this book. To make this case, I not only argued for the causal adequacy of *an* intelligent agent as the best explanation for the origin of biological information; I also provided reasons to affirm the unique causal adequacy of a *transcendent* intelligence as the best explanation for the origin of the universe and its fine tuning.

For example, I argued that any cause capable of explaining the origin of the universe and its fine tuning must in some way stand causally separate from the universe or, in philosophical terms, *transcend* matter, space, time, and energy. Materialists themselves have tacitly conceded the need for a transcendent explanatory entity by positing universes beyond our universe and abstract nonmaterial mathematical entities (such as those in quantum cosmology) as explanations for the origin of the universe and its fundamental attributes.

As previously explained, these either fail as specifically *materialistic* explanations, or they have science-destroying consequences. On the other hand, God, as conceived by theists, possesses the attribute of transcendence and can be invoked without science-destroying consequences. Therefore, the God hypothesis qualifies as a causally adequate explanation for the origin of the material universe, more so than any explanation expressing a materialistic or pantheistic worldview.

There is another aspect to this. In Chapter 12, I explained how positing the action of a free agent could resolve the otherwise seemingly intractable dilemma for naturalism posed by the beginning of the material universe. The choice of an intelligent free agent made it possible to explain the origin of the universe without invoking either an uncaused material event (in violation of the principles of sufficient reason and causality) or an infinite regress of material causes at odds with observations about the universe having a beginning.[17] By highlighting the ability of agents with free will to generate abrupt changes of state uncompelled by a suite of necessary and sufficient material conditions, Chapter 12 provided an additional justification for affirming the *causal adequacy* of a transcendent, intelligent, conscious agent with free will—a.k.a. God—as an explanation for the origin of the universe. Moreover, positing such an agent provides a causally adequate explanation without undermining the basis of rationality or contradicting relevant evidence.

Chapters 8 and 16 reinforced that argument. There I explained how the fine tuning of the universe exhibits two properties—extreme improbability *and* functional specification—that we routinely associate with the activity of intelligent agents. We have observed intelligent agents (and *only* intelligent agents) producing highly improbable systems or events that exemplify a set of functional requirements, whether finely tuned Swiss watches, digital computers, engines, recipes, or coded messages. Consequently, we have empirical evidence of the sufficiency (and necessity) of intelligent agency as the cause of systems exhibiting fine tuning. Moreover, since the fine tuning of the universe originated at the beginning of the universe itself, this class of evidence suggests the need for a transcendent intelligence to most adequately explain it.

In addition, I argued for the superior causal adequacy of theism over deism as an explanation for the origin of the functional information necessary to produce life, because deism assigns *only* transcendence, not immanence or involvement in the universe after its creation, to the concept of God. Yet life first arose long after the beginning of the universe.

Theoretical Justifications of Causal Adequacy

Of course, as David Hume famously fretted, we do not have "uniform and repeated" experience of God creating universes.[18] So, admittedly, we cannot certify the causal adequacy of the God hypothesis in exactly the same way we would a generic intelligent design hypothesis or many other historical-scientific hypotheses. Nevertheless, architects of the method of inference to the best explanation and historical-scientific reasoning, also called reconstructive causal analysis, allow that, absent the ability to observe cause-and-effect relationships directly, we may have "theoretical reasons"[19] for regarding a postulated cause as causally adequate.

For example, a particle physicist might postulate the existence of an unobserved particle with certain attributes (e.g., spin, mass, momentum, and/or charge) and deduce from those attributes certain causal powers, thus providing a theoretical justification for the causal adequacy of that entity with respect to certain expected observable phenomena. The physicist might then infer the particle's *actual* existence upon observing those and perhaps other expected phenomena. Remember, the method of inference to the best explanation does not require us to know in advance that a posited entity actually exists. Instead, using this method, scientists and philosophers infer the cause, among a group of possible causes, that would, *if true*, *existent*, *or actual*, best explain the evidence in question.

In much the same way, philosophers or scientists who postulate the existence of God understand God to have certain attributes (e.g., transcendence, intelligence, creativity, omnipresence, and free will) and corresponding causal powers that would allow God to produce certain kinds of effects expected on the basis of those powers. In other words, the concept of God as advanced by philosophers and theologians, or as revealed in scripture, entails particular attributes, properties, and corresponding causal powers. By articulating those attributes, properties, and powers, philosophers can provide a theoretical rationale for affirming the causal adequacy of a God hypothesis—as I did in Chapters 12 to 14—*with respect to specific observable phenomena or kinds of events.*

This in turn allowed us to formulate a God hypothesis with empirical content and corresponding expectations about what we should see in the world if such a God existed and had acted. Since scientists have evidence confirming those expectations in, for example, the discovery of the fine tuning and temporal beginning of the universe and the digital code and information-processing systems in living cells, evidence from the world provides confirmation of the theistic hypothesis and *positive* support for its causal adequacy as an explanation.

In addition, recall that historical scientists often justify the causal adequacy of a postulated entity by extrapolating from the effects produced by a relevantly similar entity. Darwin, for example, saw that artificial selection could produce modest changes in organisms over a short period of time, and he extrapolated from those effects to propose that natural selection and random variation operating over a long period of time could produce fundamental transformations in the morphology of organisms. In this way, he sought to establish that natural selection qualified as a causally adequate explanation of morphological innovation in the history of life.

Proponents of a theistic design hypothesis can, in a similar way, extrapolate from the creative power of human agents. Since human agents, uncompelled by a set of necessary and sufficient material conditions, can create new structures by arranging preexisting matter and energy at discrete points in time, we might reasonably postulate an *omnipotent* divine intelligence as the cause of an abrupt change of state that resulted in the creation of matter and energy at the beginning of time in the first place.[20] Similarly, theists might reasonably extrapolate from the known abilities of intelligent human agents to produce finely tuned or information-rich terrestrial artifacts and systems to posit a "supermind" or "superintellect" as an adequate cause of the fine tuning of the universe or the information necessary to produce the first life.

Indeed, the concept of God has inherent in it precisely those attributes—transcendence, omnipotence, creative power, free will, and intelligence—that confirm its adequacy as a cause of the origin of the universe, its fine tuning, and the information necessary for life. Thus, a theistic God would, if existent, provide a more causally adequate explanation for the origin of life and the universe than any entity affirmed in competing worldviews (such as materialism or pantheism) that deny a transcendent reality and intelligent agent separate from the material universe.

By contrast, arguments that commit the GOTG fallacy exemplify a specific type of argument from ignorance—one in which the evidence against a proposed material cause leads immediately to the affirmation of a God hypothesis without any presentation of positive reasons for affirming the causal adequacy of that hypothesis. Since the argument presented here does offer positive reasons for affirming the causal adequacy of God as an explanatory entity, along with a critique of materialistic alternatives, it is not a GOTG argument—even if it does offer an alternate explanation that fills gaps or causal discontinuities in materialistic accounts of the origin of the universe, its fine tuning, and life. Instead, the argument presented here represents an inference to the best (metaphysical) explanation of those key classes of evidence.

A Prohibition and a Circular Justification

The origin of the material universe, like the origin of the fine tuning and the origin of novel forms of life, is an indicator of some past causal action. But what kind? What kind of cause best explains the events in question? By rejecting all explanations that posit a transcendent or intelligent agent as fallacious GOTG arguments, scientific materialists and theistic evolutionists effectively require scientists and philosophers to explain all events in the history of the universe materialistically.

Of course, those concerned about the God-of-the-gaps fallacy have their reasons for limiting acceptable explanations in this way. They assume that some material process or some law of nature *will eventually* provide an adequate explanation of *every* event in the history of the universe. But do we have good reason for believing this will necessarily occur?

We have already seen that neither the origin of the universe nor the fine tuning of the laws of physics and initial conditions of the universe are the kinds of events that laws of nature or materialistic processes are likely to, or can *in principle*, explain. Similarly, as noted in Chapters 9 and 14, the origin of complex and specified information present in DNA cannot in principle be explained by forces of chemical attraction or underlying laws of physics and chemistry. Such considerations alone might alert us to a problem with the idea that all events can be explained by natural laws or materialistic processes.

Yet there is a deeper logical problem with the blanket prohibition against intelligent agency as an explanation in natural history. Whether motivated by concerns about "gaps" or by a commitment to methodological naturalism (a convention requiring scientists to consider only materialistic explanations), the prohibition against explanations involving creative intelligence ends up assuming the very point at issue in the debate about origins.

To see why, consider the following illustration.[21] Imagine someone mistakenly enters an art gallery expecting to find croissants for sale. That is, he thinks the gallery is actually a fancy bakery. Observing the absence of pastries and rolls, such a person may think that he has encountered a "gap" in the services provided by the gallery. He may even think that he has encountered a gap in the staff's knowledge of what must definitely be present *somewhere* in the gallery. Based on his assumptions, the visitor may stubbornly cling to his perception of a gap, badgering the gallery staff to "bring out the croissants already," until with exasperation they show him to the exit.

The moral of this vignette? The gallery visitor's perception of a gap in service or in knowledge of the location of the croissants derives from a false assumption about the nature of this establishment or about art galleries in general and what they typically offer to visitors.

In a similar way, *perceived gaps* in our knowledge of the materialistic processes responsible for key events in natural history are based on our background assumptions about *the kind of processes or entities that ought to have been working in nature*. In the debate about biological origins, theistic evolutionists and mainstream evolutionary biologists alike assume that all living systems necessarily were produced by some materialistic process and that their origin will, thus, ultimately have a completely adequate materialistic explanation. The assumption implicit in, for instance, the question "What chemical processes first produced life?" implies a gap in our scientific knowledge when it becomes apparent (as it has; see Chapter 9) that no materialistic chemical process has been discovered that can generate the information necessary to produce the first living cells. Nevertheless, our present lack of knowledge of any such chemical process entails a "gap" in our knowledge of the *actual* process by which life arose *only if* some materialistic chemical evolutionary process *actually did* produce the first life.

Yet if life did *not* evolve via a strictly materialistic process but was, for example, intelligently designed, then our absence of knowledge of a materialistic process does not represent "a gap" in knowledge of an actual process. It only represents a gap in materialistic accounts of the origin of life. In that case, the perceived gap in our knowledge would merely reflect a false assumption about what *must have* happened or about the existence of a certain kind of process—a completely materialistic one—with the creative power to generate life.

But what if such a strictly materialistic process did not produce either the first life or the universe and its fine tuning? Then our assumption about the ultimate sufficiency of materialistic explanations would be false. Consequently, prohibitions against explanations invoking creative intelligence, based upon an aversion to explaining events that generate such "gaps" in materialistic accounts, might cause us to miss the true cause and best explanation for the event (or discontinuities) in question. It might cause us to ask: "So where *are* those croissants?"

A more intellectually rigorous approach to the challenge of explaining crucial events in the history of life and the universe would permit scientists and philosophers to consider competing possible explanations even if they posit the activity of a creative intelligence. The critical question is not "Which *materialistic* or *naturalistic* hypothesis best explains the origin of life and the universe?" but rather, "What actually caused life, the universe, and its fine tuning to arise?"

Seen in this light, the GOTG objection fades into insignificance. To make their case for the adequacy of a strictly materialistic approach to explanation in science and philosophy, defenders of this approach must first show that "gaps" in our knowledge of the materialistic causes of key events in the history of life and the universe can be filled with knowledge of an actual materialistic process capable of producing the events in question. But as I've shown in Chapters 4 through 19, this is exactly what scientific materialists have failed to do—and, indeed, look unlikely to do. Indeed, if scientific materialists had discovered materialistic processes with the demonstrated creative power to explain the origin of life and the universe, they would not need to use the God-of-the-gaps objection to counter intelligent design arguments or science-based arguments for the existence of God.

On the other hand, if there are positive reasons to consider creative intelligence as a crucial causal factor, and if, in addition to evidence against competing materialistic explanations, those reasons suggest creative and/or transcendent intelligence as a causally adequate and best explanation for the events in question, then so be it. Let the evidence and standard methods of scientific and philosophical reasoning based on metaphysically neutral criteria for assessing hypotheses determine the conclusion of the investigation, not a question-begging prohibition that smuggles in the answer to the question at issue from the outset.

Yet currently that is precisely how scientific materialists and others use the God-of-the-gaps objection. Critics of intelligent design or theistic arguments assert that scientists shouldn't invoke creative intelligence to explain events that leave gaps in our materialistic accounts of the origin of life and the universe. Why? Because we know (or assume) that natural laws or materialistic processes will eventually explain those events and close those gaps. How do we know that materialistic processes will eventually explain those events, considering that such processes haven't yet explained them and look unlikely to do so? Because the only alternative to explaining such events materialistically would be to invoke creative intelligence and that would commit—you guessed it—a GOTG fallacy.

More succinctly, we cannot allow God as an explanation for events that leave gaps in our materialistic accounts of the origin of life and the universe, because we know that scientists will eventually develop adequate materialistic explanations of those events. How do we know that? Because the only alternatives to materialistic explanations commit the God-of-the-gaps fallacy. And around and around we go.

An Urban Legend

As I noted above, critics who warn about the God of the gaps often cite Isaac Newton as a cautionary tale. Supposedly, Newton invoked *specific acts* of God (or angels) to occasionally fix the orbits of the planets and to compensate for Newton's inability to describe the regular motion of those planets accurately. Critics cite this alleged episode in the history of science as the kind of GOTG reasoning that concerns them. And, indeed, a second way that a scientist could commit a genuine God-of-the-gaps blunder would be to invoke a *singular act of divine agency* to answer a question about what *nature ordinarily does.* Clearly, when scientists ask questions of the form "What does nature ordinarily do," any answer of the form "God did it" simply fails to address the question at hand. Invoking a *singular* idiosyncratic act of God (as opposed to merely affirming that ongoing divine action *lies behind* all fundamental natural laws or that God acted to establish those regularities in the first place) could well function as a placeholder for ignorance (or a gap in our knowledge) of the actual regularities that nature manifests. In so doing, it could also

FIGURE 20.3
Neil deGrasse Tyson, astrophysicist and charismatic science popularizer.

impede scientific progress in developing a mathematical description of those regularities, thus functioning "as a science stopper."

But did Newton actually do this?

In a 2010 lecture critiquing intelligent design, Neil deGrasse Tyson cited Newton as the main example of a scientist who used such science-stopping reasoning (Fig. 20.3).

According to Tyson, after Newton "solved his equations for gravity," he realized that the interaction between Jupiter and earth's gravitation would render the solar system unstable. As Tyson tells the story, Newton knew "that not only is [*sic*] the sun and earth pulling on each other, [but] Jupiter is tugging on earth every time we come around the back stretch [of the earth's orbit]." Thus, according to Tyson, these competing gravitational "tugs" led Newton to conclude, "this system is unstable. You keep this up, the orbits will get distorted beyond recognition, and earth would fly off into space." Then Tyson asserted that in the *Principia* "for the first time in his entire record of the discovery of the laws of mechanics and the laws of gravity . . . Isaac Newton says God must step in and fix things."[22]

Over the years, I've heard or read many similar versions of this story, but based upon my reading of the *Principia*, they always seemed suspect. So I decided to dig back into the primary sources. My suspicions were confirmed. First, Newton *did* indeed believe that God sustains the orderly concourse of nature in what we call the laws of nature. Thus, he stated in the General Scholium of the *Principia*: "In him [God] are all things contained and moved."[23]

Second, Newton also believed that God could act, and had acted, in more discrete and special ways at specific times in the past history of the universe and of life. He argued that both living organisms and the solar system exhibited evidence of special creative acts distinct from the constant exercise of the divine power that, he thought, maintains the laws of nature. Thus, in the General Scholium of the *Principia*, Newton argued that these laws could *preserve* the stability of the planetary orbits in the solar system, but only the design of "an intelligent and powerful Being" could have at first established the "position of the orbits."[24] Newton made similar design arguments in a later book, the *Opticks*, based upon the qualities of light and the exquisite functional integration of the many parts of the eye.[25]

Thus, in his work Newton affirmed what theologians since the Middle Ages had conceived of as two complementary but distinct powers of God: (1) the *potentia ordinata*, God's ordinary power, by which God sustains the order of nature, and (2) the *potentia absoluta*, the absolute or fiat power of God, by which God accomplishes special acts of creation or design or initiates events in human history at discrete times for special purposes—not in opposition, but as a complement to the laws of nature.[26]

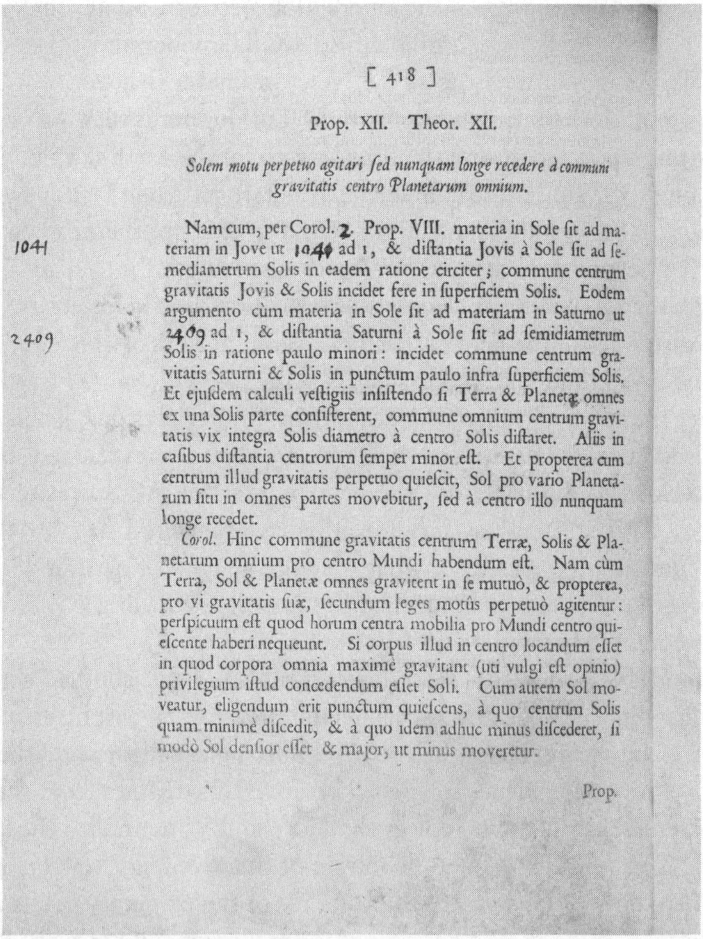

[418]

Prop. XII. Theor. XII.

Solem motu perpetuo agitari ſed nunquam longe recedere à communi gravitatis centro Planetarum omnium.

Nam cum, per Corol. 2. Prop. VIII. materia in Sole ſit ad materiam in Jove ut 1041 ad 1, & diſtantia Jovis à Sole ſit ad ſemediametrum Solis in eadem ratione circiter; commune centrum gravitatis Jovis & Solis incidet fere in ſuperficiem Solis. Eodem argumento cùm materia in Sole ſit ad materiam in Saturno ut 2409 ad 1, & diſtantia Saturni à Sole ſit ad ſemidiametrum Solis in ratione paulo minori: incidet commune centrum gravitatis Saturni & Solis in punctum paulo infra ſuperficiem Solis. Et ejuſdem calculi veſtigiis inſiſtendo ſi Terra & Planetæ omnes ex una Solis parte conſiſterent, commune omnium centrum gravitatis vix integra Solis diametro à centro Solis diſtaret. Aliis in caſibus diſtantia centrorum ſemper minor eſt. Et propterea cum centrum illud gravitatis perpetuo quieſcit, Sol pro vario Planetarum ſitu in omnes partes movebitur, ſed à centro illo nunquam longe recedet.

Corol. Hinc commune gravitatis centrum Terræ, Solis & Planetarum omnium pro centro Mundi habendum eſt. Nam cùm Terra, Sol & Planetæ omnes gravitent in ſe mutuò, & propterea, pro vi gravitatis ſuæ, ſecundum leges motûs perpetuò agitentur: perſpicuum eſt quod horum centra mobilia pro Mundi centro quieſcente haberi nequeunt. Si corpus illud in centro locandum eſſet in quod corpora omnia maximè gravitant (uti vulgi eſt opinio) privilegium iſtud concedendum eſſet Soli. Cum autem Sol moveatur, eligendum erit punctum quieſcens, à quo centrum Solis quam minimè diſcedit, & à quo idem adhuc minus diſcederet, ſi modò Sol denſior eſſet & major, ut minus moveretur.

Prop.

FIGURE 20.4

A passage from Book III, Proposition XII, of the *Principia*, where Newton affirms the stability of the solar system and explains that the sun "never recedes far from the common centre of gravity of all the planets."

Third, though Newton affirmed these powers of God, he did *not* postulate occasional, special, or singular acts of God in place of a law-like description of planetary motion or to remedy irregularities in the laws of nature or to fix an unstable planetary system. Newton thought that God was responsible on an ongoing basis for the mathematical regularities evident in nature, not fixing irregularities or rectifying instabilities. His *Mathematical Principles of Natural Philosophy* sought to use mathematics to describe the orderly concourse of nature as a way of demonstrating the rationality of God, the divine geometer. Indeed, for Newton to posit episodic divine action to fix the laws of nature would have implied that God worked at odds with himself.

Book III of the *Principia*[27] (Fig. 20.4) definitely shows that Newton did not posit episodic divine action to fix an unstable solar system. There Newton shows how the mutual gravitational attraction between the outer planets Saturn and Jupiter *will* result in discernible orbital perturbations when these planets come into conjunction. Nevertheless, he argues in that same section that "the common centre of gravity of the earth, the sun, and all the planets, is immovable"[28] and that though "the sun is agitated by a continual motion," it "never recedes far from the common centre of gravity of all the planets."[29] Indeed, contrary to the false story we often hear, Newton demonstrated the *stability* of the solar system *despite* small periodic perturbations in the orbits of the larger planets. Further, he calculated that the solar system will remain stable for an "immense tract of time."[30] Newton's analysis implies that the solar system *does not require* any singular divine "intervention" to compensate for perturbations. In other words, the often repeated story of Newton's God-of-the-gaps blunder is completely false.[31]

Given the number of times that I'd heard the story about Newton invoking periodic divine intervention to fix the solar system, I was stunned upon reading and rereading these most relevant sections of the *Principia*.[32] Yet those who are interested can read these passages for themselves.[33] Clearly, science popularizers like Tyson and even many historians of science who have perpetrated this myth have not done so.[34]

The God Hypothesis: A Science Starter

One additional aspect of the GOTG objection warrants comment. The main concern about invoking creative intelligence as a scientific explanation concerns the possibility that such explanations will stop the advance of science. Nevertheless, as historian of science Stephen Snobelen, a renowned Newton scholar at the University of King's College, has pointed out,[35] Newton's way of understanding God's interaction with nature did not at all inhibit his scientific work. Instead, his understanding of God as (1) the source and sustainer of mathematically describable order in the universe and (2) the intelligent designer of living systems and the solar system actually *inspired* his scientific endeavor. The *Principia* was a theologically inspired mathematical treatise in which Newton sought to bring glory to God by discovering, as its title indicates, the "mathematical principles" that governed the universe.[36] Similarly, Newton's conviction in the intelligibility of nature led him to investigate how natural systems work, so that he could appreciate and bear witness to their ingenious design. For Newton, postulating divine action as part of his natural philosophy was anything but a science stopper.

Newton not only formulated the theory of universal gravitation as an explicit expression of his theology of nature, but he also discovered the three laws of motion, invented the calculus, constructed[37] the first reflecting telescope, developed the generalized binomial theorem, inferred the oblate shape of the sphere of the earth, and conducted the most detailed scientific study of light undertaken to that point in the history of science.

For Newton, nature not only provided evidential support for belief in God, but his God hypothesis functioned as a hugely productive science starter. There is no reason to think that updating that hypothesis will threaten scientific advance today. On the contrary, there is good reason to expect that it will inspire deeper interest in discovering more about the intricacy, order, and design of the universe, just as it did for Newton himself.

21

The Big Questions and Why They Matter

As I was completing this book, Stephen Hawking spoke to me from the grave. Of course, I wasn't the only one to whom his message was addressed. In his posthumously published bestseller, *Brief Answers to the Big Questions*, he made a beautifully impassioned appeal to confront the fundamental questions of human existence—an appeal that would please any philosopher. As advertised, his new book also gave succinct answers to those big questions, including the question, "Is there a God?"[1]

Hawking's answer didn't surprise me, but the way he justified it prompted me to think about the question motivating this book: What can science—or the evidence from nature—tell us about the existence of God?

Hawking affirmed that he thought science could help answer this question, and he reiterated a claim that he had made in his book *The Grand Design*. There he had said, "Because there is a law such as gravity, the universe can and will create itself from nothing. . . . Spontaneous creation is the reason there is something rather than nothing, why the universe exists, why we exist."[2]

Oddly, the latter part of this statement represents what logicians call a tautology—a vacuous statement that simply states the same thing twice in two different ways. "Spontaneous creation" is not "the reason there is something rather than nothing." The phrase "spontaneous creation" simply refers to something coming into existence from nothing. By invoking "spontaneous creation," Hawking did not identify a cause of the universe, still less a materialistic one.[3] Nevertheless, in *The Grand Design* he asserted that spontaneous creation made it unnecessary "to invoke God to . . . set the universe going."[4] Now in his new book Hawking doubled down, saying, "The universe was spontaneously created out of nothing, according to the laws of science."[5] Consequently, for him that meant "the simplest explanation is that there is no God."[6]

In this final message to his readers, Hawking (Fig. 21.1) dispensed with any appeal to complicated mathematics. But this mode of presentation laid the logic of his argument bare. As we have seen in previous chapters, Hawking's statements about the laws of nature explaining how the universe originated betray a confusion of categories—a philosophical misunderstanding about what the laws of nature do and don't do. The laws of nature are *our descriptions*, typically framed in mathematical terms, of what nature ordinarily does.

FIGURE 21.1
A mature, pensive Stephen Hawking.

Unless those equations exist in the mind of God and reflect his way of actively ordering the universe, a possibility that Hawking rejected, they have no objective existence in the universe independent of our minds. Indeed, unless the laws express God's ideas and action, they are not "things" or entities in nature that exist independently of the universe and certainly not things that can cause events in the world; still less would they cause the origin of the universe itself. Saying they do is like saying that the longitude and latitude lines on the map explain how the Hawaiian Islands popped up in the middle of the Pacific Ocean.

Alas, Hawking's mistaken philosophy of science led him to claim that he had explained the origin of the universe from nothing—"the ultimate free lunch," as he memorably put it.[7] As a consequence, he also claimed that he had rendered the God hypothesis unnecessary, despite his own earlier proof of the singularity theorems and his acknowledgment of the exquisite fine tuning of the laws and constants of physics.

Reading Hawking's final reflections saddened me. I knew from my time in Cambridge and from friends who still lived there that Hawking thought seriously and deeply about the big questions. So I knew Hawking was serious about his plea to consider life's big questions and equally well considered in his rejection of the existence of God. Yet his rationale was philosophically confused, and *this* confusion, not any scientific evidence, led him to reject the God hypothesis. All this made his final words ring with poignancy, especially since he so squarely faced the logical consequences of atheism. "No one created the universe and no one directs our fate," he wrote. "This leads me to a profound realisation: there is probably no heaven and afterlife either. I think belief in an afterlife is just wishful thinking. . . . When we die, we return to dust."[8]

A Bigger Picture

Reading Hawking's final words saddened me not only for Hawking, but also for the many millions of people who have long labored under the misimpression that the testimony of nature renders belief in God untenable. Hawking grappled admirably and authentically with this question. But philosophically flawed thinking led him, and others by virtue of his scientific authority, to miss the clear implications of the extraordinary and unexpected scientific discoveries of the last century, one of which he himself helped to establish: the universe had a beginning.

Similarly, though I admire his facility for clearly framing big worldview issues, Richard Dawkins has had a hugely unwarranted influence on the perspective of millions of people, especially young people. Dawkins has acknowledged that the "machine code of the genes is uncannily computerlike" and that neither he nor anyone else has explained the origin of life, which depends upon such digital information.[9] Yet he has persisted in asserting that "the universe we observe has precisely the properties we should expect if there is, at bottom, no design, no purpose . . . nothing but blind, pitiless indifference."[10] Many prominent scientists have echoed this perspective. As Steven Weinberg, a Nobel laureate in physics at the University of Texas, has lamented, "The more the universe seems comprehensible, the more it also seems pointless."[11]

The prominence of this atheistic, indeed nihilistic, perspective, advanced as a consequence of science, deeply troubles me. It does so not only because the scientific evidence points to the opposite conclusion, but also because I have seen how this perspective has affected the lives of many people, especially college-aged students.

In my first year of teaching, I had an exceptionally bright freshman student, a religious agnostic, who happened also to star as a defensive lineman on the football team. The student, who went on to get a PhD in computer science and philosophy of mind, began during his freshman year to investigate faith questions. To his great frustration, however, he found conversations with believing students intellectually unsatisfying. He came to these discussions assuming that science had undermined the credibility of theistic belief. When he posed critical questions about why—in light of such challenges—he should consider belief in God, other students repeatedly told him, "I don't know. You just have to have faith."

After one of these conversations with a fellow undergraduate, he barged into my office to report in a loud voice about his disappointment with religious believers. He exclaimed in complete frustration, "Why can't someone give you *reasons* for faith?" Perhaps acting out of self-defense (given his size and intensity), I offered to do a class on the topic.

His incredulous reaction still haunts me. He asked, "You mean you think *there are* some?"

I then told him, by now with somewhat more conviction, "Of course. I wouldn't be a Christian if I didn't."

As his agitation began to recede, he replied, "I'd be very interested in hearing about *that*."

This interaction inspired me to develop a class called "Reasons for Faith," one that I've now taught many times at various colleges and universities, often to skeptical students like my former student who have heard so many times that science and faith in God conflict.

Polling shows that many Americans share that perception of conflict. According to a Pew poll, 55 percent of Americans believe that, in general, science and religion often conflict.[12] That number rises to 60 percent among Americans who rarely attend religious services.[13] The poll I mentioned in the Prologue showed that more than two-thirds of self-described atheists believe "the findings of science make the existence of God less probable." And those rejecting religious belief cite scientific theories of unguided chemical and biological evolution more frequently than any other reason for their loss of faith.[14]

On a speaking tour of New Zealand several years ago, I met a young science journalist who had grown up in a religious home but lost his faith after college upon reading a book by a prominent *theistic* evolutionist. The author of the book presented evolutionary mechanisms such as natural selection and random mutation as God's way of creating new forms of life. The author also denied that life manifested any evidence of intelligent design. The student concluded that if the evolutionary process could produce new forms of life without intelligent guidance or design, then there was no need to posit the existence of God at all.

Often the distressed parents of such young people approach me at conferences. They are usually looking for help or advice, reporting on how a son or daughter lost faith in God as the result of the pervasive atheistic ethos on a college campus or in a science class. One mother of an MIT student has come every year to the same East Coast conference on science and faith where I lecture, in part to update me on the story of her previously devout son's conversion to a hostile and condescending form of scientific atheism.

Many students from other universities or parts of the world have shared their own, similar stories with me. Some tell of science professors who openly ridicule theistic belief as part of an overt attempt to dissuade them from their belief in God. Others describe how the unstated presumption of materialism simply defines the range of plausible beliefs under consideration in the classroom and consequently excludes the relevance of theism for understanding reality. For many students who enter college believing in God, these learning environments precipitate a loss of faith or induce an acute cognitive dissonance and even despair.

In talking to students I've discovered that these issues often evoke an intense emotional response precisely because they understand the devasting implications of scientific materialism for their own sense of ultimate hope and meaning. I've long had special empathy for young people searching for answers to the big questions and for meaning and purpose in their lives. Their stories of cognitive dissonance and doubt, of angst and lost faith, move me in part because they remind me of my own experience of troubling questions and acute anxiety as a teenager.

Questions About Meaning

When I was fourteen years old, I began to have recurring and unwelcome questions that I couldn't answer. They began in April 1972, which happened unfortunately to be just before I broke my leg in a skiing accident. When I awoke after surgery, I found myself in a full-length cast reaching from my toes to the top of my left thigh. The cast, complete with a pin through my tibia and fibula, confined me to bed for several weeks and then severely limited my mobility for four months after that.

When I was still in the hospital, to while away the time, my dad gave me a book about the history of baseball. The great American pastime is not associated with existential angst. Yet as I convalesced, devouring the stories of the great players of yesteryear—the statistics they amassed, the records they broke, the victories they achieved—my young mind began to formulate an unwelcome question. It was this: "But what's it going to matter in a hundred years?"

The problem of human significance began to torment me. My worry about the futility of human achievements wasn't just a consequence of thinking that the achievements of the players from the early days of baseball—Honus Wagner, Tris Speaker, and Ty Cobb—were for most people long forgotten. Nor was it that the players popular at the time—Joe Rudi, Al Oliver, Tom Seaver, Johnny Bench, and Joe Morgan—would also one day be forgotten. It wasn't even that the protagonists in the stories played a game for a living. At that time in my life I could think of no more ennobling or glamorous profession than playing baseball, especially for the New York Yankees. Yet reading about the achievements of the greatest baseball players of all time left me with an increasingly empty feeling.

Each story had a similar trajectory. A scout would discover the player in his youth, full of promise. The player would rise to the big leagues, experiencing success measured statistically in an earned run or batting average, a home run record, or wins in big games. Eventually the player would retire with his statistical achievements to show for his career. Some would later be inducted into the Baseball Hall of Fame, achieving an immortality of sorts. Then the player would live out the rest of his life, enjoying the fame and fortune that came with the achievements of an athletic youth. Finally, like all other human beings, the player would die. Then what? What did any of those numbers measuring his achievements mean after that?

Years later I remember walking through the Louvre in Paris and looking at the images of the Olympic athletes from ancient Greece. The sculptures were sublime but, as far as the museum visitor was concerned, the men who inspired them might as well not have existed. How spectacular their victories must have seemed at the time, but how insignificant today. At some point in the future, the exploits of the greatest athletes of our time will be similarly lost or so lacking in context that people will regard them with the same indifference that I felt about the athletes of the Greek Olympics.

Nor did this sense of human futility apply only to athletics. It seemed to me that all of us were destined for a similar end. If a great surgeon saved many lives during her career, that would certainly benefit those people she saved. But ultimately those people would also die, and so would she. How could any achievement in this life have any lasting meaning if in the end no one—no *person*—was here to value or remember it?

As I tried in vain to make sense of these questions, I read a book called *Psycho-Cybernetics*, published in 1960 by a popular physician and self-help author, Maxwell Maltz. To find meaning and satisfaction in life, the book recommended setting goals every day. It called for "steering your mind to a productive, useful goal so you can reach the greatest port in the world, peace of mind."[15]

I began to follow the advice in the book. Some days I would meet my homework goal. Other days I would plan and execute a practical joke on one of my siblings, another goal. But setting goals didn't satisfy my quest for meaning, because the goals themselves seemed ultimately meaningless.

Another thought began to pop up into my mind: "Maybe this is what it means to be insane." That thought frightened me. Soon I began to dread the questions that elicited this fear. I developed a fear of a fear. And soon a fear of the worry about my sanity and thus a fear of a fear of a fear. A mental maelstrom ensued. I was a mess. It seemed a dark cloud followed me throughout my day. I remember thinking that, at fourteen, my life was over.

This all sounds terribly melodramatic now, but it didn't feel that way at the time. Probably if I had seen a therapist, I would have been diagnosed with an anxiety disorder. I was a socially unpopular and somewhat neurotic "smart kid" with a hyperactive brain and clearly prone to anxious thoughts. In addition, my parents were experiencing a rough patch in their marriage at the time, and I could sense the sadness and discord in our home. These were significant factors in my distress.

But I was also experiencing a specific kind of panic, a metaphysical panic, a fear of the meaninglessness of life. It seemed to me that no improvement in my circumstances or in the human condition generally could satisfy my real concern. At some point, as everything in the universe wound down and the last flicker of subjective awareness or human consciousness expired, no one—*no person*—capable of valuing anything would remain on the scene. In that case, how could any human achievement or love or kindness have any lasting value or meaning? The impersonal cosmos guaranteed that they could not.

Since my teenage years, I've encountered many other people, particularly students, who have experienced a similar metaphysical anxiety about whether their lives or human existence generally has any ultimate purpose. I suspect this hopelessness about the human condition, despite the affluence of our society, has contributed to the epidemic levels of suicide among young people, including many who appear for all the world to have everything going for them. Of course, feelings like these are not limited to the young. I've wondered if the plague of opioid addiction, much in the news now and crossing the generations, reflects a wish on the part of many people to numb themselves against a gnawing despair about their own significance.

Philosophy and the Death of God

My own story ultimately took a much more hopeful turn. In college, I learned that the questions I had been asking were actually philosophical and that I wasn't the only person ever to have asked them. I learned that a group of philosophers, the existentialists, identified the "death of God" at the hands of materialistic science as the principal reason for the angst that many modern people felt—and that they felt. In my junior year in college, I remember encountering a thought attributed to the French philosopher Jean-Paul Sartre: "No finite point has any meaning unless it has an infinite reference point."[16] This dictum captured my worry in philosophical terminology: human mortality meant that eventually every person would die and that every person and human action or achievement would ultimately be forgotten. Nothing finite could have any lasting meaning.

I later encountered the work of the British atheist and analytical philosopher Bertrand Russell, who understood the connection between the loss of belief in God and humankind's existential predicament. As he explained in 1910:

> That Man is the product of causes which had no prevision of the end they were achieving; that his origin, his growth, his hopes and fears, his loves and his beliefs, are but the outcome of accidental collocations of atoms; that no fire, no heroism, no intensity of thought and feeling, can preserve an individual life beyond the grave; that all the labours of the ages, all the devotion, all the inspiration, all the noonday brightness of human genius,

> are destined to extinction in the vast death of the solar system, and that the whole temple of Man's achievement must inevitably be buried beneath the debris of a universe in ruins—all these things, if not quite beyond dispute, are yet so nearly certain, that no philosophy which rejects them can hope to stand.[17]

As I considered the logical consequences of an impersonal universe devoid of God, I began to understand my own questions better. "That's it. That's what had been bothering me," I remember telling myself. "I wasn't insane. I was just a philosopher!" That brought some relief until one of my professors tempered my relief by noting that "there is a fine line between philosophy and insanity."

During my last two years of high school, I had begun to read the Bible and found, in its implicit worldview, possible answers to some of my existential questions. This had for a time made my questions and the anxiety they generated bearable. If an eternal personal God created the universe, and if after human death that God could offer human beings continued conscious existence and eternal life, then a person capable of valuing people and their deeds would continue. An eternally existing personal God provided Sartre's infinite reference point and suggested the basis for continued and ultimate meaning.

Later, as I reflected in my college philosophy classes on the questions of my distressed adolescence, I realized why belief in God dissolved the fear of meaninglessness. Only personal agents with subjective conscious awareness can value or confer significance or meaning on something or someone. Nothing can *mean* anything to a molecule or an energy field or a rock. If, as Bertrand Russell wrote, the "vast death of the solar system" and the ruin of the universe ensures that eventually no personal agents will exist, then the existentialist philosophers were right. On the other hand, theism, if true, affirmed the existence—indeed, the eternal and self-existence—of a personal being who loves and regards as meaningful to himself both his creation and the persons within it.

The Epistemological Necessity of Theism

While taking college philosophy classes, I realized that theism solved other fundamental philosophical problems. For example, since the Enlightenment, philosophers have found it difficult to justify a belief in the reliability of human knowledge of the physical world. Oddly, I worried about this too as teenager. I remember looking at a windowsill in my bedroom and wondering if the impression of it in my mind accurately represented the actual object in the world. I worried, "How do I know that my perceptions of reality are accurate?" You can probably imagine that I wasn't much fun at parties!

The problem of epistemology, the basis and justification of human knowledge, has commanded the attention of philosophers for centuries, many of whom doubted our perceptions and our ability to understand the workings of nature. Many philosophers have adopted various forms of skepticism or "antirealism" that deny the reliability of the human mind or our ability to form accurate representations of a mind-independent world around us.

The Scottish empiricist philosopher David Hume initiated much of this skepticism in the eighteenth century. In his famous argument against inductive reasoning, Hume showed that our understanding of the laws of nature depends upon inductive reasoning, the kind of reasoning by which we infer a general principle or law of nature by observing a finite number of instances of the same kind of thing occurring.

For example, if I repeatedly observe a ball falling to the earth after lifting it and then letting it drop, I will soon infer that *all* unsuspended bodies fall. But Hume pointed out that such inferences are inevitably based upon a limited number of observations, since we can't possibly observe all unsuspended bodies to see if they will fall. Thus, inductive inferences, upon which much of science depends, necessarily must assume that "the future resembles the past" or more generally, as the British literary scholar C. S. Lewis put it, "that Nature when we are not watching her behaves in the same way as when we are."[18]

This assumption expresses what philosophers call the principle of the uniformity of nature. That principle affirms that the basic regularities (e.g., unsuspended bodies fall) and patterns of cause and effect (e.g., eating bread nourishes) at work in nature will remain constant through space and time. This assumption seems unproblematic enough until we try to justify it by empirical observation.

For example, we might try to do so by pointing to several different places or different times in the past in which nature seemed to exhibit the same basic laws or patterns of cause and effect that we observe now. We might then be tempted to infer from those *several observed instances* of the constancy of the order of nature that nature *always* exhibits such uniformity. But as Hume pointed out, this conclusion depends upon inductive reasoning—the very kind of reasoning that the uniformity of nature was invoked to justify. In other words, the attempt to justify inductive reasoning by reference to the uniformity of nature involves circular reasoning. Justifying inductive reasoning requires assuming the principle of the uniformity of nature. But justifying the principle of the uniformity of nature, by observing instances of it, inevitably requires using inductive reasoning.

In college, I took philosophy courses from a professor named Norman Krebbs, who became an influential mentor to me. He argued that the problem of knowledge, identified by Hume, raised the fundamental question of the reliability of the human mind. By showing that the uniformity of nature could not be justified by an appeal to repeated sense experience, Hume actually showed that the human mind *assumed* such uniformity. But since Hume showed this assumption could not be justified empirically, he doubted our ability to know the world around us.

Nevertheless, Krebbs pointed out that Hume himself and all other epistemological skeptics *acted* as though they believed in the uniformity of nature. Moreover, all of us, skeptics included, demonstrate that belief every time we walk through a door rather than a window or every time we push the brake in a car. All of us act as though we believe the world, in its most fundamental regularities, will behave in the future the same way that it has behaved in the past. Boarding an airplane, we expect air foils to produce lift tomorrow just as they did yesterday, even though we cannot yet observe how air foils will work in the future. Thus, we all act as though the principle of uniformity holds and that inductive reasoning is generally reliable even if, based on Hume's skeptical argument, we may say we doubt uniformity and induction.

But what justifies the belief that our actions betray? Krebbs argued that though the reliability of the human mind—and the assumptions it makes about the world—could not be justified empirically, it could be justified *theologically*. If one presupposed the existence of a benevolent God, one had good reason to trust in the design of the mind and the reliability of its built-in assumptions about the world. Theists assume the uniformity of nature, because they believe that God is a God of order who sustains the regularities that we describe as the laws of nature. Moreover, theists also believe that God designed human beings with their cognitive capacities. Therefore, they have reason to think that the assumption we all necessarily make about the uniformity of nature matches the way the world actually works. That assumption, in other words, is objectively true as well as subjectively necessary to our everyday functioning. As you will recall, this way of thinking led to the idea of the intelligibility of nature that provided a foundation for the scientific revolution.

This type of argument—known as a presuppositional argument for the existence of God—did not prove the existence of God. But it did suggest that positing God's existence allowed one to live consistently—such that one's stated philosophy would match one's implicit beliefs as expressed in action. Since we all live as though we believe that nature will exhibit the same basic laws and regularities in the future as it has in the past, and since only belief in a benevolent God provides an adequate explanation for the reliability of that and other such necessary assumptions, only theists have a belief system that matches the way they act. Krebbs argued, further, that when people act as if they accept the reliability of the mind and its built-in assumptions about the world, they are essentially acting as if they believe that God exists, even if they deny as much in their explicitly stated philosophies.

I found this line of reasoning persuasive, and find it more so now after encountering the work of the Notre Dame philosopher Alvin Plantinga (Fig. 21.2) and other philosophers of science such as Robert Koons (Fig. 21.3) of the University of Texas. Both Plantinga and Koons have critiqued what is sometimes called evolutionary or naturalistic epistemology—the attempt to justify the reliability of the human mind by reference to a strictly naturalistic and evolutionary account of the origin of our belief-forming faculties.[19]

FIGURE 21.2
Alvin Plantinga, the Notre Dame philosopher and author of *Warrant and Proper Function*.

In his book *Warrant and Proper Function*, Plantinga argues that having what he calls warranted beliefs depends upon having reliable cognitive faculties. He argues further that having such reliable belief-forming capabilities depends upon those capabilities being well designed for attaining knowledge in the environment in which they operate.[20] He acknowledges that the design of our cognitive apparatus (what he calls

FIGURE 21.3
University of Texas philosopher Robert Koons, a critic of naturalistic evolutionary epistemology.

its "design plan") did not necessarily have to originate from a conscious designing agent such as God. Instead, it might have formed by a purely naturalistic evolutionary mechanism of some kind. Nevertheless, Plantinga and Koons have argued that, given the conjunction of evolutionary theory and a naturalistic worldview—and thus a completely *undirected* evolutionary process—we have significant reason to doubt the reliability of our cognitive faculties.[21]

Plantinga notes that for a belief or belief-forming apparatus to affect the evolutionary process, it must affect our behavior in a way that would also influence our prospects for survival (or reproductive success). That's because natural selection favors behaviors that enhance survival; it doesn't favor or "care about" whether the beliefs associated with the behaviors are true or not.[22] Robert Koons amplifies this point. He notes we have no reason to trust the reliability of our faculties in matters that would have had little or nothing to do with survival *when those cognitive structures and beliefs were being formed*.[23] For example, belief-forming mechanisms that might have helped early human beings survive by enabling them to spear mastodons or avoid saber-toothed tigers might not be suited to forming reliable beliefs about particle physics, cosmology, or the evolution of life on earth.

In his autobiography, Darwin worried about a closely related problem. He wondered how we could trust the reliability of our cognitive faculties if they had evolved from the minds of lower animals. As he explained in a letter to a friend, "But then with me the horrid doubt always arises whether the convictions of man's mind, which has been developed from the mind of the lower animals, are of any value or at all trustworthy. Would any one trust in the convictions of a monkey's mind, if there are any convictions in such a mind?"[24]

Plantinga uses a clever thought experiment to tease apart the distinction between a behavior that aids survival and a true belief, showing how the two could easily diverge given a naturalistic evolutionary explanation of our belief-forming capabilities. He asks his readers to imagine a saber-toothed tiger chasing a hypothetical prehistoric man called Paul. Oddly, Paul likes the prospect of being eaten, but mistakenly thinks this particular tiger lacks the ability to eat him. So he runs away in hopes of finding another tiger that will eat him. As Plantinga comments wryly, "This [belief] will get his body parts in the right place so far as survival is concerned, without involving much by way of true belief."[25] To illustrate the point further he asks his readers to imagine that Paul mistakenly "thinks the tiger is a large, friendly, cuddly pussycat and wants to pet it; but he also believes that the best way to pet it is to run away from it."[26] Again, his belief may aid survival without being even approximately true.

Richard Dawkins has also acknowledged, if unintentionally, that survival value and true belief do not necessarily correlate given a naturalistic evolutionary account of our belief-forming faculties. For example, based on studies of the beneficial health effects of religion, Dawkins concedes that it's "perfectly plausible" that religious belief "could indeed have highly beneficial effects upon health."[27] Yet he also argues that belief in God is false and delusional. Thus, he tacitly concedes that natural selection can preserve grotesquely false (from his point of view) beliefs.[28]

Dawkins's concession highlights a deep problem for a strictly naturalistic and evolutionary account of the origin of our cognitive equipment. If naturalism accurately depicts reality, then human beings have overwhelmingly failed to perceive this fact. Though theistic belief has declined in the increasingly secular West, especially among college-educated millennials, a recent Pew Research study shows that 84 percent of the world's population identifies with religious systems of belief, most of which contradict strict naturalism and many of which affirm some form of theism. Moreover, multiple studies across many populations indicate that human beings are hardwired for religious belief.[29] Acceptance of the supernatural appears to be deeply built into the foundations of our cognition, evident even among young children. As Berkeley psychology professor Alison Gopnik observes, "By elementary-school age, children start to invoke an ultimate God-like designer to explain the complexity of the world around them—even children brought up as atheists."[30]

In addition, the Pew Research model projects that over the next three decades, as a percentage of the total population of the world, religiously affiliated populations will likely increase in comparison to religiously unaffiliated populations.[31] The Pew survey projects this increase mainly due to higher birth rates among religiously affiliated than among unaffiliated populations. Evidently, religious belief correlates with higher rates of reproductive success, precisely what the evolutionary process favors. But that means the evolutionary process seems to select or favor human populations with false beliefs—at least as defined by evolutionary naturalists.

It is with considerations like this in mind[32] that Plantinga and Koons argue that *given evolutionary naturalism*, we have significant reason to doubt the reliability of our minds. They conclude that the probability of human beings having reliable belief-forming cognitive faculties given evolutionary naturalism is, as Plantinga puts it, "either low or inscrutable."[33] Consequently, according to Plantinga, a naturalistic evolutionary account is ultimately self-defeating, since it induces a justifiable skepticism about our belief-forming faculties and the beliefs those faculties may form—including beliefs about a naturalistic evolutionary origin of human beings and their minds.[34]

On the other hand, Plantinga argues that the probability of our having a reliable belief-forming apparatus is much, much higher given theism. Since theism holds that a benevolent God possessing a rational intellect created our minds, we have good reason to expect the basic reliability of our belief-forming cognitive equipment. Since most versions of theism hold that the same benevolent and rational God who created our minds created them in God's image and also created a rationally ordered universe, we have good reason to think we can understand and perceive the order built into that universe. Thus, Plantinga argues, those "who believe in God" need not doubt the reliability of their minds or their ability to know the world around them—and for good reason.[35] As he explains, "If God has created us in his image, then even if he fashioned us by some evolutionary means, he would presumably want us to resemble him in being able to *know*; but then most of what we believe might be true even if our minds have developed from those of the lower animals."[36]

Plantinga concludes that the probability of human beings having reliable cognitive equipment is much greater given theism than given evolutionary naturalism. Or, as he expresses this judgment using Bayesian probabilistic formalism: $P(R \mid T) \gg P(R \mid N + E)$, where R symbolizes the reliability of the mind, T symbolizes theism, and N + E represents the conjunction of naturalism and evolutionary theory. It follows that if one assumes the reliability of the mind, a belief that almost all people betray by their actions, the hypothesis of theism provides a better explanation for that presumed fact than does naturalism—or in symbolic terms, $P(T \mid R) \gg P(N + E \mid R)$.[37]

A Presupposition That Solves Philosophical Problems

The power of these various theistic arguments from epistemology came home to me several years ago on a visit to New York. In the fall of 2013, several years after the controversy broke out in the *Times Literary Supplement* over Thomas Nagel's review of *Signature in the Cell*, I contacted Professor Nagel before a scheduled trip to New York. I asked if he would like to get together in person. Given all the trouble his review of my book had caused him, I anticipated that he might not want any direct association. But to my surprise he replied and suggested that I join him for lunch.

Over our meal, Nagel and I talked about his just then recently published book, which I mentioned in Chapter 15, *Mind & Cosmos: Why the Materialist Neo-Darwinian Conception of Nature Is Almost Certainly False.*[38] We also discussed the vicious reaction to his review of my book. We talked about the concept of "naturalistic teleology," which he had formulated as an alternative to both intelligent design, on the one hand, and reductionism and evolutionary materialism, on the other. Nagel asked me to explain how I had come to believe in God. I began to recount how I had become convinced as an undergraduate in philosophy courses that theism best accounted for the reliability of the mind. As I started to explain the argument from epistemological necessity, he politely interrupted me.

"Oh yes," he said. "There is no question that theism solves a lot of philosophical problems."

He then told me a bit more about his own philosophical sensibilities, explaining that he simply lacked an innate *sensus divinitatis*, a "sense of the divine," and that, further, he actually didn't want the universe to be the kind of place in which God existed.[39] His explanation of his views over lunch echoed a famous admission he'd made in print years earlier. As he explained in 1997: "I want atheism to be true and am made uneasy by the fact that some of the most intelligent and well-informed people I know are religious believers. It isn't just that I don't believe in God and, naturally, hope that I'm right in my belief. It's that I hope there is no God! I don't want there to be a God; I don't want the universe to be like that."[40]

Nagel's candor about his philosophical predilections impressed me and reminded me of some of my own internal struggles. During college, I was particularly conflicted in my philosophical inclinations and desires. As I perceived how theism answered many philosophical questions, some of my existential angst abated. At the same time, the sense of accountability that theistic belief placed on me put me in the awkward position of believing in God for philosophical reasons, but not wanting theism (and specifically Christianity) to be true for other, more personal reasons.

Of course, I realized that my wishes and motivations didn't determine the truth about reality. As Lawrence Krauss has correctly pointed out, "The universe is the way it is, whether we like it or not. The existence or nonexistence of a creator is independent of our desires. A world without God or purpose may seem harsh or pointless, but that alone doesn't require God to actually exist."[41] Or, as he put it in another context with characteristic pith, "The universe doesn't exist to make you happy."[42]

Recognizing that our subjective preferences can prove unreliable as guides to truth, philosophers have long looked for other more rational criteria by which to adjudicate worldview issues and to answer the big questions. One such criterion is logical coherence. This criterion seemed to me to favor theism over other worldviews. Our actions betray a belief in objective moral values and moral principles. A belief in God's existence justifies such a belief in a Supreme Valuer. Our actions betray a belief in our ability to know the world around us. A prior belief in the existence of a benevolent God justifies belief in the reliability of the human mind and in the truth of the assumptions we make about the world that make knowledge possible.

What the Oxford philosopher Richard Swinburne has called the "coherence of theism" is no trivial consideration in its favor. Philosophers of science regard coherence as one of the great "theoretical virtues," a key indicator of the truth of a theory. In science and philosophy, as in a court of law, a lack of coherence is routinely considered a fatal defect. Nothing disqualifies a witness in court as quickly as inconsistent or contradictory testimony. That theism provides a coherent account of ethics and epistemology commends it as a more rational worldview than many competing systems of thought.

Nevertheless, in my mid-twenties, I was exposed to another way of assessing the big questions—namely, by examining scientific evidence, especially concerning the crucial worldview-shaping questions of cosmological and biological origins. In this book, I have told the story of my investigation of that evidence and why I've concluded that theism provides a better explanation of the evidence from nature than do competing worldviews.

The Uninvented God

The French philosopher Voltaire once said, "If God did not exist, it would be necessary to invent him."[43] If by that he meant that we need the concept of God to build a coherent, internally consistent worldview, then in my college years I reluctantly came to that same conclusion. I realized that presupposing the existence of God did indeed "solve a lot of philosophical problems."

Yet if Voltaire instead meant that we need to invent the concept of God to cope emotionally—to use religion as an opiate—then I would agree with Professor Krauss. It would be better to face that reality honestly than to live a delusion for the sake of comfort.

Some have argued, of course, that acute angst about the death of God is unnecessary. They have suggested that optimism is still possible in a universe without God and without ultimate meaning. Krauss, for one, emphasizes that the scientific understanding of our place in the universe makes us "insignificant on a scale that Copernicus never would have imagined."[44] He avers that "we're an accident in a remote corner of the universe" and that "the universe doesn't care about us."[45]

Nevertheless, despite the insignificance of our planet and the ultimate meaningless of human life, Krauss argues that we still have cause for optimism and joy. As he explains, "So in a purposeless universe that may have a miserable future you may wonder, 'Well how can I go about each day?' And the answer is we make our own purpose. We make our own joy. We are here by a cosmic accident as I've tried to show, but it's a remarkable accident that's allowed you and I to be here to talk, to think and appreciate the beauty and splendor of the universe."[46]

The existentialist philosophers would have agreed with Krauss's advice about the need to create our own meaning. They too argued that we need to create our own values and morality. But they would have regarded Krauss's claim to have found a basis for genuine optimism as "inauthentic." Jean-Paul Sartre argued that a universe without a transcendent personal God—an infinite reference point—left people in a state of "anguish, forlornness, and despair"[47]—anguish because we can never know if we have chosen the right values, since there is no ultimate standard by which to judge such choices; forlornness because we are truly alone in our choices and no external source or transcendent God can confer lasting meaning on our existence; and despair because the world around us may not cooperate with the choices we make.[48] Sartre also famously describes the existential mood of "nausea" that results from recognizing the ultimate meaningless of the world.

Even so, I readily acknowledge the logic of what Professor Krauss advises. It is wonderful to be alive; the universe is a beautiful as well as an orderly and mysterious place, and our planet is a unique and exquisite blue jewel in the vastness of—as far as we know—an otherwise uninhabited universe. And if we know that everything we love will one day dissolve in the heat death of the universe and that we ourselves will die long before then, why not enjoy everything about our lives and our loves as long as we can? Why not admit the inevitable and be merry while we are here?

Perhaps how we respond to our existential predicament—whether by celebrating our fleeting existence or descending into "anguish, forlornness, and despair"—reflects our own personal temperaments or our frame of mind on a given day. I am not arguing that only one response to the human condition without God makes sense. After all, if God does not exist, there is no standard by which to judge any such choice. Each of us must then decide for ourselves how to respond to a future without hope of ultimate survival or meaning.

Nevertheless, this book has better news: neither of the widely offered responses to the death of God—angst or Sisyphean resistance—is in fact necessary. Not only does theism solve a lot of philosophical problems, but empirical evidence from the natural world points powerfully to the reality of a great mind behind the universe. Our beautiful, expanding, and finely tuned universe and the exquisite, integrated, and informational complexity of living organisms bear witness to the *reality* of a transcendent intelligence—a personal God.

The press of this evidence upon our scientific awareness suggests that we do not need to "invent" God or even to accept God's existence as a mere philosophical necessity. Instead, reflecting on this evidence can enable us to discover—or rediscover—the reality of God. And that discovery is good news indeed. We are not alone in a vast impersonal and meaningless universe—the product of "blind, pitiless indifference." Instead, the evidence points to a personal intelligence behind the physical world that we observe.

This realization has inspired and can continue to inspire deep scientific investigation into the underlying order, beauty, and design of life and the universe. But it has another implication as well. As psychologist Viktor Frankl noted in his classic book *Man's Search for Meaning*,[49] human beings cannot help but ask questions about the meaning of their own existence. But since meaning can only be recognized and conferred by persons, and is arguably found best in relationship between persons, the return of the God hypothesis also revives a hopeful possibility—that our search for ultimate meaning need not end in vain.

Epilogue

Response to Critics

Since the initial publication of this book, critics have raised a few scientific objections to its main arguments—one of which was thoroughly addressed in the book[1] and all of which have been addressed at returnof thegodhypothesis.com.

Nevertheless, one common objection deserves specific mention here. Writing opposite me in the journal *Inference*, physicist Lawrence Krauss recently challenged the idea that the physical parameters of our universe were fine-tuned to make life possible. Rather, he argued "life evolved on earth because it could adapt itself to prevailing [cosmic] circumstances."[2] Or as he put it, "Life on earth is fine-tuned [to] the universe, not the other way around." Consequently, he concluded, we should be no more surprised to find ourselves in a universe fine-tuned for life than we should be to find our legs "remarkably fine-tuned to touch the ground."

To support this view, he suggests that life could exist in some other form—one that could have evolved in accord with other seemingly (but not actually) finely tuned parameters. As he speculated, "Could silicon-based rather than carbon-based life, like the Horta in Star Trek, exist elsewhere? . . .When one imagines such possibilities, the connection between small values of the cosmological constant and the inference to design . . . disappears."

Though ingenious, Krauss's argument fails for several reasons. First, many physical parameters must be fine-tuned not only to make life on earth possible, but also to make possible a universe with stable galaxies, rocky planets, and even basic chemistry—all of which are necessary to establish any biologically relevant process of evolution.[3] Consider: absent the fine tuning of the cosmological constant (estimated at 1 part in 10^{90}), the universe would have either collapsed or blown apart. Absent the hyper-exponential fine tuning of the initial distribution of mass-energy (or "initial entropy"), the universe would have been dominated by life-prohibiting black holes. Unless quarks have precise masses within exceedingly narrow tolerances, atoms more complex than helium (including silicon and carbon), and most basic chemistry, would be impossible. Thus, the precise fine tuning of many critical factors needed to arise *first* before any conceivable form of life could have begun to evolve from necessary chemical elements in a stable galaxy and on a fit planetary platform. Consequently, appeals to the way life evolved to match preexisting physics do not render the improbability of the fine tuning unremarkable. Instead, improbable fine tuning of fundamental physics is necessary to make the origin and evolution of life even possible.

But could *a different set* of physical parameters have enabled the evolution of radically different forms of life? Krauss believes so. Although biologists have found it difficult to agree on a comprehensive definition of life, they do agree that anything worth being called alive must (a) maintain a boundary between itself and its environment, (b) perform integrated sets of complex chemical reactions within that boundary, (c) convert and direct energy, (d) store and process information, and (e) maintain itself in a stable state far from thermodynamic equilibrium. All these functions lie beyond the power of simple hydrogen, helium, or beryllium atoms. Yet, producing atoms more complex than helium or beryllium requires significant fine tuning not only of the cosmological constant and the masses of quarks, but also of the ratio between the strong nuclear and electromagnetic force, the ratio between the electromagnetic and gravitational force, and many other factors. Again, considerable fine tuning must *precede* any biologically relevant process of evolution—and, thus, does indeed require explanation apart from appeals to the evolutionary process.

Another type of potential objection warrants consideration. Several new cosmological models have appeared in technical literature that, once again, attempt to portray the universe as infinitely old. These models could, therefore, be construed as challenging the cosmological arguments developed here. Speculative cosmological models, however, do not constitute *evidence* of an infinitely old universe. Further, two of these models—Roger Penrose's conformal cyclic cosmology and Paul Steinhardt's cyclic cosmology—represent variants of the oscillating universe model critiqued in Chapter 5. Recall that this model was subject to the problem of steadily increasing entropy (and decreasing energy available to do work) with each cycle of postulated expansion.

To address this problem, Penrose's model invokes a hypothetical Phantom Field with powers associated with no known physical field (but, instead, only with god-like agency).[4] He invokes the *unexplained activation* of this field at the right time and in just the way in order to generate—*de novo*—just the right amount of mass to drive a new cycle of expansion at just the right rate from the dying embers of a previous universe. Though he may, thus, envision an infinite series of big bang–like events, he does so only by invoking a finely tuned amount of unexplained mass in the activation of his "Phantom Field." Similarly, Steinhardt's model must invoke multiple forms of *unexplained fine tuning* to ensure repeated expansions of the universe after each postulated recollapse in the infinitely repeating cycles he envisions. Thus, both models inadvertently imply the need for intelligent agency to solve the problem of universe generation. Moreover, physicists Will Kinney and Nina Stein have demonstrated that the Borde-Guth-Vilenkin (BGV) theorem applies to Steinhardt's model, so, even assuming his model, the universe must have ultimately had a beginning.

A third model, known as causal set cosmology, uses a form of mathematics that can describe infinite sets. Causal set cosmology mistakenly assumes, however, that because this form of mathematics can describe infinite sets, applying it to model our universe necessarily implies the universe could be infinite. Causal sets can, however, model finite *and* infinite sets. Thus, absent *evidence* of an infinite universe, the mathematics of causal sets implies nothing about the age of our universe, however useful it might be in modeling hypothetical universes.

Articles on returnofthegodhypothesis.com critique these new cosmological models in more detail. Readers will also find an article there critiquing the false claim, circulating in the media, that discoveries made using NASA's new James Webb Space Telescope challenge the big bang theory and its implication of a cosmic beginning. In fact, they do not. I recommend these articles for further study.

Acknowledgments

Interdisciplinary works benefit greatly from the critical review of experts in different fields. The chapters in this book have benefited from such critical review and deliberative discussion and I'm indebted to the scholars and scientists who have provided that. These include Steve Fuller, Stephen Stobelen, and Michael Newton Keas (history of science); Robert Sheldon and Guillermo Gonzalez (astronomy); Brian Miller, Brian Pitts, David Snoke, Frank Tipler, and Luke Barnes (physics and quantum mechanics); David Berlinski and William Dembski (mathematics), Paul Nelson, Douglas Axe, Richard Sternberg, Michael Denton, and Günter Bechly (biology and paleontology); Bruce Gordon (philosophy of physics); Timothy McGrew, J. P. Moreland, and Robert Koons (philosophy of science). I'd also like to thank Brian Miller for his thorough research in support of the book's physics chapters and for many deep conversations about the issues arising from that research. Thanks to Andrew McDiarmid for his invaluable assistance in entering sources, managing chapter drafts, coordinating the internal peer review and editing of the manuscript and for spearheading the acquisition of permissions. I greatly appreciate the work of our Discovery Institute editors—David Klinghoffer, Jonathan Witt, and Elaine Meyer—who have helped make the manuscript more readable. I'm also grateful to Ray Braun for his delightful pen and ink drawings. Thanks to John West for creating space for me to write by assuming additional burdens of management and for wise counsel. I offer special thanks to Joseph Condeelis, Steve Dilley, Mark Montie, Thomas Marseille, Carolyn Siebe, John and Georgia Wiester, and Elaine Meyer for behind-the-scenes inspiration and support. I'd also like to acknowledge the HarperOne production staff, particularly Lisa Zuniga, Ann Moru, and Chantal Tom, for their professionalism and exquisite attention to detail, and David Fassett for his beautiful design work. Finally, I thank my editor, Kathryn Hamilton, for her skillful guidance in shaping this project, for her deliberative approach to managing it, and for her faith that it would ultimately come to fruition.

Notes

Prologue

1 "Krauss, Meyer, Lamoureux: What's Behind It All? God, Science and the Universe," March 2016, https://youtu.be/mMuy58DaqOk.
2 Because materialists claim that science supports this view, scholars refer to this philosophy as *scientific* materialism or sometimes *scientific* naturalism. Scientific naturalism refers to the closely related idea that the natural world—and the matter and energy out of which it is made—is *all* that ultimately exists.
3 See "Lost Between Immensity and Eternity: The Best of Carl Sagan's *Cosmos* (Part 2)," https://youtu.be/vIVsDg6U0LU?t=41.
4 Dawkins, *The Blind Watchmaker*, 1.
5 Nye, *Undeniable*, 46.
6 Dawkins, *The Blind Watchmaker*, 7.
7 Dennett, *Darwin's Dangerous Idea*, 63.
8 Krauss, *A Universe from Nothing*.
9 Hawking and Mlodinow, *The Grand Design*, 180.
10 West, *Darwin's Corrosive Idea*, 3–7.
11 Michael Lipka, "A Closer Look at America's Rapidly Growing Religious 'Nones,'" *Pew Research Center*, May 13, 2015, http://www.pewresearch.org/fact-tank/2015/05/13/a-closer-look-at-americas-rapidly-growing-religious-nones.
12 Dawkins, *The Blind Watchmaker*, 1, emphasis added.

Chapter 1: The Judeo-Christian Origins of Modern Science

1 Barash, "God, Darwin and My College Biology Class."
2 Dawkins, *River Out of Eden*, 133.
3 Albert B. Paine, *Mark Twain: A Biography, 1835–1910* (1912), chap. 197, https://www.gutenberg.org/files/2988/2988-h/2988-h.htm#link2H_4_0102.
4 See Dawkins, *The God Delusion*, *River Out of Eden*, *The Blind Watchmaker*; Dennett, *Darwin's Dangerous Idea*; Harris, *The End of Faith*, *Letter to a Christian Nation*; Hitchens, *God Is Not Great*.
5 For a refutation of a corollary to this thesis, namely, the view that science has caused the secularization of society, see Brooke, "Myth 25."
6 Tyson, *Cosmos*, Episode 3.
7 Tyson, *Cosmos*, Episode 3, pg. 45 in the transcript, found at http://investigacion.izt.uam.mx/alva/cosmos2014_01-1.pdf. Also available in video form at https://www.fox.com/watch/cebd37976d6172a938ec77621fb35699/, timestamp 28:46.
8 Tyson, *Cosmos*, Episode 3, pg. 45 of the transcript, available at http://investigacion.izt.uam.mx/alva/cosmos2014_01-1.pdf. Also available in video form at https://www.fox.com/watch/cebd37976d6172a938ec77621fb35699/, timestamps 21:11 and 28:16.
9 See also Ungureanu, *Science, Religion, and the Protestant Tradition*.
10 Larson, *Summer for the Gods*, 21.

11 Draper, *History of the Conflict Between Religion and Science*, 363.
12 Russell, *Inventing the Flat Earth*, 38.
13 Larson, *Summer for the Gods*, 22.
14 Russell, "The Conflict of Science and Religion," 4.
15 Russell, "The Conflict of Science and Religion," 4.
16 Butterfield, *The Origins of Modern Science*.
17 Crombie, *The History of Science from Augustine to Galileo*, vol. 2.
18 Foster, *Creation, Nature, and Political Order in the Philosophy of Michael Foster (1903–1959)*.
19 Describing the origin of modern science, Loren Eiseley gave credit to "the sheer act of faith that the universe possessed order and could be interpreted by rational minds. . . . The philosophy of experimental science . . . began its discoveries and made use of its method in the faith, not the knowledge, that it was dealing with a rational universe controlled by a Creator who did not act upon whim nor interfere with the forces He had set in operation" (*Darwin's Century*, 62).
20 Lindberg, "Medieval Science and Religion."
21 Gingerich, *God's Universe*.
22 Hooykaas, *Religion and the Rise of Modern Science*.
23 Merton, "Science, Technology and Society in Seventeenth Century England."
24 Duhem, *The System of World*.
25 Russell, "The Conflict of Science and Religion"; *Cross-Currents*.
26 Whitehead, *Science and the Modern World*.
27 Hodgson, "The Christian Origin of Science"; *The Roots of Science and Its Fruits*; *Theology and Modern Physics*.
28 Barbour, *Religion and Science*.
29 Kaiser, *Creation and the History of Science*.
30 Rolston, *Science and Religion*.
31 Fuller, *Science vs. Religion?*
32 Harrison, *The Bible, Protestantism, and the Rise of Natural Science*; *The Fall of Man and the Foundations of Science*.
33 Stark, *For the Glory of God*.
34 Hodgson, "The Christian Origin of Science," *Occasional Papers*, 1. See also "The Christian Origin of Science," *Logos*, esp. 138.
35 Hodgson, "The Christian Origin of Science," *Occasional Papers*, 1. For an amplifying discussion and extended quotations, see Chapter 1, n. a, at www.returnofthegod hypothesis.com/extendedresearchnotes.
36 Barbour, *Religion and Science*, 27, emphasis in original.
37 Butterfield, *The Origins of Modern Science*, 16–17.
38 Lois Kieffaber, "Christian Theism Alive and Well in the Physics and Astronomy Classroom," in Arlin C. Migliazzo, ed., *Teaching as an Act of Faith: Theory and Practice in Church-Related Higher Education* (New York: Fordham Univ. Press, 2002), 121.
39 Hodgson, "The Christian Origin of Science," *Logos*, 142; Zilsel, "The Genesis of the Concept of Physical Law," 255.
40 Heraclitus and the Stoics represent the most famous of the Greeks to believe in the *logos*, but a similar concept, the *nous*, is also present in the works of Plato and Aristotle. Both "*logos* and *nous*," Richard Tarnas writes, "were variously employed to signify mind, reason, intellect, organizing principle, thought, word, speech, wisdom, and meaning, in each case relative to both human reason and a universal intelligence" (*The Passion of the Western Mind*, 47).
41 See Aristotle, *On the Heavens*; Ptolemy, *Almagest*.
42 Hooykaas, *Religion and the Rise of Modern Science*, 12.
43 The historian and philosopher of science Steve Fuller, of the University of Warwick, offers a different interpretation of the origin of the necessitarian thinking that Bishop Tempier condemned in 1277. In personal correspondence with me about this chapter, he notes that "by modern standards, most Greek philosophers—with the possible exception

of Plato—were quite modest in what they thought 'science' of any sort could ultimately accomplish. (Consider the atomists. They definitely did not have an overblown conception of human reason.)" Instead, Fuller attributes the origin of necessitarian thinking less to Aristotle or Greek science generally and more to the influence of Islamic scholarship on the *interpretation* of Aristotle as his works came into currency in the Christian West during the twelfth and thirteenth centuries. For an amplifying discussion, see Chapter 1, n. b, at www.returnofthegodhypothesis.com/extendedresearchnotes.

44 See Boethius of Dacia, "On the Supreme Good; On the Eternity of the World; On Dreams"; Siger of Brabant, "On the Eternity of the World"; and Dales, *Medieval Discussions of the Eternity of the World*, for a thorough discussion of this subject.

45 Chaberek, *Aquinas and Evolution*, 48–59.

46 Crombie, *The History of Science from Augustine to Galileo*, 2:57. For an amplifying discussion and extended quotations, see Chapter 1, n. c, at www.returnofthegod hypothesis.com/extendedresearchnotes.

47 For example, Albertus Magnus (Crombie, *The History of Science from Augustine to Galileo*, 2:54), Pierre d'Auvergne, Jean Buridan, Marsilius of Inghen (2:54), Thierry of Chartres (1:49), and Averroës (1:72–73).

48 For example, Thierry of Chartres (Crombie, *The History of Science from Augustine to Galileo*, 1:49) and Averroës (1:72–73).

49 Klima, Allhoff, and Vaidya, eds., "Selections from the Condemnation of 1277."

50 Torrance, *Divine and Contingent Order*, 3.

51 For an amplifying discussion and extended quotations, see Chapter 1, n. d, at www .returnofthegodhypothesis.com/extendedresearchnotes.

52 Boyle, Royal Society, Miscellaneous MS 185, fol. 29, cited in Davis, "The Faith of a Great Scientist." For an amplifying discussion and extended quotations, see Chapter 1, n. e, at www.returnofthegodhypothesis.com/extendedresearchnotes.

53 Barbour, *Religion and Science*," 28, emphasis in original.

54 I often use the term "scientist" or "early modern scientist" anachronistically to describe those who, during the sixteenth and seventeenth centuries, did what we would call scientific research today. At that time they were actually called "natural philosophers," "experimental philosophers," or "mechanical philosophers." The term "scientist" was not used to describe those who systematically study nature until 1833, when the philosopher of science William Whewell first coined it.

55 Whitehead, *Science and the Modern World*, 3–4, emphasis in original.

56 Whitehead, *Science and the Modern World*, 12.

57 Fuller, *Science vs. Religion?*, 15.

58 Rolston, *Science and Religion*, 39.

59 Kepler, Letter to Herwart von Hohenburg.

60 Hodgson, "The Christian Origin of Science," *Logos*, 145. For an amplifying discussion and extended quotations, see Chapter 1, n. f, at www.returnofthegodhypothesis.com/ extendedresearchnotes.

61 Fuller, Foreword, in *Theistic Evolution*, 30. See also Harrison, *The Bible, Protestantism, and the Rise of Natural Science*; *The Fall of Man and the Foundations of Science*.

62 Fuller, Foreword, in *Theistic Evolution*, 30.

63 Fuller, Foreword, in *Theistic Evolution*, 30.

64 Harrison, *The Bible, Protestantism, and the Rise of Natural Science*; *The Fall of Man and the Foundations of Science*.

65 Fuller, Foreword, in *Theistic Evolution*, 31.

66 Crombie, *Robert Grosseteste and the Origins of Experimental Science 1100–1700*, 139–62.

67 Crombie, *Robert Grosseteste and the Origins of Experimental Science 1100–1700*, 52–57, 81–90. Grosseteste's method of "Resolution and Composition" involved induction from particulars to universals, and vice versa (see pp. 52–57), whereas his method of "Verification and Falsification" described the gradual elimination of variables to find the true cause of a given phenomenon and to affirm the correct hypothesis (see pp. 81–90; or chap. 4 for more detail).

68 Crombie, *The History of Science from Augustine to Galileo*, 2:27.
69 Lewis, "Robert Grosseteste."
70 See McGrade, "Natural Law and Moral Omnipotence," 273–301.
71 Spade, "Ockham's Nominalist Metaphysics: Some Main Themes," 101.
72 Aristotle, *Metaphysics* 1.3, 7-8.
73 Moliére uses this phrase in his play *The Imaginary Invalid* to make fun of a group of physicians explaining the dormitive powers of opium in this way.
74 See Spade, "Ockham's Nominalist Metaphysics."
75 See Freddoso, "Ockham on Faith and Reason," 328–31.
76 Ockham, Sent. I.30.1 (290), in Spade, "Ockham's Nominalist Metaphysics," 104.
77 Dawkins, *River Out of Eden*, 133.

Chapter 2: Three Metaphors and the Making of the Scientific World Picture

1 Hess, "God's Two Books"; Howell, *God's Two Books*. For an amplifying discussion and extended quotations, see Chapter 2, n. a, at www.returnofthegodhypothesis.com/extendedresearchnotes.
2 Socrates Scholasticus, *Historia Ecclesiastica*, IV, 23 (67, 518), cited in Tanzella-Nitti, "The Two Books Prior to the Scientific Revolution."
3 Basil of Caesarea, *Homilia de Gratiarum Actione*, 2 (PG 31, 221C–224A), cited in Tanzella-Nitti, "The Two Books Prior to the Scientific Revolution."
4 See Augustine, *Enarrationes in Psalmos*, *Sermones*, *Confessiones*; Maximus the Confessor, *Ambigua*; Aquinas, *Super Epistolam ad Romanos*, *Summa Theologiae*, cited in Tanzella-Nitti, "The Two Books Prior to the Scientific Revolution."
5 Ps. 19:1.
6 Ps. 19:2, emphasis added.
7 Rom. 1:20.
8 Ray, *The Wisdom of God Manifested in the Works of the Creation*.
9 Boyle, "Of the Study of the Book of Nature," 154–55, cited in Davis, "The Faith of a Great Scientist."
10 Boyle, "Of the Study of the Book of Nature," 154–55, cited in Davis, "The Faith of a Great Scientist."
11 The metaphor of nature as a book lingered in common usage among scientists long after it first gained currency during the scientific revolution. Albert Einstein used an adapted version of the metaphor in which he refers to the universe as a library of books intelligently arranged in a definite but, to us, mysterious order. As he explained: "We are in the position of a little child, entering a huge library whose walls are covered to the ceiling with books in many different tongues. The child knows that someone must have written those books. It does not know who or how" (quoted in Viereck, *Glimpses of the Great*, 372–73). For the extended quotation, see Chapter 2, n. b, at www.returnofthegodhypothesis.com/extendedresearchnotes.
12 Boyle, "Usefulness of Natural Philosophy," cited in Davis, "The Faith of a Great Scientist."
13 Boyle, "Usefulness of Natural Philosophy," cited in Davis, "The Faith of a Great Scientist."
14 Oresme, *Le Livre du ciel et du monde*.
15 Boyle, "A Free Enquiry into the Vulgarly Receiv'd Notion of Nature" (emphasis in original), cited in Davis, "The Faith of a Great Scientist."
16 Newton, in his *Opticks*, writes: "To tell us that every Species of Things is endow'd with an occult specifick Quality by which it acts and produces manifest Effects, is to tell us nothing" (401). For an amplifying discussion and extended quotations, see Chapter 2, n. c, at www.returnofthegodhypothesis.com/extendedresearchnotes.
17 Corpuscularian theories were proposed by, besides Boyle, Descartes (*The World*), Gassendi (*Opera Omnia*), Newton (*Opticks*), and Locke (*An Essay Concerning Human Understanding*).

18 Boyle, *New Experiments Physico-Mechanical.*
19 Boyle, *A Defense of the Doctrine Touching the Spring and Weight of the Air.*
20 Boyle, "A Free Enquiry into the Vulgarly Receiv'd Notion of Nature," 450–60, cited in Davis, "The Faith of a Great Scientist."
21 Boyle, "A Free Enquiry," 448.
22 Boyle, "Disquisition about the Final Causes of Natural Things," 150–51, cited in Davis, "The Faith of a Great Scientist."
23 Boyle, "Final Causes," 150–51.
24 Boyle, "Final Causes," 150–51, emphasis in original.
25 Boyle, "Final Causes," 150–51, emphasis in original.
26 Davis, "The Faith of a Great Scientist."
27 Davis, "The Faith of a Great Scientist."
28 For an alternate view of the effect of Darwin's theory on the reception of Paley's design argument, see Shapiro, "Myth 8."
29 Boyle, Royal Society, Boyle Papers, vol. 5, fol. 105, cited in Davis, "The Faith of a Great Scientist."
30 Boyle, Royal Society, Boyle Papers, vol. 5, fol. 105, cited in Davis, "The Faith of a Great Scientist."
31 Brooke, "Science and Theology in the Enlightenment," 9.
32 Calvin, *Chemical Evolution*, 258.
33 Calvin, *Chemical Evolution*, 258.
34 Calvin, *Chemical Evolution*, 258.
35 Zilsel, "The Genesis of the Concept of Physical Law," 245–47. For more recent scholarship exploring this connection, see Harrison, "The Development of the Concept of Laws of Nature." See also Gingerich, "Kepler and the Laws of Nature," 17–23; and Daston and Stolleis, *Natural Law and Laws of Nature in Early Modern Europe.*
36 Zilsel, "The Genesis of the Concept of Physical Law," 247–49.
37 Zilsel, "The Genesis of the Concept of Physical Law," 247–49.
38 Job 28:25–26: "When he gave to the wind its weight and apportioned the waters by measure, when he made a decree for the rain and a way for the lightning of the thunder."
39 Zilsel, "The Genesis of the Concept of Physical Law," 247–48. Ps. 104:9: "The Lord has set a boundary to the waters that they may not pass over"; Prov. 8:29: "The Lord gave his decree to the sea that the waters should not pass his commandment"; Jer. 5:22: "The Lord has placed the sand for the bound of the sea by a perpetual decree"; Job 26:10: "The Lord made a boundary to the water, until light and darkness come to an end."
40 Zilsel, "The Genesis of the Concept of Physical Law," 248.
41 Zilsel, "The Genesis of the Concept of Physical Law," 248.
42 Zilsel, "The Genesis of the Concept of Physical Law," 249–53. For an amplifying discussion and extended quotations, see Chapter 2, n. d, at www.returnofthegodhypothesis.com/extendedresearchnotes.
43 Zilsel, "The Genesis of the Concept of Physical Law," 250.
44 Zilsel, "The Genesis of the Concept of Physical Law," 250.
45 Zilsel, "The Genesis of the Concept of Physical Law," 251. For an extended quotation, see Chapter 2, n. e, at www.returnofthegodhypothesis.com/extendedresearchnotes.
46 Zilsel, "The Genesis of the Concept of Physical Law," 255.
47 Gingerich, "Kepler and the Laws of Nature," 17–23.
48 See Zilsel, "The Genesis of the Concept of Physical Law," 258ff.
49 Zilsel, "The Genesis of the Concept of Physical Law," 254.
50 Zilsel, "The Genesis of the Concept of Physical Law," 254. For an amplifying discussion and extended quotations, see Chapter 2, n. f, at www.returnofthegodhypothesis.com/extendedresearchnotes.

51 Oakley, "Christian Theology and the Newtonian Science," 435.

52 These ideas may, interestingly, be suggested by the biblical terminology. The Hebrew word *chok*, or "decree," shares a root with the verb that means "to carve," which is "to impress" words or images on a substance. It has a different connotation from another word used to mean something similar, *mishpat*, often translated as "judgment." Divine decrees of the first type have what may seem like a willful, even arbitrary quality, their reasons not necessarily known (Rashi [Rabbi Shlomo Yitzchaki, 1040–1105] and Rabbeinu Bachya [Bahya ibn Paquda, 1050–1120] on Numbers 19:1). On the other hand, the rationales behind God's "judgments" are clearer, and some could potentially be predicted as logically necessary to a justly ordered society.

53 Oakley, "Christian Theology and the Newtonian Science," 436–37.

54 Newton held this view. He wrote in the preface to the *Principia* that the natural philosophers of the seventeenth century were rightly "rejecting substantial forms and occult qualities" so they could "reduce the phenomena of nature to mathematical laws" (*Mathematical Principles of Natural Philosophy*, 381).

55 Descartes, *Principles of Philosophy*.

56 Huygens, "Discours de la Cause de la Pesanteur."

57 Davis, "Newton's Rejection of the 'Newtonian World View.'"

58 As Leibniz argued: "The Assertors of them [miraculous gravitational attractions] must suppose to be effected by *Miracles*, or else have recourse to Absurdities, that is, to the occult Qualities of the schools; which some Men begin to revive under the specious Name of *Forces*; but they bring us back again into the Kingdom of Darkness" (*A Collection of Papers, Which Passed Between the Late Learned Mr. Leibnitz* [*sic*] *and Dr. Clarke, in the Years 1715 and 1716*, 265, emphasis in original).

59 Leibniz, *New Essays on Human Understanding*, 66.

60 For an amplifying discussion and extended quotations from Leibniz, see Chapter 2, n. g, at www.returnofthegodhypothesis.com/extendedresearchnotes.

61 Leibniz, *Die Philosophischen Schriften*, 358.

62 Leibniz, *The Leibniz-Clarke Correspondence*, 94.

63 Leibniz, *The Leibniz-Clarke Correspondence*, 94.

64 Leibniz, *The Leibniz-Clarke Correspondence*, 92.

65 Leibniz, "New Essays on Human Understanding," 66; Allen, *Mechanical Explanations and the Ultimate Origin of the Universe According to Leibniz*, 8; Buchdahl, *Metaphysics and the Philosophy of Science*, 426; Leibniz, *The Leibniz-Clarke Correspondence*, 19.

66 Allen, *Mechanical Explanations and the Ultimate Origin of the Universe According to Leibniz*, O.

67 Allen, *Mechanical Explanations and the Ultimate Origin of the Universe According to Leibniz*, W.

68 Newton, *Cambridge Manuscript Add. 3970.3, 478v.*

69 Newton, Letter from Isaac Newton to Richard Bentley, February 25, 1692/3, 189.R.4.47, 7r–7v.

70 Newton, Letter from Isaac Newton to Richard Bentley, February 25, 1692/3, 189.R.4.47, W.

71 See Bentley's seventh and eighth Boyle lectures, in Bentley, *Eight Sermons Preach'd at the Honourable Robert Boyle's Lecture, in the First Year 1692*.

72 Bentley, Letter from Richard Bentley to Isaac Newton, February 18, 1692/3, 189.R.4.47, ff.3–4.

73 Bentley, Letter from Richard Bentley to Isaac Newton, February 18, 1692/3, 189.R.4.47, 3r, emphasis added.

74 Newton, Letter from Isaac Newton to Richard Bentley, February 25, 1692/3, 189.R.4.47, 7r, emphasis added.

75 Newton, *Mathematical Principles of Natural Philosophy*, 941, emphasis added.

76 Collingwood, *The Idea of Nature*.

77 As Samuel Clarke, Newton's frequent spokesman, wrote, the course of nature is "nothing else but the will of God producing certain effects in a continued, regular, constant and uniform manner" (Yenter, "Samuel Clarke"; citations within are from Clarke, *The Works*, 698; *A Demonstration of the Being and Attributes of God and Other Writings*, 149).
78 Davis, "Newton's Rejection of the 'Newtonian World View.'"
79 Schaffer, "Occultism and Reason," 129; Davis, "Newton's Rejection of the 'Newtonian World View.'"
80 Gillespie, "Natural History, Natural Theology, and Social Order."
81 Kepler, *Mysterium Cosmographicum*, 93–103; *Harmonies of the World*, 170, 240; Gingerich, "Kepler and the Laws of Nature," 17–23.
82 Boyle, *Selected Philosophical Papers of Robert Boyle*, 172.
83 As historian of science James Larson has noted: "Rational inquiry must inevitably, in Linné's [Linnaeus's] opinion, lead, not to skepticism or disbelief, but to the acknowledgement of and respect for an omniscient and omnipotent creator" (*Reason and Experience*, 151).
84 Newton, *Opticks*, 369–70.
85 Newton, *The Mathematical Principles of Natural Philosophy*, trans. Motte, 388. The more recent Cohen and Whitman translation renders this passage as: "This most elegant system of the sun, planets, and comets could not have arisen without the design and dominion of an intelligent and powerful being." *Mathematical Principles of Natural Philosophy*, 942.
86 Brooke, "Science and Theology in the Enlightenment," 8–9.
87 Newton, *Opticks*, 370.

Chapter 3: The Rise of Scientific Materialism and the Eclipse of Theistic Science

1 Intellectual historians recognize different strands of Enlightenment philosophy, some more secularizing than others. French Enlightenment figures such as Voltaire tended to advance a more atheistic and anticlerical perspective, while British Enlightenment figures such as John Locke saw both human reason and scriptural revelation as valid and complementary sources of knowledge. The Scottish Enlightenment figure David Hume, though British, is well known for his skepticism about arguments for the existence of God and the possibility of miracles. Ironically, he also expressed profound skepticism about the use of inductive reasoning in scientific investigation (see p 52).
2 Ross, "Scientist: The Story of a Word," 72.
3 Hume further insisted that there *must* "be a uniform experience against every miraculous event, otherwise the event would not merit that appellation. And as uniform experience amounts to a proof, there is here a direct and full proof, from the nature of the fact, against the existence of any miracle" ("An Enquiry Concerning Human Understanding," 114).
4 Markie, "Rationalism vs. Empiricism."
5 Hume, "An Enquiry Concerning Human Understanding," 114. Many philosophers today think that Hume's skeptical argument against miracles depends upon a circular justification (begs the question), because he dismisses all alleged historical evidence for the occurrence of miracles on the grounds that miracles do not happen and then justifies the claim that miracles do not happen by asserting that they have never been observed. His critique of the possibility of miracles also assumes that they violate the laws of nature, yet in other parts of "An Enquiry Concerning Human Understanding" (e.g., 46) he argues that the laws of nature lack objective existence, but instead represent mere habits of cognition in our minds. But if that is so, it is hard to see how miracles "violate" them. See McGrew, "Miracles."
6 See Comte, "The Positive Philosophy of Auguste Comte."
7 For more on Kant's view of the cosmological argument, see Kant, *The Critique of Pure*

Reason, *The Critique of Practical Reason and Other Ethical Treatises*, *The Critique of Judgement*, 368–70, 440–56; for Kant on the design argument, see 523.

8 Craig, *Reasonable Faith*, 79–83.

9 Craig, *The Cosmological Argument from Plato to Leibniz*, x, xi, 48–126, 158–204, 282–95; Craig, *Reasonable Faith*, 79–80; Swinburne, *The Existence of God*, 116–32.

10 Bonaventure. *Commentaries on the Sentences of Peter Lombard*, Book II, Distinction 1, Part 1, Article 1, Question 1.

11 Kant, *The Critique of Pure Reason*, *The Critique of Practical Reason and Other Ethical Treatises*, *The Critique of Judgement*, 368–70.

12 Kant, *The Critique of Pure Reason*, *The Critique of Practical Reason and Other Ethical Treatises*, *The Critique of Judgement*, 440–56.

13 The Kalām cosmological argument attempts to argue for the existence of God as a necessary first cause for the origin of a finite universe. The Kalām argument is not the only version of the cosmological argument, however. Thomas Aquinas argued for God as a necessary first cause of the universe, not in a temporal sense, but in an ontological sense (Craig, *Reasonable Faith*, 80–83). Gottfried Leibniz championed another version of the cosmological argument in which he postulated God as the only "sufficient reason" for the contingent causal structure of the universe as a whole ("The Monadology," 235–38). For additional discussion of the status and impact of these versions of the cosmological argument, see Chapter 3, n. a, at www.returnofthegodhypothesis.com/extendedresearchnotes.

14 Newton, *The Correspondence of Isaac Newton*, vol. 3, letter 398.

15 Hawking, *A Brief History of Time*, 9.

16 As H. S. Thayer writes, "Newton speaks, in the *Optics*, of space as the *divine sensorium*; space is that in which the power and will of God directs and controls the physical world. Space is not to be identified with God. . . . Newton says in this scholium that God 'governs all things, and knows all that are or can be done. He is not eternity or infinity, but eternal and infinite; He is not duration or space, but he endures and is present. He endures forever, and is everywhere present; and by existing always and everywhere, he constitutes duration and space.' God constitutes duration and space since 'by the same necessity [as he exists] he exists *always* and *everywhere*'" (*Newton's Philosophy of Nature*, 185–86).

17 Newton likely believed that the material contents of the universe were created a finite time ago, but that time had existed infinitely far back. Gorham, "Newton on God's Relation to Space and Time," 281–320.

18 Even so, few physicists and astronomers during the nineteenth century articulated this view explicitly. Rather, with the rise of scientific materialism, many scientists and philosophers simply seemed to assume that the universe must be eternal and self-existent. Consequently, the discovery of evidence supporting a temporally finite universe during the early part of the twentieth century induced significant cognitive dissonance among many physicists and astronomers. Their reaction clearly revealed that during the nineteenth century the assumption of an infinitely old universe had become deeply entrenched. The most prominent example of this adverse reaction was Einstein's decision to select a precise value for the cosmological constant to ensure that he could depict the universe as neither expanding nor contracting but static. Sir Arthur Eddington and Sir Fred Hoyle also reacted negatively to the idea of a temporally finite universe. See Chapter 5, where I discuss each of these examples in greater detail.

19 Hume, "Dialogues Concerning Natural Religion," 51.

20 Hume, "Dialogues Concerning Natural Religion," 72–83.

21 Hume, "Dialogues Concerning Natural Religion," 72–83. In Hume's "Dialogues," his character Philo says: "I affirm that there are other parts of the universe (besides the machines of human invention) which bear still a great resemblance to the fabric of the world, and which therefore afford a better conjecture concerning the universal origin of this system. These parts are animals and vegetables. The world plainly resembles more

an animal or a vegetable, than it does a watch or a knitting-loom. Its cause, therefore, it is more probable, resembles the cause of the former. The cause of the former is generation or vegetation. The cause, therefore, of the world, we may infer to be some thing similar or analogous to generation or vegetation" (78).

22 Kant sought to limit the scope of the design argument, but did not reject it wholesale. Though he rejected the argument as proof of the transcendent and omnipotent God of Judeo-Christian theology, he still accepted that it could establish the reality of a powerful and intelligent author of the world. In his words, "Physical-theological argument can indeed lead us to the point of admiring the greatness, wisdom, power, etc., of the Author of the world, but can take us no further" (Kant, *The Critique of Pure Reason, The Critique of Practical Reason and Other Ethical Treatises, The Critique of Judgement*, 523).
23 Reid, *Lectures on Natural Theology*, 59.
24 Paine, *The Age of Reason*, 6.
25 Paley, *Natural Theology*, 8–9. For a different view that challenges the idea that Darwin decisively refuted Paley, see Shapiro, "Myth 8." For additional discussion of the status of the design argument during this period, see Chapter 3, n. b, at www.returnofthegodhypothesis.com/extendedresearchnotes.
26 Paley, *Natural Theology*, 8–9.
27 Bridgewater, *The Bridgewater Treatises.*
28 Adam Shapiro takes an even more sanguine view of the durability of the design argument, pointing out that Darwin did not answer all the questions that motivated Paley's case for design in his *Natural Theology* ("Myth 8").
29 Darwin, *On the Origin of Species* (1964), 453.
30 Laplace, *The System of the World*, 354–75.
31 Laplace, *The System of the World*, 354–75.
32 Quoted in Kaiser, *Creation and the History of Science*, 267.
33 For a more extensive discussion of the factual status of this incident, see Kaiser, *Creation and the History of Science*, 267. See also Chapter 3, n. c, at www.returnofthegodhypothesis.com/extendedresearchnotes.
34 For a more extensive discussion of the factual status of this report, see Chapter 3, n. c, at www.returnofthegodhypothesis.com/extendedresearchnotes.
35 Lyell, *Principles of Geology*.
36 Darwin, *On the Origin of Species* (1964), 130–72.
37 Ayala, "Darwin's Revolution," 4.
38 Sire, *The Universe Next Door*, 18.
39 For a discussion of this methodological shift, see Gillespie, *Charles Darwin and the Problem of Creation*, 41–66, 82–108.
40 Gillespie, *Charles Darwin and the Problem of Creation*, 67–81.
41 Gillespie, *Charles Darwin and the Problem of Creation*, 36, 75, and notes therein. See also Agassiz, "The Primitive Diversity and Number of Animals in Geological Times."
42 Darwin, *The Origin of Species* (2003), 414.
43 For a more extensive discussion of the shifting conventions concerning the acceptability of invoking creative intelligence as a scientific explanation during the nineteenth century, see Chapter 3, n. d, at www.returnofthegodhypothesis.com/extendedresearchnotes.
44 Marx discusses the concept of dialectical materialism in various works, but the most notable is *Das Kapital*, in which Marx applied Hegel's concept of dialectical idealism to focus more on the material world (Marx, *Capital*.)
45 Freud's works are numerous, and a full summary is not relevant here. Some important points here, however, are his ideas regarding sexuality, in particular infantile sexuality (see *The Interpretation of Dreams* and *Three Essays on the Theory of Sexuality*), and his model of the human psyche, constituted by the id, ego, and superego (see *Beyond the Pleasure Principle* and *The Ego and the Id*).

46 Some have noted that Marx disliked the label "atheist" and did not apply it to himself, because he regarded people who defined themselves in this way as trying too hard to define themselves in relation to what they did not believe. Nevertheless, he clearly did not believe in God, nor was he merely agnostic in the popular sense. Instead, as a committed dialectical materialist, he rejected any notion of God as a reality and so, as a matter of philosophical categorization, the label clearly applied to him. See Marx and Engels, *Economic and Philosophic Manuscripts of 1844*, 72.
47 Marx, *A Contribution to the Critique of Hegel's Philosophy of Right*.
48 Freud, *The Future of an Illusion*.
49 Draper, *History of the Conflict Between Religion and Science*; White, *A History of the Warfare of Science with Theology in Christendom*.
50 Ayala, "Darwin's Revolution," 4–5.
51 Provine, "Evolution and the Foundation of Ethics."
52 Futuyma, *Evolutionary Biology*.
53 Dawkins, *The Blind Watchmaker*.
54 Simpson, *The Meaning of Evolution*.
55 Simpson, *The Meaning of Evolution*, 344–45.
56 Ayala, "Darwin's Greatest Discovery."
57 See Bingham, "Richard Dawkins"; HuffPost.com. "Bill Nye On Belief In God: Explains How And Why He Is Agnostic (VIDEO)." 1/22/2014. Accessible at: https://www.huffpost.com/entry/bill-nye-on-belief-in-god_n_4645891.
58 See Dawkins, "The Improbability of God."
59 Futuyma, *Evolutionary Biology*, 3.
60 Michael Peterson et al. provide a helpful threefold typology of perceived relationships between science and religion (*Reason and Religious Belief*, 196–214). They discuss the conflict, compartmentalism, and complementarity models of the interaction between science and religion. They do not, however, consider the possibility that scientific evidence might support theistic belief, though that remains a logical possibility. I have, therefore, proposed (and am defending here) a fourth model called "qualified agreement" or "epistemic support." See Meyer, "The Demarcation of Science and Religion," 18–26. See also Dembski and Meyer, "Fruitful Interchange or Polite Chitchat?"
61 Gould, *Rocks of Ages*.
62 MacKay, *The Clockwork Image*, 51–55; Van Till, *The Fourth Day*, 208–15; Van Till, Young, and Menninga, *Science Held Hostage*, 39–43, 127–68. See Gruenwald, "Science and Religion: The Missing Link," for a different interpretation of complementarity, which affirms the methodological autonomy of science and religion, but conjoins their findings.
63 Pond, "Independence."
64 See Russell, *Michael Faraday: Physics and Faith*.
65 See Guyot, *Memoir of Louis Agassiz: 1807–1873*. I also discuss Agassiz and his work elsewhere (Meyer, *Darwin's Doubt*, 7–25).
66 Indeed, it was Maxwell who insisted on placing the inscription *Magna opera Domini, Exquista in omnes voluntates ejus* on the archway of the Cavendish Laboratory in Cambridge. The inscription, quoting Psalm 111, reads "Great are the works of the Lord, sought out by all who take pleasure therein." See also Hutchinson, "James Clerk Maxwell and the Christian Proposition."

Chapter 4: The Light from Distant Galaxies

1 Dowden, "Time."
2 Aristotle, *Physics* III.
3 Augustine discusses the concept of creation *ex nihilo* in his *Confessions*: "Thus it was that in the beginning, and through thy Wisdom which is from thee and born of thy substance, thou didst create something and that out of nothing" (bk. 12, chap. 7; *Confessions and Enchiridion*, 175).

4 Aquinas, *Summa contra Gentiles* 2:17.
5 Maimonides, *Guide*, 2.13.
6 Bonaventure, *Commentaries on the Sentences of Peter Lombard*, Book II, Distinction 1, Part 1, Article 1, Question 2.
7 Craig, *The Kalām Cosmological Argument*, 1–49.
8 Moreland, *Scaling the Secular City*, 31.
9 See Jonathan Jacobs, "Maimonides (1138–1204)," *Internet Encyclopedia of Philosophy*, ed. James Fieser and Bradley Dowden, https://www.iep.utm.edu/maimonid.
10 For a more extensive discussion of Bonaventure's argument for a finite universe, see Chapter 4, n. a, at www.returnofthegodhypothesis.com/extendedresearchnotes.
11 Harrison, *Darkness at Night*, 15–27.
12 Harrison, *Darkness at Night*, 50–60.
13 For a detailed historical overview, see Jaki, *The Paradox of Olbers' Paradox.*
14 Jaki, *The Paradox of Olbers' Paradox*, 26–28.
15 Perceptive readers will realize that Poe's solution to Olbers's paradox implied that the universe had not only expanded in the past, but that it must have expanded faster than the speed of light. Though Einstein's theory of relativity stipulates that objects within a local frame of reference cannot move faster than the speed of light, it does not preclude the possibility that space itself might expand faster than the speed of light. And, indeed, astronomers and cosmologists today think that the rapid expansion of space in the early "inflationary" phase of the history of the universe pushed particles of matter away from each other at rates faster than the speed of light by many orders of magnitude.
16 Poe, *Eureka*, 62.
17 For more background on what Poe and astronomers at the time knew about the speed of light, see Chapter 4, n. b, at www.returnofthegodhypothesis.com/extendedresearchnotes.
18 Poe, *Eureka*, 18.
19 Hawking, *A Brief History of Time*, 6.
20 Recall that though Newton believed that the material contents of the universe had been created a finite time ago, he assumed that space extended infinitely far in every direction and time extended infinitely back into the past.
21 Singh, *Big Bang*, 79.
22 Singh, *Big Bang*, 190–94.
23 Astronomers still use the term "nebula" (and "nebulae," plural) to describe any celestial object that looks cloudy, but now recognize that many objects in the night sky that look (or looked) cloudy represent galaxies far beyond the Milky Way.
24 Singh, *Big Bang*, 190–94.
25 For background on how William Herschel laid the foundation for the discovery that our solar system resides in what we now call the Milky Way galaxy, see Chapter 4, n. c, at www.returnofthegodhypothesis.com/extendedresearchnotes.
26 See Hockey, *The Biographical Encyclopedia of Astronomers.*
27 Cepheid variables were so named because the first such variable star was identified as the fourth brightest in the constellation Cepheus. It was called Delta Cephei—"Delta," for the fourth letter in the Greek alphabet. The periods in question for different Cepheids ranged from about one to one hundred days (Johnson, *Miss Leavitt's Stars*, 34–44, esp. 38).
28 Johnson, *Miss Leavitt's Stars*, 38. Leavitt established a precise mathematical relationship between the periods of variation and apparent brightnesses of Cepheids in the Small Magellanic Cloud. Since all the Cepheids in the galaxy that she observed were approximately the same distance from the earth, the observed differences in the apparent brightness of these stars were proportional to the differences in the absolute brightness as well. Leavitt established the relationship between apparent brightness and the period of pulsation by plotting one variable against the other. Her data would also allow astronomers to plot the absolute brightnesses of those stars against their pulsation

periods. Since the apparent and absolute brightnesses of these stars *are* proportional, the two lines plotting brightness (absolute or apparent) against period necessarily correspond to each other by a predictable factor. In fact, they are plotted as parallel lines on a logarithmic scale, or what is known as a "log-log" graph. Logarithmic scales are typically used to measure exponential, rather than linear, changes. The units that these scales use to plot change on a graph are powers, or "logarithms," of a base number, often base ten.

29 Singh, *Big Bang*, 206–12.

30 A parsec is defined by reference to a simple trigonometric formula. One parsec equals the distance at which one astronomical unit of perpendicular motion relative to an observer generates an angular displacement of one arcsecond, which corresponds to $648000/\pi$ astronomical units (where an astronomical unit is 93,000,000 miles, the distance from the earth to the sun).

31 Dust in the galactic plane can reduce apparent brightness and introduce a complicating factor in calculating absolute brightness and distance to Cepheid stars from apparent brightness. Astronomers have attempted to develop methods for compensating for this problem in their calculations but have not yet developed methods that do so entirely reliably (Mandel et al., "The Type Ia Supernova Color–Magnitude Relation and Host Galaxy Dust").

32 Leavitt did not know the factor for converting apparent to absolute brightness, or what astronomers call the "zero point constant." Astronomers report apparent and absolute brightness in magnitudes. Magnitudes are measured with a logarithmic scale rather than a linear scale. Consequently, astronomers talk about the "zero point," which refers to the additive term in a magnitude equation using a logarithmic measure of apparent brightness.

33 For a more extensive discussion of parallax-based methods of measuring astronomical distances, including Hertzsprung's statistical parallax method, see Chapter 4, n. d, at www.returnofthegodhypothesis.com/extendedresearchnotes.

34 For just a bit more on how Hertzsprung used statistical parallax and how he also built on Henrietta Leavitt's results, see Chapter 4, n. e, at www.returnofthegodhypothesis .com/extendedresearchnotes.

35 Technically, Hertzsprung used Leavitt's graph plotting the *logarithm* of the period of pulsation against the *logarithm* of the apparent brightness of Cepheids in the Small Magellanic Cloud. He did this in the process of determining the apparent brightness of a star in the Small Magellanic Cloud with the same period of pulsation as the average of the *logarithm* of the Cepheid variables of the group near the sun.

36 Recall Leavitt's graph plotted the period of pulsation versus the apparent brightness, and the period of pulsation versus the absolute brightness, of Cepheids in the Small Magellanic Cloud.

37 Hearnshaw, *The Measurement of Starlight*, 349.

38 Johnson, *Miss Leavitt's Stars*, 55.

39 Fernie, "The Period-Luminosity Relation."

40 See "Hubble's Famous M31 VAR! Plate."

41 The actual distance to the Andromeda galaxy as determined by the most current astronomical measurements is 2,500,000 light-years. Virginia Trimble documents the long sequence of errors in calculation that led to the underestimation of that distance ("Extragalactic Distance Scales").

42 The emission and absorption lines and the spacing and relative intensity of these lines are plotted as a function of wavelength.

43 For an amplifying discussion of how astronomers use spectral lines and spectroscopy more generally to study objects in the night sky, see Chapter 4, n. f, at www.returnof thegodhypothesis.com/extendedresearchnotes.

44 Trimble, "Anybody but Hubble!"

45 Singh, *Big Bang*, 247–49.

46 Trimble, "H_0: The Incredible Shrinking Constant."

47 Oddly, Hubble may have never believed the obvious implication of his observations—that the universe was expanding (Sandage, "Edwin Hubble 1889–1953"). In addition, as we'll see in the next chapter, the Belgian astronomer Georges Lemaître had discovered the existence of a linear relationship between recessional velocity and distance two years earlier, though he published his conclusion in less prominent scientific publications.
48 It's important to note that there are gravitational effects that swamp the distancing effect of the expansion rate of the universe. These are called "peculiar motions." Thus, for instance, due to the close proximity of the Andromeda galaxy to us, the Milky Way and Andromeda are approaching each other. Consequently, light coming from a few nearby galaxies exhibits a Doppler blueshift, even if the vast majority of galaxies show a red shift.
49 Hubble, "A Relation Between Distance and Radial Velocity Among Extra-Galactic Nebulae."
50 The age of the universe approximates the inverse of the Hubble constant (Liddle, *An Introduction to Modern Cosmology*, 61–66).
51 Science writer Fred Heeren has a nice explanation of how an expanding balloon illustrates the uniform expansion of the universe; see *Show Me God*, 152.

Chapter 5: The Big Bang Theory

1 Einstein, "Die Feldgleichungen der Gravitation"; "Die Grundlage der allgemeinen Relativitätstheorie." The English translation is in Lorentz et al., *The Principle of Relativity*, 109–64; Chaisson and McMillan, *Astronomy Today*, 604–5.
2 Krane, *Modern Physics*, 31–43.
3 Singh, *Big Bang*, 109–16.
4 Krane, *Modern Physics*, 486–92.
5 As quoted in Sutton, "Review of *Einstein's Universe*."
6 Eddington, "The Deflection of Light During a Solar Eclipse."
7 Harrison, *Darkness at Night*, 73. Even so, Newton's proposed homogeneous distribution of mass solution didn't really solve the problem. The hypothetical universe that he described would have itself been vulnerable to the emergence of any slight "lumpiness" or inhomogeneity. Theoretically, an ant's sneeze would have caused an imbalance in the equipoised gravitational forces, causing the whole system to congeal.
8 Einstein, "Kosmologische Betrachtungen zur allgemeinen Relativitätstheorie."
9 In this paper he argued that his equations allowed for a static universe if two assumptions held, namely, (1) that the curvature of space was positive (like the surface of a sphere) and (2) that the field equations included an additional term known as the cosmological constant (with a precisely calibrated value). For a historical overview, see O'Raifeartaigh, "Einstein's Greatest Blunder?"
10 Einstein's field equations as applied to cosmology represent the volume and curvature of the space in the universe at a given time as a function of the mass-energy density within space. According to Einstein's theory, the mass-energy of the universe always tends to curve space in on itself. Consequently, if gravity were the only force at work in the universe, the mass-energy of the universe would eventually cause space to collapse. Thus, to account for the existence of the space in our universe, Einstein needed to invoke something else that could plausibly counteract the force of gravitational contraction due to the mass-energy in the universe. To do this, Einstein proposed his cosmological constant to represent the energy *inherent in space itself*—energy that causes space itself to expand. In order to depict a static—neither expanding nor contracting—universe, he further assigned an extremely precise value to this negative vacuum energy, so as to balance its repulsive force precisely against the gravitational attraction produced by the mass-energy contained in space.
11 Though the *value* for the cosmological constant that he chose was arbitrary in the sense of being unmotivated by any physical consideration other than his assumption that the

universe must be static, the constant itself did appear naturally in the derivation of the field equations as a constant of integration.

12 In the field equations of general relativity, the radius of the universe (or the radius of curvature) is sometimes represented with a term called "the scale factor." The scale factor provides a measure of the radius of the universe relative to (or in proportion to) the universe's current radius.

13 Einstein expressed this assumption mathematically by setting the derivatives of density and radius (i.e., the rate of change of density and the rate of change of radius) in his field equations to zero.

14 Einstein's assumptions about the radius of curvature (analogous to the radius of a sphere) and the mass density of the universe functioned as what physicists call "initial conditions" in his equations. In physics, an initial condition specifies the value of a variable term at some point in the past deemed as the starting point in the mathematical analysis of some physical process. Typically, physicists need to know initial conditions to determine the value of variable terms in the future. Though Einstein's static universe rejected a beginning to the universe, his assumption that neither the density nor the radius of the universe had changed over time allowed him to assign specific values to these (otherwise) variable terms for all times in the past. In particular, Einstein stipulated that the universe would have had the same radius of curvature and density at any time in the past as it has now.

15 Singh, *Big Bang*, 116–43.

16 Nussbaumer, "Einstein's Conversion from His Static to an Expanding Universe."

17 Even with the precise value of the cosmological constant that Einstein had chosen, Einstein's equations—like Newton's—implied an unstable universe, subject to the slightest perturbations (and resulting inhomogeneities) in the distribution of matter. Yet such perturbations would almost certainly occur in a universe of infinite duration.

18 Luminet, "The Rise of Big Bang Models, from Myth to Theory and Observations."

19 Hawking, *A Brief History of Time*, 49. See also Friedmann, "On the Curvature of Space." Hawking actually used these words to describe Friedmann's model of the universe, but his description is more accurate of Lemaître's, since Friedmann did not attempt to decide which model of the universe he thought actually best described its origin. Hawking's description of the early universe as converging on a singularity also makes more sense written after, as it was, his own proofs of various singularity theorems in the late 1960s and early 1970s (see Chapter 6). Before that, many cosmologists entertained the idea that the universe in the distant past could have "necked down" or compressed into a small but finite volume rather than ultimately beginning from a true spatial singularity (Hawking and Ellis, "The Cosmic Black-Body Radiation and the Existence of Singularities in Our Universe"; see also Hawking, "Properties of Expanding Universes," 105; Hawking and Penrose, "The Singularities of Gravitational Collapse and Cosmology"; Hawking and Ellis, *The Large Scale Structure of Space-Time*).

20 For an excellent discussion of some of this history, see Farrell, *The Day Without Yesterday*.

21 Einstein, "Note on the Work of A. Friedmann 'On the Curvature of Space'" (1922), cited in Luminet, "Lemaître's Big Bang," n. 4.

22 Einstein, "Note on the Work of A. Friedmann 'On the Curvature of Space'" (1923), cited in Luminet, "Lemaître's Big Bang," n. 5.

23 The Solvay conference was sponsored by the International Solvay Institutes for Physics and Chemistry.

24 Nussbaumer, "Einstein's Conversion from His Static to an Expanding Universe," 4. R. W. Smith, "E. P. Hubble and The Transformation of Cosmology," 57.

25 Luminet, "Lemaître's Big Bang," 10.

26 Douglas, "Forty Minutes with Einstein."

27 Nussbaumer, "Einstein's Conversion from His Static to an Expanding Universe," esp. 4–6. See also Douglas, "Forty Minutes with Einstein."

28 Nussbaumer, "Einstein's Conversion from His Static to an Expanding Universe," 5. For an extended quotation from Eddington describing why he came to the conclusion that Einstein's static universe was unstable, see Chapter 5, n. a, at www.returnofthgod hypothesis.com/extendedresearchnotes.
29 "Prof. Einstein Begins His Work at Mt. Wilson"; actual quote: "New observations by Hubble and Humason concerning the red shift of light in distant nebulae make it appear likely that the general structure of the Universe is not static."
30 "Redshift of Nebulae a Puzzle, Says Einstein," referenced in O'Raifeartaigh and McCann, "Einstein's Cosmic Model of 1931 Revisited."
31 Luminet, "Lemaître's Big Bang," 10.
32 Dicke et al., "Cosmic Black-Body Radiation," 415.
33 Kragh, *Cosmology and Controversy*, 179–87.
34 Bondi and Gold, "The Steady-State Theory of the Expanding Universe"; Hoyle, "A New Model for the Expanding Universe."
35 Singh, *Big Bang*, 343.
36 Luminet, "Lemaître's Big Bang," 2. In fairness, Hoyle's creation field was no more ad hoc than Einstein's cosmological constant. To see why, see Chapter 5, n. b, at www.return ofthegodhypothesis.com/extendedresearchnotes.
37 The age of the solar system has been estimated by dating the age of meteorites that have fallen to earth. See Singer, "The Origin and Age of the Meteorites."
38 Kragh, *Cosmology and Controversy*, 79.
39 Singh, *Big Bang*, 283–85; See also: Gamow, George. "Expanding Universe and the Origin of the Elements." *Physical Review* 70 (1946): 572–73.
40 Singh, *Big Bang*, 333.
41 Kragh, *Cosmology and Controversy*, 132–35.
42 Liddle, *An Introduction to Modern Cosmology*, 13–16.
43 Singh, *Big Bang*, 376–83.
44 Baade, "Problems in the Determination of the Distance of Galaxies," 207.
45 Sandage, "Current Problems in the Extragalactic Distance Scale."
46 Planck Collaboration, "Planck 2015 Results."
47 Kragh, *Cosmology and Controversy*, 299–305. See also Burbidge et al., "Synthesis of the Elements in Stars," 547.
48 It is important to note that elements heavier than iron aren't made by the fusion of two nuclei, but instead by neutron capture in supernovae or, as has been more recently argued, by neutron star disruption in merging binary stars. See Wallerstein et al., "Synthesis of the Elements in Stars: Forty Years of Progress."
49 Singh, *Big Bang*, 422–35.
50 Penzias and Wilson, "A Measurement of Excess Antenna Temperature at 4080 Mc/s."
51 Alpher, "Ralph A. Alpher, George Antonovich Gamow, and the Prediction of the Cosmic Microwave Background Radiation," 17–26.
52 Singh, *Big Bang*, 440; Kragh, "The Steady State Theory," 403. For an excellent history of the steady-state theory, see also http://www.astro.ucla.edu/~wright/stdystat.htm.
53 Guth and Sher, "The Impossibility of a Bouncing Universe," 505–7. More recent oscillating universe models have been proposed by such physicists as Paul Steinhardt and Paul Frampton. These models commonly assert that the expansion of the universe was preceded by a contraction phase and then a bounce that initiated the subsequent expansion. These models posit that during the contraction phase the patch of space corresponding to our visible universe experienced a smoothing process that explains the observed homogeneity and isotropy ("Big Bounce Simulations Challenge the Big Bang | Quanta Magazine," accessed October 9, 2020, https://www.quantamagazine.org/big -bounce-simulations-challenge-the-big-bang-20200804/). The models that allow for an eternal universe with infinite cycles then also invoke mechanisms that would reset the entropy to an extremely low value after each cycle. Nevertheless, no evidence for such

exotic mechanisms has ever been discovered (or likely could be). In addition, even the models that simply describe a single bounce face significant theoretical challenges such as instabilities resulting from the bounce violating the null energy condition (Diana Battefeld and Patrick Peter, "A Critical Review of Classical Bouncing Cosmologies," *Physics Reports* (Elsevier B.V., April 1, 2015)).

54 "What Is the Ultimate Fate of the Universe?" *National Aeronautics and Space Administration*, June 29, 2015, https://map.gsfc.nasa.gov/universe/uni_fate.html.

55 A significant minority of astronomers contest the idea that the universe is accelerating in its expansion. See, e.g., Billings, "Cosmic Conflict."

56 Peebles and Ratra, "The Cosmological Constant and Dark Energy."

57 Luminet, "Dodecahedral Space Topology as an Explanation for Weak Wide-Angle Temperature Correlations in the Cosmic Microwave Background."

58 The quote is from 1992 press conference. See Krehl, *History of Shock Waves, Explosions and Impact*, 787.

59 Though Sandage did confirm the linear relationship between the rate of galactic recession and distance, he did, as noted above, also recalibrate distance measurements to many of those galaxies.

60 Allan Sandage quotes are from my own transcript of a private film of Sandage's remarks at "Christianity Challenges the University: An International Conference of Theists and Atheists," Dallas, Texas, February 7–10, 1985. See also Sandage, "A Scientist Reflects on Religious Belief."

61 Willford, "Sizing up the Cosmos."

62 My transcript of Allan Sandage's remarks; see n. 60. See also Sandage, "A Scientist Reflects on Religious Belief."

63 Gingerich, "Scientific Cosmology Meets Western Theology"; Meyer, "Owen Gingerich."

64 Jastrow, *God and the Astronomers*, 116.

Chapter 6: The Curvature of Space and the Beginning of the Universe

1 Here's the full (unsimplified) quotation: "Therefore, by equations 1 and 5, any time-like or null irrotational geodesic must have a singular point on each geodesic within a finite affine distance. If the flow lines form an irrotational geodesic congruence, there will be a physical singularity at the physical points of the congruence, where the density and hence the curvature are infinite" (Hawking, "Properties of Expanding Universes," 105).

2 Hawking and Penrose, "The Singularities of Gravitational Collapse and Cosmology."

3 Hawking and Ellis, *The Large Scale Structure of Space-Time*.

4 Hawking and Penrose, "The Singularities of Gravitational Collapse and Cosmology."

5 Luminet, "The Rise of Big Bang Models," 2.

6 As Senovilla and Garfinkle comment: "Of course, this singular behaviour could be due to an excess of symmetry (spherical) [i.e., homogeneity], which, as exact, would not be realistic. Very remarkably he gave up spherical symmetry and studied the spatially homogeneous but anisotropic models that today we call Bianchi I models. The conclusion was unambiguous: the singularity is still there, 'anisotropy can no more prevent the vanishing of space'" ("The 1965 Penrose Singularity Theorem").

7 Friedmann's and Lemaître's assumption of homogeneity has proved to be a relatively accurate description of the universe today (especially when comparing the density of large volumes of space in different quadrants of the universe); it did prove an inaccurate assumption about the universe in its earliest stages of development. Thus, their assumption of homogeneity cast doubt on extrapolations (based upon their mathematical models) back to the beginning of the universe.

8 Hawking and Ellis, *The Large Scale Structure of Space-Time*, 6; Luminet, "The Rise of Big Bang Models."

9 Confirmed in a personal interview with George Ellis, Cap Estel, France, June 12, 2018.

10 Lemaître did eventually consider solutions to the field equations with nonhomogeneous distributions of matter, but his work did not have the mathematical rigor of that of Hawking, Penrose, and Ellis as expressed in their later proofs of the singularity.
11 Hawking and Ellis, *The Large Scale Structure of Space-Time*, 6; see also their discussion on 261–62.
12 Hawking and Ellis, *The Large Scale Structure of Space-Time*, 261–62.
13 Hawking and Ellis, *The Large Scale Structure of Space-Time*, xi.
14 Hawking and Ellis, *The Large Scale Structure of Space-Time*, 8.
15 Hawking and Ellis, *The Large Scale Structure of Space-Time*, 78.
16 In particular, they argued unless there are singularities, "focal points" would emerge on the time-like and light-like lines of trajectory (what they called "longest curves") that would deny the finite nature of these light/time-like lines ("longest curves") back into the past. Focal points allow light rays to travel through them and to continue traveling indefinitely (and thus infinitely far back into the past). But this result would contradict the proof that they had already established of the incompleteness of the time-like geodesics. Thus, it follows that there must have been a spacetime singularity in the past.
17 Hawking and Ellis, *The Large Scale Structure of Space-Time*, 256.
18 Davies, "Spacetime Singularities in Cosmology," 78–79.
19 Hawking and Ellis, *The Large Scale Structure of Space-Time*, 364.
20 Ross, *The Creator and the Cosmos*, 66–67; Vessot et al., "Test of Relativistic Gravitation with a Space-Borne Hydrogen Maser," 2081–84.
21 Today most physicists think that quantum gravitational effects would begin to manifest themselves inside the so-called Planck length of 10^{-35} of a meter corresponding to the so-called Planck time of 10^{-43} of a second after the big bang.
22 For more technical definition of these various energy conditions, see Chapter 6, n. a, at www.returnofthegodhypothesis.com/extendedresearchnotes. Most important, see also Hawking and Ellis, *The Large Scale Structure of Space-Time*, 88–96; Curiel, "A Primer on Energy Conditions."
23 Hawking and Penrose, "The Singularities of Gravitational Collapse and Cosmology," 529, 531. (Recall that Hawking and Penrose just call this strong energy condition "the energy condition.") See also Hawking and Ellis, *The Large Scale Structure of Space-Time*, 95–96. Physicist Frank Tipler has argued that the strong energy condition can be replaced by the weak energy condition if the strong energy condition holds on average ("Energy Conditions and Spacetime Singularities").
24 Hawking and Ellis, *The Large Scale Structure of Space-Time*, 96, 363.
25 Guth, "Inflationary Universe." See also Linde, "A New Inflationary Universe Scenario"; Albrecht and Steinhardt, "Cosmology for Grand Unified Theories with Radiatively Induced Symmetry Breaking." In Guth's original model he proposed that the rapid initial expansion of space was generated by what he called an "inflationary field" that he equated with the "Higgs field." Later proponents of eternal chaotic inflation refer more generically to the field responsible for the early rapid outward expansion of space as an "inflaton field." They do not associate it, as Guth did his inflationary field, with the Higgs field—the field that is responsible for giving particles their masses according to the standard model of particle physics.
26 Linde, "Eternally Existing Self-Reproducing Chaotic Inflationary Universe."
27 Borde and Vilenkin, "Violation of the Weak Energy Condition in Inflating Spacetimes."
28 Personal interview with George Ellis, Cap Estel, France, June 12, 2018.
29 Guth, "Eternal Inflation and Its Implications."
30 Liddle, *An Introduction to Modern Cosmology*, 80–82.
31 Perfect flatness is only possible if the universe exhibits both critical mass density and perfect homogeneity.
32 Note that in my description of the expansion of the universe at this point in its history, I have begun to focus on mass rather than mass-energy. That's because between about

50,000 and 100,000 years after the beginning of the universe, mass rather than radiation begins to dominate the dynamics and geometry of the universe. See Carroll and Ostlie, *An Introduction to Modern Astrophysics*, 1194. In the distant future dark energy will become the dominant factor. See Frieman, Turner, and Huterer, "Dark Energy and the Accelerating Universe."

33 The curvature of space can be estimated based on certain observed gravitational effects.

34 Inflation also explains the range and distribution of the wavelengths in the cosmic background radiation, though a discussion of how it does so would require too much explanation given the scope and focus of this chapter.

35 For a discussion of the earlier papers by Borde and Vilenkin that concluded, first (in 1994), that inflationary cosmology *could not* and, then (in 1997), that it *could* avoid an initial singularity, see Chapter 6, n. b, at www.returnofthegodhypothesis.com/extendedresearchnotes.

36 Borde, Guth, and Vilenkin, "Inflationary Spacetimes Are Incomplete in Past Directions."

37 Vilenkin, *Many Worlds in One*, 175.

38 Regions of space far apart will recede away from each other with a recession velocity (V_{space}) proportional to the separation distance (d). The constant of proportionality is the Hubble constant (H), which yields the equation $V_{space} = Hd$. Moreover, the observed velocity (V_{ob}) is equal to the velocity of the object, such as a ship (V_{ship}) with respect to the surrounding space minus the recession velocity of that region of space: $V_{ob} = V_{ship} - V_{space}$. I am indebted to Robert J. Spitzer, SJ, for this excellent illustration of the Borde-Guth-Vilenkin theorem. Spitzer, "Evidence for God from Physics and Philosophy," 13–15.

39 As Borde, Guth, and Vilenkin explain in more technical language: "Our argument shows that null and time-like geodesics are, in general, *past-incomplete in inflationary models*, whether or not energy conditions hold, provided only that the averaged expansion condition $H_{av} > 0$ holds along these past-directed geodesics. This is a stronger conclusion than the one arrived at in previous work." ("Inflationary Spacetimes Are Incomplete in Past Directions," 3–4, emphasis added).

40 As Borde, Guth, and Vilenkin summarize the implications of the BGV theorem for the inflationary string landscape multiverse model: "Our argument can be straightforwardly extended to cosmology in higher dimensions. For example, . . . brane worlds are created in collisions of bubbles nucleating in *an inflating higher-dimensional bulk spacetime*. Our analysis implies that the inflating bulk cannot be past-complete" ("Inflationary Spacetimes Are Incomplete in Past Directions," 4, emphasis added).

41 The only proposed cosmologies that avoid the BGV theorem entail physically unrealistic features. For a discussion of these unrealistic cosmologies, see Chapter 6, n. c, at www.returnofthegodhypothesis.com/extendedresearchnotes.

42 Vilenkin, *Many Worlds in One*, 176. See also: Vilenkin, "The Beginning of the Universe"; Grossman, "Why Physicists Can't Avoid a Creation Event," 7. In response, cosmologists now look to quantum cosmological models to describe or explain the beginning of the universe from nothing. As Vilenkin indicates: "What can lie beyond this boundary? Several possibilities have been discussed, one being that the boundary of the inflating region corresponds to the beginning of the Universe in a quantum nucleation event" (Borde, Guth, and Vilenkin, "Inflationary Spacetimes Are Incomplete in Past Directions," 4). I will critique and evaluate the implications of these proposals in Chapters 17–19.

43 Guth, "Inflation," 19 (pdf).

44 Guth, "Inflation," 19 (pdf).

Chapter 7: The Goldilocks Universe

1 Hoyle, "The Expanding Universe."

2 Hoyle, "The Expanding Universe."

3 Harvard astrophysicist Owen Gingerich commented, "I am told that Fred Hoyle said that nothing shook his atheism as much as this discovery" (*God's Universe*, 57).

4 Other parameters require "one-sided" fine tuning. One-sided fine-tuning parameters impose a single condition on the existence of life by ensuring that life can only exist if the parameter in question has a value either greater than or less than some particular threshold. Often in these cases of one-sided fine tuning the value of the parameter in question falls just near the edge of the life-permitting region.

5 In addition to the values of constants within the laws of physics, the fundamental laws themselves have specific mathematical and logical structures that could have been otherwise—that is, the laws themselves have contingent rather than logically necessary features. Yet the existence of life in the universe depends on the fundamental laws of nature having the precise mathematical structures that they do. For example, both Newton's universal law of gravitation and Coulomb's law of electrostatic attraction describe forces that diminish with the square of the distance. Nevertheless, without violating any logical principle or more fundamental law of physics, these forces could have diminished with the cube (or higher exponent) of the distance. That would have made the forces they describe too weak to allow for the possibility of life in the universe. Conversely, these forces might just as well have diminished in a strictly linear way. That would have made them too strong to allow for life in the universe. Moreover, life depends upon the existence of various different kinds of forces—which we describe with different kinds of laws—acting in concert. For example, life in the universe requires: (1) a long-range attractive force (such as gravity) that can cause galaxies, stars, and planetary systems to congeal from chemical elements in order to provide stable platforms for life; (2) a force such as the electromagnetic force to make possible chemical reactions and energy transmission through a vacuum; (3) a force such as the strong nuclear force operating at short distances to bind the nuclei of atoms together and overcome repulsive electrostatic forces; (4) the quantization of energy to make possible the formation of stable atoms and thus life; (5) the operation of a principle in the physical world such as the Pauli exclusion principle that (a) enables complex material structures to form and yet (b) limits the atomic weight of elements (by limiting the number of neutrons in the lowest nuclear shell). Thus, the forces at work in the universe itself (and the mathematical laws of physics describing them) display a fine tuning that requires explanation. Yet, clearly, no *physical* explanation of this structure is possible, because it is precisely physics (and its most fundamental laws) that manifests this structure and requires explanation. Indeed, clearly physics does not explain itself. See Gordon, "Divine Action and the World of Science," esp. 258–59; Collins, "The Fine-Tuning Evidence Is Convincing," esp. 36–38.

6 Denton, *Nature's Destiny*, 101–16.

7 Dicke, "Dirac's Cosmology and Mach's Principle."

8 Some might object to the wording of this statement, since elements on the periodic table are defined not by the number of neutrons they have but by the number of protons. Nevertheless, since new elements cannot be built without neutrons (as well as protons) for stability, it is entirely correct to think of building new elements one proton or neutron at a time. Even so, new elements cannot be built just by adding new neutrons (or protons). Indeed, new elements need both types of nucleons, even if different isotopes of those elements exist with different numbers of neutrons.

9 Alpher, Bethe, and Gamow, "The Origin of Chemical Elements."

10 Singh, *Big Bang*, 323–25.

11 Hoyle actually calculated the differences in *mass* between the beryllium and the helium combined, on the one hand, and the mass of known carbon atoms, on the other. That difference between the two also constituted—by Einstein's energy-to-mass conversion equation of $E = mc^2$—a calculable difference in energy. As Simon Singh notes, "The combined mass of a helium nucleus and a beryllium nucleus is very slightly greater than the mass of a carbon nucleus, so if they did fuse to form carbon then there would be the problem of getting rid of the excess mass. Normally nuclear reactions can dissipate any excess mass by converting it into energy [in accord with $E = mc^2$], but the greater the mass difference, the longer the time required for the reaction to happen. And time is something

the beryllium-8 nucleus does not have" (Singh, *Big Bang*, 392–93). Consequently, Hoyle had to propose a state of carbon with an energy excitation level of exactly the right magnitude to make possible the immediate fusion of the beryllium and the helium within the time dictated by the beryllium half-life.

12 Lewis and Barnes, *A Fortunate Universe*, 113–20.

13 In addition, life in the universe depends upon roughly comparable abundances of carbon and oxygen. Had this energy level in oxygen (at 7.1 MeV) been just a little higher, most of the carbon would be consumed to make oxygen inside stars. Yet both carbon and oxygen are required for life in comparable amounts. See Denton, *Nature's Destiny*, 11–12.

14 Burbidge et al., "Synthesis of the Elements in Stars."

15 Csoto, Oberhummer, and Schlattl, "Fine-Tuning the Basic Forces of Nature Through the Triple-Alpha Process in Red Giant Stars," 560. Epelbaum et al., "Dependence of the Triple-Alpha Process on the Fundamental Constants of Nature." See also Adams and Grohs, "Stellar Helium Burning in Other Universes." Against this, Adams and Grohs have argued that they can explain away the fine tuning of the strong nuclear force. They note that an increase in the strong nuclear force by a small amount (within the range where carbon and oxygen are still produced in comparable amounts) will allow beryllium-8 to be stable. In that case, carbon and oxygen can be produced by two-collision reactions instead of the three-step triple-alpha reaction. Nevertheless, this possibility only eliminates fine tuning on one side. Indeed, even in a two-step reaction, the strong nuclear force cannot be set more than a few percent smaller, even if it can be larger and still allow beryllium production. Moreover, other factors also constrain the value of the SNF on the upper side. For instance, if the SNF were 50 percent larger, the majority of hydrogen would have turned into helium in the early universe, which would have hindered star formation. See MacDonald and Mullan, "Big Bang Nucleosynthesis."

16 Barnes, "The Fine-Tuning of the Universe for Intelligent Life," 548–50.

17 The assumed upper bound for the masses is the Planck mass. The Planck mass is the unit of mass in the system of natural units known as Planck units. The Planck units normalize the speed of light in a vacuum (c), the gravitational constant (G), the reduced Planck constant (ħ), the Coulomb constant (k_e), and the Boltzmann constant (k_B) to 1. The Planck mass is defined as follows: $m_p = \sqrt{\hbar c/G}$. It equals approximately 0.02 milligrams.

18 The mass of the up quark is approximately 1.6×10^{-22} of the Planck mass, and the mass of the down quark is approximately 3.9×10^{-22} of the Planck mass. For the universe to permit life, the up and down quarks must have a mass roughly between 10^{-22} and 10^{-21} times the Planck mass. (S. M. Barr and Almas Khan, "Anthropic Tuning of the Weak Scale and of M_u/M_d in Two-Higgs-Doublet Models," *Phys. Rev. D* 76, no 4: 045002.)

19 In theory, alterations to the gravitational force constant (G) could be partially compensated for by alternations to the process of nucleosynthesis inside stars resulting from variations in the strength of electromagnetism and the size of the stars in question. But even taking the most extreme case, where nuclear reactions are extremely favorable to stellar burning, stars in universes with larger values of G would burn out much faster than stars in our universe. As physicist Luke Barnes has shown, regardless of the strengths of the forces, all stars of all sizes burn out in less than a million years unless values of G are extremely finely tuned—in particular, G for all stars of all sizes must fall within a range that is less than 1 part in 10^{30} of the value of the strong nuclear force, the upper bound defining the range of expected possible values of G. (Specifically, this case concerns the ratio of the proton mass to the Planck mass, which depends on G.) See Barnes, "Binding the Diproton in Stars." Barr and Khan, "Anthropic Tuning of the Weak Scale and of the Mu/Md in Two-Higgs' Doublet Models.

20 Lewis and Barnes, *A Fortunate Universe*, 108.

21 Lewis and Barnes, *A Fortunate Universe*, 108.

22 Specifically, if gravity were increased by a factor of 3000, stars would be too short-lived for terrestrial planets to develop sufficiently to support complex life. Conversely, if the gravitational force constant were set to zero, neither planetary atmospheres nor planets

could exist. This requirement corresponds to two-sided fine tuning of 1 part in ~10^{36} (or $3000/10^{40}$). See Collins, "Evidence for Fine-Tuning," 189–90.

23 In addition, if the gravitational force pulls too weakly, planets would not be able to hold down an atmosphere, making respiration impossible for living organisms. Conversely, if the gravitational force pulls too strongly, planets would retain noxious gases in their atmosphere. Of course, compensatory factors could mitigate the degree of this fine tuning as well. For example, for much smaller planets, gravity could be much larger and not cause the retention of noxious gases in the planetary atmosphere even for a larger G value. Nevertheless, smaller planets are subject to other constraints. Smaller planets have a larger surface area-to-volume ratio, and that leads to more rapid cooling of the planet's interior, making volcanism and plate tectonics impossible—both of which are necessary for life for other reasons. A larger surface area-to-volume ratio and more rapid cooling also lead to a weakened magnetic field, depriving potential life forms on a planet of protection against incoming solar radiation. In addition, a stronger gravitational force constant would have led to *a universe* composed of pure helium. See Lewis and Barnes, *A Fortunate Universe*, 78.

24 It might be possible that gravity could vary over an even larger range, but physicists tend to estimate the expected range more conservatively by defining the "comparison range" to be an actual observed range of the force strengths.

25 The range for the possible values of the constant is set between 0 and 10^{40}G. A universe that can support life must have a gravitation constant less than 10^5G. Therefore, the degree of fine tuning is $10^5G/10^{40}G = 10^5/10^{40} = 1$ in 10^{35}. See Lewis and Barnes, *A Fortunate Universe*, 109. Physicist Sabine Hossenfelder has recently argued that many fine-tuning parameters cannot in fact be quantified [Hossenfelder, "Screams for Explanation: Finetuning and Naturalness in the Foundations of Physics," 1–19]. On this basis, she contests the reality of fine tuning as a feature of nature that has to be explained. To support her claim, she points out that many physicists calculate the degree of fine-tuning associated with different parameters by assuming that all possible values of different physical constants, for example, within a given range are equally probable. She then argues that physicists have no way of knowing whether or not this assumption is true.

Perhaps, she suggests, some universe generating mechanism (see Chapter 16) exists that produces universes with, for example, certain gravitational force constants more frequently than universes with other gravitational force constants. Taking such biasing into account would clearly change the calculated degree of fine tuning (or the probability) associated with any given range of values that correspond to a life-permitting universe. Thus, she argues that the possibility of such biasing in the generation of universes implies that we cannot make accurate assessments of fine tuning—and, therefore, that we cannot be sure that the universe actually is fine tuned for life.

Nevertheless, Hossenfelder's objection has an obvious problem. The allowable ranges of many physical constants and parameters are so incredibly narrow within the vast array of other possible values for those different constants and parameters that any universe generating mechanism capable of favoring the production of those specific and tiny ranges would itself need to be finely tuned in order to produce those values in those ranges with high probability. In other words, her universe generating mechanism would require fine tuning to ensure that biasing that would allow her to explain away fine tuning in our universe.

For a critique of other challenges to the quantifiability of the fine tuning associated with different physical parameters see Chapter 7, n. a., at www.returnofthegodhypothesis.com/extendedresearchnotes.

26 Hoyle was apparently not influenced explicitly by anthropic considerations in predicting the carbon resonance levels in 1953, as many writers have claimed. Instead, as noted above, he predicted the precise resonance levels based on his understanding of what would be necessary to produce carbon and thus to account for its abundance in the universe. Later he realized the anthropic significance of the discovery of the precise value of the resonance level. Indeed, he realized that his discovery implied that the universe had been finely tuned for the production of carbon and thus life. He first made the explicit connection between

the finely tuned coincidences necessary for carbon and oxygen production and the presence of life in the universe in his 1965 book *Galaxies, Nuclei, and Quasars.* In it he writes, "The whole balance of the elements of carbon and oxygen is critical not only for the chemistry of living organisms but for the distribution of the planets" (147).

27 Davies, *The Accidental Universe*, 71–73.
28 Carr and Rees, "The Anthropic Principle and the Structure of the Physical World."
29 Hoyle, "The Universe."
30 Lewis and Barnes, *A Fortunate Universe.*
31 Davies, *The Cosmic Blueprint*, 203.
32 Lewis and Barnes, *A Fortunate Universe*, 323, 320.
33 In describing the fine tuning of the laws of physics, physicists may, however, also be describing the fine tuning of the mathematical structures of the laws themselves, not just their constants. See n. 5 above.
34 "The Anthropic Principle," May 18, 1987, Episode 17, Season 23, *Horizon* series, BBC.
35 Hawking, *A Brief History of Time*, 26.
36 Probably ten to twelve out of thirty-one total constants exhibit significant fine tuning.
37 Ekström et al., "Effects of the Variation of Fundamental Constants on Population III Stellar Evolution"; Epelbaum et al., "Dependence of the Triple-Alpha Process on the Fundamental Constants of Nature."
38 Csoto, Oberhummer, and Schlattl, "Fine-Tuning the Basic Forces of Nature Through the Triple-Alpha Process in Red Giant Stars," 560.
39 Rees, "Large Numbers and Ratios in Astrophysics and Cosmology." Also Lewis and Barnes, *A Fortunate Universe*, 78.
40 Davies, *The Accidental Universe*, 71–73.
41 Rees, *Just Six Numbers*, 22.
42 Personal interview with Sir John Polkinghorne, Portland, Oregon, 1992. For more information on Polkinghorne's arguments, see Polkinghorne, *Belief in God in an Age of Science.*
43 Personal interview with Sir Brian Josephson, Yale University, March 2, 1986.
44 David Klinghoffer, "Brian Josephson, Nobel Laureate in Physics, Is '80 Percent' Confident in Intelligent Design," *Evolution News & Science Today*, June 28, 2017, https://evolutionnews.org/2017/06/brian-josephson-nobel-laureate-in-physics-is-80-percent-confident-in-intelligent-design.
45 Personal interview with Henry Margenau, Yale University, March 2, 1986.
46 Greenstein, *The Symbiotic Universe*, 27.
47 Longley, "Focusing on Theism."

Chapter 8: Extreme Fine Tuning—by Design?

1 Bonnie Azab Powell, "'Explore as Much as We Can': Nobel Prize Winner Charles Townes on Evolution, Intelligent Design, and the Meaning of Life," *UC Berkeley NewsCenter*, June 17, 2005, https://www.berkeley.edu/news/media/releases/2005/06/17_townes.shtml.
2 Lewis and Barnes, *A Fortunate Universe*, 120–28.
3 Physicists have debated whether equating entropy with disorder is the best approach to communicating the idea to the general public. Technically, entropy relates to the logarithm of the number of configurations accessible to a given state of a system. Some prefer connecting entropy to the ideas of uncertainty or the Shannon measure of information. Nevertheless, the term "order" is usually associated with a more precise and defined arrangement of entities and, therefore, a smaller number of configurations. Consequently, the connection provides an intuitively accurate picture in the context of the early universe.
4 Calculating the entropy of water also requires knowing the number of possible energy states associated with the water molecules as well as knowing the number of possible configurations of those molecules, that is, their relative positions in relation to one another in space.

5 In fact, physicists don't really know how to count or describe the possible states or configurations of matter and energy that might characterize black holes. Indeed, they can't envision the different configurations of matter, energy, and spacetime inside a black hole as they can when thinking about atoms in a gas. They don't know how to characterize the "microstates" in black holes, since they lack a theory of quantum gravity. Strictly speaking, therefore, physicists use thermodynamic considerations, rather than considerations of statistical mechanics, to justify their conclusion that black holes exhibit high entropy. Since, they argue, black holes are in an equilibrium state, they must also be in a state of maximum entropy. Indeed, although a galaxy could collapse (eventually) into a black hole, a black hole will not spontaneously release all its matter to create a galaxy.

6 Our normal ideas about entropy seem a bit counterintuitive when applied to black holes. In his book *The Road to Reality* Roger Penrose explains why: "Gravitation is somewhat confusing, in relation to entropy, because of its universally attractive nature. We are used to thinking about entropy in terms of an ordinary gas, where having the gas concentrated in small regions represents low entropy . . . and where in the high-entropy state of thermal equilibrium, the gas is spread uniformly. But with gravity, things tend to be the other way about. A uniformly spread system of gravitating bodies would represent relatively low entropy (unless the velocities of the bodies are enormously high and/or the bodies are very small and/or greatly spread out, so that the gravitational contributions become insignificant), whereas high entropy is achieved when the gravitating bodies clump together" (706). See also Carroll, *From Eternity to Here*, 287–314.

7 Penrose, "Time-Asymmetry and Quantum Gravity." See also Penrose, *The Road to Reality*, 757–65; Gordon, "Divine Action and the World of Science," 259–61, 267.

8 Penrose used a formula known as the Bekenstein-Hawking formula to estimate the entropy per elementary particle, or "baryon" (i.e., neutrons and protons), in the universe. Using that formula, he obtained a value of 10^{43} per baryon. Since the visible universe contains 10^{80} baryons, he calculated the total entropy of the universe as 10^{123} (i.e., 10^{43} times 10^{80}). As noted, Penrose assumed this number defined the upper range of possible entropy values for the universe. In physics, entropy measures can be unitless (or tendered in "natural units"). Penrose calculated entropy in this context using natural units. Entropy is determined by taking the logarithm of the number of possible configurations consistent with a given state (or the logarithm of the number of possible "microstates" consistent with a given "macrostate"). For information on the Bekenstein-Hawking formula, see Sfetsos and Skenderis, "Microscopic Derivation of the Bekenstein-Hawking Entropy Formula for Non-Extremal Black Holes." See also Jacob D. Bekenstein, "Bekenstein-Hawking Entropy," *Scholarpedia*, 2008, http://www.scholarpedia.org/article/Bekenstein-Hawking_entropy.

9 To calculate the entropy of the present universe, he estimated that each galaxy had a black hole of a million solar masses at its center, which yields a value of 10^{21} natural entropy units per baryon. By multiplying 10^{21} by 10^{80} baryons (the total number in the universe) he calculated a total entropy for the present universe of 10^{101} natural entropy units. He then assumed that the early universe would have an entropy no greater than that of the present universe.

10 Entropy is proportional to the logarithm of the number of possible configurations of particles for a given state. Physicists refer to the number of such possible configurations as the "phase space volume." Moreover, mathematically, the number of configurations equals 10 to the power of the entropy. Penrose used the logarithmic relationship between number of configurations and entropy to calculate the probability of a universe having an entropy as low as ours. To do that he computed the number of configurations from the entropy value of the universe. This calculation yielded a number of configurations (or a phase space volume, V_0) consistent with how the universe could have started of 10 to the power of 10^{123}. Similarly, the number of configurations or phase space volume

associated with the actual early universe, V_a, equates to 10 to the power of 10^{101}. The precision associated with the choice of the early conditions for the universe is then the number of configurations associated with the entropy of the early universe divided by the total number that could have been possible. This value is the ratio of the phase space volumes (V_a/V_o), which approximates to 1 in 10 to the power of 10^{123}. The smaller exponent is, again, swallowed up by the massively larger one. See also Penrose, *Emperor's New Mind*, 444–45.

11 It turns out the initial entropy of our universe was even lower than is necessary to allow the universe to sustain life. Oddly, however, this extremely and extravagantly low initial entropy has made possible extensive astronomical observation and investigation of the universe. In theory at least, we could have lived in a universe in which only our local galactic environment exhibited such low entropy. The rest of the universe could have been characterized by high entropy, which, as noted, corresponds to vast spaces filled with black holes. In that case, life would have been possible in the local galactic environment, but nowhere else. Nevertheless, in our visible universe the whole of space exhibits low-entropy conditions just like those in our local galaxy. Indeed, the rest of the visible universe contains over 100 billion highly ordered (low-entropy) galaxies. Since the rest of the universe is not dominated by black holes, we can observe and investigate it and learn about its history by observing other galaxies and stars beyond our Milky Way. (I'm indebted to my colleague Brian Miller for this insight.)

Guillermo Gonzalez and Jay Richards developed a similar insight concerning our local planetary system in their book *The Privileged Planet*. They argue that the earth was intelligently designed not only to host intelligent life, but also to serve as a platform from which to observe the broader universe. They note, for example, that the earth's geological processes provide us with the basic materials to do science and that its transparent atmosphere allows us to study the planets, stars, and galaxies. In a similar way, the more of the universe we can observe, the better we can determine its properties, structure, and history, and the extremely low entropy of the universe as a whole allows us to do just that. Thus, the way the universe is extravagantly finely tuned beyond what is necessary for life suggests a universe designed for discovery. In support of this conclusion, physicist Brian Miller, a colleague of mine at Discovery Institute, has estimated the degree of entropy fine tuning needed for life to exist in just our low-entropy galaxy as 1 part in $10^{10^{98}}$ as opposed to the fine tuning that would be necessary (1 part in $10^{10^{123}}$) to have low-entropy galaxies suffused throughout the cosmos. The difference in these two measures of fine tuning represents the extent to which the universe is fine-tuned beyond what is necessary just for life to exist and for, arguably, scientific discovery of the cosmos. For a detailed discussion of how Miller made this estimate, see Chapter 8, n. a, at www.returnofthegod hypothesis.com/extendedresearchnotes.

12 For an extended quotation from Davies on this point, see Chapter 8, n. b, at www .returnofthegodhypothesis.com/extendedresearchnotes. See also Schaefer, *Science and Christianity*, 31; Penrose, *The Road to Reality*, 728–30, 762–64; Lewis and Barnes, *A Fortunate Universe*, 125–26, 318–19.

13 Douglas Ell nicely illustrates how absurdly small Penrose's probability is in *Counting to God*, 79–81.

14 Bruce Gordon notes: "Two proposals have been suggested by way of trying to mitigate this entropic fine tuning: (1) the inflationary multiverse overcomes the probabilistic obstacles; and (2) there is some special law that requires a perfectly uniform gravitational field at the beginning of time, thus giving rise to maximally low entropy" ("Divine Action and the World of Science," 261). Nevertheless, he shows that neither of these proposals solves the problem. In the first place, he notes: "The inflationary multiverse proposal has massive fine-tuning problems of its own, as well as creating conditions that undermine the very possibility of scientific rationality" (261). In Chapters 16 and 19 of this book, I show why this is the case in detail. Gordon also notes: "The second proposal,

that there is a special law requiring a perfectly uniform gravitational field . . . merely shifts the locus of fine tuning from the big bang itself to the gravitational field associated with it." Moreover, he notes that this proposal has been "unpopular among naturalistically minded physicists for a different reason: it requires a genuine singularity at the beginning of time at which all the laws of physics break down" (261).

15 Davies, *Superforce*, 184.

16 For quantitative estimates of the fine tuning of the expansion rate, see Hawking, *A Brief History of Time*, 125. Hawking states that the expansion could not have been smaller by 1 part in 10^{17}, but he does not quantify how much larger the expansion rate could have been without precluding a life-friendly universe. As he stated, "Why did the universe start out with so nearly the critical rate of expansion that separates models that recollapse from those that go on expanding forever, that even now, ten thousand million years later, it is still expanding at nearly the critical rate? If the rate of expansion one second after the big bang had been smaller by even one part in a hundred thousand million million [10^{17}], the universe would have recollapsed before it ever reached its present size." Meanwhile, Paul Davies affirms a two-sided fine tuning of the expansion rate of 1 part in 10^{18} (*Superforce*, 184).

17 The precise value of these different physical factors varies depending upon which of several contending inflationary cosmological models physicists decide to affirm. Since presently physicists do not agree on a standard inflationary model, it's difficult to know whether each of these physical factors are in fact independently finely tuned, though it looks likely that at least some are, again depending upon the model in question. Luke Barnes has argued that, irrespective of these considerations, the expansion rate of the universe does represent a separate parameter of the universe that requires precise fine tuning ("The Fine-Tuning of the Universe for Intelligent Life," 545).

18 Lewis and Barnes, *A Fortunate Universe*, 167.

19 For an explanation on how physicists calculate the lower bound of the fine tuning associated with the cosmological constant, see Chapter 8, n. c, at www.returnofthegod hypothesis.com/extendedresearchnotes.

20 For an extended discussion of why physicists now commonly agree that the degree of fine tuning for the cosmological constant is *no less than* 1 part in 10^{90}, see Chapter 8, n. d, at www.returnofthegodhypothesis.com/extendedresearchnotes.

21 Maria Temming, "How Many Stars Are There in the Universe?" *Sky & Telescope*, July 15, 2014, https://www.skyandtelescope.com/astronomy-resources/how-many-stars-are-there.

22 Lewis and Barnes, *A Fortunate Universe*, 50–51.

23 Lewis and Barnes, *A Fortunate Universe*, 79.

24 Lewis and Barnes, *A Fortunate Universe*, 177.

25 Carter, "Large Number Coincidences and the Anthropic Principle in Cosmology."

26 Brandon Carter defines two different versions of what is now commonly referred to as *the* weak anthropic principle, one of which he calls the "strong anthropic principle." His weak anthropic principle affirms that our local area in the cosmos (our planet, solar system, and galaxy) exhibits fine-tuning parameters that are necessarily consistent with our existence. His strong anthropic principle affirms that *the universe* exhibits fine-tuning parameters that are necessarily consistent with our existence. Since both of his anthropic principles cite necessary conditions of our existence rather than proposing a causal explanation for the fine tuning itself (either local or universal), both are subject to the same critique offered here. For simplicity's sake, I've combined Carter's two slightly different versions of the anthropic principle for the purpose of this critique. See also n. 28 below.

27 Leslie, "Anthropic Principle, World Ensemble, Design." See also Craig, "The Teleological Argument and the Anthropic Principle."

28 Barrow and Tipler, *The Anthropic Cosmological Principle*, 21. Brandon Carter similarly described the strong anthropic principle as the idea that "the Universe (and hence

the fundamental parameters on which it depends) must be as to admit the creation of observers within it at some stage" ("Large Number Coincidences and the Anthropic Principle in Cosmology," 294). As Lewis and Barnes make clear, however, Carter did not propose this as a causal explanation of the fine tuning, as Barrow and Tipler described the strong anthropic principle in their 1988 book. As Lewis and Barnes explain, "Carter's SAP is easily misunderstood; the source of most confusion is the word *must*. The sense is not logical or metaphysical, that is, that a universe without observers is impossible. Neither is it causal, as if we made the Universe. Rather, this *must* is consequential, as in 'there is frost on the ground, so it must be cold outside.' *Given that we exist*, the Universe (and its laws) must allow observers" (*A Fortunate Universe*, 19).

29 Physicists define the "strong anthropic principle" differently. By the strong anthropic principle I am referring to the maximalist version of the principle as articulated by Barrow and Tipler in their classic 1988 work, *The Anthropic Cosmological Principle*. That version of the principle not only affirms that "the Universe must have those properties which allow life to develop within it at some stage in its history" and that "there exists one possible Universe 'designed' with the goal of generating and sustaining 'observers,'" but also that "observers are necessary to bring the Universe into being." This version of the principle (also called the participatory anthropic principle) affirms observers acting *after* the establishment of the fine-tuning parameters as the cause of the origin of the fine tuning.

Some physicists also use the term "strong anthropic principle" to designate more minimalist concepts (see notes 26 and 28 above), but those versions of the anthropic principle do not specify a cause for the origin of the fine tuning and thus do not present themselves as explanations for it; instead, they are simply facts about our universe. Thus, they do not warrant separate critique here except insofar as they are mistakenly interpreted as causal explanations of the fine tuning as, for example, the "weak anthropic principle" and Brandon Carter's version of the SAP are. Thus, definitional differences notwithstanding, I have attempted to critique all versions of the anthropic principle (weak or strong) that purport to offer causal explanations for the origin of the fine tuning. See Barrow, "Anthropic Definitions"; Barrow and Tipler, *The Anthropic Cosmological Principle*, 16–25.

30 Barrow and Tipler, *The Anthropic Cosmological Principle*, 16–25, esp. 21–22.

31 Gardner, "WAP, SAP, FAP & PAP."

32 Dembski, *The Design Inference*, 33–36. In my book, *Signature in the Cell*, I explained in a more technical way why a "set of functional requirements" represents a kind of "independent pattern" and why Dembski's concept of specification thus subsumes both "independently recognized patterns" and "functional specifications." There I point out that systems that exemplify a set of functional requirements can be thought of as hitting a small independently definable target within a much larger combinatorial space of possibilities. See: *Signature in the Cell*, 360–363. See also Chapter 8, n. e., at: www.returnofthegodhypothesis.com/extendedresearchnotes.

33 As it happens the overwhelming majority of leading physicists accept that the universe does exhibit an extraordinary degree of fine tuning and that the fine tuning is real, not just apparent. In the recent book *A Fortunate Universe*, Lewis and Barnes give a partial list of such leading physicists that includes John Barrow, Bernard Carr, Brandon Carter, Paul Davies, Stephen Hawking, David Deutsch, George Ellis, Brian Greene, Alan Guth, Edward Harrison, Andrei Linde, Donald Page, Roger Penrose, John Polkinghorne, Martin Rees, Allan Sandage, Lee Smolin, Leonard Susskind, Max Tegmark, Frank Tipler, Alexander Vilenkin, Steven Weinberg, John Wheeler, and Frank Wilczek. As Barnes has explained, the scientists on this list "all agree that there is enough evidence for fine tuning that we should do something about it. The list is a roughly equal mix of theist, non-theist and unknown. The non-theists often reach for the multiverse. The theists are divided between those who think that the multiverse is a good scientific

solution (especially Page) and those who think that God is required." In Chapters 13 and 16, I evaluate these competing interpretations. See Lewis and Barnes, *A Fortunate Universe*, 243; Luke Barnes, "Carroll's Five Replies to the Fine-Tuning Argument: Number 1," *Letters to Nature*, August 17, 2014, https://letterstonature.wordpress.com/2014/08/17/carrolls-five-replies-to-the-fine-tuning-argument-number-1.

34 Polkinghorne, "So Finely Tuned a Universe," 16.

Chapter 9: The Origin of Life and the DNA Enigma

1 Ayala, "Darwin's Greatest Discovery," 8567.
2 Dawkins, *The Blind Watchmaker*, 1, emphasis added.
3 Ayala, "Darwin's Greatest Discovery," 8567.
4 Crick, *What Mad Pursuit*, 138.
5 Kenyon and Steinman, *Biochemical Predestination*, 199–211.
6 "They [Thaxton, Bradley, and Olsen] believe, and I now concur, that there is a fundamental flaw in all current theories of the chemical origins of life" (Kenyon, Foreword to *The Mystery of Life's Origin*, vii).
7 Meyer, "Of Clues and Causes," 143–61.
8 Kamminga, "Studies in the History of Ideas on the Origin of Life," 222–45.
9 Shannon, "A Mathematical Theory of Communication."
10 Dretske, *Knowledge and the Flow of Information*, 6–10.
11 Moreover, information increases as improbabilities multiply. The probability of getting four heads in a row when flipping a fair coin is $1/2 \times 1/2 \times 1/2 \times 1/2$, or $(1/2)^4$. Thus the probability of attaining a specific sequence of heads and/or tails decreases exponentially as the number of trials increases. The quantity of information increases correspondingly. Even so, information theorists found it convenient to measure information additively rather than multiplicatively. Thus the common mathematical expression ($I = -\log_2 p$) for calculating information converts probability values into informational measures through a negative logarithmic function, where the negative sign expresses an inverse relationship between information and probability.
12 Dembski, *The Design Inference*, 1–35, 136–74.
13 Crick, "On Protein Synthesis," esp. 144, 153.
14 Dawkins, *River Out of Eden*, 17.
15 Gates, *The Road Ahead*, 188.
16 Hood and Galas, "The Digital Code of DNA."
17 Wald, "The Origin of Life," 44–53.
18 Lehninger, *Biochemistry*, 782.
19 Shapiro, *Origins*, 121; Kamminga, "Studies in the History of Ideas on the Origin of Life," 303–4.
20 Prigogine, Nicolis, and Babloyantz, "Thermodynamics of Evolution," 23.
21 De Duve, "The Constraints of Chance"; Crick, *Life Itself*, 89–93.
22 De Duve, "The Beginnings of Life on Earth," 437.
23 De Duve, "The Beginnings of Life on Earth," 437. Of course, examples of such reasoning abound in our ordinary experience. If, for instance, someone repeatedly rolls a pair of dice and turns up a sequence such as 9, 4, 11, 2, 6, 8, 5, 12, 9, 6, 8, and 4, no one would suspect anything but the interplay of random forces. Yet rolling ten (or, say, one hundred) consecutive sevens *in a game that rewards sevens* will justifiably arouse suspicion that something more than random forces are at work. Indeed, as we saw in the previous chapter, when a highly improbable event occurs that also conforms to a "functionally significant pattern," our uniform experience justifiably leads us to reject chance as the best explanation. Though most origin-of-life researchers have not been open to considering design as an option, they have rejected the chance hypothesis because they recognize that producing the genetic information necessary to synthesize proteins (and thus the

first life) would require just such a conjunction of (a) an extremely improbable series of events with (b) a functionally significant outcome. Indeed, origin-of-life scientists recognize that the critical problem is not just generating an improbable series of molecular interactions that might result in, say, an improbable arrangement of bases in DNA. Instead, the problem is relying on a random shuffling of molecular building blocks to generate one of the very rare arrangements of bases in DNA (or amino acids in proteins) that also *performs a biological function*.

24 In 1977, the physicists Ilya Prigogine and Grégoire Nicolis proposed another theory of self-organization based on their observation that open systems driven far from equilibrium often display self-ordering tendencies (*Self-Organization in Nonequilibrium Systems*, 339–53, 429–47). For example, gravitational energy will produce highly ordered vortices in a draining bathtub, and thermal energy flowing through a liquid or viscous medium will generate distinctive convection currents, or "spiral wave activity." In his 1993 book *The Origins of Order*, Stuart Kauffman attempted to explain the origin of life based upon thermodynamic considerations as well (285–341; see also *At Home in the Universe*, 47–92). I critique these and other theories invoking *external* self-organizing forces (and nonequilibrium thermodynamics) in *Signature in the Cell* (253–70). In brief, I show that these theories do a good job of explaining symmetric or redundant order, but they do not explain the origin of the specified complexity or specified information that characterizes living systems. Physicist Brian Miller has also developed a powerful critique of the recent use of thermodynamic "fluctuation theorems" to explain—or at least render a bit less mysterious—the origin of life (see "Hot Wired: The Thermodynamics of Life," *Inference*, vol. 5, iss. 2).

25 For other arguments supporting this conclusion, see Chapter 9, n. a (as well as Chapter 14), at www.returnofthegodhypothesis.com/extendedresearchnotes.

26 Yockey, *Information Theory and Molecular Biology*, 274–81.

27 Dobzhansky, "Discussion of G. Schramm's Paper," 310.

28 De Duve, *Blueprint for a Cell*, 187.

29 Shapiro, "Prebiotic Cytosine Synthesis."

30 Wolf and Koonin, "On the Origin of the Translation System and the Genetic Code in the RNA World by Means of Natural Selection, Exaptation, and Subfunctionalization," 14.

31 Johnston et al. "RNA-Catalyzed RNA Polymerization."

32 De Duve, *Vital Dust*, 23.

33 Crick, *Life Itself*, 88.

34 Stein, *Expelled*. The Stein-Dawkins interview begins at time stamp 1:26:32.

35 Stein, *Expelled*. The Stein-Dawkins interview begins at time stamp 1:26:32.

36 Crick and Orgel, "Directed Panspermia."

37 Thaxton, Bradley, and Olson, *The Mystery of Life's Origin*, 211.

38 As Thaxton, Bradley, and Olson put it: "We have observational evidence in the present that intelligent investigators can (and do) build contrivances to channel energy down nonrandom chemical pathways to bring about some complex chemical synthesis, even gene building. May not the principle of uniformity then be used in a broader frame of consideration to suggest that DNA had an intelligent cause at the beginning?" (*The Mystery of Life's Origin*, 211).

39 Dean Kenyon, "Going Beyond a Naturalistic Approach to the Origin of Life," presentation at "Christianity Challenges the University: An International Conference of Theists and Atheists," Dallas, TX, February 9, 1985.

40 Peirce, "Deduction, Induction, and Hypothesis," 375.

41 Gould, "Evolution and the Triumph of Homology," 61.

42 Chamberlain, "The Method of Multiple Working Hypotheses," 754–59.

43 Lipton, *Inference to the Best Explanation*, 1.

44 Lyell, *Principles of Geology*, 75–91.

45 Kavalovski, "The Vera Causa Principle," 78–103.

46 Scriven, "Explanation and Prediction in Evolutionary Theory," 480.
47 Meyer, "Of Clues and Causes," 96–108.
48 Quastler, *The Emergence of Biological Organization*, 16.
49 Thaxton, Bradley, and Olson, *The Mystery of Life's Origin*, 42–172; Shapiro, *Origins*; Dose, "The Origin of Life"; Yockey, *Information Theory and Molecular Biology*, 259–93; Thaxton and Bradley, "Information and the Origin of Life"; Meyer, *Signature in the Cell*.
50 Of course, the phrase "large amounts of specified information" raises a quantitative question, namely, "How much specified information would a biomacromolecule (or a minimally complex cell) have to possess before that specified information implied intelligent design?" In *Signature in the Cell*, I give and justify a precise quantitative answer to this question. I show that the *presence* of roughly 500 or more bits of specified information reliably indicates intelligent design in a prebiotic context. For the basis of these calculations, see chaps. 8–10 of *Signature in the Cell* and esp. Chapter 9, n. b, at www.returnofthegodhypothesis.com/extendedresearchnotes. Suffice to say, for now, that many of the information-bearing biomacromolecules present in even the simplest one-celled organisms easily exceed this (specified) informational threshold.
51 Protein folds constitute the smallest unit of *structural* innovation in the history of life. See Chapter 9, n. c, at www.returnofthegodhypothesis.com/extendedresearchnotes; Meyer, *Darwin's Doubt*, 189–98, 219–27.
52 McDonough, *The Search for Extraterrestrial Intelligence*.
53 Dawkins, *River Out of Eden*, 133.

Chapter 10: The Cambrian and Other Information Explosions

1 Koonin, "The Biological Big Bang Model for the Major Transitions in Evolution."
2 Darwin, *On the Origin of Species* (1964), 129–30.
3 Darwin writes: "There is another and allied difficulty, which is much graver. I allude to the manner in which numbers of species of the same group, suddenly appear in the lowest known fossiliferous rocks. . . . To the question why we do not find records of these vast primordial periods, I can give no satisfactory answer." (*On the Origin of Species* (1964), 396–97).
4 For an extended discussion of, and relevant quotations from, paleontologists who have questioned the gradualistic Darwinian picture of the history of life, see Chapter 10, n. a, at www.returnofthegodhypothesis.com/extendedresearchnotes.
5 Bechly and Meyer, "The Fossil Record and Universal Common Ancestry."
6 Bechly and Meyer, "The Fossil Record and Universal Common Ancestry."
7 According to John Maynard Smith: "The fact of evolution was not generally accepted until a theory had been put forward to suggest how evolution had occurred, and in particular how organisms could become adapted to their environment; in the absence of such a theory, adaptation suggested design, and so implied a creator. It was this need which Darwin's theory of natural selection satisfied" (*The Theory of Evolution*, 42).
8 Huxley, "The Evolutionary Vision," 249, 253.
9 Eden, "Inadequacies of Neo-Darwinian Evolution as a Scientific Theory," 11.
10 For example, in 1980 the Harvard paleontologist and evolutionary biologist Stephen Jay Gould declared neo-Darwinism "effectively dead, despite its persistence as textbook orthodoxy" ("Is a New and General Theory of Evolution Emerging?" 120). See also Müller and Newman, "Origination of Organismal Form," esp. 7.
11 Müller, "The Extended Evolutionary Synthesis."
12 Müller and Newman, "Origination of Organismal Form," esp. 7.
13 Gilbert, Opitz, and Raff, "Resynthesizing Evolutionary and Developmental Biology." Andreas Wagner, *The Arrival of the Fittest* (New York: Penguin, 2014). The historical origins of this catchphrase are surprisingly complicated. Most who cite the phrase (e.g., evolutionary biologist Andreas Wagner) credit it to Hugo de Vries, but de Vries himself, in the last sentence of his 1904 monograph *Species and Varieties*, credits Arthur Harris:

"Or, to put it in the terms chosen lately by Mr. Arthur Harris in a friendly criticism of my views: 'Natural selection may explain the survival of the fittest, but it cannot explain the arrival of the fittest'" (*Species and Varieties: Their Origin by Mutation*, 2nd ed., ed. Daniel Trembly MacDougal (Chicago: Open Court, 1906), http://www.gutenberg.org/files/7234/7234-h/7234-h.htm).

14 Peterson and Müller, "Phenotypic Novelty in EvoDevo," 328.

15 Denton, *Evolution: A Theory in Crisis*, 308–25.

16 Berlinski, *The Deniable Darwin*, 41–64.

17 Meyer, *Darwin's Doubt*, 169–84.

18 Axe determined the rarity of "function-ready" protein folds in a region of sequence space close to the wild-type beta-lactamase enzyme. His analysis showed that folds capable of performing beta-lactamase functions were extremely rare (i.e., 1 in 10^{77}) even in this specific region of amino acid sequence space. His result, therefore, implied an even greater rarity for functional folds in the vastly larger sequence space of possible amino acid combinations. This conclusion follows even taking into account the most optimistic estimates for the number of existing folds populating that space. Clearly, the probability of finding function-ready protein folds is higher close to known targets than elsewhere in sequence space. Axe, "Estimating the Prevalence of Protein Sequences Adopting Functional Enzyme Folds." For an earlier estimate also derived from mutagenesis experiments, see Reidhaar-Olson and Sauer, "Functionally Acceptable Solutions in Two Alpha-Helical Regions of Lambda Repressor." Weizmann Institute protein scientist Dan Tawfik has performed studies showing that as mutations accumulate, protein folds—first gradually and then increasingly quickly—lose their structural stability. For a discussion of how these and other recent results reinforce Axe's quantitative estimates of the rarity of protein folds, see the discussion in Chapter 15. See also Tokurik et al., "The Stability Effects of Protein Mutations Appear to Be Universally Distributed"; Tokuriki and Tawfik, "Stability Effects of Mutations and Protein Evolvability"; Bershtein et al., "Robustness–Epistasis Link Shapes the Fitness Landscape of a Randomly Drifting Protein"; Lundin et al., "Experimental Determination and Prediction of the Fitness Effects of Random Point Mutations in the Biosynthetic Enzyme HisA"; Bechly, Miller, and Berlinski, "Right of Reply"; Miller, "A Dentist in the Sahara: Doug Axe on the Rarity of Proteins Is Decisively Confirmed."

19 Darwin, *The Life and Letters of Charles Darwin*, 278–79.

20 Mayr, Foreword, in Ruse, *Darwinism Defended*.

21 Meyer, *Darwin's Doubt*, 271–335. See also Meyer, Gauger, and Nelson, "Theistic Evolution and the Extended Evolutionary Synthesis." See also Chapter 15 of this book.

22 Quastler, *The Emergence of Biological Organization*, 16.

23 Again, the phrase "large amounts of specified information" elicits a quantitative question, namely, "How much specified information would a system have to have before it implied design?" As I explain in *Signature in the Cell* and *Darwin's Doubt*, the answer to that question depends upon the available probabilistic resources—the number of opportunities for solving a relevant search problem—in a given context. (See also Chapter 9, n. 50). In the context of biological evolution, the search for a novel protein fold defines the relevant search, and the duration of the evolutionary history of life on earth (roughly 3.85 billion years) and the number of organisms that have existed determines the maximum number of mutational trials that could have occurred—i.e., the available probabilistic resources.

Recall that a novel protein fold represents the smallest unit of innovation in the history of life. Since building fundamentally new forms of life requires structural innovation, mutations must generate new protein folds for natural selection to have an opportunity to preserve and accumulate structural or morphological innovations. Thus, the ability to produce new protein folds represents a *sine qua non* of macroevolutionary innovation.

Douglas Axe's discovery that protein folds are extremely rare in amino acid sequence space poses a formidable challenge to the creative power of the random mutation and natural selection mechanism. Though random mutations may produce slight changes in protein function within the structure of a common, preexisting fold, Axe's work showing the extreme rarity of "function-ready" folds (estimated at no better than 1 out of every 10^{77} amino acid sequences for a sequence of a relatively modest length) implies that generating new folds requires more information than could be reasonably expected to arise given the probabilistic resources available to earth's evolutionary history. Recall that there have been "only" 10^{40} organisms and replication events in the history of life on earth.

Since I also show that proposed alternative evolutionary mechanisms also fail to explain the origin of the amount of information necessary to generate proteins folds (*Darwin's Doubt*, 291–355), and that intelligent agents routinely solve informational search problems of much greater magnitudes (360–63), I argue that intelligent design provides the best explanation for the origin of the information necessary to account for fundamental innovation in the history of life.

For examples of failed attempts to demonstrate that mutation and natural selection can generate novel protein folds, see Berlinski and Hampton, "Hopeless Matzke "; Axe, "Answering Objections from Martin Poenie"; Axe, "More on Objections from Martin Poenie"; Gauger, "Protein Evolution"; Axe, "Show Me"; Meyer, *Darwin's Doubt*, 221–27. See also Miller, "A Dentist in the Sahara: Doug Axe on the Rarity of Proteins Is Decisively Confirmed," and Chapter 15 in this book.

24 Agassiz, "Evolution and the Permanence of Type," 444.

25 Dawkins, *The Blind Watchmaker*, 47–49; Küppers, "On the Prior Probability of the Existence of Life"; Schneider, "The Evolution of Biological Information"; Lenski et al., "The Evolutionary Origin of Complex Features." For a critique of these genetic algorithms and claims that they simulate the ability of random mutation and natural selection to generate new biological information apart from intelligent activity, see Meyer, *Signature in the Cell*, 281–95.

26 Rodin, Szathmáry, and Rodin, "On the Origin of the Genetic Code and tRNA Before Translation."

27 Denton, *Evolution: A Theory in Crisis*, 309–11.

28 Polanyi, "Life Transcending Physics and Chemistry"; "Life's Irreducible Structure."

29 Dawkins, *River Out of Eden*, 133.

Chapter 11: How to Assess a Metaphysical Hypothesis

1 The quote is also found in Sagan's book by the same title; see Sagan, *Cosmos*, 4.

2 Niiler, "Maybe You're Not an Atheist."

3 Carroll, "Turtles Much of the Way Down."

4 Carroll, *The Big Picture*, 11.

5 Dawkins, *River Out of Eden*, 133.

6 Peirce, "Deduction, Induction, and Hypothesis," 375.

7 In his essay "Deduction, Induction, and Hypothesis" Peirce outlined the differences between these three forms of inference. Deductive inferences apply a general rule to a particular case and yield logically necessary results. As he stated, for deductive inference, "the major premise lays down [the] rule, . . . the minor premise states a case under the rule, . . . [and] the conclusion applies the rule to the case and states the result." By contrast, Peirce defined induction as an inferential process in which rules are *generated* from knowledge of particular cases and results. As he stated, "Induction infers a rule." Peirce's third form of inference, which he called hypothesis or abduction, occurs when a particular case is postulated from knowledge of a rule and a fact or result. "Hypothesis," he stated, "infers from one set of facts of one kind to facts of another."

8 Meyer, "Of Clues and Causes," 25.

9 Gingerich, "The Galileo Affair"; see also *The Galileo Affair*, 110.

10 Peirce, "Deduction, Induction, and Hypothesis," 375.

11 Peirce, "Deduction, Induction, and Hypothesis," 375.
12 Work in the philosophy of science suggests that predictive success constitutes a special case of explanatory power in which a theory's ability to predict an event stands as evidence of its ability to explain it (Lipton, *Inference to the Best Explanation*). In addition, other work has shown that scientists can often explain events after the fact that they could not have predicted before the fact (Scriven, "Explanation and Prediction in Evolutionary Theory"). Still other work in the history of science has shown that the explanation of previously known facts often accounts more for the success of a theory than does a theory's ability to predict previously unknown events (Brush, "Prediction and Theory Evaluation"). All these results have suggested the primacy of explanation as an indicator of theory success. See also n. 21 below.
13 Chamberlain, "The Method of Multiple Working Hypotheses."
14 Meyer, "Of Clues and Causes," 90–97; Lipton, *Inference to the Best Explanation*, 1–5, 6–8, 56–74; Sober, *The Philosophy of Biology*, 27–46.
15 Cleland, "Historical Science, Experimental Science, and the Scientific Method"; "Methodological and Epistemic Differences Between Historical Science and Experimental Science."
16 Lipton, *Inference to the Best Explanation*; Meyer, "The Methodological Equivalence of Design and Descent," 67–112, 300–312, esp. 88–94.
17 For another homespun example of the way both explanatory power and considerations of simplicity contribute to the evaluation of competing possible explanations, see Chapter 11, n. a, at www.returnofthegodhypothesis.com/extendedresearchnotes.
18 In my books making the case for an intelligent design and in Chapters 9 and 10 of this book, I have shown how the case for intelligent design—which begins as an abductive inference—has been strengthened by just such a process of elimination, rendering the argument for intelligent design not just an abductive inference, but an abductive inference to the *best* explanation. See Meyer, *Darwin's Doubt; Signature in the Cell.*
19 Meyer, "Of Clues and Causes," 99–108, esp. 102.
20 Scriven, "Causes, Conditions and Connections in History," 249–50.
21 See Kline, "Theories, Facts, and Gods," 37–44. Kline argues that in cases where there is no known or observable cause of the effect or event in need of explanation, historical scientists may posit a novel causal theory by extrapolating from the powers of a cause known to be capable of producing a "relevantly similar" effect or event. Such extrapolation will generally need to be justified on some theoretical grounds. As Kavalovski has shown, Darwin used such a general strategy to establish the causal adequacy of natural selection. By drawing an analogy between artificial and natural selection, Darwin suggested that the latter could produce morphological change just as the former could. By invoking the theoretical consideration that natural selection would have more time in which to achieve its results, Darwin suggested that it was legitimate to expect (i.e., to extrapolate) that natural selection could produce more morphological change than artificial selection—enough to produce new species. Technically this method of reasoning did not meet the strict requirements of *vera causa*, because historical scientists cannot observe natural selection producing the amount of morphological change required by the fossil record and the extant diversity of life. Nevertheless, as Kavalovski notes, the use of analogy and extrapolation (justified theoretically) was widely accepted by influential philosophers even before Darwin as a valid strategy for establishing causal adequacy ("The Vera Causa Principle," 104–29).
22 For a more extensive primer on Bayesian probability calculus, see Chapter 11, n. b, at www.returnofthegodhypothesis.com/extendedresearchnotes.
23 For a discussion of an objection known as the "problem of old evidence" to the use of the Bayesian formalism to evaluate hypotheses, see Chapter 11, n. c, at www.returnofthegodhypothesis.com/extendedresearchnotes.
24 There is a common objection to the use of both inference to the best explanation and Bayesian analysis to test hypotheses. Specifically, philosophers of science worry that inference to the best explanation as a method and the Bayesian formalism used in

support of it treat already known evidence and new evidence equally when assessing the effect of evidence on the strength of a hypothesis. (This objection is also related to the problem of "old evidence"; see n. 23 above.) In other words, in inference to the best explanation and the Bayesian formalism, the ability to explain already known evidence counts just as much in support of a hypothesis as predicting a previously unknown event, phenomenon, or piece of evidence (see Talbott, "Bayesian Epistemology").

Indeed, in these methods of hypothesis testing, the relationship between hypothesis and evidence is unaffected by the way it comes to an observer in time. Some philosophers of science worry, therefore, that when evidence presents itself before scientists predict it, scientists can "accommodate" or gerrymander the features of their proposed explanations to match the evidence in question. Such explanations, so goes the worry, will not explain anything other than the evidence at hand and will fail to bring deeper understanding about the world—that is, understanding not provided already by the data themselves. In other words, such explanations will lack broader explanatory power or depth. Historians and philosophers of science characterize such explanations as ad hoc. They also note such explanations can become extremely convoluted in order to match the data at hand as, for example, those of Ptolemaic astronomy did with their use of epicycles to describe planetary motions. When hypotheses require complex interactions between multiple theoretical entities to explain the evidence, historians and philosophers of science think that reliance on these explanations may obscure deeper regularities or patterns of cause and effect at work in nature. In short, what gerrymandered explanations after the fact may gain in empirical adequacy, they may lose in parsimony, explanatory depth, and prior plausibility.

Most historians and philosophers of science acknowledge that explaining events after the fact does create more of an opportunity for scientists to contrive explanations in ways that can diminish parsimony, explanatory depth, and coherence. Nevertheless, *post hoc* hypotheses with explanatory power need not lack these and other explanatory virtues. Indeed, the recognition of the presence or absence of other explanatory virtues often tacitly complements assessments of causal adequacy and explanatory power and figures into determinations about which among a competing set of explanations qualifies as best. Thus, there is no reason to reject explanations of already known facts simply because they may not also make predictions, unless such explanations also lack coherence or parsimony, for example. Instead, there may be good reasons to accept such explanations, especially if in addition to explanatory power and causal adequacy they exhibit other explanatory virtues such as coherence, parsimony, explanatory depth, and breadth.

As the University of Maryland historian of science Stephen Brush has shown, many theories in physics were initially accepted because of their ability to explain already known facts and anomalies better than previously dominant theories. Brush shows, in particular, that physicists accepted Einstein's theory of general relativity more because of its immediate ability to explain known facts than because of its later successful predictions. (See Brush, "Prediction and Theory Evaluation." For specifically Bayesian responses to this problem, see Horwich, *Probability and Evidence*; Maher, "Prediction, Accommodation, and the Logic of Discovery.") In any case, I will show in later chapters that the God hypothesis not only exhibits greater causal adequacy than competing metaphysical hypotheses, but that it also exhibits other explanatory virtues such as simplicity/parsimony, explanatory breadth and depth, internal consistency and coherence, and fruitfulness. (See Keas, "Systematizing the Theoretical Virtues.")

25 I'm indebted to philosopher of science Tim McGrew for this excellent illustration.

26 Meyer, "The Return of the God Hypothesis."

27 Richard Dawkins, "Why There Almost Certainly Is No God," *Edge*, October 25, 2006, https://edge.org/conversation/richard_dawkins-why-there-almost-certainly-is-no-god.

Chapter 12: The God Hypothesis and the Beginning of the Universe

1 For a more complete explication of the logical structure of the Kalām cosmological argument for God's existence, see Chapter 12, n. a, at www.returnofthegodhypothesis .com/extendedresearchnotes.

2 This argument was known as the trademark argument. Some philosophers questioned its first premise, that finite human beings have a clear and distinct idea of a perfect, infinitely powerful or infinitely wise God. Others questioned the second premise, the idea that only God could cause the idea of a perfect being in our minds.

3 Dembski and Meyer, "Fruitful Interchange or Polite Chitchat?," 418–22.

4 See Moser, *The Elusive God*, 243–45.

5 McMullin, "How Should Cosmology Relate to Theology?," 39.

6 McMullin, "How Should Cosmology Relate to Theology?," 39.

7 Quoted in Browne, "Clues to Universe Origin Expected."

8 Gen. 1:1.

9 Isa. 51:16; Isa. 45:12; Ps. 104:2; Jer. 10:12; Zech. 12:1; 2 Tim. 1:9; Titus 1:2.

10 As quoted in Sutton, "Review of *Einstein's Universe.*"

11 Of course, in addition to this primary sense of "bringing into being that which did not previously exist," many versions of theism also affirm that God sustains the universe in existence moment by moment. Some theists think of this sustaining power as part of God's role as the creator; others think of it as a separate power. Even those who think of it as part of the way God functions as creator also typically think of God's sustaining the universe as a secondary sense of creating, the need for which follows from the primary act of God's having brought the universe into existence in the first place.

Some theists, typically analytical philosophers, do deny that God *necessarily* acted to create the universe at a point in time in the finite past. For example, a famous argument for God's existence, the "argument from contingency," offers God as the sufficient reason and best explanation for the existence of the universe and its contingent features whether the universe had a beginning or not. Philosophers who offer such proofs often conceive of God, at least for the sake of argument, as continually creating and sustaining the universe on a moment-by-moment basis rather than having created it at a specific time in the finite past. They think of God as "the ground of all being," who continuously sustains the universe in existence, and they suggest that God has possibly done so for an infinitely long time.

Even so, there are reasons for theists to prefer the idea that God both creates and sustains the universe over the view that God only sustains it. First, creation without a beginning has no precedent in our experience. Instead, in our experience, creators with powers of deliberation will bring various "creations" (bridges, paintings, cars, cell phones, etc.) into existence that did not exist before. Indeed, creation implies a new entity coming into being and thus temporal sequence and beginning.

In addition, as many theistic philosophers from the Middle Ages to the present have argued, the idea that God continually creates time as well as space, matter, and energy but has been doing so for an infinitely long time generates various absurdities.

Similarly, St. Bonaventure, for one, argued that the universe could no more have had an infinite past than a man could have climbed out of a hole infinitely deep (Moreland, *Scaling the Secular City*, 31). Just as a man, climbing up from an infinitely deep hole one step at a time, would never reach the top because he would have an infinite distance to traverse, a universe that began an infinitely long time ago would never reach the present through a series of temporal events, because an infinitely long time would have had to have transpired before that series of events could reach the present moment.

As William Lane Craig has argued, potential infinities (or the idea of approaching infinity as a mathematical limit) make sense, but "a collection of things" or events "formed by adding one member after another can't be actually infinite" in reality. Craig explains why. A collection formed by adding one member to another can never actually be infinite because no matter how many members might exist in the collection, they could be numbered *and* one more could always be added before reaching an infinite number. And since a series of events in time is a "collection formed by adding one member to another," it follows that a series of events in time cannot form an actual infinite either. That means that the universe could not have begun an infinitely long time ago even if God existed to create it "then" (Craig, *Reasonable Faith*, 98–99).

12 See n. 11 above for a discussion of the implausibility of positing an actual infinite, including an actual temporally infinite universe, and thus the implausibility of positing that God created such a temporally infinite universe.

13 Indeed, though some versions of theism expect a temporally finite universe and other versions might not expect or at least require it, naturalism (at least, *basic* naturalism) would not expect the physical universe to have a beginning at all. Thus, the evidence that we actually have of a temporally finite universe is better explained by theism than by basic naturalism.

14 Dicke et al., "Cosmic Black-Body Radiation," 415.

15 Eddington, "The End of the World," 450. Eddington was raised a Quaker and may have retained some religious sensibilities or even theistic belief into his adult life. Nevertheless, in his work as an astronomer he was a functional materialist, accepting methodological materialism as a normative canon of method. Thus, he would have found a picture of the universe that was effectively impossible to explain materialistically "repugnant."

16 There may now be reasons to think the singularity theorems are more well-grounded than current opinion in physics suggests, since, as I show in Chapter 16 (pp. 341–42), the inflationary models that justify doubting the applicability of the singularity theorems have encountered significant explanatory difficulties.

17 Recall that a prior probability in Bayesian analysis is the probability of some hypothesis, P(H), before some body of new or relevant evidence is taken into account. Usually that means that estimates of prior probabilities are based upon the background knowledge that we have before we begin to assess a hypothesis with respect to such new evidence. Nevertheless, when assessing competing worldviews or metaphysical hypotheses, some philosophers think it entirely legitimate to assume that no worldview should be considered any more intrinsically or inherently probable than another. In the case of theism, the situation is complicated; some philosophers argue that considerations of symmetry dictate equal priors for theism and atheism, while others appeal to simplicity considerations or entailments to argue that one of these views should be given at least a modest preference over the other.

These issues are subtle and complex, but for the purposes of our argument we need not resolve them. Virtually no one argues that the ratio of the prior probabilities for theism and atheism, $P(T) \mid P(\sim T)$, is very far from 1. But if the argument of this book is correct, the ratio of the cumulative likelihoods, $P(E_1 \& \ldots \& E_n \mid T) \mid P(E_1 \& \ldots \& E_n \mid \sim T)$, is very large indeed, large enough to swamp even a hefty skeptical ratio of the priors. In the absence of a compelling reason to think that the prior probabilities are wildly skewed against theism, the empirical evidence that we marshal *a posteriori* will and should predominate in assessment of the plausibility of competing hypotheses. Insofar as the evidence considered in Chapters 4–6 is much more strongly expected given theism than materialism or naturalism, that evidence not only confers greater epistemic support on theism than materialism, but, as I show in subsequent chapters, the whole ensemble of evidence under consideration (in Chapters 4–10) also makes theism the most reasonable thing to believe, all things considered.

18 Personal interview with William Lane Craig, July 1994, Cambridge, England.

19 Hawking and Penrose, "The Singularities of Gravitational Collapse and Cosmology."

20 Moreland, *Scaling the Secular City*, 42.

21 As Moreland notes: "Naturalists like John Searle, John Bishop, and Thomas Nagel all admit that our basic concept of action [i.e., human choice or decision] is itself a libertarian one. Searle goes so far as to say that our understanding of [physical or material] event

causality is conceptually derived from our first-person experience of our own causation. There is a major tradition in philosophy that agent causation is clearer and more basic than event [i.e., physical or material] causation, and it may actually be that if any sort of causation is inscrutable, it is [such] event causation" (Moreland, "The Explanatory Relevance of Libertarian Agency as a Model of Theistic Design," 273–74).

22 Moreover, considerable neurophysiological evidence now supports the reality of human libertarian agency or some form of mind-body dualism. See, for example, Custace, *The Mysterious Matter of Mind*; Beauregard and O'Leary, *The Spiritual Brain: A Neuroscientist's Case for the Existence of the Soul.* Nevertheless, whatever one thinks about the debate between mind-body dualists and physicalists, it remains the case that simply positing libertarian free agency to explain the beginning of the universe does circumvent the explanatory conundrum confronting naturalists or materialists. As J. P. Moreland explains, "The only way for the first event to arise spontaneously from a timeless, changeless, space-less state of affairs, and at the same time be caused, is this—the event resulted from the free act of a person or agent. In the world, persons or agents spontaneously act to bring about events. I myself raise my arm when it is done deliberately. There may be necessary conditions for me to do this (e.g., I have a normal arm, I am not tied down), but these are not sufficient. The event is realized only when I freely act. Similarly, the first event [i.e., the beginning of the universe] came about when an agent freely chose to bring it about, and this choice was not the result of other conditions which were sufficient for that event to come about" (Moreland, *Scaling the Secular City*, 42).

23 Haldane, *Possible Worlds and Other Essays*, 209.

24 As J. P. Moreland has explained in response to materialists who deny the intelligibility of a personal agent cause as the best explanation for the beginning of the universe: "The Divine creation of the initial singularity is precisely analogous to human libertarian acts; for example, both involve first movers who initiate change. There is nothing particularly mysterious or inscrutable about the latter, so in the absence of some good reason to think that there is some specific problem with the initial Divine creation, the charge of inscrutability is question begging. Moreover, we understand exercises of [will] power primarily from introspective awareness of our own libertarian acts, and we use the concept of action so derived to offer third-person explanations of the behavior of other human persons. There is nothing obscure about such explanations for the effects produced by other finite persons, and I see no reason to think that this approach is illicit in the case of Divine initial creation" (Moreland, "The Explanatory Relevance of Libertarian Agency as a Model of Theistic Design," 273–74). See also: Moreland, J.P. "Agent Causation and the Craig/Grünbaum Debate about Theistic Explanation of the Initial Singularity," 539–54.

25 Anthony Aguirre and John Kehayias, "Quantum Instability of the Emergent Universe."

26 Spinoza, like many Eastern philosophers, equates God and nature, but, unlike many Eastern pantheistic philosophers, does regard God as possessing rationality as opposed to simply constituting the impersonal unity or oneness of all reality. See Kaufmann and Baird, *Philosophic Classics: From Plato to Nietzsche*, 478, 479–86.

27 Ferm, *An Encyclopedia of Religion*, 557–58.

28 See Sire, *The Universe Next Door*, 118–35.

Chapter 13: The God Hypothesis and the Design of the Universe

1 "*The Abrams Report* for September 29, 2005."

2 Dawkins, *River Out of Eden*, 133.

3 Carroll, "Turtles Much of the Way Down."

4 Some object to the fine-tuning argument by asking why God would need to fine-tune a universe at all. "Surely," they ask, "if God wanted a life-permitting universe, God wouldn't perch it on a razor's edge." The argument presented here circumvents this objection by showing that the inference to a transcendent designer better explains the evidence we have (based upon what we know about the features that designed systems

typically exhibit) without speculating about why a designing intelligence would have chosen to fine-tune the universe the way it did.

5 Physicist Luke Barnes formulates the argument slightly differently. Rather than focusing on the probability of the fine tuning per se given either theism or naturalism, he focuses on the probability of a *life-permitting universe* given either theism or naturalism (given what we know about the fine tuning). He articulates the argument as follows:

Premise One: For two theories T_1 and T_2, in the context of background information B, if it is true of evidence E that $P(E \mid T_1B) >> P(E \mid T_2B)$, then E strongly favors T_1 over T_2.

Premise Two: The likelihood that a life-permitting universe exists on [given] naturalism is vanishingly small.

Premise Three: The likelihood that a life-permitting universe exists on [given] theism is not vanishingly small.

Conclusion: Thus, the existence of a life-permitting universe strongly favors theism over naturalism.

Barnes has written an excellent article titled "A Reasonable Little Question: A Formulation of the Fine-Tuning Argument" in which he develops this argument by defending each of the above premises. He especially focuses on using what physicists know about the quantitative precision of the physical constants to support Premise 2 above. I draw on his work to support a slightly modified version of that premise (in my version of the fine-tuning argument) in this chapter. Barnes then uses his formulation of the argument to respond to many common objections to the fine-tuning argument for the existence of God. Among others, he addresses such objections as: (a) deeper physical laws explain the fine tuning; (b) the multiverse explains the fine tuning; and (c) we can't know whether God would be inclined to create a finely tuned or a life-permitting universe since what God would do is inscrutable.

I find both Barnes's argument as he formulates it and his responses to these objections persuasive and compelling. Nevertheless, I have chosen to formulate the argument from fine tuning slightly differently in this chapter. I have emphasized what we know from our uniform and repeated experience about characteristic features of designed objects to suggest that we do have a strong empirically based reason to expect that a designing mind would fine-tune the parameters necessary for life. Recall that fine tuning represents (a) a highly improbable set of conditions or values that (b) exemplify a set of functional requirements, making possible a functional or significant outcome. Recall also that intelligently designed objects and systems often exemplify precisely these features in combination. Indeed, producing finely tuned systems is one of the things that intelligent agents frequently and uniquely do. I argue further that, since fine tuning has been present from the beginning of the universe, the evidence of fine tuning points to a *transcendent* intelligence rather than an immanent one.

By arguing this way, I do not need to justify the idea that we have reason to expect God would have produced a *life-permitting* universe. Instead, I only need to justify the idea that we have reason to expect that an intelligent agent would produce *a finely tuned system*, since, again, we have ample evidence of agents doing just that. I prefer this way of making the argument, because Bayesian likelihoods (i.e., assessments of the probability of the evidence E given the hypothesis H) are determined largely by considerations of causal adequacy—that is, by reference to our knowledge of cause and effect. Making the argument this way allows us to employ our empirically based knowledge of cause and effect to suggest that the probability/expectation of the fine-tuning evidence given a design hypothesis is high, and certainly higher than the probability of that evidence given naturalism. In other words, the fine tuning *would be expected* given the activity of a preexisting designing intelligence—one that I argue on other grounds must possess the attribute of transcendence.

By contrast, Barnes must defend the proposition that a *life-permitting* universe (rather than the fine tuning necessary to produce it) would be expected given theism—or, as he puts it, "the probability that a life-permitting universe exists on theism is not vanishingly small." He gives a perfectly good justification for that proposition that turns, first, on the attributes associated with the concept of God. Thus, he also rejects the idea that the intentions of God (so conceived) would be utterly inscrutable. In any case, he shows that however much we might be uncertain about whether God would be inclined to create a life-permitting universe, we certainly have greater reason to expect such a universe given theism than naturalism.

Barnes's defense of his third premise, and his argument as a whole, is compelling. Nevertheless, I prefer to make the argument by focusing on the probability of the *fine tuning itself*, rather than the probability of a *life-permitting* universe, given theism or naturalism. I do so, because this way of making the argument appeals directly to our uniform and repeated experience, rather than just to our concept of God, to generate the Bayesian likelihoods. Indeed, whereas we have observed intelligent agents generating finely tuned systems (highly improbable arrangements of parts or conditions that exemplify a functional specification), we do not have a similar direct observation of God producing life.

Even so, I don't deny the force of Barnes's argument since, as I explained in Chapter 11, theoretical considerations can justify claims of causal adequacy in other ways. Indeed, I think he answers the "God's intentions are inscrutable" objection persuasively. Thus, I regard his argument and the one presented in this chapter as complementary ways of reaching the same conclusion.

Several other ways of making the fine-tuning argument also have formidable force. See, e.g., Swinburne, *The Existence of God*; Leslie, *Universes*; Craig, "Design and the Anthropic Fine-Tuning of the Universe"; Collins, "The Teleological Argument." For a philosopher of science who makes the argument by assessing the probability of the fine tuning given theism as opposed to naturalism (as do I), see Roberts, "Fine-Tuning and the Infrared Bull's-Eye."

6 Crick, *Life Itself*, 88, 95–166. See also Crick and Orgel, "Directed Panspermia."
7 Hoyle and Wickramasinghe, *Evolution from Space*, 35–50.
8 Stein, *Expelled.*
9 Crick, *Life Itself*, 88.
10 Dawkins, "Ben Stein vs. Richard Dawkins Interview."
11 See also Sober, "Intelligent Design Theory and the Supernatural —The 'God or Extraterrestrial' Reply," 1–12. Sober, a philosophical naturalist who rejects the case for intelligent design, argues that *if* one does accept the argument for intelligent design in biology (from irreducible complexity), it makes more sense to affirm a supernatural designer than an extra-terrestrial one. He argues that the "minimalist case" for intelligent design when supplemented with a few additional and plausible premises (such as, for example, "the universe is finite") leads logically to the conclusion that a transcendent intelligent designer must exist.
12 Another metaphysical hypothesis that posits an immanent form of intelligence as the prime or ultimate reality is known as panpsychism. Panpsychism holds that a universal mind or ubiquitous consciousness present in the universe, and partially present in each part of the material universe, underlies all of reality (Goff, Seager, and Allen-Hermanson, "Panpsychism"). Critics of panpsychism worry that it fails to give an account of how one conscious mind—for example, my mind—differs from another, say, yours. They ask: if all matter is part of the same universal consciousness, what makes one mind different than another? On what basis can our ordinary experience of having individual minds separate from each other be affirmed if all that ultimately exists is a single universal mind? A popular form of panpsychism among some analytical philosophers known as emergent panpsychism addresses this dilemma by arguing that the smallest constitutive parts of the material universe have little "droplets" of proto-consciousness, but as more complex material arrangements emerge more developed forms of consciousness arise. This makes it possible to affirm a universe in which many minds evolve within one emerging universal mind. Whatever advantages panpsychism may

offer over materialism as a philosophical concept, both versions of it—i.e. straight panpsychism and emergent panpsychism—lack promise as explanations for the *origin of the material* universe and the origin of its fine tuning. Indeed, all forms of panpsychism, and especially popular emergent panpsychism, deny the existence of any transcendent conscious agent existing outside of, or prior to, matter coming into being. Instead, since consciousness and matter (or mass-energy) are co-extensive, panpsychism necessarily must affirm that mind and matter would have begun coterminously with the beginning of the universe—if, indeed, the universe had a beginning as the evidence suggests that it did. Thus, panpsychism necessarily denies any entity separate from the universe that could explain its origin or fine tuning, two of the three classes of evidence about cosmological and biological origins under examination here. And yet, as I have argued, adequately explaining the origin of the material universe and its fine tuning require positing just such a transcendent and intelligent entity.

13 Carroll, "Turtles Much of the Way Down."

14 Ugural and Fenster, "Hooke's Law and Poisson's Ratio."

15 Polanyi, "Life Transcending Physics and Chemistry," 61.

16 Polanyi, "Life's Irreducible Structure," esp. 1309.

17 Barnes, "The Fine-Tuning of the Universe for Intelligent Life," 530; Halliday and Resnick, *Physics: Part Two*, appendix B, A23.

18 Susskind, *The Cosmic Landscape*.

19 Dawkins, *River Out of Eden*, 133.

20 Recall from Chapter 8, n. 11, that the entropy required for life in a stable galaxy is not as extreme as the initial entropy required to produce the low-entropy universe that we actually have. Whereas Roger Penrose has calculated the entropy fine tuning necessary to generate our low-entropy universe as $10^{10^{123}}$ my colleague Brian Miller calculates the initial entropy to produce a life-permitting galaxy at "just" $10^{10^{98}}$. For the basis of Miller's calculations see, again, Chapter 8, n. a, at www.returnofthegodhypothesis.com/extendedresearchnotes.

21 Recall from n. 5 above that Barnes takes a slightly different tact than I do in *what* he calculates and in how he makes use of his calculation in his version of the fine-tuning argument. He calculates the probability of a *life-permitting universe* given naturalism, whereas I calculate the probability of the fine tuning given naturalism. But since a life-permitting universe also depends precisely and directly upon the fine tuning of the constants of physics and the initial conditions of the universe, the precise quantitative degree of the fine tuning also allows me to calculate the probability of observing *the fine tuning itself* given naturalism. And, of course, the two probabilities are the same. In addition, rather than arguing, as I do, that the observation of the exquisite fine tuning of the universe for life "confirms precisely what we might well expect if a purposive intelligence . . . had acted to design the universe and life," he argues that "the likelihood that a *life-permitting* universe exists on theism is *not* vanishingly small." He focuses on the probability given theism of a life-permitting universe as opposed to the fine tuning that makes a life-permitting universe possible. He also makes a more modest claim about what we have reason to expect based upon theism than I do, in part because he bases his argument on the properties associated with God, whereas I base my assessment of likelihoods on our repeated experience of the attributes (small probability specifications) of designed objects and systems that relevantly similar intelligent agents are known to produce. Using Bayesian analysis we both come to similar conclusions. He argues that the probability of a *life-permitting universe* (given the high degree of fine tuning we observe) is much less expected (and less probable) given naturalism than theism. I argue that the probability of observing the *extreme degree of fine tuning* that we do in the universe is much less expected (and less probable) given naturalism than theism. Consequently, we both agree the fine tuning provides greater evidential support for theism than naturalism.

22 See n. 5.

23 See Sire, *The Universe Next Door*, 119–35.

24 Indeed, according to some forms of Eastern pantheism (for example, the Sankara school of Vedanta Hinduism), even our own awareness of ourselves as conscious minds separate from the oneness of nature (*brahman*) represents an illusion or false consciousness. For an amplifying discussion of this point, see Chapter 13, n. a, at www.returnofthegodhypothesis/extendedresearchnotes; and Kohler, *Asian Philosophies*, 81.
25 Sean Carroll and others object to this by arguing that there is no reason for the concept of a cause to be extended beyond the physical universe. Yet holding that position denies the principle of sufficient reason.

Chapter 14: The God Hypothesis and the Design of Life

1 See Lamoureux, "Evolutionary Creation." In the following notes, I will refer to this online essay. See also his recent book-length treatment, *Evolutionary Creation: A Christian Approach to Creation.*
2 Lamoureux, "Evolutionary Creation," 2.
3 Quoted in Woodward, "The End of Evolution," 33.
4 Lamoureux, "Evolutionary Creation," 2.
5 Lamoureux, "Evolutionary Creation," 3.
6 Lamoureux, "Evolutionary Creation," 2.
7 Lamoureux, "Evolutionary Creation," 1n1.
8 The information in other stretches of DNA—specifically, that in the nonprotein-coding regions—helps to regulate the timing of expression of the information contained in the coding regions. In addition, higher levels of "ontogenetic" (beyond the genes) information are also stored in cytoskeletal arrays, the distribution of membrane targets, and supracellular structures, tissues, and organs. These more structural forms of information also play crucial roles in the regulation and expression of specifically genetic information. Ontogenetic information necessary for animal development is also transmitted between cells by sugar-signaling molecules via the "sugar code." See Wells, "Membrane Patterns Carry Ontogenetic Information That Is Specified Independently of DNA"; Meyer, *Darwin's Doubt*, 271–87.
9 Dawkins, *River Out of Eden*, 17; Gates, *The Road Ahead*, 228; Hood and Galas, "The Digital Code of DNA."
10 Clearly, these two possibilities are contradictory. If *all* the information necessary to produce either the first life or new forms of life was present at the start of the universe, then the evolutionary process would not need to generate any new genetic information or novelty. Instead, it would just be unfolding what was already "in" the initial conditions of the universe. But if *new* information was generated, then clearly the initial and boundary conditions must have lacked at least some of the information needed to generate life or build new forms of life.
11 Lamoureux, "Evolutionary Creation," 1.
12 For a definition of specification, see Dembski, *The Design Inference*, 1–66, 136–74.
13 Eigen, *Steps Towards Life*, 12. Eigen's statement also contradicts what is known as algorithmic information theory. Algorithmic information theory states that the amount of information or data that a system outputs *cannot* exceed the amount input into the system or the amount in the algorithm that operates upon the system (Chaitin, *Algorithmic Information Theory*).
14 Dretske, *Knowledge and the Flow of Information*, 12.
15 Alberts et al., *Molecular Biology of the Cell*, 105.
16 Polanyi, "Life Transcending Physics and Chemistry"; see also "Life's Irreducible Structure," esp. 1309.
17 Küppers, *Information and the Origin of Life*, 170–72; also Thaxton and Bradley, "Information and the Origin of Life"; also Thaxton, Bradley, and Olson, *The Mystery of Life's Origin*, 24–38.
18 Kok, Taylor, and Bradley, "A Statistical Examination of Self-Ordering Amino Acids in Proteins."
19 Lamoureux, "Evolutionary Creation," 3.

20 Tompa and Rose, "The Levinthal Paradox of the Interactome," 2074.

21 Tompa and Rose themselves think that the cell must have emerged as the result of some "preferred pathways" or by an "iterative hierarchic assembly of its component sub-assemblies" as opposed to either a straightforward deterministic (or self-organizational) process or a random process. Nevertheless, they posit no specific process that could overcome the combinatorial complexity that they describe. Instead they state: "The central biological question of the twenty-first century is: how does a viable cell emerge from the bewildering combinatorial complexity of its molecular components? Here, we estimate the combinatorics of self-assembling the protein constituents of a yeast cell, a number so vast that the functional interactome could only have emerged by iterative hierarchic assembly of its component sub-assemblies. *We surmise that this non-deterministic temporal continuum could not be reconstructed de novo under present conditions*" ("The Levinthal Paradox of the Interactome," 2074, emphasis added).

22 Recall that I also showed in Chapter 10, and in *Darwin's Doubt* (chaps. 8–16), that all currently proposed evolutionary processes fail to account for the large increases in genetic and epigenetic forms of information necessary to build new forms of life after the beginning of the universe.

23 Shannon describes this process of error correction using a correction channel as follows: "We consider a communication system and an observer (or auxiliary device) who can see both what is sent and what is recovered (with errors due to noise). This observer notes the errors in the recovered message and transmits data to the receiving point over a 'correction channel' to enable the receiver to correct the errors." He then proposes his tenth theorem: "*Theorem 10:* If the correction channel has a capacity equal to $H_y(x)$ it is possible to so encode the correction data as to send it over this channel and correct all but an arbitrarily small fraction ε of the errors. This is not possible if the channel capacity is less than $H_y(x)$." His accompanying diagram, Figure 8, makes clear that the correction channel supervenes over the transmission channel and often depends upon an observer to detect deviations from the original transmission of information ("A Mathematical Theory of Communication").

24 W. Ross Ashby's "law of requisite variety" advanced a similar (indeed, mathematically "isomorphic") principle that he discovered in the context of "self-organization" theory. Ashby's principle states that the control or design of an informational process depends on a correction channel that has a capacity *equal to or greater than* all the possible states that a system can adopt. For even small physical systems the number of possible states can be hyperastronomical (a problem known in control theory as the "curse of dimensionality"; "Requisite Variety and Its Implications for the Control of Complex Systems").

25 Some physicists have argued against an indeterministic and probabilistic interpretation of quantum mechanics. Consequently, they regard quantum indeterminacy as only apparent and not real. The small minority of physicists who hold to the Bohmian interpretation of quantum mechanics, for example, argue that "hidden variables" follow deterministic laws that drive the evolution of quantum states (Vaidman, "Quantum Theory and Determinism"). Therefore, on this view, measurements that appear to result from random events actually stem from the hidden variables changing with time according to some law or algorithm. This view, if true, could be used to challenge the argument presented against front-loaded design in this chapter. Some might suggest, for example, that an omniscient God could have set all of the hidden variables in some region of space at the start of the universe to the specific values needed to ensure that natural processes would generate a cell billions of years in the future. Therefore, the information required to build the first cell would not need to enter the biosphere as the result of a later direct action or "intervention" of an intelligent agent. This way of formulating the front-loaded design idea might seem reasonable at first, but it is

implausible due to the chaotic dynamics that govern the interactions of large systems of particles. For a more complete explanation as to why, see Chapter 14, n. a, at www.returnofthegodhypothesis/extendedresearchnotes. See also Dellago and Posch, "Kolmogorov-Sinai Entropy and Lyapunov Spectra of a Hard-Sphere Gas").

26 In classical theism, the omniscience and omnipotence of God are closely related doctrines. God is omniscient in part because God is omnipotent. Consequently, some theists have argued that God might be causing the collapse of the wave function as a way of understanding both God's omniscience and the basis of the regularity of natural law despite the underlying stochastic nature of quantum processes. By contrast, a deistic God who does not exercise omnipotence over nature after the beginning would seem to lack the attributes (immanence and omnipotence) necessary to omniscience, at least, in a world of quantum fluctuations and indeterminacy such as ours.

27 There may be an even deeper problem with this whole line of thinking. The front-loaded design hypothesis of Denis Lamoureux and others seems to assume that life could be generated from an essentially computational process. In short, it seems to assume the validity of what is known as the "Church-Turing conjecture" in computer science, which asserts that natural laws and processes can be represented as a computational process. For a discussion of why this tacit assumption of front-loaded design models fails, see Chapter 14, n. b, at www.returnofthegodhypothesis/extendedresearchnotes.

28 According to modern quantum theory, the interactions and evolution of subatomic particles and energy in the universe do not operate like large-scale objects such as billiard balls, which follow clear trajectories and interact predictably according to deterministic laws. At microscopic levels, a physical system must be described quantum mechanically using probability distributions describing the probability of a given state of affairs arising from some prior state or condition. For example, unlike billiard balls interacting deterministically in accord with the law of conservation of momentum, the angle at which a subatomic particle deflects off of another much larger particle cannot be exactly known beforehand. Instead, physicists can only calculate the probability of that particle adopting a particular angle of refraction. Likewise, an atom in an excited energy state will eventually drop to a lower energy state and release a photon. The time required for the event to occur, the specific final energy level, and the direction of the released photon cannot be determined or predicted, only the probabilities of the allowed outcomes.

29 Briggs, "Science, Religion Are Discovering Commonality in Big Bang Theory."

30 Sandage, "A Scientist Reflects on Religious Belief," 53.

31 In addition to panpsychism (see Ch. 13, n. 12), I'm often asked about whether the worldview or metaphysical hypothesis known as panentheism can explain the evidence concerning cosmological and biological origins discussed in this book. Panentheism comes in different varieties, but it's most commonly associated with the American philosopher and theologian Charles Hartshorne (see Hartshorn, *The Divine Relativity*). Like theism, panentheism, as developed by Hartshorne, holds that a personal God exists and that the physical universe depends upon God and can't exist without God. Nevertheless, unlike classical or biblical theism, Hartshorne's panentheism also affirms that God depends in some sense upon the universe and can't exist without it. Indeed, Hartshorne envisions the physical world and God as simultaneously "co-evolving."

Clearly, panentheism, as articulated by Hartshorn, would fail as an explanation for the origin of the universe itself. If God's existence depends upon the universe, then until the universe comes into existence, no God of the panentheistic variety would have yet existed. But since the universe appears to have come into existence a finite time ago, a panentheistic God could not have acted to cause the origin of that universe, since God's own existence depends upon the universe itself already existing.

Similarly, since the fine tuning of the universe has existed from the beginning of the universe, and since God as conceived by Hartshorn has no existence independent of the universe, a panentheistic God cannot be invoked as either a logically or temporally prior entity capable of causing or selecting the fine-tuning parameters that apply to the laws and constants of physics and the initial conditions of the universe. Instead, since God's existence, again, depends upon the universe, it is not clear that it could explain either the temporal beginning of the universe or the features of the universe that were set from the beginning.

A panentheistic God might be posited as part of a co-evolutionary process that produces new forms of life. Nevertheless, one could also argue that such a thesis fails the test of experience. We have a great breadth of experience showing that intelligent agents can and do generate specified information of the kind that is present in living systems. Nevertheless, we do not have experience of designing agents changing in their fundamental nature as the result of generating such information or designing technological objects. This may be more debatable, but once what I mean by "in their fundamental nature" is tightly defined, it appears to me to be a quite defensible statement.

In any case, panentheism as conceived by Hartshorn clearly cannot invoke a truly independent or transcendent intelligence as the cause of the origin and fine tuning of the universe and thus lacks explanatory power with respect to at least these two key facts in need of explanation. For a thorough exposition and critique of Hartshorn's panentheism, also known as process theology, see Richards, *The Untamed God*, 172–94. For a discussion of some contemporary classical theists who also use the term "panentheism" to describe their view of God, see Chapter 14, n. c, at www.returnofthegodhypothesis.com /extendedresearchnotes.

32 One objection to all theistic arguments in support of God's existence is the well-known problem of evil. Atheistic critics of theism pose this as what philosophers call a "defeater" argument. They contend that the existence of evil, both human moral evil and so-called natural evil, renders belief in the existence of God, or at least a benevolent God, logically incoherent—thus, "defeating" theistic arguments for God's existence. Atheists pose a familiar dilemma to support this claim: A benevolent and all-powerful God would not have allowed evil in the world. Since there *is* evil in the world, God is either not good, not all powerful or—more likely—does not exist.

Since at least the time of St. Augustine (or the writing of the book of Job), Christians, Jews, and other theists have answered this objection with the classical free will defense. They have insisted that the existence of human moral evil in the world is consistent with the existence of God if one considers, first, that God wanted to create human beings in God's own image with genuine free will; and second, that God clearly thought it better to make a world in which human beings could exercise their freedom, even if they might use it badly, rather than to create a world in which human beings were compelled as mere puppets to do only what God thought best. There is much to say about this philosophical and theological issue. Nevertheless, the free will defense has seemed to me and many theists a satisfactory response to the philosophical problem of evil. It certainly defeats the atheistic defeater argument of evil by showing that it is possible to reconcile belief in the existence of the omnipotence and benevolence of God with the presence of human moral evil in the world.

But what about the problem of natural evil or what is sometimes called "malevolent design" in nature? This argument has often seemed more troubling for theists and more difficult to answer. Atheists and scientific materialists have often pointed to the existence of virulent strains of bacteria or killer viruses as inconsistent with the existence of an intelligent designer, or at least a benevolent designer or creator. Answering this objection completely would take another book and lies beyond the scope of this work.

Nevertheless, I offer a few thoughts that I think can establish a framework for addressing the objection of natural evil and for showing that the existence of natural evil is not necessarily inconsistent with the theory of intelligent design, a larger God hypothesis, or even a belief in the existence of a benevolent designer or creator.

Clearly, the problem of natural evil only poses a problem for those who want to affirm, as I do, the benevolence of the designing intelligence responsible for life or a God such as the one the Judeo-Christian scriptures affirm. Nevertheless, those same Judeo-Christian scriptures, and what they teach about God and the created order, provide explanatory resources for reconciling the presence of natural evil in the world with the existence of a benevolent designer or creator. In other words, Judeo-Christian proponents of intelligent design have a framework for answering this objection that purely secular or nonreligious proponents of the theory of intelligent design may not.

Based on the Judeo-Christian scriptures, one should expect to find not one, but two classes of phenomena in nature. Indeed, one should expect to find evidence of intelligent design and goodness in the creation, but also evidence of subsequent decay and degradation.

Concerning the first expectation, the Judeo-Christian scriptures clearly affirm that God's original design of the universe and life was "good" and even beautiful. And, of course, there are many such evidences of good design in living systems and the universe (see Chapters 7–10) and much beauty to enjoy in the natural world. Thus, a significant body of evidence supports the hypothesis that a benevolent intelligent creator designed the natural world.

Nevertheless, there are aspects of nature, particularly in the living realm, such as virulent strains of bacteria or viruses, that do not promote human flourishing, but instead disease and suffering. Yet, this too is not unexpected from the standpoint of a specifically Judeo-Christian version of theism or by proponents of intelligent design (or a larger God hypothesis) who hold this worldview. The Judeo-Christian scriptures not only teach that God created the world and pronounced it good; they also teach that something went wrong that adversely affected both the human moral condition and the natural order. The scriptures also provide a backstory, whether understood mytho-poetically or more strictly historically, explaining in part why and how this disruption to the original created order occurred.

In any case, based on the Judeo-Christian scriptures we should not only expect to see evidence of an intelligent and good original design, but also evidence of subsequent decay in nature and living systems. The entropy-maximizing (order-destroying) processes to which all physical systems are subject may well be considered evidence confirming this expectation. Moreover, at the molecular level in living systems, biologists are increasingly discovering evidence of both elegant aboriginal design—in, for example, the information-bearing biomacromolecules and information-processing systems in cells as well as the miniature machines and circuitry in cells and of the decay of those systems, often via mutations.

Intriguingly, microbiologists who study virulence increasingly recognize mutational degradation and loss of genetic information, or the lateral transfer of genetic information out of its original context, as the mechanisms by which virulent strains of bacteria emerge. [See, for example, Monday et al., "A 12-base-pair Deletion in the Flagellar Master Control Gene flhC Causes Nonmotility of the Pathogenic German Sorbitol-fermenting Escherichia coli O157:H-strains," 2319–27; Minnich and Rohde, "A Rationale for Repression and/or Loss of Motility by Pathogenic *Yersinia* in the Mammalian Host," 298–310.] Moreover, virulence experts document that such informational losses or transfers—losses or mutations that, from an intelligent design perspective, reverse or alter the original creative acts that made life possible—are responsible for the emergence of the harmful bacteria that cause human suffering. For example, *Yersinia pestus*, the microorganism that caused the plague, arose as the result of four or five identifiable mutations of various kinds during human history, altering an innocuous bacterium for which humans had an in-built immune response into a killer bug [Rasmussen et al., "Early Divergent Strains of *Yersinna pestus* in Eurasia 5000 Years

Ago." 571–82]. As University of Idaho microbiologist Scott Minnich explained to me in a 2020 personal interview, "With molecular techniques and DNA sequencing we have in the last 10 years shown that the plague 'evolved'—or rather devolved—from an innocuous progenitor strain of bacteria."

Thus, just as the bursts of novel biological information that occur in the generation of new forms of life give evidence of the activity of a designing intelligence, the mutations that degrade or alter that information show subsequent processes of decay at work in living systems after their original design. That we see evidence of both good design and subsequent decay, and that we further recognize that processes of decay, not the aboriginal design of living systems, are responsible for human suffering, is precisely what we should expect to see based on a Judeo-Christian understanding of the natural world—a natural world that, as one biblical book puts it, is in "bondage to decay" (Rom. 8:21). Indeed, if the Judeo-Christian account is correct, we should positively *expect* to find tragic natural evils in the world around us. That expectation should temper any surprise we might otherwise have felt when, in fact, we do. Thus, our encounter with such natural evil actually provides evidential support for the Judeo-Christian understanding of nature considered as a kind of metaphysical hypothesis. It certainly shows that the existence of natural evil is not logically incompatible with belief in God. Those who argue otherwise fall into common logical fallacy. For a discussion of that fallacy, and how understanding it helps answer the atheistic argument from natural evil, see Chapter 14, n. d, at www.returnofthegodhypothesis.com/extendedresearchnotes.

Chapter 15: The Information Shell Game

1 Nagel, "Books of the Year."
2 Stephen Fletcher (December 4, 2009), *The Times Literary Supplement* 5566 (letter to the editor): 6.
3 Marshall, "When Prior Belief Trumps Scholarship."
4 Pera, *The Discourses of Science.*
5 Fletcher (December 4, 2009), *The Times Literary Supplement* 5566 (letter to the editor): 6.
6 Nagel, "Books of the Year."
7 Fletcher (December 4, 2009), *The Times Literary Supplement* 5566 (letter to the editor): 6.
8 Thomas Nagel (December 11, 2009), *The Times Literary Supplement* 5567 (letter to the editor): 6.
9 John Walton (December 11, 2009), *The Times Literary Supplement* 5567 (letter to the editor): 6.
10 Stephen Fletcher (December 18, 2009), *The Times Literary Supplement* 5568 (letter to the editor): 6.
11 Stephen C. Meyer (January 15, 2010), *The Times Literary Supplement* 5572 (letter to the editor): 6.
12 De Duve, *Blueprint for a Cell*, 187.
13 Powner, Gerland, and Sutherland, "Synthesis of Activated Pyrimidine Ribonucleotides in Prebiotically Plausible Conditions."
14 Lincoln and Joyce, "Self-Sustained Replication of an RNA Enzyme."
15 Stephen Fletcher (February 5, 2010), *The Times Literary Supplement* 5575 (letter to the editor): 6.
16 In February of 2010, I wrote another letter in response to Fletcher's third letter; the *TLS* did not publish it. They had understandably had enough!
17 Berlinski, "Responding to Stephen Fletcher's Views in *The Times Literary Supplement* on the RNA World."
18 Tour, "Animadversions of a Synthetic Chemist;" Tour, "Time Out"; see also Tour, "An Open Letter to My Colleagues."
19 Marshall, "When Prior Belief Trumps Scholarship."

20 Specifically, Marshall writes, "But today's GRNs have been overlain with half a billion years of evolutionary innovation (which accounts for their resistance to modification), whereas GRNs at the time of the emergence of the phyla were not so encumbered" ("When Prior Belief Trumps Scholarship").
21 According to Eric Davidson: "There is always an observable consequence if a dGRN subcircuit is interrupted. Since these consequences are always catastrophically bad, flexibility is minimal, and since the subcircuits are all interconnected, the whole network partakes of the quality that there is only one way for things to work. And indeed the embryos of each species develop in only one way" ("Evolutionary Bioscience as Regulatory Systems Biology," 40). See also the discussion in *Darwin's Doubt*, 264–70.
22 Peter and Davidson, *Genomic Control Processes.*
23 For a diagrammed schematic of the dGRN circuitry in the purple sea urchin, *Strongylocentrotus purpuratus*, see Fig. 13.4 in *Darwin's Doubt*, 266.
24 Davidson, "Evolutionary Bioscience as Regulatory Systems Biology," 38.
25 Davidson, "Evolutionary Bioscience as Regulatory Systems Biology," 38.
26 Davidson and Erwin, "An Integrated View of Precambrian Eumetazoan Evolution," esp. 72.
27 As Davidson notes, "Contrary to classical evolution theory, the processes that drive the small changes observed as species diverge cannot be taken as models for the evolution of the body plans of animals" (*The Regulatory Genome*, 195; "Evolutionary Bioscience as Regulatory Systems Biology," 35–36).
28 Davidson, "Evolutionary Bioscience as Regulatory Systems Biology," 40.
29 Marshall, "Nomothetism and Understanding the Cambrian 'Explosion'."
30 Marshall and Valentine, "The Importance of Preadapted Genomes in the Origin of the Animal Bodyplans and the Cambrian Explosion," esp. 1195–96.
31 Marshall, "Explaining the Cambrian 'Explosion' of Animals," 366.
32 Marshall and Valentine, "The Importance of Preadapted Genomes," 1189.
33 Meyer, *Darwin's Doubt*, 191.
34 See Shannon, "A Mathematical Theory of Communication."
35 Reidhaar-Olson and Sauer, "Functionally Acceptable Solutions in Two Alpha-Helical Regions of Lambda Repressor"; Axe, "Estimating the Prevalence of Protein Sequences Adopting Functional Enzyme Folds."
36 As it happens, recent comparative studies of the genetic diversity of the animal phyla have confirmed my original contention rather than Marshall's proposal. These studies have established that many thousands of novel genes did arise abruptly during the Cambrian explosion in order to build the first animals. As Jordi Paps and Peter Holland, the authors of one study, put it: "Contrary to the prevailing view, this [study] uncovers an unprecedented increase in the extent of *genomic novelty* during the origin of the metazoans," that is, during the period of or just before the appearance of the disparate body plans in the Cambrian explosion (emphasis added). The authors concluded that "internal genomic changes were as important as external factors in the emergence of animals" ("Reconstruction of the Ancestral Metazoan Genome Reveals an Increase in Genomic Novelty"). A similar study has recently confirmed the same pattern of explosive gene origination just before or coincident with the origin of land plants; see Bowles, Bechtold, and Paps, "The Origin of Land Plants Is Rooted in Two Bursts of Genomic Novelty."
37 Charles Marshall and Stephen Meyer debate on *Unbelievable* with Justin Brierley, November 30, 2013, https://www.youtube.com/watch?v=6yOCpb0wBPw. For a transcript of the relevant portions of this dialogue, see Chapter 15, n. a, at www.returnofthegodhypothesis/extendedresearchnotes.
38 Haarsma makes this claim in response to me in "Response from Evolutionary Creation" (224), her essay in the book *Four Views on Creation, Evolution, and Intelligent Design*, to which both she and I contributed.

39 Durston et al., "Measuring the Functional Sequence Complexity of Proteins"; Reidhaar-Olson and Sauer, "Functionally Acceptable Solutions in Two Alpha-Helical Regions of Lambda Repressor"; Taylor et al., "Searching Sequence Space for Protein Catalysts"; Yockey, "A Calculation of the Probability of Spontaneous Biogenesis by Information Theory."
40 Axe, "Estimating the Prevalence of Protein Sequences Adopting Functional Enzyme Folds."
41 Tokuriki and Tawfik, "Stability Effects of Mutations and Protein Evolvability"; Tokuriki et al., "The Stability Effects of Protein Mutations Appear to Be Universally Distributed"; see also Bershtein et al., "Robustness–Epistasis Link Shapes the Fitness Landscape of a Randomly Drifting Protein"; Lundin et al., "Experimental Determination and Prediction of the Fitness Effects of Random Point Mutations in the Biosynthetic Enzyme HisA."
42 Bechly, Miller, and Berlinski, "Right of Reply." See also Miller, "Protein Folding and the Four Horsemen of the Axocalypse." Miller, "A Dentist in the Sahara: Doug Axe on the Rarity of Proteins Is Decisively Confirmed."
43 Chiarabelli et al., "Investigation of De Novo Totally Random Biosequences, Part II"; Ferrada and Wagner, "Evolutionary Innovations and the Organization of Protein Functions in Sequence Space."
44 Venema, in: Venema and McKnight, *Adam and the Genome*, 85.
45 Venema, in: Venema and McKnight, *Adam and the Genome*, 85.
46 Venema, in: Venema and McKnight, *Adam and the Genome*, 85.
47 Negoro et al., "X-ray Crystallographic Analysis of 6-Aminohexanoate-Dimer Hydrolase."
48 Indeed, the close sequence identity between nylonase and its cousin suggests the genes for both proteins arose from a common ancestral gene, which also would have coded for a protein with nylonase activity. It follows that the mutations that produced the gene for nylonase did not generate a "brand-new" functional gene and protein, but instead merely optimized a *preexisting* function in a similar protein using the same fold. Kato et al., "Amino Acid Alterations Essential for Increasing the Catalytic Activity of the Nylon-Oligomer-Degradation Enzyme of *Flavobacterium* sp."
49 Negoro et al., "X-ray Crystallographic Analysis of 6-Aminohexanoate-Dimer Hydrolase."
50 It is worth pointing out that a close reading of Venema's critique shows that he does not understand protein structure. To see why, see Chapter 15, n. b, at www.returnofthe godhypothesis/extendedresearchnotes.
51 Dawkins, comment 14 in Coyne, "God vs. Physics."
52 For Moran's position, see Moran, "You Need to Understand Biology If You Are Going to Debate an Intelligent Design Creationist"; for Myers's postmortem, see Myers, "A Suggestion for Debaters."
53 Axe, "Estimating the Prevalence of Protein Sequences."
54 See Meyer, *Darwin's Doubt*, 292–335, and Meyer, *Signature in the Cell*, 272–323.

Chapter 16: One God or Many Universes?

1 In inflationary cosmology, the production of bubble universes was a natural consequence of the quantum character of the proposed inflationary mechanism. Once this was realized, however, this feature of the model was put to use to explain away initial-condition fine tuning. String theory had a similarly innocuous origin as well—first as an attempt to develop a theory of the strong nuclear interaction and then as a promising candidate for a "theory of everything"—before being appropriated as an explanation for the fine tuning of the laws and constants of nature.
2 I've chosen, by the way, to defer addressing these other more abstract possible explanations till now for two reasons. First, I wanted to give readers a chance to see just how unexpected the main discoveries about the complexity of life and the origin and fine tuning of the universe really are from a standard naturalistic or materialistic point of view—and thus to feel the force of the core case for theism as a better explanation than

the most "natural" forms of naturalism. Second, I wanted to leave some of the more difficult concepts and technical material to this final section of the book to allow readers who feel they've already gone deep enough the opportunity to skim or skip these chapters and other, more technically minded readers the opportunity to dig deeper and evaluate the strength of my case for theism against even the most exotic naturalistic cosmologies and theories.

3 *Lexico*, https://en.oxforddictionaries.com/definition/exotic.

4 Recall that physicists disagree about how precisely to define these different models. See my discussion about these semantic differences in Chapter 8, nn. 26, 28, and 29. See also Barrow, "Anthropic Definitions," 150; Barrow and Tipler, *The Anthropic Cosmological Principle*, 16–25; Carter, "Large Number Coincidences and the Anthropic Principle in Cosmology," 291–98; Lewis and Barnes, *A Fortunate Universe*, 19.

5 Science writer Clifford Longley explains the concept this way: "There could have been millions and millions of different universes created each with different dial settings of the fundamental ratios and constants, so many in fact that the right set was bound to turn up by sheer chance" ("Focusing on Theism").

6 A few physicists have proposed that if our bubble universe bumped into another bubble universe, it would leave detectable patterns in the CMBR (Sokol, "A Brush with a Universe Next Door"). Roger Penrose has made a similar claim for his conformal cyclic cosmology (CCC) model in which the universe goes through infinitely many cycles with the future time-like infinity of each earlier iteration being identified with the big bang singularity of the next (for a popular account, see his book *Cycles of Time: An Extraordinary New View of the Universe*). He argues that observed "hot spots" in the CMBR represent evidence of interaction between the different modes of the universe in its collapsing and expanding phases. Specifically, he sees hot spots in the CMBR as evidence of the collapse of black holes prior to the beginning of our universe in its present expansion phase ("On the Gravitization of Quantum Mechanics 2"). Even so, his model does not, strictly speaking, represent a multiverse model, since the universes exist in succession, not in parallel.

7 Linde, "A New Inflationary Universe Scenario"; Guth, "Inflationary Universe"; Albrecht and Steinhardt, "Cosmology for Grand Unified Theories with Radiatively Induced Symmetry Breaking."

8 Linde, "Eternally Existing Self-Reproducing Chaotic Inflationary Universe."

9 Stenger, "Fine-Tuning and the Multiverse."

10 Physicists and philosophers call this an "observer selection effect." By this they mean we necessarily must observe a universe with features compatible with complex life forms and thus should not be surprised to find ourselves in such a universe—especially if the multiverse correctly depicts reality and various universe-generating mechanisms will eventually produce some life-permitting universe somewhere.

11 String theory was first proposed in the late 1960s to describe the strong nuclear force. Another approach, known as quantum chromodynamics, eventually proved more effective for that task, however. Then, in the 1970s, Caltech physicist John Schwarz and others noticed that string theory held promise for reconciling general relativity with quantum mechanics. That realization generated renewed interest in developing the theory.

12 Manoukian, "Introduction to String Theory."

13 The earliest version of string theory offered only a description of the bosons that carry the strong nuclear force, and it required twenty-six-dimensional spacetimes in order to work. So as initially formulated, string theory was bosonic and twenty-six-dimensional and could not account for the existence of matter! What Schwarz and his collaborators discovered as they continued to work on the theory in the 1980s was a way to extend string theory to include all matter and radiation. For a short discussion of how they did this, see Chapter 16, n. a, at www.returnofthegodhypothesis/extendedresearchnotes.

14 Dimopoulos, "Splitting Supersymmetry in String Theory."

15 Susskind, "The Anthropic Landscape of String Theory."
16 Bena and Graña, "String Cosmology and the Landscape."
17 Bousso and Polchinski, "The String Theory Landscape."
18 This postulation is highly dubitable, since there is no way of knowing how much of the string landscape will get explored by such a means. There is no *a priori* reason to suppose the process of exploring the landscape will be complete. But if it isn't significantly complete, it's unlikely that cascading down the energy landscape will generate a universe like ours. In addition, Baylor University engineering professor and information theorist Robert Marks has recently challenged the idea that the inflationary string-theory multiverse produces enough universes to generate enough "contradistinctions" to render the fine tuning in our universe probable. See Marks, "Diversity Inadequacies of Parallel Universes."
19 Ellis, "Cosmology." Readers familiar with my previous work in the philosophy of science will know that I don't think a bright line of demarcation between science and metaphysics can be drawn. Consequently, I don't think it's justified to disregard or reject a hypothesis simply because it may invoke philosophical or metaphysical ideas. We may by convention classify such hypotheses as metaphysical, but that does not mean they are necessarily false, insignificant, untestable, or beyond rational evaluation. For an extended discussion of the so-called demarcation issue and its applicability to assessing an intelligent design and/or a God hypothesis, see Chapter 16, n. b, at www.returnofthe godhypothesis/extendedresearchnotes. See also Meyer, "Sauce for the Goose"; "The Scientific Status of Intelligent Design"; "The Demarcation of Science and Religion."
20 As George Ellis has argued, "So one can motivate multiverse hypotheses as plausible, but they are not observationally or experimentally testable—and never will be. It is easy to support your favourite model over others because no one can prove you wrong—you can simply adjust its parameters to fit the latest information" ("Cosmology," 295).
21 Swinburne, *The Existence of God*, 185.
22 Swinburne, *The Existence of God*, 185.
23 For a popular account of this process, see Bousso and Polchinski, "The String Theory Landscape." For a more extended popular treatment, see Susskind, *The Cosmic Landscape*.
24 Gordon, "Postscript to Part One"; "Balloons on a String."
25 These points are explicit in a set of unpublished lecture notes that Gordon has shared with me, but implicit in a variety of Gordon's publications, for example, "Balloons on a String" and "Divine Action and the World of Science."
26 One version of string theory—known as the "cyclic ekpyrotic model"—does attempt to explain the fine tuning of both the initial conditions and the laws and constants of physics without invoking inflation. Yet it too offers a bloated ontology measured by the number of entities it must invoke to explain these two different kinds of fine tuning. For an explanation of this defect in the "cyclic ekpyrotic model," see Chapter 16, n. c, at www.returnofthegodhypothesis/extendedresearchnotes.
27 Collins, "The Fine-Tuning Design Argument."
28 Interview with Robin Collins in Strobel, *The Case for a Creator*, 145.
29 The energy associated with the inflaton field—in particular, something called the "inflation-preheating coupling parameters" required to convert inflationary energy to normal mass-energy—is also reverse-engineered (fine-tuned) by physicists modeling the origin of the universe to produce a universe similar to ours in which life would be possible (see, e.g., Kofman, "The Origin of Matter in the Universe"; DeCross et al., "Preheating after Multifield Inflation with Nonminimal Couplings."
30 Carroll and Tam, "Unitary Evolution and Cosmological Fine-tuning."
31 Rees, *Just Six Numbers*, 115.
32 Page, "Inflation Does Not Explain Time Asymmetry."
33 Personal interview with Bruce Gordon, Seattle, July 18, 2019.
34 As allowed by quantum mechanics, individual bubble universes may occasionally

"tunnel" through a potential energy barrier to a higher-energy universe that will in turn expand and then either decay or tunnel, generating yet more universes. Nevertheless, such tunneling events are extremely improbable, or "exponentially suppressed," as some theoretical physicists put it (see Linde, "Sinks in the Landscape, Boltzmann Brains and the Cosmological Constant Problem"). For a popular account of the whole process of "exploring the landscape," see Bousso and Polchinski, "The String Theory Landscape." For a more extended popular treatment, see Susskind, *The Cosmic Landscape*.

35 Smolin, *The Trouble With Physics*, xiv.

36 Gordon, "Balloons on a String," 580–81.

37 Kallosh, Kofman, and Linde, "Pyrotechnic Universe."

38 Kallosh, Kofman, and Linde, "Pyrotechnic Universe."

39 Collins, "The Fine-Tuning Design Argument," 61; see also "The Multiverse Hypothesis." Some cosmologists might argue that even the prior sources of fine tuning presupposed in the inflationary string landscape model can be explained simply by positing a mechanism for generating an infinite number of universes with different inflaton fields and shutoff parameters. To do so, they might first envision each string vacua in the landscape producing an inflaton field. They could then envision that each of these different inflaton fields would be subject to random quantum fluctuations that will produce different fields with different shutoff energies and intervals. Each such fluctuation would then produce a new universe, though in all probability not a life-conducive one. Nevertheless, if (1) an *infinite* number of such fluctuations occurred in (2) an infinite space produced from (3) an infinite singularity within either a hyperbolic or flat universe, *then* an actually infinite number of different universes would emerge, some of which would have correct inflaton shutoff energies and intervals to ensure the production of many life-conducive universes. Thus, some might argue that such an "infinite-verse" could explain the prior fine tuning of the inflaton field—if, again, one posited an infinite number of random quantum fluctuations producing an infinite number of universes with different inflaton fields and shutoff parameters. If an infinite number of universes and inflaton fields will inevitably arise, then the fine tuning required for a life-conducive universe will eventually emerge.

This speculative scenario depends upon several contestable assumptions (enumerated as 1-3 above) and does not, in any case, actually circumvent the need for prior fine tuning. Indeed, the inflaton field necessary to the universe-generating mechanism of the inflationary string landscape requires several sources of built-in fine tuning that *precede* the mechanism for producing any new bubble universes.

For example, proponents of the inflationary multiverse (and the combined string inflationary multiverse) make a number of gratuitous assumptions about the structure of our universe in order to get inflationary cosmology to mesh with general relativity. Moreover, they *must* do this because the mechanism that produces bubble universes presupposes general relativity. Thus, proponents of these models have to make specific assumptions about the nature of spacetime and reject others (Penrose, "Difficulties with Inflationary Cosmology"; The Road to Reality, 757). That's in part because there's no guarantee that any given inflaton field, when conjoined with general relativity, will actually produce inflation (Hawking and Page, "How Probable Is Inflation?"). Consequently, physicists have to select some inflationary models and exclude others based on whether they would allow inflation to occur and bubble universes to form. But this implies fine tuning in the structure of spacetime, a fine tuning that precedes the operation of any specific mechanism that could generate new universes.

In addition, explaining the homogeneity of the universe using inflaton fields also requires built-in fine tuning. To explain the homogeneity of the universe using inflationary cosmology physicists have to make assumptions about the singularity from which everything came. As Roger Penrose has pointed out, however, if the singularity were perfectly generic, expansion from it could yield many different kinds of irregular (inhomogeneous) universes, even *if* inflation had occurred ("Difficulties with Inflationary Cosmology"). Thus inflation alone, without additional assumptions about the singularity (and a corresponding "spacetime metric"), does not solve the homogeneity problem.

(For a discussion of the cosmic "chicken and egg" problem this objection to my argument creates, see Chapter 16, n. d, at www.returnofthegodhypothesis/extended researchnotes.)

Further, though the inflaton field may be conceived to generate an infinite number of universes, it doesn't generate enough of the right kind. As noted, though the decay of the inflaton field may produce bubble universes with many new initial conditions, it does not produce new universes with new laws and physical constants. Consequently, the inflationary string landscape model must rely on the string-theoretic generating mechanism to do that.

Nevertheless, the process of "exploring the landscape" will not itself produce an infinite number of new universes, but instead only a finite number corresponding to—at most—the number of solutions to the string-theoretic equations that have a positive cosmological constant. Moreover, nothing in string theory guarantees an exhaustively random search of that finite number of possible universes in the landscape. At best, the process of "cascading down the landscape" will explore a large number of those possible universes but only if the process starts with an initial high-energy compactification. But such a condition implies exquisite initial-condition fine tuning, as noted on page 339. And that fine tuning—fine tuning that may still only make a limited search through the landscape possible—necessarily precedes such a search (and precedes the generation of new bubble universes as envisioned in the combined model).

In addition, significant additional fine tuning is built into string theory itself, implying—if string theory accurately represents the universe—the existence of additional sources of fine tuning in the universe. In the late 1990s, string theorists found in their modeling that if they wrapped lines of flux around the compactified dimensions of space, they could stabilize them and ensure their continued compactification. They also found that lines of flux could only be wrapped around the compactified dimensions of space a limited number of times before they became unstable again, but that "tying them down" in specific ways ensured that the corresponding compactifications had a positive cosmological constant, thus matching a key physical feature of our universe. That string theory requires such a precise selection of parameters in order for its solutions to match the physics of our universe shows that the universe as described by string theory has contingent features that must be finely tuned to produce a universe like ours. And this implies that if string theory accurately depicts the universe, additional kinds of unexplained fine tuning must be built into it and the universe it putatively describes.

Consequently, even if the inflationary string multiverse could produce an infinity of universes, it still leaves unexplained many significant sources of prior fine tuning, a fine tuning that precedes the operation of a specific mechanism for generating new universes.

Beyond all this, inflationary cosmology presupposes fine tuning in the structure of the laws of physics themselves, a fine tuning that turns out to be a necessary condition of an efficacious inflaton field. As Robin Collins and Bruce Gordon have pointed out, the inflaton field depends upon many specific laws of physics that could exhibit different mathematical structures or relationships. For example, the mechanism for generating bubble universes depends upon a mechanism for translating energy into mass. Thus, it presupposes a universe operating in accord with Einstein's famous equation $E = mc^2$. Yet conceivably many other such mathematical relationships (or none whatsoever) might govern the relationship between mass and energy, many of which would preclude the operation of the kind of universe-generating mechanism that inflationary cosmology envisions.

Similarly, both Gordon and Collins point out that the inflationary universe-generating mechanism depends upon a larger built-in and finely tuned law structure that includes: Einstein's field equations of general relativity, something like the Pauli exclusion principle (to allow the formation of complex chemical structures), and a principle of quantization governing all physical fields to permit the stability of matter (see n. 5, Chapter 7, p. 469). Though we rarely think about the possibility of different laws and physical principles governing our universe, such built-in mathematical laws and structures represent a type of fine tuning that would have to precede the operation of the inflationary universe-generating mechanism. Indeed, since inflationary cosmology's universe-creating mechanism does not generate universes with new laws or constants of physics, but instead only new initial conditions, it does not explain the fine tuning of the law structure of the universe. See Collins, "The Teleological Argument," 264; Gordon, "Postscript to Part One," esp. 97; "Balloons on a String."

40 Ijjas, Steinhardt, and Loeb, "Pop Goes the Universe."

41 For example, the uniform distribution of the wavelengths of the cosmic background radiation may be a consequence of inflation. But physicists can just as easily explain the uniformity of this distribution on straightforward mathematical grounds without reference to any cosmological model whatsoever. As Bruce Gordon notes, "The Gaussian (normal) distribution prediction of inflation is a straightforward consequence of the Central Limit Theorem, which states that the mean of a sufficiently large iteration of random variables with well-defined means and variances will have a near-normal distribution" ("Divine Action and the World of Science," 270). For an elaboration, see Peacock, *Cosmological Physics*, 342, 503.

42 Jogalekar, "Why the Search for a Unified Theory May Turn Out to Be a Pipe Dream."

43 For an extensive discussion of the key predictions of inflationary cosmology and why they fail, see Chapter 16, n. c, at www.returnofthegodhypothesis/extendedresearchnotes. See also Ijjas, Steinhardt, and Loeb, "Pop Goes the Universe," 37.

44 Oddly, inflationary cosmology also suffers from the opposite problem as well. Many of the evidences it explains or the predictions it makes can be explained or have been predicted on the basis of other models. For a discussion of how other cosmological models make the same predictions as the inflationary multiverse, see Chapter 16, n. e, at www.returnofthegodhypothesis/extendedresearchnotes.

45 Several leading physicists have suggested that postulating an inflaton field seems an increasingly contrived explanation for a range of cosmological evidence, in part because the field has to be highly gerrymandered to account for recent anomalies and failed predictions and in part because such fields represent purely hypothetical entities with idiosyncratic attributes. Indeed, inflaton fields, with their uncanny ability to activate the rapid expansion of space and then decay at just the right time in one model (between 10^{-37} to 10^{-35} seconds after the big bang) and in just the right measure, have properties associated with no other physical fields.

In addition, to accommodate recent failed predictions about gravity waves and the cosmic background radiation, inflationary cosmologists have had to revise their models of inflaton fields in extremely idiosyncratic ways, casting further doubt on the existence of these fields. Proponents of inflation now posit sudden, discontinuous, and/or irregular changes in the energy density of space as well as in the other parameters that affect the overall strength of the inflaton field. Proponents of inflation have also arbitrarily made adjustments to the mathematical function that relates the energy density of space and the strength of the field. These functions no longer define smooth curves as they did in the original models of Guth, Linde, and Steinhardt, but instead freakishly irregular curves that Steinhardt now describes as "arcane" and "contrived." The choice of inflation (energy) potential is essentially reverse-engineered to fit the data and then put forward as an "explanation" for what is observed (Ijjas, Steinhardt, and Loeb, "Pop Goes the Universe"). As the theoretical physicist William Unruh observed, "I'll fit any dog's leg that you hand me with inflation" (referenced in Holder, *God, the Multiverse, and Everything*, 130).

46 Ijjas, Steinhardt, and Loeb, "Pop Goes the Universe," 36.

47 Note that supersymmetry cuts both ways: it is not just that regular bosons have

fermionic superpartners, but if they do, then the regular fermions have bosonic superpartners as well. What string theory requires is not just fermionic supersymmetric particles, but bosonic ones as well. In any case, experimental evidence for supersymmetry (whether by discovery of a supersymmetric boson or fermion) is a necessary but insufficient condition for the correctness of string theory. Yet *neither* supersymmetric bosons *nor* supersymmetric fermions have yet been detected in expected energy ranges, even though the energy scales have been adjusted upward multiple times. Given that supersymmetry is a necessary condition of the correctness of string theory, failure to detect either (by *modus tollens*) strongly disconfirms the theory.

48 Horgan, "Why String Theory Is Still Not Even Wrong."

49 Hooft, *In Search of the Ultimate Building Blocks*, 163–64. Or as physicist Lee Smolin has noted, "If string theory is to be relevant at all for physics, it is because it provides evidence for the existence of a more fundamental theory. This is generally recognized, and the fundamental theory has a name—M-theory—even if it has not yet been invented" (*The Trouble with Physics*, 182).

50 Quoted in Gefter, "Is String Theory in Trouble?"

51 Carr, "Introduction and Overview," 16.

52 Lewontin, "Billions and Billions of Demons," 31.

Chapter 17: Stephen Hawking and Quantum Cosmology

1 Hawking and Penrose, "The Singularities of Gravitational Collapse and Cosmology."

2 Hawking and Ellis, *The Large Scale Structure of Space-Time.*

3 Hawking and Ellis, *The Large Scale Structure of Space-Time*, 363.

4 "Has Hawking Explained God Away?"

5 Hawking, *A Brief History of Time*, 138.

6 Hartle and Hawking, "Wave Function of the Universe," 2960–75.

7 In fact, it's a bit more accurate to say that Hawking *effectively* introduced the $i\tau$ term into the spacetime metric because he first introduced the $i\tau$ term into a functional integral that includes the spacetime metric within its mathematical structure.

8 Calculating the probabilities for different states of the universe required him to construct integrals that could not be solved using real time, but could be solved using imaginary time. Wiltshire, "An Introduction to Quantum Cosmology," 488.

9 The time variable in complex analysis is plotted on an axis for imaginary time that has no physical meaning.

10 Hawking, "The Beginning of Time."

11 Hawking, *A Brief History of Time*, 140–41.

12 Another version of the cosmological argument known as the "cosmological argument from contingency" does not depend upon the universe having a beginning. It affirms, first, that the universe has many contingent features that could be otherwise, and one of those contingent facts about the universe is that it exists. Proponents of this argument contend that the existence of the universe is a contingent fact because it is logically possible that the universe might *not* exist. They also argue that every contingent fact must have a sufficient reason for its existence. Proponents further contend that the universe itself (and in some versions, the contingent relationships within it) cannot depend for its (or their) existence on any contingent fact within the universe. Instead, the universe must depend upon some necessarily existing cause—some cause that must exist independent of the universe, whether the universe began a finite time ago or not. The eighteenth-century German philosopher and mathematician Gottfried Leibniz summarized the argument as follows: "the sufficient or final reason must be outside of the succession or *series* of this diversity of contingent things [i.e., in the universe], however infinite it may be. Thus the final reason of things must be in a necessary substance . . . and this substance we call *God.*" (Leibniz, *The Principles of Philosophy Known as Monadology*). Prominent proponents of versions of the argument from contingency have not only included Leibniz, but also Thomas Aquinas and

the capable contemporary philosophers Alexander Pruss of Baylor University and Andrew Loke of Hong Kong Baptist University. Pruss, *The Principle of Sufficient Reason*; "The Leibnizian Cosmological Argument"; Loke, "God and Ultimate Origins."

13 Craig and Sinclair, "The Kalām Cosmological Argument," p. 177–78, 179, Craig, *Reasonable Faith*, 109–113; Peacock, *A Brief History of Eternity*.

14 Hawking, *A Brief History of Time*, 139.

15 See n. 9 above.

16 Hawking does acknowledge that imaginary numbers have no real-world referent. Nevertheless, he also seems to make a weak attempt at justifying his decision to treat his mathematical depiction of the universe using imaginary time as if it told us something about the real universe—in particular, that the universe had no temporal beginning. In his 2001 book *The Universe in a Nutshell*, Hawking observes that "one might think . . . that imaginary numbers are just a mathematical game having nothing to do with the real world. From the viewpoint of positivist philosophy, however, one cannot determine what is real. All one can do is find which mathematical models describe the universe we live in" (p. 56). His argument here seems to be that since science doesn't *ever* tell us what is real about the world, but only produces models of it, it is perfectly acceptable to model the universe with a form of mathematics that can have no possible application to the real world. Perhaps so, but since Hawking has already affirmed that nothing in mathematical physics tells us about the real universe but only gives us models (positivism), he undercuts any claim to have produced a specific model that accurately depicts the real universe. If models in general don't tell us what is real, then his specific model using imaginary time doesn't either. Indeed, by proclaiming himself a positivist in his philosophy of science, he eschews all realist interpretations of mathematical physics, including any that depict a universe without a beginning in time.

17 Hawking, *A Brief History of Time*, 139.

18 Time has an inherent directionality in our universe; thus our use of descriptive words to describe it such as "before" and "after." By "spatializing time," Hawking's mathematical transformation rendered time directionless mathematically and thus inapplicable to spacetime in our universe.

19 Hawking, *A Brief History of Time*, 136.

20 Page, "Susskind's Challenge to the Hartle-Hawking No-Boundary Proposal and Possible Resolutions," 4.

21 Craig, "The Ultimate Question of Origins." See also Spitzer, *New Proofs for the Existence of God*; Copan and Craig, *Creation Out of Nothing*; Craig, *The Kalām Cosmological Argument*; *The Cosmological Argument from Plato to Leibniz*.

22 Vilenkin, "Quantum Cosmology and the Initial State of the Universe"; *Many Worlds in One*.

23 Mlodinow, "The Crazy History of Quantum Mechanics."

24 Vilenkin and Yamada, "Tunneling Wave Function of the Universe," 066003.

25 Huygens, *Treatise on Light*.

26 Young, "Bakerian Lecture."

27 Hertz, "Über den Einfluss des ultravioletten Lichtes auf die electrische Entladung."

28 "Photoelectric Effect," *The Physics Hypertext*, https://physics.info/photoelectric.

29 Einstein, "Über einen die Erzeugung und Verwandlung des Lichtes betreffenden heuristischen Gesichtspunkt."

30 To read more about what Einstein had to say about the particle-like quality of light, see Chapter 17, n. a, at www.returnofthegodhypothesis/extendedresearchnotes. See also Einstein, "Über einen die Erzeugung und Verwandlung des Lichtes betreffenden heuristischen Gesichtspunkt."

31 Taylor, "Interference Fringes with Feeble Light."

32 Taylor, "Interference Fringes with Feeble Light."

33 Davisson, "The Diffraction of Electrons by a Crystal of Nickel." (Don't forget the experiments of G. P. Thomson as well: "Experiments on the Diffraction of Cathode Rays.")

34 "What Is the Schrödinger Equation, Exactly?" *YouTube*, July 6, 2018, https://www.youtube.com/watch?v=QeUMFo8sODk&t=7s. See also Schrödinger, "An Undulatory Theory of the Mechanics of Atoms and Molecules."
35 Born, "Zur Quantenmechanik der Stoßvorgänge." See N. P. Landsman, "The Born Rule and Its Interpretation," http://www.math.ru.nl/~landsman/Born.pdf.
36 The known constant *e* is the base of the natural logarithm.
37 The function ψ is called a *wave* function because it determines a unitless amplitude that yields a probability distribution that, in turn, describes the likelihood of the photon being observed at a particular location (or having a particular momentum) over time.
38 The interpretation that the wave function does not represent a real entity is the most popular interpretation. Nevertheless, some theoretical physicists have proposed interpretations in which it does represent a physical wave (Matzkin, "Realism and the Wavefunction").
39 Feynman, *The Character of Physical Law*, 129.
40 Though the observer-caused interpretation of the collapse of the wave function is now associated with Niels Bohr and called "the Copenhagen interpretation" in his honor, John von Neumann and Eugene Wigner originally proposed this interpretation. Moreover, Bohr himself believed that the formalism of quantum mechanics presupposed a classical world picture. Consequently, he did not actually advance the observer-induced collapse of the wave function idea that has been attributed to him. For more on how Bohr himself interpreted quantum phenomena (i.e., the collapse of the wave function), see Chapter 17, n. b, at www.returnofthegodhypothesis/extendedresearchnotes. See also Faye, "Copenhagen Interpretation of Quantum Mechanics"; Halvorson, "Complementarity of Representations in Quantum Mechanics"; "The Quantum Experiment That Broke Reality."
41 Quoted in Heisenberg, *Physics and Beyond*, 206.
42 In its mathematical analogy to the Schrödinger equation, the Wheeler-DeWitt equation uses Paul Dirac's procedure for quantizing the dynamical equations of spacetime geometry in general relativity (called a Hamiltonian constraint) in an effort to describe the quantum evolution of geometries in superposition.
43 Cooke, "An Introduction to Quantum Cosmology."
44 In reality, the standard universal wave function includes "matter fields" that will ultimately result in the production of different possible configurations of matter and energy. For simplicity's sake, however, I will refer to different possible curvatures and configurations of matter (or mass-energy) when discussing the possible pairings that determine the gravitational fields of different possible universes in quantum cosmology.
45 In ordinary quantum mechanics the wave function assigns probabilities, via a relationship called the Born Rule, named for the German physicist Max Born, to each possible value in the probability distribution described by the wave function ψ. Similarly, the solution to the Wheeler-DeWitt equation—the *universal* wave function ψ—assigns, via the Born Rule, probabilities to each possible value in the probability distribution described by ψ. The Born Rule specifies that probabilities for each possibility in superspace are calculated by squaring the absolute value of the probability distribution ψ. Once a variety of restrictive and simplifying assumptions are made, ψ allows physicists to calculate the probabilities associated with the different possible spatial geometries and configurations of mass-energy—the different possible universes—as they exist in superposition as described by ψ. In practice, the calculation of probabilities can be more complex. (Wiltshire, "An Introduction to Quantum Cosmology," 496–98.)
46 Butterfield and Isham, "Spacetime and the Philosophical Challenge of Quantum Gravity."
47 In reality, Hawking and Hartle first calculated a "ground-state wave function" that does not even include universes such as ours, as they acknowledge in their technical paper. In ordinary quantum mechanics, the ground-state wave function describes an electron in its lowest energy state. Knowing this wave function allows physicists to calculate the probability that an electron in its lowest orbital resides at a specific place, including at

the center of the atom. By analogy, in quantum cosmology, the ground-state universal wave function allows quantum cosmologists to calculate the probability that any given universe, including a universe with zero spatial volume, will emerge out of the superposition of possible universes described by the wave function. As it turns out, Hawking and Hartle's ground-state wave function only describes *closed* universes, while our universe is an open universe (meaning it will expand indefinitely into the future). Thus, to explain how our universe arose from the universal wave function that they calculate, Hawking and Hartle had to first calculate an "excited-state" wave function from their original "ground-state" wave function. That excited-state wave function also initially included only closed universes, some of which could "tunnel" into continuously expanding universes such as ours. In this way, they "explained" the origin of our universe.

48 Technically, Hawking and Hartle's initial solution to the Wheeler-DeWitt equation did not itself include a universe like ours. See n. 47.

49 Their model does not eliminate a spatial singularity, however. Hawking and Hartle still assume a spatial singularity in what they call a "zero three-geometry" ("Wave Function of the Universe," 2961).

50 Indeed, prior to their use of the Wick rotation to solve the Wheeler-DeWitt equation (or rather the specific path integral they solved in its place), Hawking and Hartle first presuppose an actual spacetime singularity out of which various possible universes could have emerged. As they noted in their technical paper, "One can interpret the functional integral over all compact four-geometries bounded by a given three-geometry as giving the amplitude for that three-geometry to arise from a zero three-geometry, that is, a single point. In other words, the ground state is the amplitude for the Universe *to appear from nothing*" ("The Wave Function of the Universe," 2961).

51 Note, again, that our universe actually represents a continually expanding universe, but Hawking and Hartle's initial solution to the Wheeler-DeWitt equation included only closed universes. See n. 47.

52 See n. 47 for important qualifications.

Chapter 18: The Cosmological Information Problem

1 Krauss, "Ultimate Issues Hour: How Does Something Come from Nothing?," interview on *The Dennis Prager Show*, January 29, 2013.

2 "Ultimate Issues Hour: Evolution Revolution," interview on *The Dennis Prager Show*, October 16, 2018.

3 Vilenkin, "Creation of Universes from Nothing."

4 Borde and Vilenkin, "Eternal Inflation and the Initial Singularity"; Borde and Vilenkin, "Violation of the Weak Energy Condition in Inflating Spacetimes"; Borde, Guth, and Vilenkin, "Inflationary Spacetimes Are Incomplete in Past Directions."

5 For a very recent discussion and expansion of Vilenkin's model, see Vilenkin and Yamada, "Tunneling Wave Function of the Universe."

6 Krauss, *A Universe from Nothing*, 159. The statement is not actually original to Krauss. It can be traced back to a remark made by the Nobel laureate Frank Wilczek in a *Scientific American* article on matter-antimatter symmetry that he wrote back in 1980: "The answer to the ancient question 'Why is there something rather than nothing?' would then be that 'nothing' is unstable" ("The Cosmic Asymmetry Between Matter and Antimatter"). This quote has also appeared repeatedly in new atheist literature (see, e.g., Stenger, *God: The Failed Hypothesis*, 133).

7 Hawking and Mlodinow, *The Grand Design*, 180.

8 Krauss, *A Universe from Nothing*, 161–70.

9 Krauss, *A Universe from Nothing*, 169.

10 Krauss, *A Universe from Nothing*, 142.

11 For a good, concise, and accessible introduction to these topics, see Baggott, *The Quantum Story*, 361–71. For a discussion of the ill-defined, intractable nature of the

Wheeler-DeWitt equation and its interpretation, see Butterfield and Isham, "On the Emergence of Time in Quantum Gravity," esp. 43–62 (secs. 5.2–5.5). Comprehensive technical introductions can be found in Bojowald, *Quantum Cosmology*, and Rovelli, *Quantum Gravity*. An examination of some of the philosophical implications of quantum gravity and quantum cosmology can be found in Callender and Huggett, eds., *Physics Meets Philosophy at the Planck Scale*. See also Bojowald, Kiefer, and Moniz, "Quantum Cosmology for the XXIst Century: A Debate"; Gordon, "Balloons on a String," esp. 563–69; Kiefer, "The Need for Quantum Cosmology"; "Conceptual Problems in Quantum Gravity and Quantum Cosmology"; and Rovelli, "The Strange Equation of Quantum Gravity."

12 And, of course, there are even deeper questions here. Why is conservation of momentum one of the "laws" of nature in the first place, and what does it mean for something to be a natural "law"? It is certainly *not* a "law" on the grounds of it being a logical or metaphysical necessity—there is no absurdity that would result from its denial—so why is this regularity there in the first place and what maintains it?

13 Actually, there are two ways of understanding ψ, and neither specifies a material cause of the universe. One implies that ψ does not represent a definite state of affairs with material existence prior to the emergence of one of these possible universes. The universal wave function is *not a material entity* or even an energy-rich field in a real space. It only represents an information-rich mathematical expression describing different possibilities. The universal wave function, in this view, is just a mathematical expression that makes possible assigning differing probabilities to different possible universes. These probabilities are just artifacts of the mathematical calculating procedure that generates them. They do not correspond to actual material universes existing simultaneously or "in superposition" with one another. Consequently, ψ does not describe or specify any *thing* that could act as the cause of any particular outcome—that is, any particular universe. Nor does any material antecedent precede the universal wave function that could do so.

In another interpretation of the universal wave function, ψ does represent an actually existent universe, as opposed to a merely mathematical reality, albeit one in which many different possible universes with different possible gravitational fields exist simultaneously "in superposition." Nevertheless, even in this interpretation, nothing about that universe (or those universes) in that indeterminate state causes one, rather than another, of the possibilities to emerge as our universe. Instead, each of those different possible universes exists in parallel and in isolation from the others, and none of them in any way causes the others to materialize. Nor does any material antecedent determine the probability of the possible universes described by ψ. Instead, how the *physicist chooses* to solve the Wheeler-DeWitt equation determines the resulting ψ function and, thus, those probabilities. Thus, this interpretation of the wave function does not specify a material antecedent as a cause of our universe either. Nor does it provide a causal explanation for the origin of any of the universes that ψ describes as putatively existing in superposition.

14 Put another way, ψ describes a *necessary* condition for a universe to come into being—namely, that a particular universe is one of the possibilities included in the universal wave function. It doesn't describe a *sufficient* condition for that universe's arising, however. Thus, ψ doesn't specify the cause of the origin of any particular universe, much less actually cause its ultimate origin. Even so, another popular interpretation of the universal wave function tries to solve this problem. It holds that every spatial geometry and configuration of matter (matter field) described by the universal wave function does exist as a separate universe "in some possible world." This many-worlds interpretation (MWI) of quantum mechanics and quantum cosmology does not, however, specify a cause of the existence of these universes or, for that matter, our own. Instead, it represents an interpretation of what the wave function describes without addressing the underlying question of causation. See the next chapter, Chapter 19, for an extensive discussion of the MWI.

15 In practice, quantum cosmologists often solve a functional integral that stands in place of the Wheeler-DeWitt equation. It happens that mathematicians have proven that solutions to a specific path integral, a type of functional integral, correspond to certain solutions to the Wheeler-DeWitt equation. Because it is often easier to solve that specific path integral, quantum cosmologists often, but not always, solve that path integral rather than the Wheeler-DeWitt equation itself. Not surprisingly, the Wheeler-DeWitt equation can be derived from this path integral. For the mathematically curious, the path integral that quantum cosmologists often use is depicted in Figure 18.2.

16 Hawking, *A Brief History of Time*, 174.

17 Vilenkin, *Many Worlds in One*, 205, emphasis added.

18 "Platonism about mathematics (or *mathematical Platonism*) is the metaphysical view that there are abstract mathematical objects whose existence is independent of us and our language, thought, and practices" (Linnebo, "Platonism in the Philosophy of Mathematics").

19 Gage, "Darwin, Design & Thomas Aquinas."

20 Kline, "Theories, Facts, and Gods"; Meyer, "Of Clues and Causes," 91.

21 As Vilenkin and Yamada state, "The resulting wave function can be interpreted as describing *a universe originating at zero size*, that is, from 'nothing'" ("Tunneling Wave Function of the Universe"; emphasis added).

22 Drees, "Interpretation of 'The Wave Function of the Universe.'"

23 Hawking and Hartle, for example, claim that the "ground state of the wave function" allows them to calculate the probability of the universe coming into existence "from nothing." As mentioned in n. 47 on page 506, in ordinary quantum mechanics, knowing the ground-state wave function allows the physicist to calculate the probability that an electron in its lowest orbital will be found at a specific place. By analogy, in quantum cosmology, the ground-state wave function allows quantum cosmologists to calculate the probability that any given universe with a given geometry will be the one we observe out of the superposition of possible universes described by the wave function. Nevertheless, Hawking and Hartle do not explain what causes the universe to begin from the singularity that they presuppose in their mathematical procedure.

Instead, Hawking and Hartle calculate the ground state by summing possible paths from an initial singularity to different possible geometries and mass-energy configurations (i.e., different possible universes) in a restricted mini-superspace. They claim that "the ground state is the amplitude for the universe to appear from nothing," but what they really mean is that the ground-state wave function allows them to calculate the probability of a specific universe *evolving* from the singularity. Indeed, by their own logic, the ground-state cannot explain a universe appearing out of nothing. Instead, it represents the probability of observing a universe with a given geometry evolving or arising out of a true singularity of zero spatial volume.

24 In the case of quantum tunneling, the tunneling occurs not only after the quantum cosmologists have already presupposed the existence of a universe, but also after they have assumed that universe to be expanding in opposition to a potential energy barrier—one produced by a matter field associated with space that they have also simply presupposed to exist.

25 Of course, those who interpret the universal wave function instrumentally and interpret its description of different possible configurations of matter and spatial geometries as *merely* an abstract mathematical description of different possibilities do not beg the question, since they do not presuppose that these mathematical constructs, or the possibilities they describe, correspond to an *actual* universe. They merely regard these mathematical constructs as describing possible universes with different properties. Nevertheless, the universal wave function then cannot, by the same token, offer a causal explanation for the origin of the universe, because it does not posit an actual material state with causal powers that could conceivably generate a universe. Attempting to

conjure a material reality out of a purely mathematical description in this way would, again, commit the fallacy of "reification."

26 Halliwell, "Introductory Lectures on Quantum Cosmology," 38–39.

27 As Vilenkin has noted: "Thus, to explain the initial conditions of the universe, all we need to do is find the wave function ψ from [the Wheeler-DeWitt] eq. (9). However, as any differential equation, it has an infinite number of solutions" ("Quantum Cosmology," 7).

28 Vilenkin, "Quantum Cosmology," 7.

29 Vilenkin and Yamada, "Tunneling Wave Function of the Universe."

30 As Vilenkin nicely summarizes his procedure: "The role of the Schrödinger equation [in quantum cosmology] is played by the Wheeler-DeWitt equation, which is a functional differential equation on superspace. Since one does not know how to solve such an equation, *one restricts the infinite number of degrees of freedom of* $g_{\mu\nu}$ *and* ϕ [spatial geometries and matter fields] *to a finite number*; the resulting finite-dimensional manifold is called *mini-superspace*. Here we shall employ a simple mini-superspace model in which we restrict the 3-geometry to be homogeneous, isotropic and closed, so that it is described by a single scale factor *a*" ("Quantum Origin of the Universe," 144).

31 Hartle and Hawking, "Wave Function of the Universe," 2967; emphasis added.

32 In jargon of the discipline, the Hawking-Hartle model utilized a Euclidean rather than a Lorentzian metric. In a Euclidean metric, time is treated like a dimension of space ($ds^2 = dx^2 + dy^2 + dz^2 + c^2dt^2$), whereas a Lorentzian metric distinguishes between the status of spatial and temporal dimensions in its spatiotemporal structure ($ds^2 = dx^2 + dy^2 + dz^2 - c^2dt^2$). For an accessible discussion of oscillating versus nonoscillating regions of superspace in the context of quantum cosmological models, see Isham, "Theories of the Creation of the Universe," esp. 68–81.

33 Hartle and Hawking, "Wave Function of the Universe," 2967.

34 Hartle and Hawking, "Wave Function of the Universe," 2967.

35 James Hartle, "What Is Quantum Cosmology?" *Closer to Truth*, https://www.closertotruth.com/series/what-quantum-cosmology.

36 More precisely, they arbitrarily restricted degrees of mathematical freedom to ensure that the paths through superspace that they summed would produce a "ground-state" wave function that could be used to calculate another "excited-state" wave function that included universes *capable of tunneling into universes* with *geometries* similar to our own. See n. 50 for additional details Chapter 17. See also Gordon, "Balloons on a String," 563–69.

37 Theoretical physicist Jonathan Halliwell makes this same point using more technical terminology. Hawking and Hartle, he writes, "impose initial conditions on the histories [their mathematical path-integral procedure] which ensure that (i) the four-geometry closes, and (ii) the saddle-points of the functional integral *correspond to metrics and matter fields which are regular solutions to the classical field equations* matching the prescribed data on the bounding three-surface B" ("Introductory Lectures on Quantum Cosmology," 41).

38 Hartle and Hawking, "Wave Function of the Universe," 2960, emphasis added.

39 Interestingly, a recently proposed competing theory of quantum gravity known as loop quantum gravity (LQG) is subject to the same problem. For an extensive discussion of loop quantum gravity and why it neither explains nor attempts to explain the origin and the fine tuning of the universe, see Chapter 18, n. a, at www.returnofthegodhypothesis/extendedresearchnotes. See also Baggott, *Quantum Space*, xii–xiii; Rovelli, "Loop Quantum Gravity"; Date and Hossain, "Genericness of Inflation in Isotropic Loop Quantum Cosmology"; Mithani and Vilenkin, "Collapse of Simple Harmonic Universe"; Carroll, "Against Bounces."

40 Isham, "Theories of the Creation of the Universe," 72.

41 Halliwell, "Introductory Lectures on Quantum Cosmology," 46.

42 Halliwell, "Introductory Lectures on Quantum Cosmology," 46, emphasis added.

Halliwell made this point in relation to the Hawking-Hartle model, but it applies just as much to Vilenkin's. As Vilenkin acknowledges, he not only needed to impose boundary conditions on the Wheeler-DeWitt equation in order to solve it, he also needed to add an extra boundary term to prevent generating many unstable solutions to the wave function. He needed to choose this term carefully to match the required boundary conditions that he had already chosen to make the Wheeler-DeWitt equation solvable. These constraints also constitute an external infusion of information into the mathematical procedure by which Vilenkin modeled the origin and development of the universe.

43 See Gordon, "Balloons on a String," 568–69.

44 Dawkins, *The Blind Watchmaker*, 46–49. See the discussion in Ewert et al., "Efficient Per Query Information Extraction from a Hamming Oracle."

45 Vilenkin, "Quantum Cosmology," 7. Emphasis added.

46 For a technical article summarizing the result of this research effort see, Meyer, "Mind Before Matter: The Unexpected Implications of Quantum Cosmology."

Chapter 19: Collapsing Waves and Boltzmann Brains

1 See Chapter 17, n. 40.

2 The models described earlier focused on a single scalar matter field, but additional fields corresponding to ordinary matter and energy can be included as variations in the models (Kiefer, "Emergence of a Classical Universe from Quantum Gravity and Cosmology").

3 For more information on the controversy, see Collins, "The Many Interpretations of Quantum Mechanics." For attempts to explain the collapse of the universal wave function, see Kiefer, "Emergence of a Classical Universe from Quantum Gravity and Cosmology."

4 Craig, "What Place, Then, for a Creator?"

5 Kiefer, "On the Interpretation of Quantum Theory—from Copenhagen to the Present Day." The philosopher Robert Koons, of the University of Texas, has developed a similar view that invokes the chemical and thermodynamic properties of large macroscopic objects, including measuring devices, to account for the collapse of the wave function; see "The Many-Worlds Interpretation of Quantum Mechanics."

6 See also Faye, "Copenhagen Interpretation of Quantum Mechanics." For a discussion of technical problems with the standard Copenhagen view, see Bell, "Against 'Measurement'"; Wallace, "The Quantum Measurement Problem."

7 Quantum cosmologists have eschewed other non-Copenhagen interpretations of quantum cosmology such as the Bohmian (named for the late physicist David Bohm) interpretation and the GRW (named for the physicists Giancarlo Ghirardi, Alberto Rimini, and Tullio Weber) interpretation. Both of these interpretations presuppose preexisting matter upon which the universal wave function ψ acts. Consequently, in neither interpretation does ψ *produce* matter and energy upon its collapse. Consequently, in neither interpretation can ψ be invoked to explain the origin of the material universe. In the Bohmian interpretation, ψ directs or dictates the movements of preexisting particles. In this interpretation, ψ doesn't produce particles; it just moves them around. Similarly, in the GRW interpretation, the wave function describes how preexisting matter spread (in a wavelike structure) across some extended space and then "collapses" in the specific sense of becoming more densely concentrated. Neither of these interpretations of the quantum mechanics has any utility for quantum cosmologists who want to explain the origin of the universe. Since these interpretations treat ψ and the mass-energy of the universe as distinct entities (rather than mass-energy as a manifestation of ψ), they offer no explanation of how matter (or energy) emerged out of ψ.

8 Everett, "The Theory of the Universal Wave Function."

9 For an excellent defense of the principle of sufficient reason, see Pruss, *The Principle of Sufficient Reason*; Koons and Pruss, "Skepticism and the Principle of Sufficient Reason."

10 Vilenkin, *Many Worlds in One*, 205.
11 Moniz, "A Survey of Quantum Cosmology."
12 Tegmark, *The Multiverse Hierarchy*, 99–126, especially 110.
13 Tegmark, "Is 'The Theory of Everything' Merely the Ultimate Ensemble Theory?" 5.
14 The Baylor University philosopher of science Alexander Pruss develops this point in his critique of philosopher David Lewis's modal realism; see *Actuality, Possibility, and Worlds*, 117–19.
15 Recall that inflationary cosmology putatively generates new universes with different initial conditions, but not different laws and constants. Thus, this cosmology would not have the problem of new irregular laws generating unpredictable events. Nevertheless, inflationary cosmology envisions random quantum fluctuations in the inflaton field generating an infinite number of new bubble universes as well as random events within different bubble universes. Thus, we could be in a universe in which virtually anything could occur as the result of such random quantum fluctuations (albeit roughly within the framework of the known physical laws of this universe).
16 Tegmark, "What Scientific Idea Is Ready for Retirement?"
17 For a good overview of the problem of Boltzmann brains, see Carroll, "Why Boltzmann Brains Are Bad." See also Page, "Is Our Universe Likely to Decay Within 20 Billion Years?"; "Return of the Boltzmann Brains"; "Susskind's Challenge to the Hartle-Hawking No-Boundary Proposal and Possible Resolutions"; Bousso and Freivogel, "A Paradox in the Global Description of the Multiverse"; Koberlein, "Can Many-Worlds Theory Rescue Us from Boltzmann Brains?"
18 Carroll, "Why Boltzmann Brains Are Bad."
19 See Linde, "Towards a Gauge Invariant Volume-Weighted Probability Measure for Eternal Inflation"; and Bousso, Freivogel, and Yang, "Boltzmann Babies in the Proper Time Measure."
20 Tegmark's mathematical universe hypothesis does not lend itself to the strategy that proponents of inflationary cosmology tried to use to demonstrate the rarity of Boltzmann brains. Recall that inflationary cosmology proponents attempted to establish that the ratio of Boltzmann brains to natural brains is low in one quadrant of space and then tried to show that the frequency of Boltzmann brains would decrease toward a limit as they extended their analysis to larger and larger parts of that same space. In Tegmark's mathematical universe hypothesis the different universes exemplifying different mathematical structures do not reside in the same physical space or share a common time parameter, so no such extrapolation is possible. In more technical terms, his model rules out any plausible way of "managing" Boltzmann brains by applying a "measure function."
21 Tegmark, "Infinity Is a Beautiful Concept—and It's Ruining Physics." As he says, "Not only do we lack evidence for the infinite but we don't need the infinite to do physics. Our best computer simulations, accurately describing everything from the formation of galaxies to tomorrow's weather to the masses of elementary particles, use only finite computer resources by treating everything as finite. So if we can do without infinity to figure out what happens next, surely nature can, too—in a way that's more deep and elegant than the hacks we use for our computer simulations. Our challenge as physicists is to discover this elegant way and the infinity-free equations describing it—the true laws of physics. To start this search in earnest, we need to question infinity. I'm betting that we also need to let go of it."
22 Tegmark, "Our Mathematical Universe," 101–50.
23 Someone could wonder whether quantum cosmology accurately describes the origin of the universe at all. Quantum cosmology is based upon an *analogy* with ordinary quantum mechanics. But quantum mechanics uses mathematics to describe an actual physical system that already exists. Quantum cosmology attempts to use analogous mathematics to conjure a universe into existence. In ordinary quantum mechanics, the Schrödinger equation and its solution, the wave function, describes an actual physical system: a light

source, photons, a barrier with slits, and a detector. But what in quantum cosmology corresponds to the light source? Or the detector? Or the barrier or the screen? Quantum cosmologists would say that the light source and the screen with a detector are analogous to the boundary conditions that they impose on the Wheeler-DeWitt equation. But they have to impose those boundary conditions *arbitrarily* precisely because *no physical systems yet* exist to which their mathematical descriptions apply.

Consequently, as I argued in the previous chapter, quantum cosmologists have to first presuppose the existence of a universe in order to construct a wave function that allegedly explains its origin. But in ordinary quantum mechanics that would be like calculating the wave functions for a hypothetical photon and experimental apparatus and then expecting the equations to cause the apparatus and the photon to pop into existence. The mathematical apparatus of quantum mechanics describes *something*. The mathematical apparatus of quantum cosmology must *presuppose* something in order for the mathematics to apply to anything. Thus, it may be that the mathematics of quantum cosmology and quantum mechanics are analogous, but the physical situations to which the math is being applied are not. That at least raises a question about whether the math of the one (quantum mechanics) applies to the physics of the other (the universe) in a meaningful way.

Chapter 20: Acts of God or God of the Gaps?

1 Dawkins, *The Blind Watchmaker*, 1, emphasis added.
2 Dawkins, *The Blind Watchmaker*, 7.
3 Hawking, *Brief Answers to the Big Questions*, 38.
4 Gould, "Nonoverlapping Magisteria"; *Rocks of Ages*, 14.
5 Galileo and Finocchiaro, *The Essential Galileo*, 119. This is in Galileo's letter to the Grand Duchess Christina (1615). Galileo used this aphorism, but claimed it originated with Cardinal Cesare Baronio.
6 MacKay, *The Clockwork Image*, 51–55; Van Till, *The Fourth Day*, 208–15; Van Till, Young, and Menninga, *Science Held Hostage*, 39–43, 127–68. For a different interpretation of complementarity that affirms the methodological autonomy of science and religion, but conjoins their findings, see also Gruenwald, "Science and Religion."
7 Newton, *The Mathematical Principles of Natural Philosophy*, 391–92. The more recent Cohen and Whitman translation renders this passage as: "the discussion of God, and to treat of God from phenomena is certainly a part of natural philosophy." Newton, *Mathematical Principles of Natural Philosophy*, 943.
8 For a more nuanced reading of this passage in conjunction with related sayings in some of Newton's unpublished manuscripts, see Snobelen, "Isaac Newton."
9 Not all scholars who have written for the BioLogos website oppose intelligent design. For example, historian of science Ted Davis neither supports nor opposes intelligent design. Even so, the most prominent leaders of BioLogos have strenuously opposed and critiqued intelligent design. See, e.g., Collins, *The Language of God*, 181–96; Falk, "Thoughts on *Darwin's Doubt*, Part 1"; "Further Thoughts on 'Darwin's Doubt' after Reading Bishop's Review"; Haarsma, "Reviewing 'Darwin's Doubt'"; "Response from Evolutionary Creation"; Venema, "Intelligent Design and Nylon-Eating Bacteria"; Venema and Kuebler, "Biological Information and Intelligent Design"; Venema and McKnight, *Adam and the Genome*, 67–92.
10 "Are Gaps in Scientific Knowledge Evidence for God?"
11 Recall that in Newton's time scientists (as we would call them today) were called natural philosophers.
12 Here's another version of this often told story published in a more scholarly volume: "Perhaps the most famous example of the God-of-the-gaps argument came about when Isaac Newton considered the question of the long-term stability of the solar system. He was not able to calculate whether the small gravitational forces between pairs of planets would cancel on the average or accumulate. He considered that in the latter case the

unstable behavior would be avoided by the gentle action of God, applying small forces at the right times and places. A century later, Pierre Simon de Laplace showed that the solar system is indeed stable against such perturbations. When his former student, Napoléon Bonaparte, asked why Laplace's treatise on celestial mechanics did not mention God, Laplace answered, 'I did not need that hypothesis'" (Albright, "God of the Gaps," 955).

13 Shermer, "ID Works in Mysterious Ways."

14 McMullin, "The Virtues of a Good Theory."

15 Robert Larmer has argued that theistic arguments based upon gaps in the natural order do not necessarily commit the fallacy of arguing from ignorance. He argues that it is simply dogmatic to insist that God can never be the cause of an event in nature. Instead, he argues that some *apparent* cases of "arguments from ignorance" have considerable epistemic force. He observes that "presumed examples of the fallacy of *argumentum ad ignorantiam* can often be redescribed in a positive way that makes them seem not to be arguments from ignorance at all" ("Is There Anything Wrong with 'God of the Gaps' Reasoning?", esp. 131).

16 "Are Gaps in Scientific Knowledge Evidence for God?"

17 Recall this argument is based upon our own introspective awareness of our conscious minds. We all intuitively sense that we have the ability to produce abrupt changes of state (free will) uncompelled by material causes. I noted that, though positing the uncaused act of an agent did represent an exception to the rule that "all events have causes," it did not undermine scientific rationality in so doing. On the other hand, positing or allowing the possibility of uncaused *material* events not only violates the principle of sufficient reason; it does so in a way that undermines our confidence in the intelligibility of nature and scientific rationality. See the discussion in Chapter 12.

18 See Hume's *Dialogues Concerning Natural Religion*, Part II: "And will any man tell me with a serious countenance, that an orderly universe must arise from some thought and art like the human, because we have experience of it? To ascertain this reasoning, it were requisite that we had experience of the origin of worlds; and it is not sufficient, surely, that we have seen ships and cities arise from human art and contrivance."

19 Scriven, "Causes, Connections and Conditions in History," 249–50.

20 Critics of the God hypothesis could argue that we have no experience of minds creating matter *ex nihilo*, and further that the power to create matter from nothing is a qualitatively different power than the power to create a new structure by arranging or reconfiguring preexisting matter. Consequently, they could argue that positing a God with the power to generate matter itself *ex nihilo* does not qualify as a reasonable extrapolation from the causal powers of a known entity.

Nevertheless, we do have experience of minds choosing to actualize specific states out of a larger ensemble of possibilities, thereby using and/or generating information. Moreover, since the advent of quantum mechanics, we now understand that a material particle (matter) results from the informative actualization of a possible state from among a much larger ensemble of states described by a quantum wave function. In other words, in quantum mechanics matter results from an informational input as an observation or interaction with a larger macroscopic object results in the actualization of a specific material state from an ensemble of possible states described by a wave function (i.e., the collapse of the wave function). Moreover, we know that intelligent agents have demonstrated the ability to actualize specific states out of a larger ensemble of possibilities, thus using or generating information. Thus, theists could argue that human minds have demonstrated a "relevantly similar" causal power to that required to actualize specific states of affairs described by quantum wave function—that is, they can choose among possibilities, thus using or generating information. Since human choices can actualize possibilities and generate information, it is reasonable to extrapolate and postulate that a divine mind using a relevantly similar but greater causal power could choose among possibilities described by quantum (or universal) wave functions to actualize specific states of affairs resulting in the production of matter or the universe itself.

21 I'm indebted to Paul Nelson for thinking of this illustration. For a thorough critique of the use of the God-of-the-gaps objection to prohibit the use of intelligent causes in explanations of the history of life, see Meyer and Nelson, "Should Theistic Evolution Depend on Methodological Naturalism."

22 Tyson continued at great length in his 2010 lecture to indict Newton for his fallacious reasoning. For Tyson's badly misinformed history of science on display in an extended excerpt from the transcript of his 2010 lecture, see Chapter 20, n. a, at www.returnofthegodhypothesis/extendedresearchnotes. Notice there that Tyson asserts, among many other errors, that Newton thought that the solar system was unstable.

23 Newton, *Mathematical Principles of Natural Philosophy*, 941.

24 Newton, *Mathematical Principles of Natural Philosophy*, 940. In the *Principia*, Newton also developed four methodological principles or "rules of reasoning" in natural philosophy, including a version of the *vera causa* principle. He articulated this principle as follows: "No more causes of natural things should be admitted than are both true and sufficient to explain their phenomena" (794). For a discussion of how Newton applied his *vera causa* principle to justify the God hypothesis without making a God-of-the-gaps argument, see Chapter 20, n. b, at www.returnofthegodhypothesis/extendedresearchnotes.

25 As Newton wrote there: "How came the Bodies of Animals to be contrived with so much Art, and for what ends were their several parts? Was the Eye contrived without Skill in Opticks, and the Ear without Knowledge of Sounds? . . . And these things being rightly dispatch'd, does it not appear from Phænomena that there is a Being incorporeal, living, intelligent, omnipresent?" (*Opticks*, 369–70).

26 Courtenay, "The Dialectic of Omnipotence in the High and Late Middle Ages," 243–69; Kaiser, *Creational Theology and the History of Physical Science*, 53–55.

27 Newton, *Mathematical Principles of Natural Philosophy*, 815–20 (Book III, Propositions X–XV).

28 Newton, *Mathematical Principles of Natural Philosophy*, 816 (Book III, Proposition XI).

29 Newton, *Mathematical Principles of Natural Philosophy*, 816–17 (Book III, Proposition XII).

30 Newton, *Mathematical Principles of Natural Philosophy*, 815–16 (Book III, Proposition X).

31 Newton, *Mathematical Principles of Natural Philosophy*, 940. The passage commonly cited to justify the claim that Newton specifically invoked episodic divine acts to adjust the motions of the planets appears in the General Scholium of the *Principia*. There Newton argues that though the planetary bodies may "persevere in their orbits by the mere laws of gravity, yet they could by no means have at first derived the regular position of the orbits themselves from those laws," and he goes on to argue that "this most beautiful System of Sun, planets and comets could only proceed from the counsel and dominion of an intelligent and powerful being." Science popularizers and historians often misinterpret these passages. [See for example, the BioLogos staff-written article, "Are Gaps in Scientific Knowledge Evidence for God?"] Notice that in the passages from the General Scholium, Newton does not say that God intervenes to *fix* the planetary orbits or to stabilize the system. Instead, he's talking about the *origin* of the solar system and its manifest order and stability. He recognizes that the laws of nature describe regularities, but also that they do not explain the origin of *specific initial conditions* of systems that make those regularities possible.

Thus, later in the General Scholium he amplifies that argument by explaining that "no variation in things arises from blind metaphysical necessity [i.e., the laws of nature], which must be the same always and everywhere." Instead, he argues that "All the diversity of created things, each in its place and time, could only have arisen from the ideas and the will of a necessarily existing being" (942). Newton here displays a sophisticated understanding of what the laws of nature do—and don't do. Specifically, in this case, he realizes that his universal law of gravitation *can* describe the regularities in the planetary motions but can't *in principle* determine the specific and irregular initial conditions of the solar system (or any system) to which the law applies. Yet since the present stability of the system depends upon a highly specific and irregular (or complex) positioning of "the Sun, planets and comets," he infers the activity of a designing intelligence as an explanation for the origin of the system itself. Nevertheless, he does not claim that God, after establishing this system, periodically intervenes to adjust irregularities in the system. Instead, he makes an initial-condition *fine-tuning* argument based on a correct understanding of what agents can do (arrange matter in highly specific and complex ways to accomplish desired ends) and what laws can't.

Moreover, in the BioLogos article cited above, the authors do distinguish between Newton's interest in "the ongoing motion of the planets" and "the origin of the motions." But the authors provide no citations from the *Principia* to show that Newton posited God's singular, episodic action to adjust ongoing planetary motions. They do cite the passage from the General Scholium noted above in which Newton credits God with *the origin* and design of the solar system. Nevertheless, because they give no supporting quotes for their claim that Newton postulated singular divine interventions into the *ongoing* workings of the solar system, the quote attributing the initial design of the solar system to God gives the false impression that Newton also proposed episodic and singular acts of God to fix (alleged) irregularities and perturbations. Indeed, they repeat the same false story as Tyson without any direct attribution. As they put it: "Newton suspected that these gravitational perturbations would accumulate and slowly disrupt the magnificent order of the solar system. To counteract these and other disruptive forces, Newton suggested that God must necessarily intervene occasionally to tune up the solar system and restore the order. Thus, God's periodic *special* actions were needed to account for the ongoing stability of the solar system [emphasis added]."

32 Stephen Snobelen, an excellent Newton scholar at the University of King's College in Halifax, Nova Scotia, also rejects the claim that Newton made God-of-the-gaps arguments. He emphasizes that because Newton thought God continuously upheld the laws of nature, he did not think that nature could have gaps in its lawlike regularities in need of filling. As he put it, "The 'God of the gaps' critique applied to Newton can imply that where God isn't filling a gap, He is not at work. But Newton believed that God is ultimately behind all operations in the cosmos." Indeed, Newton saw God constantly sustaining the orderly concourse of the universe through what we call the laws of nature. Those laws admit no gaps for God to fill with special divine interventions, since in Newton's view God sustains the orderly concourse of nature on a moment by moment basis. Moreover, since God's predictable action and character are precisely what allow us to perceive the existence of laws of nature at all, Newton's affirmation of constant divine action in sustaining the order of nature did not inhibit his scientific investigation of it. Just the opposite. His view of God's dominion over nature inspired it—a point that Snobelen has emphasized to me in correspondence.

In addition, Snobelen notes the passages often cited as evidence of Newton's invoking God to fill gaps are instead "expostulations of natural theology," that is, design arguments celebrating the wisdom of God in establishing the natural order in the first place ("Newton and the God of the Gaps"). For example, Newton also made design arguments that implied that God had acted in the past to design the integrated complexity of the eye and establish the specific material conditions that made the stability of the solar system possible. Yet Newton's invoking of divine action in this way does not constitute a GOTG fallacy for the same reason that contemporary intelligent design arguments do not constitute instances of a fallacy as explained in this chapter. Newton inferred intelligent design based upon the presence of features in nature that—given *our knowledge* of cause and effect—are best explained by the activity of an intelligent cause.

Even so, Snobelen allows how one passage in the Newton corpus might provide some support for the idea that Newton *envisioned* the need for divine action to fix the solar system at some point far into the future (as opposed to invoking God's intervention at episodic intervals on an ongoing basis, as most versions of the God-of-the-gaps story assert). Snobelen notes that in a short passage in Query 31 in the *Opticks*, Newton anticipates that "some inconsiderable Irregularities" will arise "from the mutual Actions of Comets and Planets upon one another." And that those irregularities "will be apt to increase, *till this System wants a Reformation*" (emphasis added). Even so, Snobelen views this passage neither as evidence of Newton's willingness to invoke divine action to fix irregularities in the ordinary concourse of nature nor of his having a penchant for invoking such action to the exclusion of looking for lawful regularities in nature. Instead, Snobelen argues that Newton, as a student of biblical eschatology, anticipated that the solar system, like nature itself, would one day run down, after which God would remake the heavens and the earth. As Snobelen explains: "The trajectory toward decline has its remedy in the God of dominion, who reforms and adjusts to keep the cosmos orderly, and who recreates when the time comes for a new heaven and a new earth. Newton's cosmos is not deterministic in the secular and materialistic senses often applied to him; nevertheless, its future is ultimately guided by divine action." See Snobelen, "Cosmos and Apocalypse," esp. 93.

In my view, reading Query 31 in the *Opticks* in context casts further doubt on the claim that Newton made a God-of-the-gaps argument there. First, in the following sections of this passage, Newton *does not postulate any specific act of God* or angels to rectify these anticipated irregularities. Instead, he seems only to be affirming the reality of what physicists today would call entropy, the tendency of systems to move from order to disorder over time. (Snobelen has told me in personal correspondence that he interprets this passage in much the same way, as Newton affirming a kind of proto-entropy concept.) Second, in Book III of the *Principia*, Newton shows mathematically that the solar system is stable over "a long tract of time" and consequently shows no concern whatsoever in that most relevant section of his corpus to posit direct divine action to remedy orbital irregularities or instabilities in the solar system.

Third, in Query 31 Newton is marveling at the "wonderful Uniformity in the Planetary System" and arguing that the order of the system arose "in the first Creation by the Counsel of an intelligent Agent." Thus, he does invoke divine action, but, again, only as the cause of the "Origin of the World." He does not postulate any specific divine action to stabilize the solar system, on an ongoing basis or even at some time in the future. Here's the passage in question:

> "Now by the help of these Principles, all material Things seem to have been composed of the hard and solid Particles above mention'd, variously associated in the first Creation by the Counsel of an intelligent Agent. For it became him who created them to set them in order. And if he did so, it's unphilosophical to seek for any other Origin of the World, or to pretend that it might arise out of a Chaos by the mere Laws of Nature; though being once form'd, it may continue by those Laws for many Ages. For while Comets move in very excentrick Orbs in all manner of Positions, blind Fate could never make all the Planets move one and the same way in Orbs concentrick, some inconsiderable Irregularities excepted which may have risen from the mutual Actions of Comets and Planets upon one another, and which will be apt to increase, till this System wants a Reformation. Such a wonderful Uniformity in the Planetary System *must be allowed the Effect of Choice*. And so must the Uniformity in the Bodies of Animals." (*Opticks*, 402)

33 See especially Book III of the *Principia*, where Newton specifically analyzes the perturbations in planetary orbits.

34 Snobelen points out that Newton had a providential and dynamic view of the cosmos that paralleled his interpretation of biblical eschatology. For an extensive discussion of what that implied for Newton's view of divine action in the natural world and why it did not imply that he made God-of-the-gaps arguments, see Chapter 20, n. c, at www .returnofthegodhypothesis/extendedresearchnotes.

35 Snobelen, "Newton and the God of the Gaps"; see also 'God of Gods and Lord of Lords.'

36 See also Iliffe, *Priest of Nature.*
37 The first reflecting telescope was invented by James Gregory.

Chapter 21: The Big Questions and Why They Matter

1 Hawking, *Brief Answers to the Big Questions*, 23–38.
2 Hawking and Mlodinow, *The Grand Design*, 180.
3 By invoking spontaneous creation, Hawking was drawing an analogy to a physical process known as virtual particle production (or spontaneous particle/antiparticle production). In this process, described by the Heisenberg uncertainty principle, a particle can emerge spontaneously out of an energy-rich quantum field for a time provided an antiparticle with equivalent negative energy also arises. If the net energy of this process cancels out to equal zero total energy, it does not violate the law of the conservation of energy. Nevertheless, the mathematical equation that describes how this can occur describes an actual—that is, already existing—physical situation in which particles (and their virtual complements) arise out of a preexisting energy-rich space. The particle/antiparticle production is made possible by the prior existence of the energy-rich quantum field and occurs in a preexisting space. Thus, the analogy that Hawking draws is not apt. The laws of physics that describe this process do not apply to the origin of the universe itself, because before the universe existed there was no space or energy to draw on to drive the particle (universe) production.
4 Hawking and Mlodinow, *The Grand Design*, 180.
5 Hawking, *Brief Answers to the Big Questions*, 29.
6 Hawking, *Brief Answers to the Big Questions*, 38.
7 Hawking, *Brief Answers to the Big Questions*, 33.
8 Hawking, *Brief Answers to the Big Questions*, 38.
9 Dawkins, *River Out of Eden*, 17; see also 12, 18–20.
10 Dawkins, *River Out of Eden*, 133.
11 Weinberg, *The First Three Minutes*, 154.
12 David Masci, "Public Opinion on Religion and Science in the United States," *Pew Research Center*, November 5, 2009, http://www.pewforum.org/2009/11/05/public-opinion-on-religion-and-science-in-the-united-states.
13 The Pew poll also notes that, although the majority of Americans believe science and religion often conflict *in general*, the majority also regard science as not conflicting with their *particular* religious beliefs.
14 West, *Darwin's Corrosive Idea*, 3–7.
15 Maltz, *Psycho-Cybernetics*, back cover.
16 The quote I encountered actually came from the Christian philosopher Francis Schaeffer, who appears to have paraphrased and synthesized some of Jean-Paul Sartre's key ideas (see Schaeffer, *He Is There and He Is Not Silent*, 1). Here's a passage from Sartre that expresses some of the ideas that Schaeffer may have been summarizing: "The existentialist, on the contrary, finds it extremely embarrassing that God does not exist, for there disappears with Him all possibility of finding values in an intelligible heaven. There can no longer be any good *a priori*, since there is no infinite and perfect consciousness to think it. It is nowhere written that 'the good' exists, that one must be honest or must not lie, since we are now upon the plane where there are only men. Dostoievsky [*sic*] once wrote 'If God did not exist, everything would be permitted'; and that, for existentialism, is the starting point. Everything is indeed permitted if God does not exist, and man is in consequence forlorn" (Sartre, "Existentialism and Humanism," 70–71; see also 65–76). In his novel *Nausea* (1938), Sartre expresses the idea that the death of God has specifically left humankind without ultimate meaning.
17 Russell, *Mysticism and Logic and Other Essays*, 10–11.
18 Lewis, *Miracles: A Preliminary Study*, 102; see also 100–107.

19 Other thinkers have articulated arguments of a similar ilk as well. See, e.g., Menuge, *Agents Under Fire*; Willard, "Knowledge and Naturalism," 24–48; Lewis, *Miracles*; Reppert, *C. S. Lewis's Dangerous Idea*; Crisp, "On Naturalistic Metaphysics," 61–74.

20 More precisely, as Plantinga says elsewhere, "A belief has warrant for a person *S* only if that belief is produced in *S* by cognitive faculties functioning properly (subject to no dysfunction) in a cognitive environment that is appropriate for *S*'s kind of cognitive faculties, according to a design plan that is successfully aimed at truth" (*Warranted Christian Belief*, 156).

21 For a short summary of how Plantinga summarizes his own argument, see Chapter 21, n. a, at www.returnofthegodhypothesis/extendedresearchnotes. See also, Plantinga, *Where the Conflict Really Lies*, 314, emphasis in original.

22 Plantinga, "Evolution vs. Naturalism."

23 Koons, "The General Argument from Intuition." For related arguments, see also "Epistemological Objections to Materialism" and "The Incompatibility of Naturalism and Scientific Realism."

24 Darwin Correspondence Project, "Letter no. 13230." See also Darwin, *The Autobiography of Charles Darwin 1809–1882*, 92–93.

25 Plantinga, *Warrant and Proper Function*, 225.

26 Plantinga, *Warrant and Proper Function*, 225.

27 Dawkins, Interview with Ben Wattenberg.

28 In his more recent book *The God Delusion*, Dawkins accounts for the origin of religion in terms of "memetic natural selection" rather than "genetic natural selection." In this way, he attempts to distance natural selection from the production of cognitive equipment that enabled the origin, development, and promulgation of (false) religious beliefs. Yet in Dawkins's view, "genetic natural selection" undergirds the origin of all human cognitive faculties (indeed, of *all* of flora and fauna on earth), including human brains capable of forming and passing along "memes." This leaves his account of the origin of our cognitive equipment *ultimately* resting on natural selection, in combination with other evolutionary processes, as the fundamental force that enabled the development and spread of what he regards as a false belief (*The God Delusion*, chap. 5).

29 "The Global Religious Landscape," *Pew Research Center*, December 18, 2012, https://www.pewforum.org/2012/12/18/global-religious-landscape-exec; Barrett, *Born Believers*. For more extensive documentation of these claims, see Chapter 21, n. a, at www.returnofthegodhypothesis/extendedresearchnotes.

30 Gopnik, "See Jane Evolve." For more extensive documentation of these claims, see Chapter 21, n. a, at www.returnofthegodhypothesis/extendedresearchnotes.

31 See Pew Research Center, "The Changing Global Religious Landscape," "The Future of World Religions: Population Growth Projections, 2010–2050." And as Conrad Hackett and colleagues note in other research: "The religiously unaffiliated are projected to decline as a share of the world's population in the decades ahead because their net growth through religious switching will be more than offset by higher childbearing among the younger affiliated population" (Hackett et al., "The Future Size of Religiously Affiliated and Unaffiliated Populations," 829–42 [830]).

32 Plantinga argues that to justify the reliability of the mind, evolutionary naturalism requires that adaptive beliefs need to correlate with truth. As noted, he provides many reasons to doubt that coupling. Yet he also argues that other ways of conceiving of the relationship between belief and adaptive behavior also reinforce doubts about the reliability of the mind. Indeed, he notes that there are several different mutually exhaustive ways of conceiving of the relationship between the mind and body—and thus between cognitive states and behaviors—given philosophical naturalism. For each such way of conceiving of this relationship, he argues that the probability of the reliability of our belief-forming apparatus (and consequent beliefs) is either inscrutable or low.

For example, evolutionary naturalists might hold to (a) various epiphenomenalist views of mind-body interaction. Epiphenomenalism denies that either our beliefs or the specific semantic content of those beliefs affect our behaviors. It follows in this view that our beliefs would be invisible to natural selection and the probability of our possessing reliable beliefs given naturalism and evolution—that is, P(R | N + E)—would be low or inscrutable. Indeed, in this case, the action of natural selection would certainly not give us a reason to trust (or certify) the reliability for the mind.

Naturalists might also hold the view that beliefs *do* cause behaviors, but they are either (b) maladaptive and true or (c) adaptive and false. For both these cases he argues that the probability of possessing reliable beliefs given evolutionary naturalism, P(R | N + E), is again very low. In the case of (b), where beliefs are true but maladaptive, natural selection would weed out cognitive structures that produce such beliefs. In the case of (c), where beliefs are false but adaptive, natural selection would preserve cognitive structures responsible for producing false beliefs, again, casting doubt on the reliability of our cognitive equipment. Since Plantinga also offers many reasons for doubting that adaptive beliefs will necessarily be true, he concludes that there are good reasons to doubt the reliability of our belief-forming structures—given any conceivable conjunction of evolution and a naturalistic view of the relationship between belief and behavior or mind and body.

33 Plantinga uses this phrase in a number of texts, including, for example, *Warranted Christian Belief*, 231.

34 Plantinga's refinements to his original argument along with critics' objections and his replies can be found in the following: Plantinga, *Warranted Christian Belief*, 227–40, 281–84, 350–51; Beilby, ed., *Naturalism Defeated?*; Law, "Naturalism, Evolution, and True Belief," 41–48; Fitelson and Sober, "Plantinga's Probability Arguments Against Evolutionary Naturalism," 115–29; Plantinga, "Reliabilism, Analysis and Defeators"; "Probability and Defeators"; Plantinga and Tooley, *Knowledge of God*, 31–51, 227–32; Plantinga, "Content and Natural Selection"; *Where the Conflict Really Lies*, 307–50.

35 Plantinga, "Evolution vs. Naturalism"; see also *Warrant and Proper Function*, 236–37.

36 Plantinga, "Evolution vs. Naturalism," emphasis in original.

37 In *Where the Conflict Really Lies*, Plantinga takes his argument one step farther: he applies this line of thinking, along with other considerations, to the question, "Is theism or naturalism more compatible with science?" (265–350). He contends that, if evolution is true, then naturalists have a major problem: the conjunction of naturalism and evolution undermines the reliability of naturalists' cognitive faculties (as we have already noted). By contrast, the Judeo-Christian doctrine of the *imago dei* and other conceptual and metaphysical resources of Judeo-Christian theism provide a suitable (epistemological) ground for the pursuit of scientific knowledge. Thus, Plantinga concludes, "On balance, theism is vastly more hospitable to science than naturalism" (309).

38 Nagel, *Mind & Cosmos*.

39 According to John Calvin, the *sensus divinitatis* is a natural, inborn "conviction" in all human beings "that there is some God" (McNeill, *Institutes of the Christian Religion*, bk. 1, ch. 3, 46). For a contemporary development of this doctrine, see Plantinga, *Warranted Christian Belief*.

40 Nagel, *The Last Word*, 130–31.

41 Krauss, *A Universe from Nothing*, xii.

42 "Lawrence Krauss: Atheism and the Spirit of Science." Or, as Richard Dawkins says bluntly in the Afterword to Krauss's *A Universe from Nothing*: "Reality doesn't owe us comfort" (188).

43 This line is in the third stanza of Voltaire's poetic reply to the book *The Three Impostors*. See Voltaire, "Epître à l' auteur du livre des *Trois imposteurs*," 10: 402–5. The text can also be found in French and in English at https://www.whitman.edu/VSA/trois.imposteurs.html.

44 Krauss, "Our Godless Universe Is Precious."

45 Krauss, "Our Godless Universe Is Precious."
46 Krauss, "The Universe Doesn't Give a Damn about Us."
47 Sartre, "Existentialism and Humanism," 69; see also 70–76. In this particular version, the translator uses "abandonment" rather than "forlornness," but the essential concept remains the same.
48 Sartre, "Existentialism and Humanism," 69–76.
49 Frankl, *Man's Search for Meaning*.

Epilogue: Response to Critics

1 Brian Miller, "Darrell Falk Badly Mischaracterizes RNA World Experiments . . . and Stephen Meyer," *Evolution News and Science Today* (May 28, 2021), https://evolutionnews.org/2021/05/darrell-falk-badly-mischaracterizes-rna-world-experiments-and-stephen-meyer/.
2 Lawrence Krauss, "Cosmology Without Design," *Inference: An International Review of Science* 5, no. 3 (September 2020), https://inference-review.com/article/cosmology-without-design.
3 Luke A. Barnes, "The Fine Tuning of the Universe for Intelligent Life," *Publications of the Astronomical Society of Australia* (Cambridge: Cambridge Univ. Press, 2012), http://doi.org/10.1071/AS12015.
4 Juliane Barbour, "Inside Penrose's Universe," *Physics World* (December 6, 2010), https://physicsworld.com/a/inside-penroses-universe/.

Bibliography

"The Abrams Report for September 29, 2005." *NBC News*. http://www.nbcnews.com/id/9542288/ns/msnbc-the_abrams_report/t/abrams-report-september.

Adams, Fred C., and Evan Grohs. "Stellar Helium Burning in Other Universes: A Solution to the Triple-Alpha Fine-Tuning Problem." *Astroparticle Physics* 87 (2017): 40–54.

Agassiz, Louis. "Evolution and the Permanence of Type." *Atlantic Monthly* 33 (January 1874): 430–45.

———. "The Primitive Diversity and Number of Animals in Geological Times." *American Journal of Sciences*, 2d ser. 17 (1854): 309–24.

Aguirre, Anthony, and John Kehayias. "Quantum Instability of the Emergent Universe." arXiv:1306.3232v2 [hep-th] 19 Nov 2013.

Alberts, Bruce D., et al. *Molecular Biology of the Cell*. New York: Garland, 1983.

Albrecht, Andreas, and Paul Steinhardt. "Cosmology for Grand Unified Theories with Radiatively Induced Symmetry Breaking." *Physical Review Letters* 48/17 (1982): 1220–23.

Albright, John R. "God of the Gaps." In *Encyclopedia of Sciences and Religions*, edited by Anne L. C. Runehov and Lluis Oviedo. Dordrecht: Springer, 2013.

Allen, D. *Mechanical Explanations and the Ultimate Origin of the Universe According to Leibniz*. Wiesbaden: Franz Steiner, 1983.

Alpher, R. A., H. Bethe, and G. Gamow. "The Origin of Chemical Elements." *Physical Review* 73/7 (April 1, 1948): 803–4.

Alpher, Victor S. "Ralph A. Alpher, George Antonovich Gamow, and the Prediction of the Cosmic Microwave Background Radiation." *Asian Journal of Physics*, 23/1 and 2 (2014), 17–26.

Aquinas, Thomas. *Summa contra Gentiles* 2:17. In "On Creation and Time | Inters.Org." Accessed March 12, 2020, http://inters.org/contra-gentiles-creation.

"Are Gaps in Scientific Knowledge Evidence for God?" *BioLogos*. January 19, 2019. https://biologos.org/common-questions/gods-relationship-to-creation/god-of-the-gaps.

Aristotle, *Metaphysics* 1.3 translated by C. D. C. Reeve (Indianapolis: Hackett Publishing Company, 2016): 7–8.

———. *Physics* III, Ch. 6. "Logos Virtual Library: Aristotle: Physics, III, 6." Accessed March 12, 2020, http://www.logoslibrary.org/aristotle/physics/36.html.

Ashby, W. Ross. "Requisite Variety and Its Implications for the Control of Complex Systems." In *Facets of Systems Science*, by George J. Klir, 405–17. New York: Plenum, 1991.

Augustine of Hippo. *Confessions*, bk. 12, chap. 7. In *Confessions and Enchiridion*, translated and edited by Albert C. Outler. Philadelphia: Westminster, 1955.

Axe, Douglas. "Answering Objections from Martin Poenie." In *Debating Darwin's Doubt*, edited by David Klinghoffer, 163–66. Seattle: Discovery Institute Press, 2015.

———. "Estimating the Prevalence of Protein Sequences Adopting Functional Enzyme Folds." *Journal of Molecular Biology* 341 (2004): 1295–1315.

———. "More on Objections from Martin Poenie." In *Debating Darwin's Doubt*, edited by David Klinghoffer, 167–72. Seattle: Discovery Institute Press, 2015.

———. "Show Me: A Challenge for Martin Poenie." In *Debating Darwin's Doubt*, edited by David Klinghoffer, 182–83. Seattle: Discovery Institute Press, 2015.

Ayala, Francisco. "Darwin's Greatest Discovery: Design Without Designer." *Proceedings of the National Academy of Sciences* 104 (2007): 8567–73.

———. "Darwin's Revolution." In *Creative Evolution?!* edited by J. Campbell and J. Schopf, 4–5. Boston: Jones and Bartlett, 1994.

Baade, Walter. "Problems in the Determination of the Distance of Galaxies." *Astronomical Journal* 63 (May 1958): 207–10. doi:10.1086/107726.

Baggott, Jim. *Quantum Space: Loop Quantum Gravity and the Search for the Structure of Space, Time, and the Universe.* Oxford: Oxford Univ. Press, 2018.

———. *The Quantum Story: A History in Forty Moments.* Oxford: Oxford Univ. Press, 2011.

Barash, David. "God, Darwin and My College Biology Class." *New York Times*, September 28, 2014. https://www.nytimes.com/2014/09/28/opinion/sunday/god-darwin-and-my-college-biology-class.html.

Barbour, Ian G. *Religion and Science: Historical and Contemporary Issues.* San Francisco: HarperSanFrancisco, 1997.

Barnes, Luke. "Binding the Diproton in Stars: Anthropic Limits on the Strength of Gravity." *Journal of Cosmology and Astroparticle Physics* 2015/12 (December 29, 2015): 050.

———. "The Fine-Tuning of the Universe for Intelligent Life." *Publications of the Astronomical Society of Australia* 29/4 (2013): 529–64.

Barr, S. M., and Almas Khan. "Anthropic Tuning of the Weak Scale and of the Mu/Md in Two-Higgs' Doublet Models." *Physical Review D–Particles, Fields, Gravitation, and Cosmology* 76, no. 4 (August 6, 2007), https://doi.org/10/1103?PhysRevD.76.045002.

Barrett, Justin. *Born Believers: The Science of Childhood Religion.* New York: Free Press, 2012.

Barrow, John D. "Anthropic Definitions." *Quarterly Journal of the Royal Astronomical Society* 24 (1983): 150.

Barrow, John, and Frank Tipler. *The Anthropic Cosmological Principle.* Oxford: Oxford Univ. Press, 1988.

Beauregard, Mario and Denyse O'Leary. *The Spiritual Brain: A Neuroscientist's Case for the Existence of the Soul.* San Francisco: HarperOne, 2008.

Bechly, Günter, and Stephen C. Meyer. "The Fossil Record and Universal Common Ancestry." In *Theistic Evolution: A Scientific, Philosophical and Theological Critique*, edited by J. P. Moreland et al., 323–53. Wheaton, IL: Crossway, 2017.

Bechly, Günter, Brian Miller, and David Berlinski. "Right of Reply: Our Response to Jerry Coyne." *Quillette*, September 29, 2019. https://quillette.com/2019/09/29/right-of-reply-our-response-to-jerry-coyne.

Beilby, James, ed. *Naturalism Defeated?: Essays on Plantinga's Evolutionary Argument Against Naturalism.* Ithaca: Cornell Univ. Press, 2002.

Bell, J. S. "Against 'Measurement.'" In *Sixty-Two Years of Uncertainty: Historical, Philosophical, and Physical Inquiries into the Foundations of Quantum Mechanics*, edited by Arthur I. Miller, 17–31. New York: Plenum, 1990.

Bena, Iosif, and Mariana Graña. "String Cosmology and the Landscape." *Comptes Rendus Physique* 18/3–4 (March–April 2017): 200–206.

Bentley, Richard. *Eight Sermons Preach'd at the Honourable Robert Boyle's Lecture, in the First Year 1692.* Cambridge: Cornelius Crownfield, 1724.

———. Letter from Richard Bentley to Isaac Newton, February 18, 1692/3. 189.R.4.47, ff. 3–4, Cambridge: Trinity College Library, 1693.

Berlinski, David. "Responding to Stephen Fletcher's Views in the *Times Literary Supplement* on the RNA World." https://davidberlinski.org/2010/01/15/responding-to-stephen-flethers-views-in-the-times-literary-supplement-on-the-rna-world.

———. *The Deniable Darwin.* David Klinghoffer, ed. Seattle: Discovery Institute Press, 2009.

———. "The Deniable Darwin." In *The Deniable Darwin*, edited by David Klinghoffer, 41–64. Seattle: Discovery Institute Press, 2009. Reprinted from *Commentary*, June 1996.

Berlinski, David, and Tyler Hampton. "Hopeless Matzke." In *Debating Darwin's Doubt*, edited by David Klinghoffer, 100–113. Seattle: Discovery Institute Press, 2015.

Bershtein, Shimon, et al. "Robustness–Epistasis Link Shapes the Fitness Landscape of a Randomly Drifting Protein." *Nature* 444/7121 (2006): 929–32.

Billings, Lee. "Cosmic Conflict: Diverging Data on Universe's Expansion Polarizes Scientists." *Scientific American*, May 16, 2018. Accessed March 21, 2020, https://www.scientificamerican.com/article/cosmic-conflict-diverging-data-on-universes-expansion-polarizes-scientists1/.

Bingham, John. "Richard Dawkins: I Can't Be Sure God Does Not Exist." *The Telegraph*, February 24, 2012. https://www.telegraph.co.uk/news/religion/9102740/Richard-Dawkins-I-cant-be-sure-God-does-not-exist.html.

Boethius of Dacia. *On the Supreme Good; On the Eternity of the World; On Dreams* (*De Summo Bono; De Aeternitate Mundi; De Somnis*). Translated by John F. Wippel, Mediaeval Sources in Translation. Montmagny: Les Éditions Marquis Ltée, 1987.

Bojowald, Martin. *Quantum Cosmology: A Fundamental Description of the Universe*. New York: Springer, 2011.

Bojowald, Martin, Claus Kiefer, and Paulo Vargas Moniz. "Quantum Cosmology for the XXIst Century: A Debate." In *The Twelfth Marcel Grossmann Meeting: On Recent Developments in Theoretical and Experimental General Relativity, Astrophysics and Relativistic Field Theories* (in 3 Volumes), edited by T. Damour, R. Jantzen, and R. Ruffini, Part A: 589–608. Singapore: World Scientific, 2012.

Bonaventure. *Commentaries on the Sentences of Peter Lombard*, Book II. In *Commentary on the Sentences: Philosophy of God*, Vol. 16 of Works of St. Bonaventure, translated by R. E. Houser and Timothy B. Noone, with introduction and notes. New York: Franciscan Institute Press, 2014.

Bondi, Hermann, and Thomas Gold. "The Steady-State Theory of the Expanding Universe." *Monthly Notices of the Royal Astronomical Society* 108 (1948): 252–70.

Borde, Arvind, Alan H. Guth, and Alexander Vilenkin. "Inflationary Spacetimes Are Incomplete in Past Directions." *Physical Review Letters* 90/15 (April 15, 2003): 151301.

Borde, Arvind, and Alexander Vilenkin. "Eternal Inflation and the Initial Singularity." *Physical Review Letters* 72 (1994): 3305–9.

———. "Violation of the Weak Energy Condition in Inflating Spacetimes." *Physical Review D* 56 (1997): 717–23.

Born, Max. "Zur Quantenmechanik der Stoßvorgänge (On the Quantum Mechanics of Collision Processes)." *Zeitschrift für Physik* 38 (1926): 803–27.

Bousso, Raphael, and Ben Freivogel. "A Paradox in the Global Description of the Multiverse." *Journal of High Energy Physics* 2007/06 (2007): 018.

Bousso, Raphael, Ben Freivogel, and I-Sheng Yang. "Boltzmann Babies in the Proper Time Measure." *Physical Review D* 77/10 (2008): 103514.

Bousso, Raphael, and Joseph Polchinski. "The String Theory Landscape." *Scientific American* 291/3 (September 2004): 78–87.

Bowles, Alexander M. C., Ulrike Bechtold, and Jordi Paps. "The Origin of Land Plants Is Rooted in Two Bursts of Genomic Novelty." *Current Biology* 30/3 (February 3, 2020): 530–36.

Boyle, Robert. *A Defense of the Doctrine Touching the Spring and Weight of the Air*. London: Thomas Robinson, 1662.

———. "A Disquisition about the Final Causes of Natural Things." In *The Works of Robert Boyle, vol. 11*, edited by Edward B. Davis and Michael Hunter, 14 volumes, 79–152. London: Pickering and Chatto, 1999–2000.

———. "A Free Enquiry into the Vulgarly Receiv'd Notion of Nature." In *The Works of Robert Boyle, vol. 10*, edited by Edward B. Davis and Michael Hunter, 14 volumes, 437–571. London: Pickering and Chatto, 1999–2000.

———. *New Experiments Physico-Mechanical, Touching the Spring of the Air, and Their Effects: Whereunto Is Added a Defence of the Authors Explication of the Experiments Against the Objections of Franciscus Linus and Thomas Hobbes*. Oxford: H. Hall, 1662.

———. *Selected Philosophical Papers of Robert Boyle*. Edited by M. A. Stewart. Manchester: Manchester Univ. Press, 1979.

———. "Of the Study of the Book of Nature." In *The Works of Robert Boyle, vol. 13*, edited by Edward B. Davis and Michael Hunter, 14 volumes, 147–172. London: Pickering and Chatto, 1999–2000.

———. "The Usefulness of Natural Philosophy." In *The Works of Robert Boyle, vol. 3*, edited by Edward B. Davis and Michael Hunter, 14 volumes, 189–560. London: Pickering and Chatto, 1999–2000.

———. *The Works of Robert Boyle*. Edited by Michael Hunter and Edward B. Davis. 14 vols. London: Pickering and Chatto, 1999–2000.

Bozza, Valerio, and Gabriele Veneziano. "Scalar Perturbations in Regular Two-Component Bouncing Cosmologies." *Physics Letters B* 625 (2005): 177–83.

Bridgewater, Francis Henry Egerton. *The Bridgewater Treatises on the Power, Wisdom, and Goodness of God, as Manifested in the Creation. Treatise I–VIII.* London: Pickering, 2012, 1834–39.

Briggs, David. "Science, Religion Are Discovering Commonality in Big Bang Theory." *Los Angeles Times*, May 2, 1992, B6–7. https://www.latimes.com/archives/la-xpm-1992-05-02-me-1350-story.html.

Brooke, John Hedley. "Myth 25: That Modern Science Has Secularized Modern Culture." In *Galileo Goes to Jail and Other Myths About Science and Religion*, edited by Ronald Numbers, 224–34. Cambridge, MA: Harvard Univ. Press, 2009.

———. "Science and Theology in the Enlightenment." In *Religion and Science: History, Method, Dialogue*, edited by Mark W. Richardson and Wesley J. Wildman, 7–28. New York: Routledge, 1996.

Brown, Gregory. "Is the Logic in London Different from the Logic in Hanover?" In *Leibniz and the English-Speaking World*, edited by Pauline Phemister and Stuart Brown, 145–62. Dordrecht: Springer, 2007.

Browne, Malcolm W. "Clues to Universe Origin Expected." *New York Times*, March 12, 1978, 54.

Brush, Stephen G. "Prediction and Theory Evaluation: The Case of Light Bending." *Science* 246 (1989): 1124–29.

Buchdahl, Gerd. *Metaphysics and the Philosophy of Science*. Oxford: Blackwell, 1969.

Burbidge, E. M., et al. "Synthesis of the Elements in Stars." *Reviews of Modern Physics* 29/4 (1957): 547–650.

Butterfield, Herbert. *The Origins of Modern Science*. New York: Free Press, 1957.

Butterfield, Jeremy, and Christopher Isham. "On the Emergence of Time in Quantum Gravity." In *The Arguments of Time*, edited by Jeremy Butterfield, 111–68. Oxford: Oxford Univ. Press, 1999.

———. "Spacetime and the Philosophical Challenge of Quantum Gravity." In *Physics Meets Philosophy at the Planck Scale*, edited by Craig Callender and Nick Huggett, 33–89. Cambridge: Cambridge Univ. Press, 2001.

Callender, Craig, and Nick Huggett, eds. *Physics Meets Philosophy at the Planck Scale: Contemporary Theories in Quantum Gravity.* Cambridge: Cambridge Univ. Press, 2001.

Calvin, Melvin. *Chemical Evolution*. Oxford: Clarendon, 1969.

Carr, Bernard. "Introduction and Overview." In *Universe or Multiverse?* edited by Bernard Carr, 3–28. Cambridge: Cambridge Univ. Press, 2007.

Carr, B., and M. Rees. "The Anthropic Principle and the Structure of the Physical World." *Nature* 278 (1979): 605–12.

Carroll, Bradley W., and Dale A. Ostlie. *An Introduction to Modern Astrophysics*. 2nd ed. Cambridge: Cambridge Univ. Press, 2017.

Carroll, Sean. "Against Bounces." *Discover*, July 2, 2007. https://www.discovermagazine.com/the-sciences/against-bounces.

———. *The Big Picture: On the Origins of Life, Meaning, and the Universe Itself.* New York: Dutton, 2017.

———. *From Eternity to Here*. New York: Penguin, 2010.
———. "Turtles Much of the Way Down." *Discover*, November 25, 2007. https://www.discovermagazine.com/the-sciences/turtles-much-of-the-way-down.
———. "Why Boltzmann Brains Are Bad." February 2, 2017. arXiv:1702.00850.
Carroll, Sean, and H. Tam. "Unitary Evolution and Cosmological Fine-tuning." July 8, 2010. arXiv:1007.1417.
Carter, Brandon. "Large Number Coincidences and the Anthropic Principle in Cosmology." In *Confrontation of Cosmological Theories with Observational Data*, edited by M. S. Longair, 291–98. Dordrecht: Reidel, 1974.
Chaberek, Michael Fr. *Aquinas and Evolution*. British Columbia: Chartwell Press, 2017.
Chaisson, Eric, and Steve McMillan. *Astronomy Today*. Englewood Cliffs, NJ: Prentice Hall, 1993.
Chaitin, Gregory J. *Algorithmic Information Theory*. Cambridge: Cambridge Univ. Press, 2004.
Chamberlain, Thomas C. "The Method of Multiple Working Hypotheses." *Science* (old series) 15 (1890): 92–96. Reprinted in *Science* 148 (1965): 754–59. Also reprinted in *Journal of Geology* (1931): 155–65.
Chiarabelli, C., et al. "Investigation of De Novo Totally Random Biosequences, Part II: On the Folding Frequency in a Totally Random Library of De Novo Proteins Obtained by Phage Display." *Chemistry and Biodiversity* 3/8 (August 2006): 840–59.
Clarke, Samuel. *A Collection of Papers, Which Passed Between the Late Learned Mr. Leibnitz* [*sic*] *and Dr. Clarke, in the Years 1715 and 1716*. London: James Knapton, 1717.
———. *A Demonstration of the Being and Attributes of God and Other Writings*. Edited by E. Vailati. Cambridge: Cambridge Univ. Press, 1998.
———. *The Works*. Edited by B. Hoadly. Vol. 2. London: Garland, 2002.
Cleland, Carol E. "Historical Science, Experimental Science, and the Scientific Method." *Geology* 29 (2001): 987–90.
———. "Methodological and Epistemic Differences Between Historical Science and Experimental Science." *Philosophy of Science* 69 (2002): 474–96.
Collingwood, Richard G. *The Idea of Nature*. 1945. New York: Oxford Univ. Press, 2014.
Collins, Francis. *The Language of God: A Scientist Presents Evidence for Belief*. New York: Free Press, 2006.
Collins, Graham P. "The Many Interpretations of Quantum Mechanics." *Scientific American* 297/5 (November 19, 2007): 19.
Collins, Robin. "Evidence for Fine-Tuning." In *God and Design: The Teleological Argument and Modern Science*, edited by Neil A. Manson, 178–99. New York: Routledge, 2003.
———. "The Fine-Tuning Design Argument." In *Reason for the Hope Within*, edited by Michael Murray, 60–61. Grand Rapids, MI: Eerdmans, 1998.
———. "The Fine-Tuning Evidence Is Convincing." In *Debating Christian Theism*, edited by J. P. Moreland, Chad Meister, and Khaldoun A. Sweis, 35–46. New York: Oxford Univ. Press, 2013.
———. "The Multiverse Hypothesis: A Theistic Perspective." In *Universe or Multiverse?* edited by Bernard Carr, 459–80. Cambridge: Cambridge Univ. Press, 2007.
———. "The Teleological Argument: An Exploration of the Fine-Tuning of the Universe." In *The Blackwell Companion to Natural Theology*, edited by William Lane Craig and J. P. Moreland, 202–81. Malden, MA: Wiley-Blackwell, 2009.
Comte, Auguste. *The Positive Philosophy of Auguste Comte*. Translated by Harriet Martineau. New York: Calvin Blanchard, 1858.
Cooke, Michael. "An Introduction to Quantum Cosmology." Master's thesis, Imperial College, London, September 24, 2010.
Copan, Paul, and William Lane Craig. *Creation Out of Nothing: A Biblical, Philosophical, and Scientific Exploration*. Grand Rapids, MI: Baker Academic, 2004.
Courtenay, W. "The Dialectic of Omnipotence in the High and Late Middle Ages." In *Divine Omniscience and Omnipotence in Medieval Philosophy*. Edited by T. Ruduvsky. Dordrecht, Netherlands: D. Reidel, 1985, 243–69.

Coyne, Jerry. "God vs. Physics: Krauss Debates Meyer and Lamoureux." *Why Evolution Is True* (blog), March 20, 2016. https://whyevolutionistrue.wordpress.com/2016/03/20/god-vs-physics-krauss-debates-meyer-and-lamoureaux/#comment-1316386.

Craig, William L. *The Cosmological Argument from Plato to Leibniz*. Eugene, OR: Wipf & Stock, 2001.

———. "Design and the Anthropic Fine-Tuning of the Universe." In *God and Design: The Teleological Argument and Modern Science*, edited by Neil A. Manson, 155–77. New York: Routledge, 2003.

———. *The Kalām Cosmological Argument*. Eugene, OR: Wipf & Stock, 2000.

———. *Reasonable Faith: Christian Truth and Apologetics.* 3rd ed. Wheaton, IL: Crossway, 2008.

———. "The Teleological Argument and the Anthropic Principle." In *The Logic of Rational Theism: Exploratory Essays*, edited by William Lane Craig and Mark S. McLeod, 127–53. Problems in Contemporary Philosophy 24. Lewiston, NY: Edwin Mellen, 1990.

———. "The Ultimate Question of Origins: God and the Beginning of the Universe." *Reasonable Faith*. Accessed November 23, 2018, https://www.reasonablefaith.org/writings/scholarly-writings/the-existence-of-god/the-ultimate-question-of-origins-god-and-the-beginning-of-the-universe.

———. "'What Place, Then, for a Creator?': Hawking on God and Creation." *British Journal for the Philosophy of Science* 41/4 (December 1990): 473–91.

Craig, William Lane, and James D. Sinclair. "The *Kalam* Cosmological Argument." In *The Blackwell Companion to Natural Theology*, edited by W. L. Craig and J. P. Moreland: 101–201. Oxford: Blackwell Publishing, 2009.

Crick, Francis. *Life Itself: Its Origin and Nature*. New York: Simon & Schuster, 1981.

———. "On Protein Synthesis." *Symposium for the Society of Experimental Biology* 12 (1958): 138–63.

———. *What Mad Pursuit: A Personal View of Scientific Discovery*. New York: Basic Books, 1988.

Crick, Francis, and Leslie Orgel. "Directed Panspermia." *Icarus* 19 (1973): 341–46.

Crisp, Thomas. "On Naturalistic Metaphysics." In *The Blackwell Companion to Naturalism*, edited by Kelly James Clark. Hoboken, NJ: Wiley, 2016.

Crombie, Alistair C. *The History of Science from Augustine to Galileo*, 2 vols. 1952. Reprint, Cambridge, MA: Harvard Univ. Press, 1979.

———. *Robert Grosseteste and the Origins of Experimental Science 1100–1700*. Oxford: Oxford Univ. Press, 1953.

Csoto, Attila, Heinz Oberhummer, and Helmut Schlattl. "Fine-Tuning the Basic Forces of Nature Through the Triple-Alpha Process in Red Giant Stars." *Nuclear Physics A* 688/1–2 (2001): 560–62.

Curiel, Erik. "A Primer on Energy Conditions." In *Towards a Theory of Spacetime Theories*, edited by Dennis Lehmkuhl, Gregor Schiemann, and Erhard Scholz. Einstein Studies, 13: 43–104. New York: Birkhäuser, 2017.

Custance, Arthur C. *The Mysterious Matter of Mind* (second online edition, 2001; originally published by Probe Ministries and Zondervan Publishing, 1980. http://www.custance.org/Library/MIND/.

Dales, Richard C. *Medieval Discussions of the Eternity of the World*. Leiden: Brill, 1990.

Darwin, Charles. *The Autobiography of Charles Darwin, 1809–1882*. Edited by Nora Barlow. New York: Norton, 1958.

———. *The Life and Letters of Charles Darwin, Including an Autobiographical Chapter.* Edited by Francis Darwin. 2 vols. New York: Appleton, 1898.

———. *On the Origin of Species by Means of Natural Selection.* A facsimile of the first edition, published by John Murray, London, 1859. Reprint, Cambridge, MA: Harvard Univ. Press, 1964.

———. *The Origin of Species: 150th Anniversary Edition.* New York: Signet, 2003.

Darwin Correspondence Project. "Letter no. 13230." http://www.darwinproject.ac.uk/DCP-LETT-13230.

Daston, Lorraine, and Michael Stolleis, *Natural Law and Laws of Nature in Early Modern*

Europe: Jurisprudence, Theology, Moral and Natural Philosophy. Burlington, VT: Ashgate, 2008.
Date, Ghanashyam, and Golam Mortuza Hossain. "Genericness of Inflation in Isotropic Loop Quantum Cosmology." *Physical Review Letters* 94/1 (2005): 011301.
Davidson, Eric H. "Evolutionary Bioscience as Regulatory Systems Biology." *Developmental Biology* 357 (2011): 35–40.
———. *The Regulatory Genome: Gene Regulatory Networks in Development and Evolution*. Burlington, MA: Academic, 2006.
Davidson, Eric, and Douglas Erwin. "An Integrated View of Precambrian Eumetazoan Evolution." *Cold Spring Harbor Symposia on Quantitative Biology* 74 (2009): 65–80.
Davies, Paul. *The Accidental Universe*. Cambridge: Cambridge Univ. Press, 1982.
———. *The Cosmic Blueprint*. New York: Simon & Schuster, 1988.
———. *Superforce: The Search for a Grand Unified Theory of Nature*. New York: Simon & Schuster, 1985.
Davies, P. C. W. "Spacetime Singularities in Cosmology." In *The Study of Time III*, edited by J. T. Fraser, N. Lawrence, and D. Park, 74–93. New York: Springer, 1978.
Davis, Edward B. (Ted). "The Faith of a Great Scientist: Robert Boyle's Religious Life, Attitudes, and Vocation." *BioLogos*, August 8, 2013. https://biologos.org/articles/the-faith-of-a-great-scientist-robert-boyles-religious-life-attitudes-and-vocation.
———. "Newton's Rejection of the 'Newtonian World View': The Role of Divine Will in Newton's Natural Philosophy." In *Facets of Faith and Science*. Vol. 3, *The Role of Beliefs in the Natural Sciences*, edited by Jitse M. van der Meer, 103–17. Lanham, MD: Univ. Press of America, 1996.
Davisson, C. J. "The Diffraction of Electrons by a Crystal of Nickel." *Bell System Technical Journal* 7/1 (January 1928): 90–105. doi:10.1002/j.1538-7305.1928.tb00342.x.
Dawkins, Richard. *The Blind Watchmaker: Why the Evidence of Evolution Reveals a Universe Without Design*. New York: Norton, 1986.
———. *The God Delusion*. Boston: Houghton Mifflin, 2006.
———. "The Improbability of God." Richard Dawkins Foundation for Reason & Science, June 17, 2014. https://www.richarddawkins.net/2014/06/the-improbability-of-god. Originally published in *Free Enquiry* 18/3 (Summer 1998).
———. *River Out of Eden: A Darwinian View of Life*. New York: Basic Books, 1995.
———. Interview with Ben Wattenberg. Think Tank, 1996. http://www.pbs.org/thinktank/transcript410.html.
DeCross, Matthew P., et al., "Preheating after Multifield Inflation with Nonminimal Couplings. I. Covariant Formalism and Attractor Behavior." *Physical Review* D 97/2 (January 26, 2018): 023526.
de Duve, Christian. "The Beginnings of Life on Earth." *American Scientist* 83/5 (1995): 428–37.
———. *Blueprint for a Cell: The Nature and Origin of Life*. Burlington, NC: Neil Patterson, 1991.
———. "The Constraints of Chance." *Scientific American* 274/1 (January 1996): 112.
———. *Vital Dust: Life as a Cosmic Imperative*. New York: Basic Books, 1995.
Dellago, C., and H. A. Posch. "Kolmogorov-Sinai Entropy and Lyapunov Spectra of a Hard-Sphere Gas." *Physica A: Statistical Mechanics and Its Applications* 240/1–2 (1997): 68–83.
Dembski, William A. *The Design Inference: Eliminating Chance Through Small Probabilities*. Cambridge: Cambridge Univ. Press, 1998.
Dembski, William A., and Stephen C. Meyer. "Fruitful Interchange or Polite Chitchat? The Dialogue Between Science and Theology." *Zygon* 33/3 (September 1998): 415–30.
Dennett, Daniel. *Breaking the Spell: Religion as a Natural Phenomenon*. New York: Penguin, 2006.
———. *Darwin's Dangerous Idea: Evolution and the Meanings of Life*. New York: Simon & Schuster, 1995.

Denton, Michael. *Evolution: A Theory in Crisis.* London: Adler and Adler, 1985.
———. *Nature's Destiny: How the Laws of Biology Reveal Purpose in the Universe.* New York: Free Press, 1998.
Descartes, René. *Principles of Philosophy.* Translated and edited by V. R. Miller and R. P. Miller. Dordrecht: Kluwer Academic, 1983.
———. *The World.* Translated by Michael S. Mahoney. 1633. Reprint, New York: Abaris, 1979.
De Vries, Hugo. *Species and Varieties: Their Origin by Mutation*, 2nd ed., ed. Daniel Trembly MacDougal. Chicago: Open Court, 1906, http://www.gutenberg.org/files/7234/7234-h/7234-h.htm).
Dicke, Robert H. "Dirac's Cosmology and Mach's Principle." *Nature* 192 (1961): 440–42.
Dicke, Robert H., et al. "Cosmic Black-Body Radiation." *Astrophysical Journal Letters* 142 (1965): 415–19.
Dimopoulos, Antoniadis S. "Splitting Supersymmetry in String Theory." *Nuclear Physics.* B715 (2005): 120–40.
Dobzhansky, Theodosius. "Discussion of G. Schramm's Paper." In *The Origins of Prebiological Systems and of Their Molecular Matrices*, edited by Sidney W. Fox, 309–15. New York: Academic, 1965.
Dose, K. "The Origin of Life: More Questions Than Answers." *Interdisciplinary Science Review* 13 (1988): 348–56.
Douglas, A. Vibert. "Forty Minutes with Einstein." *Journal of the Royal Astronomical Society of Canada* 50 (1956): 99–102.
Dowden, Bradley. "Time." In *Internet Encyclopedia of Philosophy*, edited by James Fieser and Bradley Dowden. http://www.iep.utm.edu/time.
Draper, John W. *History of the Conflict Between Religion and Science.* New York: Appleton, 1874.
Drees, Willem B. "Interpretation of 'The Wave Function of the Universe.'" *International Journal of Theoretical Physics* 26/10 (October 1987): 939–42. doi:10.1007/BF00670817.
Dretske, Fred I. *Knowledge and the Flow of Information.* Cambridge, MA: MIT Press, 1981.
Duhem, Pierre M. M. *The System of World: A History of Cosmological Doctrines from Plato to Copernicus (Le système du monde: histoires des doctrines cosmologiques de Platon à Copernic).* Paris: Librairie Scientifique A. Hermann et Fils, 1913.
Durston, K. K., et al. "Measuring the Functional Sequence Complexity of Proteins." *Theoretical Biology and Medical Modelling* 4 (2007): 47.
Eddington, Arthur S. "The Deflection of Light During a Solar Eclipse." *Nature* 104 (1919): 372.
———. "The End of the World: From the Standpoint of Mathematical Physics." *Nature* 127 (1956): 450.
Eden, M. "Inadequacies of Neo-Darwinian Evolution as a Scientific Theory." In *Mathematical Challenges to the Neo-Darwinian Interpretation of Evolution*, edited by P. S. Moorhead and M. M. Kaplan, 5–19. Wistar Institute Symposium Monograph No. 5. New York: Liss, 1967.
Eigen, Manfred. *Steps Towards Life.* Oxford: Oxford Univ. Press, 1992.
Einstein, Albert. "Die Feldgleichungen der Gravitation (The Field Equations of Gravitation)." *Sitzungsberichte der Königlich Preussischen Akademie der Wissenschaften* 48 (November 25, 1915): 844–47.
———. "Die Grundlage der allgemeinen Relativitätstheorie (The Foundation of the General Theory of Relativity)." *Annalen der Physik* 49 (1916): 769–822.
———. "Kosmologische Betrachtungen zur allgemeinen Relativitätstheorie (Cosmological Considerations in the General Theory of Relativity)." *Sitzungsberichte der Königlich Preussischen Akademie der Wissenschaften* (Febuary 8, 1917): 142–52.
———. "Note on the Work of A. Friedmann 'On the Curvature of Space.'" *Zeitschrift für Physik* 11/32 (1922).
———. "Note on the Work of A. Friedmann 'On the Curvature of Space.'" *Zeitschrift für Physik* 16/228 (1923).

———. "Über einen die Erzeugung und Verwandlung des Lichtes betreffenden heuristischen Gesichtspunkt (On a Heuristic Point of View About the Creation and Conversion of Light)." *Annalen der Physik* 17/6 (1905): 132–48. https://onlinelibrary.wiley.com/doi/abs/10.1002/andp.19053220607.

Eiseley, Loren. *Darwin's Century: Evolution and the Men Who Discovered It.* Garden City, NY: Doubleday, 1958.

Ekström, S., et al. "Effects of the Variation of Fundamental Constants on Population III Stellar Evolution." *Astronomy and Astrophysics* 514 (May 2010): A62.

Ell, Douglas. *Counting to God: A Personal Journey Through Science to Belief.* Attitude Media, 2004.

Ellis, George. "Cosmology: The Untestable Multiverse." *Nature* 469 (January 20, 2011): 294–95. doi:10.1038/469294a.

Epelbaum, E., et al. "Dependence of the Triple-Alpha Process on the Fundamental Constants of Nature." *European Physics Journal* A 49 (2013): id 82.

Everett, Hugh. "The Theory of the Universal Wave Function." In *The Many-Worlds Interpretation of Quantum Mechanics*, edited by Bryce DeWitt and Neil Graham, 3–140. Princeton, NJ: Princeton Univ. Press, 1973.

Ewert, Winston, et al. "Efficient Per Query Information Extraction from a Hamming Oracle." *42nd Southeastern Symposium on System Theory (SSST)*, Tyler, TX (2010): 290–97. doi:10.1109/SSST.2010.5442816.

Falk, Darrell. "Further Thoughts on 'Darwin's Doubt' after Reading Bishop's Review." *BioLogos*, September 11, 2014. https://biologos.org/articles/reviewing-darwins-doubt-darrel-falk.

———. "Thoughts on *Darwin's Doubt*, Part 1." *BioLogos*, September 9, 2014. https://biologos.org/articles/reviewing-darwins-doubt-darrel-falk.

Farrell, John. *The Day Without Yesterday: Lemaître, Einstein, and the Birth of Modern Cosmology.* New York: Basic Books, 2006.

Faye, Jan. "Copenhagen Interpretation of Quantum Mechanics." In *The Stanford Encyclopedia of Philosophy*, edited by Edward N. Zalta, Winter 2019. https://plato.stanford.edu/archives/win2019/entries/qm-copenhagen.

Ferm, Vergilius, ed. *An Encyclopedia of Religion.* New York: Philosophical Library, 1945.

Fernie, J. D. "The Period-Luminosity Relation: A Historical Review." *Publications of the Astronomical Society of the Pacific* 81/483 (December 1969): 707–31.

Ferrada, E., and A. Wagner. "Evolutionary Innovations and the Organization of Protein Functions in Sequence Space." *PLOS ONE* 5/11 (2010): e14172.

Feynman, Richard. *The Character of Physical Law.* Cambridge, MA: MIT Press, 1995.

Fitelson, Branden, and Elliott Sober. "Plantinga's Probability Arguments Against Evolutionary Naturalism." *Pacific Philosophical Quarterly* 79/2 (1998): 115–29.

Foster, Michael B. *Creation, Nature, and Political Order in the Philosophy of Michael Foster (1903–1959): The Classic Mind Articles and Others, with Modern Critical Essays.* Edited by Cameron Wybrow. Lewiston, NY: Edwin Mellen, 1993.

Freddoso, Alfred J. "Ockham on Faith and Reason." In *The Cambridge Companion to Ockham*, edited by Paul V. Spade, 326–49. Cambridge: Cambridge Univ. Press, 1999.

Frankl, Viktor. *Man's Search for Meaning.* New York: Washington Square Press, 1959.

Freud, Sigmund. *Beyond the Pleasure Principle.* Translated and edited by James Strachey. 1961. New York: Bantam, 1990.

———. *The Ego and the Id.* Translated and edited by James Strachey. 1960. New York: Norton, 1990.

———. *The Future of an Illusion.* Translated by W. D. Robson-Scott. 1927. New York: Norton, 1989.

———. *The Interpretation of Dreams.* Translated and edited by James Strachey. 1955. New York: Basic Books, 2010.

———. *Three Essays on the Theory of Sexuality.* Translated and edited by James Strachey. 1949. Eastford, CT: Martino Fine Books, 2011.

Friedmann, Aleksandr. "On the Curvature of Space." *Zeitschrift für Physik* 10 (1922): 377–86.
Frieman, Joshua A., Michael S. Turner, and Dragan Huterer. "Dark Energy and the Accelerating Universe." *Annual Review of Astronomy and Astrophysics* 46/1 (September 15, 2008): 385–432.
Fuller, Steve. Foreword. In *Theistic Evolution: A Scientific, Philosophical, and Theological Critique*, edited by J. P. Moreland, et al. Wheaton, IL: Crossway, 2017.
———. *Science vs. Religion? Intelligent Design and the Problem of Evolution.* Oxford: Polity, 2007.
Futuyma, Douglas J. *Evolutionary Biology*. 3rd ed. Sunderland, MA: Sinauer, 1998.
Gage, Logan. "Darwin, Design & Thomas Aquinas." *Touchstone*, November–December 2010. https://www.touchstonemag.com/archives/article.php?id=23-06-037-f.
Galileo, Galilei, and Maurice A. Finocchiaro. *The Essential Galileo.* Indianapolis, IN: Hackett, 2008.
Gamow, George. "Expanding Universe and the Origin of the Elements." *Physical Review* 70 (1946): 572–73.
Gardner, Martin. "WAP, SAP, FAP & PAP." *New York Review of Books* 33 (May 8, 1986): 22–25.
Gassendi, Pierre. *Opera Omnia*. Edited by Gregory Tullio. Stuttgart-Bad Cannstatt: Friedrich Frommann, 1964.
Gates, Bill. *The Road Ahead*. New York: Viking, 1995.
Gauger, Ann. "Protein Evolution: A Guide for the Perplexed." In *Debating Darwin's Doubt*, edited by David Klinghoffer, 177–81. Seattle: Discovery Institute Press, 2015.
Gefter, Amanda. "Is String Theory in Trouble?" *New Scientist*, December 14, 2005. https://www.newscientist.com/article/mg18825305-800-is-string-theory-in-trouble.
Gilbert, S. F., J. M. Opitz, and R. A. Raff. "Resynthesizing Evolutionary and Developmental Biology." *Developmental Biology* 173 (1996): 357–72.
Gillespie, Neil C. *Charles Darwin and the Problem with Creation*. Chicago: Univ. of Chicago Press, 1979.
———. "Natural History, Natural Theology, and Social Order: John Ray and the 'Newtonian Ideology.'" *Journal of the History of Biology* 20 (1987): 1–49.
Gingerich, Owen. "The Galileo Affair." *Scientific American* 247/2 (August 1982): 133–43.
———. *The Galileo Affair.* Cambridge, MA: Sky, 1992.
———. *God's Universe.* Cambridge, MA: Harvard Univ. Press, 2006.
———. "Kepler and the Laws of Nature." *Perspectives on Science and Christian Faith* 63/1 (2011): 17–24.
———. "Scientific Cosmology Meets Western Theology." *Annals of the New York Academy of Sciences* 950/1 (December 2001): 28–38.
Goff, Philip, William Seager, and Sean Allen-Hermanson. "Panpsychism." In *The Stanford Encyclopedia of Philosophy* (Winter 2017 Edition), edited by Edward N. Zalta (ed.). https://plato.stanford.edu/archives/win2017/entries/panpsychism/.
Gonzalez, Guillermo, and Jay Wesley Richards. *The Privileged Planet: How Our Place in the Cosmos Is Designed for Discovery*. New York: Simon & Schuster, 2004.
Gopnik, Alison. "See Jane Evolve: Picture Books Explain Darwin." *Wall Street Journal*, April 18, 2014. http://www.bu.edu/cdl/files/2014/04/WSJ-Teaching-Tots-Evolution-via-Picture-Books-WSJ.com_.pdf.
Gordon, Bruce. "Balloons on a String: A Critique of Multiverse Cosmology." In *The Nature of Nature: Examining the Role of Naturalism in Science*, edited by Bruce L. Gordon and William A. Dembski, 558–601. Wilmington, DE: ISI Books, 2011.
———. "Divine Action and the World of Science: What Cosmology and Quantum Physics Teach Us About the Role of Providence in Nature." *Journal of Biblical and Theological Studies* 2/2 (2017): 247–98.
———. "Postscript to Part One: Inflationary Cosmology and the String Multiverse." In *Robert J. Spitzer, New Proofs for the Existence of God: Contributions of Contemporary Physics and Philosophy*, 75–103. Grand Rapids, MI: Eerdmans, 2010.
Gorham, Geoffrey. "Newton on God's Relation to Space and Time: The Cartesian

Framework." *Archiv Fur Geschichte Der Philosophie* 93, no. 3 (September 2011): 281–320. https://doi.org/10.1515/AGPH.2011.013.

Gould, Stephen Jay. "Evolution and the Triumph of Homology: Or, Why History Matters." *American Scientist* 74 (1986): 60–69.

———. "Nonoverlapping Magisteria: Science and Religion Are Not in Conflict, for Their Teachings Occupy Distinctly Different Domains." *Natural History* 106/2 (1997): 16–22.

———. *Rocks of Ages*. New York: Ballantine, 1999.

Greenstein, George. *The Symbiotic Universe: Life and Mind in the Cosmos*. New York: Morrow, 1988.

Grossman, Lisa. "Why Physicists Can't Avoid a Creation Event." *New Scientist*, January 11, 2012. https://www.newscientist.com/article/mg21328474-400-why-physicists-cant-avoid-a-creation-event.

Gruenwald, Oskar. "Science and Religion: The Missing Link." *Journal of Interdisciplinary Studies* 6 (1994): 1–23.

Guth, Alan. "Eternal Inflation and Its Implications." *Journal of Physics A: Mathematical and Theoretical* 40/25 (June 22, 2007): 6811–26.

———. "Inflation." In *Measuring and Modeling the Universe*, edited by Wendy L. Freeman, 31–52. Carnegie Observatories Astrophysics Series, vol. 2. Cambridge: Cambridge Univ. Press, 2004. Also at https://arxiv.org/pdf/astro-ph/0404546.pdf.

———. "Inflationary Universe: A Possible Solution to the Horizon and Flatness Problems." *Physical Review D* 23/2 (January 15, 1981): 347–56.

Guth, Alan, and Marc Sher. "The Impossibility of a Bouncing Universe." *Nature* 302 (1983): 505–7.

Guyot, Arnold. *Memoir of Louis Agassiz: 1807–1873*. Paper presented to the National Academy of Sciences, April 1878.

Haarsma, Deborah. "Response from Evolutionary Creation." In *Four Views on Creation, Evolution, and Intelligent Design*, edited by J. B. Stump, 221–26. Grand Rapids, MI: Zondervan, 2017.

———. "Reviewing 'Darwin's Doubt': Introduction." *BioLogos*, August 25, 2014. https://biologos.org/articles/reviewing-darwins-doubt-introduction.

Hackett, C., et al. "The Future Size of Religiously Affiliated and Unaffiliated Populations." *Demographic Research* 32 (2015): 829–842.

Haldane, J. B. S. *Possible Worlds and Other Essays*. Philadelphia: Richard West, 1972.

Halliday, David, and Robert Resnick. *Physics: Part Two*. New York: Wiley, 1978.

Halliwell, J. J. "Introductory Lectures on Quantum Cosmology." September 14, 2009. arXiv:0909.2566 [gr-qc].

Halvorson, Hans. "Complementarity of Representations in Quantum Mechanics." *Studies in History and Philosophy of Modern Physics* 35 (2004): 45–56.

Harris, Sam. *The End of Faith*. New York: Norton, 2005.

———. *Letter to a Christian Nation*. New York: Knopf, 2006.

Harrison, Edward. *Darkness at Night: A Riddle of the Universe*. Cambridge, MA: Harvard Univ. Press, 1987.

Harrison, Peter. *The Bible, Protestantism, and the Rise of Natural Science*. Cambridge: Cambridge Univ. Press, 1998.

———. "The Development of the Concept of Laws of Nature." In *Creation: Law and Probability*, edited by Fraser N. Watts, 13–35. Minneapolis: Fortress, 2008.

———. *The Fall of Man and the Foundations of Science*. Cambridge: Cambridge Univ. Press, 2007.

Hartle, James B., and Stephen Hawking. "Wave Function of the Universe." *Physical Review D* 28/12 (1983): 2960–75.

"Has Hawking Explained God Away? God, Physics and the Beginning." *Saints and Skeptics* (blog), October 16, 2014. http://www.saintsandsceptics.org/has-hawking-explained-god-away.

Hawking, Stephen. "The Beginning of Time." *Hawking* (Stephen Hawking official website). https://www.hawking.org.uk/in-words/lectures/the-beginning-of-time.

———. *Brief Answers to the Big Questions*. New York: Bantam, 2018.

———. *A Brief History of Time: From the Big Bang to Black Holes.* New York: Bantam, 1998.
———. "Properties of Expanding Universes." PhD diss. *Cambridge Digital Library,* 1966.
———. *The Universe in a Nutshell.* New York: Bantam, 2001.
Hawking, Stephen, and G. F. R. Ellis. "The Cosmic Black-Body Radiation and the Existence of Singularities in Our Universe." *Astrophysical Journal* 152 (April 1968): 25–35.
———. *The Large Scale Structure of Space-Time.* Cambridge: Cambridge Univ. Press, 1973.
Hawking, Stephen, and Leonard Mlodinow. *The Grand Design.* London: Bantam, 2010.
Hawking, Stephen, and Don Page. "How Probable Is Inflation?" *Nuclear Physics B* 298/4 (1988): 789–809.
Hawking, Stephen, and Roger Penrose. "The Singularities of Gravitational Collapse and Cosmology." *Proceedings of the Royal Society of London,* series A, 314 (1970): 529–48.
Hartshorne, Charles. *The Divine Relativity: A Social Conception of God.* New Haven: Yale Univ. Press, 1948.
Hearnshaw, J. B. *The Measurement of Starlight: Two Centuries of Astronomical Photometry.* Cambridge: Cambridge Univ. Press, 2005.
Heeren, Fred. *Show Me God: What the Message from Space Is Telling Us About God.* Wheeling, IL: Day Star, 2000.
Heisenberg, Werner. *Physics and Beyond: Encounters and Conversations.* New York: Harper Torchbooks, 1971.
Hertz, Heinrich. "Über den Einfluss des ultravioletten Lichtes auf die electrische Entladung (On the Effect of Ultraviolet Light upon the Electrical Discharge)." *Annalen der Physik.* 267/8 (1887): 983–1000. doi:10.1002/andp.18872670827.
Hess, Peter M. "God's Two Books: Special Revelation and Natural Science in the Christian World." In *Bridging Science and Religion,* edited by Ted Peters and Gaymon Bennett, 123–40. Minneapolis: Fortress, 2003.
Hitchens, Christopher. *God Is Not Great: How Religion Poisons Everything.* New York: Twelve Books, 2007.
Hockey, Thomas, ed. *The Biographical Encyclopedia of Astronomers.* New York: Springer, 2007.
Hodgson, Peter E. "The Christian Origin of Science." *Occasional Papers,* no. 4.
———. "The Christian Origin of Science." *Logos: A Journal of Catholic Thought and Culture* 4/2 (Spring 2001): 138–59.
———. *The Roots of Science and Its Fruits: The Christian Origin of Modern Science and Its Impact on Human Society.* London: Saint Austin, 2002.
———. *Theology and Modern Physics.* New York: Routledge, 2005.
Holder, Rodney. *God, the Multiverse, and Everything: Modern Cosmology and the Argument from Design.* Aldershot, UK: Ashgate, 2004.
Hood, Leroy, and David Galas. "The Digital Code of DNA." *Nature* 421 (2003): 444–48.
Hooft, Gerard 't. *In Search of the Ultimate Building Blocks.* Cambridge: Cambridge Univ. Press, 1997.
Hooykaas, Reijer. *Religion and the Rise of Modern Science.* Grand Rapids, MI: Eerdmans, 1972.
Horgan, John. "Why String Theory Is Still Not Even Wrong: Physicist, Mathematician and Blogger Peter Woit Whacks Strings, Multiverses, Simulated Universes and 'Fake Physics.'" *Scientific American* (blog), April 27, 2017. https://blogs.scientificamerican.com/cross-check/why-string-theory-is-still-not-even-wrong.
Horwich, Paul. *Probability and Evidence.* Cambridge: Cambridge Univ. Press, 1982.
Hossenfelder, Sabine, "Screams for Explanation: Finetuning and Naturalness in the Foundations of Physics," *Synthese,* September 3, 2019, 1–19, https://doi.org/10.1007/s11229-019-02377-5.
Howell, Kenneth J. *God's Two Books: Copernican Cosmology and Biblical Interpretation in Early Modern Science.* Notre Dame, IN: Univ. of Notre Dame Press, 2001.
Hoyle, Fred. "The Expanding Universe: The Nature of the Universe, Part V." *Harper's Magazine,* April 1951. Excerpted at https://harpers.org/archive/2016/01/our-truly-dreadful-situation.
———. *Galaxies, Nuclei, and Quasars.* New York: Harper & Row, 1965.

———. "A New Model for the Expanding Universe." *Monthly Notices of the Royal Astronomical Society* 108 (1948): 372–82.

———. "The Universe: Past and Present Reflections." *Engineering & Science*, November 1981, 8–12.

Hoyle, Fred, and Chandra Wickramasinghe. *Evolution from Space: A Theory of Cosmic Creationism*. New York: Simon & Schuster, 1982.

Hubble, Edwin. "A Relation Between Distance and Radial Velocity Among Extra-Galactic Nebulae." *Proceedings of the National Academy of Sciences* 15 (1929): 168–73.

"Hubble's Famous M31 VAR! Plate." *Carnegie Science*, 2019. https://obs.carnegiescience.edu/PAST/m31var.

Hume, David. "An Enquiry Concerning Human Understanding." In *Enquiries Concerning Human Understanding and Concerning the Principles of Morals*, edited by L. A. Selby-Bigge. 1777. Oxford: Clarendon, 1902.

———. *Dialogues Concerning Natural Religion*. Buffalo, NY: Prometheus, 1989.

———. "Dialogues Concerning Natural Religion." In *Dialogues* and *Natural History of Religion*, edited by J. C. A. Gaskin, 29–130. Oxford: Oxford Univ. Press, 2008

Hutchinson, Ian. "James Clerk Maxwell and the Christian Proposition." MIT IAP Seminar: "The Faith of Great Scientists," January 1998. http://silas.psfc.mit.edu/Maxwell/maxwell.html.

Huxley, Julian. "The Evolutionary Vision." In *Evolution After Darwin: The University of Chicago Centennial*. Vol. 3, *Issues in Evolution*, edited by S. Tax and C. Callendar, 249–61. Chicago: Univ. of Chicago Press, 1960.

Huygens, Christiaan. "Discours de la Cause de la Pesanteur." In *Oeuvres complètes de Christiaan Huygens*. Vol. 21, *Cosmologie*, edited by J. A. Vollgraf, 451–462. The Hague: Société Hollandaise des Sciences, 1944.

———. *Treatise on Light*. Translated by Silvanus P. Thompson. London: Macmillan, 1912. archive.org/details/treatiseonlight031310mbp.

Ijjas, Anna, Paul J. Steinhardt, and Abraham Loeb. "Pop Goes the Universe: The Latest Astrophysical Measurements, Combined with Theoretical Problems, Cast Doubt on the Long-Cherished Inflationary Theory of the Early Cosmos and Suggest We Need New Ideas." *Scientific American* 316/2 (January 2017): 32–39.

Iliffe, Rob. *Priest of Nature: The Religious Worlds of Isaac Newton*. New York: Oxford Univ. Press, 2017.

Isham, C. J. "Theories of the Creation of the Universe." In *Quantum Cosmology and the Laws of Nature: Perspectives on Divine Action*, 2nd ed., edited by Robert John Russell, et al., 51–89. Jointly published by the Vatican Observatory and the Center for Theology and the Natural Sciences, 1993.

Jaki, Stanley. *Cosmos and Creator*. Edinburgh: Scottish Academic Press, 1980.

———. *The Paradox of Olbers' Paradox: A Startling Account of the Scientific Meaning of the Darkness of the Night Sky*. New York: Herder and Herder, 1969.

Jastrow, Robert. *God and the Astronomers*. New York: Norton, 1992.

Jogalekar, Ashutosh. "Why the Search for a Unified Theory May Turn Out to Be a Pipe Dream." *Scientific American*, May 3, 2013. https://blogs.scientificamerican.com/the-curious-wavefunction/why-the-search-for-a-unified-theory-may-turn-out-to-be-a-pipe-dream.

Johnson, George. *Miss Leavitt's Stars: The Untold Story of the Forgotten Woman Who Discovered How to Measure the Universe*. New York: Norton, 2005.

Johnston, Wendy K., et al. "RNA-Catalyzed RNA Polymerization: Accurate and General RNA-Templated Primer Extension." *Science* 292 (2001): 1319–25.

Kaiser, Christopher. *Creation and the History of Science*. Grand Rapids, MI: Eerdmans, 1991.

———. *Creational Theology and the History of Physical Science: The Creationist Tradition from Basil to Bohr*. Leiden: Brill, 1997.

Kallosh, Renata, Lev Kofman, and Andre Linde. "Pyrotechnic Universe." *Physical Review D* 64 (2001): 123523.

Kamminga, Harmke. "Studies in the History of Ideas on the Origin of Life." PhD diss., University of London, 1980.

Kant, Immanuel. *The Critique of Pure Reason, The Critique of Practical Reason and Other Ethical Treatises, The Critique of Judgement.* Chicago: Univ. of Chicago Press, 1952.

Kato, K., et al. "Amino Acid Alterations Essential for Increasing the Catalytic Activity of the Nylon-Oligomer-Degradation Enzyme of *Flavobacterium* sp." *European Journal of Biochemistry* 200/1 (August 15, 1991): 165–69.

Kauffman, Stuart A. *The Origins of Order: Self-Organization and Selection in Evolution.* New York: Oxford Univ. Press, 1993.

———. *At Home in the Universe.* Oxford: Oxford Univ. Press, 1995.

Kaufmann, Walter, and Forrest Baird, eds. *Philosophic Classics: From Plato to Nietzsche.* Englewood Cliffs, NJ: Prentice Hall, 1994.

Kavalovski, V. "The Vera Causa Principle: A Historico-Philosophical Study of a Metatheoretical Concept from Newton Through Darwin." PhD diss., University of Chicago, 1974.

Keas, Michael Newton. "Systematizing the Theoretical Virtues." *Synthese* 195/6 (June 2018): 2761–93.

Kenyon, Dean. Foreword. In *The Mystery of Life's Origin*, by Charles Thaxton, Walter L. Bradley, and Roger L. Olsen. New York: Philosophical Library, 1984.

Kenyon, Dean, and Gary Steinman. *Biochemical Predestination.* New York: McGraw-Hill, 1969.

Kepler, Johannes. *Harmonies of the World.* Translated by Charles G. Wallis. New York: Prometheus, 1995.

———. Letter to Herwart von Hohenburg, April 9/10, 1599. In Carola Baumgardt, *Johannes Kepler: Life and Letters*, 50. New York: Philosophical Library, 1951.

———. *Mysterium Cosmographicum* (*The Secret of the Universe*). Translated by A. M. Duncan. New York: Arabis, 1981.

Kiefer, Claus. "Conceptual Problems in Quantum Gravity and Quantum Cosmology." *ISRN Mathematical Physics* (2013): 509316.

———. "Emergence of a Classical Universe from Quantum Gravity and Cosmology." *Philosophical Transactions of the Royal Society A* 370/1975 (2012): 4566–75.

———. "The Need for Quantum Cosmology." In *Principles of Evolution: From the Planck Epoch to Complex Multicellular Life*, edited by H. Meyer-Ortmanns and S. Thurner, 205–17. New York: Springer, 2011.

———. "On the Interpretation of Quantum Theory—from Copenhagen to the Present Day." In *Time, Quantum and Information*, edited by Lutz Castell and Otfried Ischebeck, 291–99. Berlin, Heidelberg: Springer, 2003.

Klima, Guyla, Fritz Allhoff, and Anand J. Vaidya, eds. "Selections from the Condemnation of 1277." In *Medieval Philosophy: Essential Readings with Commentary*, 180–89. Oxford: Blackwell, 2007.

Kline, D. A. "Theories, Facts, and Gods: Philosophical Aspects of the Creation-Evolution Controversy." In *Did the Devil Make Darwin Do It?* edited by D. Wilson, 37–44. Ames: Iowa State Univ. Press, 1983.

Klinghoffer, David, ed. *Debating Darwin's Doubt: A Scientific Controversy That Can No Longer Be Denied.* Seattle: Discovery Institute Press, 2015.

Knapton, Sarah. "Stephen Hawking's Final Book: 'There Is No God or Afterlife.'" *The Telegraph*, October 15, 2018. https://www.telegraph.co.uk/science/2018/10/15/no-god-afterlife-concludes-stephen-hawking-final-book.

Koberlein, Brian. "Can Many-Worlds Theory Rescue Us from Boltzmann Brains?" *Nautilus*, April 7, 2017. http://nautil.us/blog/can-many_worlds-theory-rescue-us-from-boltzmann-brains.

Kofman, Lev. "The Origin of Matter in the Universe: Reheating After Inflation." May 24, 1996. arXiv:9605155.

Kohler, John. *Asian Philosophies.* 4th ed. Englewood, NJ: Prentice Hall, 2002.

Kok, R. A., J. A. Taylor, and W. L. Bradley. "A Statistical Examination of Self-Ordering Amino Acids in Proteins." *Origins of Life and Evolution of the Biosphere* 18 (1988): 135–42.

Koonin, Eugene V. "The Biological Big Bang Model for the Major Transitions in Evolution." *Biology Direct* 2 (2007): 1–17.

Koons, Robert C. "Epistemological Objections to Materialism." In *The Waning of Materialism*, edited by Robert C. Koons and George Bealer, 281–306. Oxford: Oxford Univ. Press, 2010.

———. "The General Argument from Intuition." In *Two Dozen (or so) Arguments for God*, edited by Jerry L. Walls and Trent Dougherty, 238–57. Oxford: Oxford Univ. Press, 2018.

———. "The Incompatibility of Naturalism and Scientific Realism." In *Naturalism: A Critical Analysis*, edited by William Lane Craig and J. P. Moreland, 49–63. London: Routledge, 2000.

———. "The Many-Worlds Interpretation of Quantum Mechanics: A Hylomorphic Critique and Alternative." In *Neo-Aristotelian Perspectives on Contemporary Science*, edited by William Simpson, Robert Koons, and Nicholas Teh, 61–104. London: Routledge, 2017.

Koons, Robert, and Alexander Pruss. "Skepticism and the Principle of Sufficient Reason." *Philosophical Studies* (June 8, 2020): 1–21; https://doi.org/10.1007/s11098-020-01482-3.

Kragh, Helge. *Cosmology and Controversy*. Princeton, NJ: Princeton Univ. Press, 1996.

———. "The Steady State Theory." In *Cosmology: Historical, Literary, Philosophical, Religious, and Scientific Perspectives*, edited by Norriss S. Hetherington, 391–404. New York: Garland, 1993.

Krane, Kenneth S. *Modern Physics*. 3rd ed. New York: Wiley, 2012.

Krauss, Lawrence. "Ultimate Issues Hour: How Does Something Come from Nothing?" *The Dennis Prager Show*, January 29, 2013, SRN.

———. *A Universe from Nothing: Why There Is Something Rather Than Nothing*. New York: Free Press, 2012.

———. "Lawrence Krauss: Our Godless Universe is Precious." Big Think video. Uploaded Dec. 26th, 2012. Accessible at https://youtu.be/SB5cBl2np-I.

———. "Lawrence Krauss: Atheism and the Spirit of Science." *The Agenda with Steve Paikin*, July 4, 2013. https://www.youtube.com/watch?v=MawwCJ5q-2Y.

———. "The Universe Doesn't Give a Damn about Us." *Big Think*. April 17, 2017. https://www.youtube.com/watch?v=fG38uXkDSyU.

Krehl, Peter O. K. *History of Shock Waves, Explosions and Impact: A Chronological and Biographical Reference*. Berlin: Springer, 2009.

Küppers, Bernd-Olaf. *Information and the Origin of Life*. Cambridge, MA: MIT Press, 1990.

———. "On the Prior Probability of the Existence of Life." In *The Probabilistic Revolution*, vol. 2, edited by Lorenz Kruger et al., 355–72. Cambridge, MA: MIT Press, 1987.

Lamoureux, Denis. "Evolutionary Creation: A Christian Approach to Evolution." https://wp.biologos.org/wp-content/uploads/2009/08/lamoureux_scholarly_essay.pdf.

———. *Evolutionary Creation: A Christian Approach to Evolution*. Eugene, OR: Wipf & Stock, 2008.

Laplace, Pierre S. *The System of the World*. Vol. 2. Translated by J. Pond. London: Richard Phillips, 1809.

Larmer, Robert. "Is There Anything Wrong with 'God of the Gaps' Reasoning?" *International Journal for Philosophy of Religion* 52/3 (2002): 129–42.

Larson, Edward J. *Summer for the Gods: The Scopes Trial and America's Continuing Debate over Science and Religion*. New York: Basic Books, 2008.

Larson, James L. *Reason and Experience: The Representation of Natural Order in the Work of Carl von Linné*. Berkeley and Los Angeles: Univ. of California Press, 1971.

Law, Stephen. "Naturalism, Evolution, and True Belief." *Analysis* 72/11 (2012): 41–48.

Lehninger, Albert L. *Biochemistry*. New York: Worth, 1970.

Leibniz, Gottfried W. *Die Philosophischen Schriften von Gottfried Wilhelm Leibniz*. Vol. 7. Edited by C. I. Gerhardt. Hildesheim: G. Olms, 1965.

———. *The Leibniz-Clarke Correspondence*. Edited by H. G. Alexander. Manchester: Manchester Univ. Press, 1956.

———. "The Monadology." In *The Monadology and Other Philosophical Writings*, translated by Robert Latta, 215–71. London: Oxford Univ. Press, 1968.

———. *New Essays on Human Understanding*. Translated and edited by Peter Remnant and Jonathan Bennett. Cambridge: Cambridge Univ. Press, 2000.

———. *The Principles of Philosophy Known as Monadology*. Translated by Robert Latta. Revised by Donald Rutherford. Available in searchable PDF at http://philosophyfaculty.ucsd.edu/faculty/rutherford/Leibniz/translations/Monadology.pdf.

Lenski, Richard, et al. "The Evolutionary Origin of Complex Features." *Nature* 423 (2003): 139–44.

Leslie, John. "Anthropic Principle, World Ensemble, Design." *American Philosophical Quarterly* 19 (1982): 150.

———. *Universes*. New York: Routledge, 1996.

Lewis, C. S. *Miracles: A Preliminary Study*. New York: Macmillan, 1978.

Lewis, Geraint F., and Luke A. Barnes. *A Fortunate Universe: Life in a Finely Tuned Cosmos*. Cambridge: Cambridge Univ. Press, 2016.

Lewis, Neil. "Robert Grosseteste." In *The Stanford Encyclopedia of Philosophy*, edited by Edward N. Zalta, Summer 2013. https://plato.stanford.edu/archives/sum2013/entries/grosseteste.

Lewontin, Richard. "Billions and Billions of Demons." *New York Review of Books*, January 9, 1997: 28–32.

Liddle, Andrew. *An Introduction to Modern Cosmology*. West Sussex, UK: Wiley, 2003.

Lincoln, Tracey A., and Gerald F. Joyce. "Self-Sustained Replication of an RNA Enzyme." *Science* 323/5918 (2009): 1229–32.

Lindberg, David C. "Medieval Science and Religion." In *Science and Religion: A Historical Introduction*, edited by Gary B. Ferngren, 57–72. Baltimore: Johns Hopkins Univ. Press, 2002.

Linde, Andrei D. "Eternally Existing Self-Reproducing Chaotic Inflationary Universe." *Physics Letters B* 175/4 (August 14, 1986): 395–400.

———. "A New Inflationary Universe Scenario: A Possible Solution of the Horizon, Flatness, Homogeneity, Isotropy and Primordial Monopole Problems." *Physics Letters B* 108/6 (1982): 389–93.

———. "Sinks in the Landscape, Boltzmann Brains and the Cosmological Constant Problem." *Journal of Cosmology and Astroparticle Physics* 2007/01 (January 24, 2007): 022. doi:10.1088/1475-7516/2007/01/022.

———. "Towards a Gauge Invariant Volume-Weighted Probability Measure for Eternal Inflation." *Journal of Cosmology and Astroparticle Physics* 2007/06 (2007): 017.

Linnebo, Øystein. "Platonism in the Philosophy of Mathematics." In *The Stanford Encyclopedia of Philosophy*, edited by Edward N. Zalta, Spring 2018. https://plato.stanford.edu/entries/platonism-mathematics.

Lipka, Michael. "A Closer Look at America's Rapidly Growing Religious 'Nones.'" *Pew Research Center*, May 13, 2015. https://www.pewresearch.org/fact-tank/2015/05/13/a-closer-look-at-americas-rapidly-growing-religious-nones.

Lipton, Peter. *Inference to the Best Explanation*. London and New York: Routledge, 1991.

Locke, John. *An Essay Concerning Human Understanding*. Edited by Roger Woolhouse. London: Penguin, 1997.

Loke, Andrew. *God and Ultimate Origins: A Novel Cosmological Argument*. Cham, Switzerland: Palgrave Macmillan, 2017.

Longley, Clifford. "Focusing on Theism." *London Times*, January 21, 1989, 10.

Lorentz, H. A., et al. *The Principle of Relativity*. Translated by W. Perrett and G. B. Jeffrey, with notes by A. Sommerfield. London: Methuen, 1923.

Luminet, Jean-Pierre. "Dodecahedral Space Topology as an Explanation for Weak Wide-Angle Temperature Correlations in the Cosmic Microwave Background." *Nature* 425 (October 9, 2003): 593–95.

———. "Lemaître's Big Bang." Lecture given at "Frontiers of Fundamental Physics 14," Aix Marseille University, Marseille, France, July 15–18, 2014. https://arxiv.org/ftp/arxiv/papers/1503/1503.08304.pdf.

———. "The Rise of Big Bang Models, from Myth to Theory and Observations." April 26, 2007. arXiv:0704.3579.

Lundin et al. "Experimental Determination and Prediction of the Fitness Effects of Random Point Mutations in the Biosynthetic Enzyme HisA." *Molecular Biology and Evolution* 35/no. 3 (March 1, 2018): 704–18. https://doi.org/10.1093/molbev/msx325.

Lyell, Charles. *Principles of Geology: Being an Attempt to Explain the Former Changes of the Earth's Surface, by Reference to Causes Now in Operation*. 3 vols. London: Murray, 1830–33.

MacDonald, J., and D. J. Mullan. "Big Bang Nucleosynthesis: The Strong Nuclear Force Meets the Weak Anthropic Principle." *Physical Review D* 80/4 (2009): 043507.

MacKay, Donald M. *The Clockwork Image: A Christian Perspective on Science*. Downers Grove, IL: InterVarsity, 1974.

Maher, Patrick. "Prediction, Accommodation, and the Logic of Discovery." *PSA: Proceedings of the Biennial Meeting of the Philosophy of Science Association* 1 (1988): 273–85.

Maimonides. *Guide*, 2.13. In *Maimonides*, edited by T. M. Rudavsky. Hoboken, New Jersey: Wiley-Blackwell, 2010.

Maltz, Maxwell. *Psycho-Cybernetics*. New York: TarcherPerigee, 2015. (An updated and expanded reprint of the 1960 original.)

Mandel, Kaisey S., et al. "The Type Ia Supernova Color–Magnitude Relation and Host Galaxy Dust: A Simple Hierarchical Bayesian Model," *Astrophysical Journal* 842/2 (June 16, 2017): 93.

Manoukian, Edouard B. "Introduction to String Theory." In *Quantum Field Theory II. Graduate Texts in Physics*, 187–321. Springer, Cham, 2016. doi:10.1007/978-3-319-33852-1_3.

Markie, Peter. "Rationalism vs. Empiricism." In *The Stanford Encyclopedia of Philosophy*, edited by Edward N. Zalta, Fall 2017. https://plato.stanford.edu/entries/rationalism-empiricism.

Marks, Robert J. "Diversity Inadequacies of Parallel Universes: When the Multiverse Becomes Insufficient to Account for Conflicting Contradistinctions." *Perspectives on Science and Christian Faith* 71/3 (2019): 1–7.

Marshall, Charles R. "Explaining the Cambrian 'Explosion' of Animals." *Annual Reviews of Earth and Planetary Sciences* 34 (2006): 355–84.

———. "Nomothetism and Understanding the Cambrian 'Explosion.'" *Palaios* 18 (2003): 195–96.

———. "When Prior Belief Trumps Scholarship." *Science* 341 (2013): 1344.

Marshall, Charles R., and James Valentine. "The Importance of Preadapted Genomes in the Origin of the Animal Bodyplans and the Cambrian Explosion." *Evolution* 64/5 (2010): 1189–201.

Marx, Karl. *Capital: Critique of Political Economy*. Translated by Samuel Moore and Edward Aveling. Edited by Frederick Engels. 1867. *Marxists Internet Archive*, 1999. https://www.marxists.org/archive/marx/works/1867-c1/index.htm.

———. *A Contribution to the Critique of Hegel's Philosophy of Right*. Introduction. 1843. *Marxists Internet Archive*, 2009. https://www.marxists.org/archive/marx/works/1843/critique-hpr/intro.htm.

Marx, Karl, and Friedrich Engels. *Economic and Philosophic Manuscripts of 1844*. Translated by Martin Milligan. Radford, VA: Wilder, 2018.

Masci, David. "Public Opinion on Religion and Science in the United States." *Pew Research Center*, November 5, 2009. https://www.pewforum.org/2009/11/05/public-opinion-on-religion-and-science-in-the-united-states.

Matzkin, Alexandre. "Realism and the Wavefunction." *European Journal of Physics* 23/3 (May 1, 2002): 307. doi:10.1088/0143-0807/23/3/307.

Mayr, Ernst. Foreword. In *Darwinism Defended*, edited by M. Ruse, xi–xii. Reading, MA: Addison-Wesley, 1982.

McDonough, T. R. *The Search for Extraterrestrial Intelligence: Listening for Life in the Cosmos.* New York: Wiley, 1987.

McGrade, A. S. "Natural Law and Moral Omnipotence." In *The Cambridge Companion to Ockham*, edited by Paul V. Spade, 273–301. Cambridge: Cambridge Univ. Press, 1999.

McGrew, Timothy. "Miracles." In *The Stanford Encyclopedia of Philosophy*, edited by Edward N. Zalta, Spring 2019. https://plato.stanford.edu/archives/spr2019/entries/miracles.

McMullin, Ernan. "How Should Cosmology Relate to Theology?" In *The Sciences and Theology in the Twentieth Century*, edited by Arthur R. Peacocke, 39. Notre Dame, IN: Univ. of Notre Dame Press, 1981.

———. "The Virtues of a Good Theory." In *The Routledge Companion to Philosophy of Science*, edited by Stathis Psillos and Martin Curd, 498–508. New York: Routledge, 2014.

McNeill, John R., ed. *Institutes of the Christian Religion*. Philadelphia: Westminster, 1960.

Menuge, Angus. *Agents Under Fire: Materialism and the Rationality of Science*. Lanham, MD: Rowman & Littlefield, 2004.

Merton, Robert K. "Science, Technology and Society in Seventeenth Century England." *Osiris* 4 (1938): 360–632.

Meyer, Stephen C. *Darwin's Doubt: The Explosive Origin of Animal Life and the Case for Intelligent Design*. San Francisco: HarperOne, 2013.

———. "The Demarcation of Science and Religion." In *The History of Science and Religion in the Western Tradition: An Encyclopedia*, edited by Gary B. Ferngren, 17–23. New York: Garland, 2000.

———. "The Difference It Doesn't Make: Why the 'Front-End Loaded' Concept of Design Fails to Explain the Origin of Biological Information." In *Theistic Evolution: A Scientific, Philosophical and Theological Critique*, edited by J. P. Moreland et al., 209–28. Wheaton, IL: Crossway, 2017.

———. "The Methodological Equivalence of Design and Descent: Can There Be a Scientific Theory of Creation?" In *The Creation Hypothesis: Scientific Evidence for Intelligent Design*, edited by J. P. Moreland, 67–112. Downers Grove, IL: InterVarsity, 1994.

———. "Mind Before Matter: The Unexpected Implications of Quantum Cosmology." In *Routledge Handbook on Idealism and Immaterialism*, edited by Joshua R. Farris and Benedikt Paul Goecke. Abington, Oxfordshire: Routledge, 2020, forthcoming.

———. "Of Clues and Causes: A Methodological Interpretation of Origin of Life Studies." PhD diss., Cambridge University, 1990.

———. "Owen Gingerich." *Eternity*, May 1986, 29.

———. "The Return of the God Hypothesis." *Journal of Interdisciplinary Studies* 6/1–2 (1999): 1–38.

———. "Sauce for the Goose: Intelligent Design, Scientific Methodology, and the Demarcation Problem." In *The Nature of Nature*, edited by Bruce L. Gordon and William A. Dembski, 95–131. Wilmington, DE: ISI Books, 2011.

———. "The Scientific Status of Intelligent Design: The Methodological Equivalence of Naturalistic and Non-Naturalistic Origins Theories." In *Science and Evidence for Design in the Universe, The Proceedings of the Wethersfield Institute*, 151–212. San Francisco: Ignatius, 2000.

———. *Signature in the Cell: DNA and the Evidence for Intelligent Design*. San Francisco: HarperOne, 2009.

Meyer, Stephen C., Ann K. Gauger, and Paul Nelson. "Theistic Evolution and the Extended Evolutionary Synthesis: Does It Work?" In *Theistic Evolution: A Scientific, Philosophical and Theological Critique*, edited by J. P. Moreland et al., 249–79. Wheaton, IL: Crossway, 2017.

Meyer, Stephen C., and Paul A. Nelson. "Should Theistic Evolution Depend on Methodological Naturalism?" In *Theistic Evolution: A Scientific, Philosophical and Theological Critique*, edited by J. P. Moreland et al., 561–92. Wheaton, IL: Crossway, 2017.

Miller, Brian. "A Dentist in the Sahara: Doug Axe on the Rarity of Proteins Is Decisively Confirmed." *Evolution News*, February 18, 2019. https://evolutionnews.org/2019/02/a-dentist-in-the-sahara-doug-axe-on-the-rarity-of-proteins-is-decisively-confirmed/.

———. "Protein Folding and the Four Horsemen of the Axocalypse." *Evolution News*, April 12, 2018. https://evolutionnews.org/2018/04/protein-folding-and-the-four-horsemen-of-the-axocalypse.

Miller, Kenneth R. *Finding Darwin's God: A Scientist's Search for Common Ground Between God and Evolution.* New York: HarperCollins, 1999.

Miller, Kenneth R., and Joseph Levine. *Biology.* 4th ed. Upper Saddle River, NJ: Prentice Hall, 1998.

Minnich, S. R., and H. N. Rohde. "A Rationale for Repression and/or Loss of Motility by Pathogenic Yersinia in the Mammalian Host." *Advances in Experimental Medicine and Biology.* 603 (2007): 298–310.

Mithani, Audrey, and Alexander Vilenkin. "Collapse of Simple Harmonic Universe." *Journal of Cosmology and Astroparticle Physics* 2012/01 (January 10, 2012): 028. doi:10.1088/1475–7516/2012/01/028.

Mlodinow, Leonard. "The Crazy History of Quantum Mechanics." TEDxJerseyCity, Febuary 8, 2016. https://www.youtube.com/watch?v=oQYsqnRYb_o.

Monday, S. R., S. A. Minnich, and P. C. Feng. "A 12-base-pair Deletion in the Flagellar Master Control Gene flhC Causes Nonmotility of the Pathogenic German Sorbitol-fermenting Escherichia coli O157:H-strains." *Journal of Bacteriology.* 186/8 (April 2004): 2319–27.

Moniz, Paulo Vargas. "A Survey of Quantum Cosmology." In *Quantum Cosmology: The Supersymmetric Perspective*, vol. 1, 13–53. Berlin, Heidelberg: Springer, 2010.

Moran, Larry. "You Need to Understand Biology If You Are Going to Debate an Intelligent Design Creationist." *Sandwalk*, March 20, 2016. http://sandwalk.blogspot.com/2016/03/you-need-to-understand-biology-if-you.html.

Moreland, J. P. "Agent Causation and the Craig-Grünbaum Debate about Theistic Explanation and the Initial Singularity," *American Catholic Philosophical Quarterly* 71 (Autumn 1997): 539–54.

———. "The explanatory relevance of libertarian agency as a model of theistic design." In Dembski, W. (ed.), Mere Creation: Science, Faith, & Intelligent Design. Downers Grove, IL: InterVarsity Press, 1998a.

———. *Scaling the Secular City.* Grand Rapids, MI: Baker, 1989.

Moser, Paul K. *The Elusive God: Reorienting Religious Epistemology.* Cambridge: Cambridge Univ. Press, 2008.

Müller, Gerd B. "The Extended Evolutionary Synthesis." Opening lecture, "New Trends in Evolutionary Biology: Biological, Philosophical, and Social Science Perspectives." The Royal Society, London, November 7–9, 2016. Audio file: https://royalsociety.org/science-events-and-lectures/2016/11/evolutionary-biology.

Müller, Gerd B., and Stuart A. Newman. "Origination of Organismal Form: The Forgotten Cause in Evolutionary Theory." In *Origination of Organismal Form: Beyond the Gene in Developmental and Evolutionary Biology*, edited by G. B. Müller and S. A. Newman, 3–10. Cambridge, MA: MIT Press, 2003.

Myers, P. Z. "A Suggestion for Debaters." *Pharyngula*, March 20, 2016. http://freethoughtblogs.com/pharyngula/2016/03/20/a-suggestion-for-debaters.

Nagel, Thomas. "Books of the Year." *The Times Literary Supplement* 5565 (27 November, 2009): 14.

———. *The Last Word.* Oxford: Oxford Univ. Press, 1997.

———. *Mind & Cosmos: Why the Materialist Neo-Darwinian Conception of Nature Is Almost Certainly False.* Oxford: Oxford Univ. Press, 2012.

Negoro, S., et al. "X-ray Crystallographic Analysis of 6-Aminohexanoate-Dimer Hydrolase: Molecular Basis for the Birth of a Nylon Oligomer-Degrading Enzyme." *Journal of Biological Chemistry* 280/47 (November 25, 2005): 39644–52.

Newton, Isaac. *Cambridge Manuscript Add. 3970.3, ff. 475r-482v*. Cambridge: Cambridge University Library, 1675.

———. *The Correspondence of Isaac Newton*. Edited by H. W. Turnbull et al. 7 vols. Cambridge: Cambridge Univ. Press, 1959–77.

———. Letter from Isaac Newton to Richard Bentley, February 25, 1692/3. 189.R.4.47, ff. 7–8. Cambridge: Trinity College Library, 1693.

———. *Mathematical Principles of Natural Philosophy*. Translated by I Bernard Cohen and Anne Whitman. Los Angeles: Univ. of California Press, 1999.

———. *The Mathematical Principles of Natural Philosophy*. Translated by Andrew Motte. (London, 1729), vol. 2. [English translation of the third Latin edition of 1726]

———. *Opticks: Or a Treatise of Reflections, Refractions, Inflections & Colours of Light*. 1718. Reprint, New York: Dover, 1952.

———. *Opticks: Or a Treatise of the Reflections, Refractions, Inflections, and Colours of Light*. 1718. Reprint, Project Gutenberg, 2010. Available at https://www.gutenberg.org/ebooks/33504.

Nicolis, Grégoire, and Ilya Prigogine. *Self-Organization in Nonequilibrium Systems*. New York: Wiley, 1977.

Niiler, Eric. "Maybe You're Not an Atheist—Maybe You're a Naturalist Like Sean Carroll." *Wired*, May 9, 2016. https://www.wired.com/2016/05/maybe-youre-not-atheist-maybe-youre-naturalist-like-sean-carroll.

Nussbaumer, Harry. "Einstein's Conversion from His Static to an Expanding Universe." *European Physics Journal—History* 39 (2014): 37–62.

Nye, Bill. *Undeniable: Evolution and the Science of Creation*. New York: St. Martin's, 2014.

O'Raifeartaigh, Cormac. "Einstein's Greatest Blunder?" Scientific American. February 21, 2017. Accessible at https://blogs.scientificamerican.com/guest-blog/einsteins-greatest-blunder/.

O'Raifeartaigh, Cormac, and B. McCann. "Einstein's Cosmic Model of 1931 Revisited: An Analysis and Translation of a Forgotten Model of the Universe." *European Physical Journal H 39* 1 (February 4, 2014): 63–85.

Oakley, Francis. "Christian Theology and the Newtonian Science: The Rise of the Concept of the Laws of Nature." *Church History* 30/4 (1961): 433–57.

Oresme, Nichole. *Le Livre du ciel et du monde*. Translated and edited by A. D. Menut and A. J. Denomy. Madison: Univ. of Wisconsin Press, 1968.

Page, Don N. "Inflation Does Not Explain Time Asymmetry." *Nature* 304/5921 (1983): 39–41.

———. "Is Our Universe Likely to Decay Within 20 Billion Years?" *Physical Review D* 78/6 (2008): 063535.

———. "Return of the Boltzmann Brains." *Physical Review D* 78/6 (2008): 063536.

———. "Susskind's Challenge to the Hartle-Hawking No-Boundary Proposal and Possible Resolutions." *Journal of Cosmology and Astroparticle Physics* 2007/01 (2007): 004.

Paine, Thomas. *The Age of Reason*. Vol. 8 in *The Life and Works of Thomas Paine*. 10 vols. Edited by William M. Van der Weyde. New Rochelle, NY: Thomas Paine National Historical Association, 1925.

Paley, William. *Natural Theology: Or Evidences of the Existence and Attributes of the Deity Collected from the Appearances of Nature*. 1802. Reprint, Boston: Gould and Lincoln, 1852.

Paps, Jordi, and Peter W. H. Holland. "Reconstruction of the Ancestral Metazoan Genome Reveals an Increase in Genomic Novelty." *Nature Communications* 9 (2018): 1730.

Peacock, John A. *Cosmological Physics*. Cambridge: Cambridge Univ. Press, 1998.

Peacocke, Roy. *A Brief History of Eternity: A Considered Response to Stephen Hawking's a Brief History of Time*. Wheaton, IL: Crossway Books, 1990.

Peebles, P. J. E., and Bharat Ratra. "The Cosmological Constant and Dark Energy." *Reviews of Modern Physics* 75/2 (April 22, 2003): 559–606.

Peirce, Charles S. "Deduction, Induction, and Hypothesis." In *Collected Papers*, edited by

Charles Hartshorne and Paul Weiss. 6 vols. 2: 372–88. Cambridge, MA: Harvard Univ. Press, 1931–35.

Penrose, Roger. *Cycles of Time: An Extraordinary New View of the Universe*. New York: Knopf, 2010.

———. "Difficulties with Inflationary Cosmology." *Annals of the New York Academy of Sciences* 571/1 (December 1989): 249–64.

———. *Emperor's New Mind*. Oxford: Oxford Univ. Press, 1989.

———. "On the Gravitization of Quantum Mechanics 2: Conformal Cyclic Cosmology." *Foundations of Physics* 44 (2014): 873–90.

———. *The Road to Reality: A Complete Guide to the Laws of the Universe*. New York: Vintage, 2004.

———. "Time-Asymmetry and Quantum Gravity." In *Quantum Gravity 2*, edited by C. Isham, R. Penrose, and D. Sciama, 245–72. Oxford: Clarendon, 1981.

Penzias, Arno, and Robert Wilson. "A Measurement of Excess Antenna Temperature at 4080 Mc/s." *Astrophysical Journal* 142 (1965): 419–21.

Pera, Marcello. *The Discourses of Science*. Translated by Clarissa Botsford. Chicago: Univ. of Chicago Press, 1994.

Peter, Isabelle S., and Eric H. Davidson. *Genomic Control Processes: Development and Evolution*. New York: Academic, 2015.

Peterson, Michael, William Hasker, Bruce Reichenbach, and David Basinger. R*eason and Religious Belief: An Introduction to the Philosophy of Religion*. Oxford: Oxford Univ. Press, 1991.

Peterson, T., and G. B. Müller. "Phenotypic Novelty in EvoDevo: The Distinction Between Continuous and Discontinuous Variation and Its Importance in Evolutionary Theory." *Evolutionary Biology* 43 (2016): 314–45.

Pew Research Center. "The Changing Global Religious Landscape." April 5, 2017. https://www.pewforum.org/2017/04/05/the-changing-global-religious-landscape.

———. "The Future of World Religions: Population Growth Projections, 2010–2050." April 2, 2015. https://www.pewforum.org/2015/04/02/religious-projections-2010–2050.

———. "The Global Religious Landscape." December 18, 2012. https://www.pewforum.org/2012/12/18/global-religious-landscape-exec.

Planck Collaboration. "Planck 2015 Results. XIII. Cosmological Parameters." *Astronomy & Astrophysics* 594 (2015): A13.

Plantinga, Alvin. "Content and Natural Selection." *Philosophy and Phenomenological Research* 83/2 (2011): 435–58.

———. "Evolution vs. Naturalism." *Books and Culture*, July–August, 2008.

———. "Probability and Defeators." *Pacific Philosophical Quarterly* 84 (2003): 291–98.

———. "Reliabilism, Analysis and Defeators." *Philosophy and Phenomenological Research* 55/2 (1995): 427–64.

———. *Warrant and Proper Function*. Oxford: Oxford Univ. Press, 1993.

———. *Warranted Christian Belief*. Oxford: Oxford Univ. Press, 2000.

———. *Where the Conflict Really Lies: Science, Religion, and Naturalism*. Oxford: Oxford Univ. Press, 2011. https://www.booksandculture.com/articles/2008/julaug/11.37.html.

Plantinga, Alvin, and Michael Tooley. *Knowledge of God*. Malden, MA: Blackwell, 2008.

Poe, Edgar Allan. *Eureka: A Prose Poem*. 1848. Andesite Press, 2015.

Polanyi, Michael. "Life's Irreducible Structure." *Science* 160 (1968): 1308–12.

———. "Life Transcending Physics and Chemistry." *Chemical and Engineering News* 45 (1967): 54–69.

———. *Personal Knowledge*. London: Routledge, 1974.

Polkinghorne, John. *Belief in God in an Age of Science*. New Haven, CT: Yale Univ. Press, 1998.

———. "So Finely Tuned a Universe: Of Atoms, Stars, Quanta & God." *Commonweal*, August 16, 1996.

Pond, Jean. "Independence: Mutual Humility in the Relationship Between Science & Christian Theology." In *Science & Christianity: Four Views*, edited by Richard F. Carlson, 67–104. Downers Grove, IL: InterVarsity, 2000.

Powner, Matthew W., Béatrice Gerland, and John D. Sutherland. "Synthesis of Activated Pyrimidine Ribonucleotides in Prebiotically Plausible Conditions." *Nature* 459/7244 (2009): 239–42.

Prigogine, Ilya, Grégoire Nicolis, and Agnes Babloyantz. "Thermodynamics of Evolution." *Physics Today* 25/11 (November 1972): 23–31.

"Prof. Einstein Begins His Work at Mt. Wilson; Hoping to Solve Problems Touching Relativity." *New York Times*, January 3, 1931, 1.

Provine, William. "Evolution and the Foundation of Ethics." *MBL Science* 3 (1988): 25–26.

Pruss, Alexander. *Actuality, Possibility, and Worlds*. London: Continuum, 2011.

———. "The Leibnizian Cosmological Argument." In *The Blackwell Companion to Natural Theology*, edited by W. L. Craig and J. P. Moreland, 24–100. Oxford: Blackwell Publishing (2009).

———. *The Principle of Sufficient Reason: A Reassessment*. New York: Cambridge Univ. Press, 2006.

"The Quantum Experiment That Broke Reality." *Space Time* series. PBS Digital Studios, July 27, 2016. https://www.youtube.com/watch?v=p-MNSLsjjdo&t=213s.

Quastler, Henry. *The Emergence of Biological Organization*. New Haven, CT: Yale Univ. Press, 1964.

Rasmussen, Simon, et al. "Early Divergent Strains of *Yersinna pestus* in Eurasia 5000 Years Ago." *Cell* 163, Issue 3, 571–82.

Ray, John. *The Wisdom of God Manifested in the Works of the Creation*. London: R. Harbin, 1717.

"Redshift of Nebulae a Puzzle, Says Einstein." *New York Times*, February 12, 1931, 2.

Rees, Martin. *Just Six Numbers: The Deep Forces That Shape the Universe*. New York: Basic Books, 2000.

———. "Large Numbers and Ratios in Astrophysics and Cosmology." *Philosophical Transactions of the Royal Society London A* 310 (1983): 317.

Reid, Thomas. *Lectures on Natural Theology*. Edited by Elmer Duncan and William R. Eakin. 1780. Washington, DC: Univ. Press of America, 1981.

Reidhaar-Olson, John, and Robert Sauer. "Functionally Acceptable Solutions in Two Alpha-Helical Regions of Lambda Repressor." *Proteins: Structure, Function, and Genetics* 7 (1990): 306–16.

Reppert, Victor. *C. S. Lewis's Dangerous Idea: In Defense of the Argument from Reason*. Downers Grove, IL: InterVarsity, 2003.

Richards, Jay. *The Untamed God: A Philosophical Exploration of Divine Perfection, Simplicity, and Immutability*. Downers Grove, IL: IVP Academic, 2003.

Roberts, John T. "Fine-Tuning and the Infrared Bull's-Eye." *Philosophical Studies* 160 (2011): 287–303.

Rodin, Andrei S., Eörs Szathmáry, and Sergei N. Rodin. "On the Origin of the Genetic Code and tRNA Before Translation." *Biology Direct* 6/1 (2011): 14.

Rolston, Holmes, III. *Science and Religion: A Critical Survey*. Philadelphia: Temple Univ. Press, 1987.

Ross, Hugh. *The Creator and the Cosmos: How the Greatest Scientific Discoveries of the Century Reveal God*. Colorado Springs, CO: NavPress, 1995.

Ross, Sydney. "*Scientist*: The Story of a Word." *Annals of Science* 18/2 (1962): 65–85.

Rovelli, Carlo. "Loop Quantum Gravity." *Living Reviews in Relativity* 11/1 (December 15, 2008): 5.

———. *Quantum Gravity*. Cambridge Monographs on Mathematical Physics. Cambridge: Cambridge Univ. Press, 2004.

———. "The Strange Equation of Quantum Gravity." *Classical and Quantum Gravity* 32 (2015): 124005. arXiv:1506.00927.

Ruse, Michael. *Darwinism Defended: A Guide to the Evolution Controversies.* London: Addison-Wesley, 1982. p. 58.
Russell, Bertrand. *Mysticism and Logic, Including a Free Man's Worship.* London: Unwin Paperbacks, 1986.
Russell, Colin A. "The Conflict of Science and Religion." In *Science and Religion: A Historical Introduction*, edited by Gary B. Ferngren, 3–12. Baltimore: Johns Hopkins Univ. Press, 2002.
———. *Cross-Currents: Interactions Between Science and Faith.* Grand Rapids, MI: Eerdmans, 1985.
———. *Michael Faraday: Physics and Faith.* New York: Oxford Univ. Press, 2000.
Russell, Jeffrey B. *Inventing the Flat Earth: Columbus and Modern Historians.* Westport, CT: Praeger, 1991.
Sagan, Carl. *Cosmos.* New York: Random House, 1980.
———. *Cosmos: A Personal Voyage.* PBS, 1980.
Sandage, Allan. "Current Problems in the Extragalactic Distance Scale." *Astrophysical Journal* 127 (May 1958): 513–27. doi:10.1086/146483.
———. "Edwin Hubble 1889–1953." *Journal of the Royal Astronomical Society of Canada* 83/6 (1989): 621.
———. "A Scientist Reflects on Religious Belief." *Truth: An Interdisciplinary Journal of Christian Thought* 1 (1985): 53–54.
Sartre, Jean-Paul. *Nausea.* New York: New Directions Books, 2007.
———. "Existentialism and Humanism." In *The Continental Philosophy Reader*, edited by Richard Kearney and Mara Rainwater. London: Routledge, 1996.
Schaeffer, Francis. *He Is There and He Is Not Silent.* Carol Stream, IL: Tyndale House Publishers, Inc., 2001.
Schaefer, Henry F. *Science and Christianity: Conflict or Coherence?* Watkinsville, GA: Apollos Trust, 2003.
Schaffer, Jonathan, "What not to Multiply Beyond Necessity," *Australasian Journal of Philosophy*, 2015, Vol. 93, No. 4, 644664, http://dx.doi.org/10.1080/00048402.2014.992447
Schaffer, Simon. "Occultism and Reason." In *Philosophy, Its History and Historiography*, edited by Alan J. Holland, 117–44. Dordrecht: Reidel, 1985.
Schneider, Thomas D. "The Evolution of Biological Information." *Nucleic Acids Research* 28 (2000): 2794–99.
Schrödinger, Erwin. "An Undulatory Theory of the Mechanics of Atoms and Molecules." *The Physical Review* 28/6 (December 1926): 1049–70. Available at https://journals.aps.org/pr/abstract/10.1103/PhysRev.28.1049.
Schroeder, Gerald L. *The Science of God: The Convergence of Scientific and Biblical Wisdom.* New York: The Free Press, 1997.
Scriven, Michael. "Causes, Conditions and Connections in History." In *Philosophical Analysis and History.* W. H. Dray, editor. New York, Harper & Row, 1966.
———. "Explanation and Prediction in Evolutionary Theory." *Science* 130 (1959): 477–82.
Senovilla, Jose M. M., and David Garfinkle. "The 1965 Penrose Singularity Theorem." *Classical and Quantum Gravity* 32/12 (June 2015). doi.org/10.1088/0264-9381/32/12/124008.
Sfetsos, K., and K. Skenderis. "Microscopic Derivation of the Bekenstein-Hawking Entropy Formula for Non-Extremal Black Holes." *Nuclear Physics B* 517/1–3 (April 27, 1998): 179–204.
Shannon, Claude E. "A Mathematical Theory of Communication." *Bell System Technical Journal* 27 (1948): 379–423, 623–56.
Shapiro, Adam R. "Myth 8: That William Paley Raised Scientific Questions About Biological Origins That Were Eventually Answered by Charles Darwin." In *Newton's Apple and Other Myths About Science*, edited by Ronald Numbers and Kostas Kampourakis, 67–73. Cambridge, MA: Harvard Univ. Press, 2015.

Shapiro, Robert. *Origins: A Skeptic's Guide to the Creation of Life on Earth*. New York City: Bantam, 1987.

———. "Prebiotic Cytosine Synthesis: A Critical Analysis and Implications for the Origin of Life." *Proceedings of the National Academy of Sciences, USA* 96 (1999): 4396–401.

Shermer, Michael. "ID Works in Mysterious Ways: A Critique of Intelligent Design." *Skeptic*, Vol. 8, No. 2, 2000: 22–24.

Siger of Brabant. *On the Eternity of the World* (*De Aeternitate Mundi*). Translated by Lottie H. Kendzierski. Milwaukee, WI: Marquette Univ. Press, 1964.

Simpson, George Gaylord. *The Meaning of Evolution*. New Haven, CT: Yale Univ. Press, 1967.

Singer, S. F. "The Origin and Age of the Meteorites." *Irish Astronomical Journal* 4/6 (June 1957): 165–80.

Singh, Simon. *Big Bang: The Origin of the Universe*. New York: HarperCollins, 2005.

Sire, James W. *The Universe Next Door.* 2nd ed. Downers Grove, IL: InterVarsity, 1988.

Smith, John Maynard. *The Theory of Evolution*. 3rd ed. 1975. Reprint, New York: Penguin, 1985.

Smith, R. W. "E. P. Hubble and The Transformation of Cosmology." *Physics Today*, April 1990, 57.

Smolin, Lee. *The Trouble with Physics: The Rise of String Theory, the Fall of a Science, and What Comes Next*. Boston: Mariner, 2007.

Snobelen, Stephen. "Cosmos and Apocalypse." *New Atlantis* 44 (2015): 76–94.

———. "Cosmos and Apocalypse: Prophetic Themes in Newton's Astronomical Physics." Unpublished manuscript.

———. "'God of Gods, and Lord of Lords': The Theology of Isaac Newton's General Scholium to the Principia." *Osiris* 16/1 (January 2001): 169–208.

———. "Isaac Newton." In *Science and Religion: A Historical Introduction*, edited by Gary B. Ferngren, 123–39. Baltimore: Johns Hopkins Univ. Press, 2017.

———. "Newton and the God of the Gaps." Unpublished lecture notes.

Sober, Elliot. "Intelligent Design Theory and the Supernatural—The 'God or Extra-terrestrials' Reply." *Faith and Philosophy: Journal of Christian Philosophers* 24/1 (January 2007): 1–12.

———. *The Philosophy of Biology*. San Francisco: Westview, 1993.

Sokol, Joshua. "A Brush with a Universe Next Door." *New Scientist* 228/3045 (October 31, 2015): 8–9.

Spade, Paul V. "Ockham's Nominalist Metaphysics: Some Main Themes." *The Cambridge Companion to Ockham*, edited by Paul V. Spade, 100–117. Cambridge: Cambridge Univ. Press, 1999.

Spitzer, Robert J. "Evidence for God from Physics and Philosophy," *Magis Center*, 2016: 13–15, accessed August 15, 2020, https://magiscenter.com/evidence-for-god-from-physics-and-philosophy/.

———. *New Proofs for the Existence of God: Contributions of Contemporary Physics and Philosophy*. Grand Rapids, MI: Eerdmans, 2010.

Stark, Rodney. *For the Glory of God: How Monotheism Led to Reformations, Science, Witch-Hunts and the End of Slavery*. Princeton, NJ: Princeton Univ. Press, 2004.

Stein, Ben. *Expelled: No Intelligence Allowed*. Directed by Nathan Frankowski. Los Angeles: Premise Media, 2008. https://youtu.be/V5EPymcWp-g.

Stenger, Victor. "Fine-Tuning and the Multiverse." *Skeptic* 19.3 (2014). https://www.skeptic.com/reading_room/fine-tuning-and-the-multiverse.

———. *God: The Failed Hypothesis: How Science Shows That God Does Not Exist*. Amhurst, NY: Prometheus, 2007.

Strobel, Lee. *The Case for a Creator.* Grand Rapids, MI: Zondervan, 2004.

Susskind, Leonard. "The Anthropic Landscape of String Theory." In *Universe or Multiverse?* edited by Bernard Carr, 247–66. Cambridge: Cambridge Univ. Press, 2007.

———. *The Cosmic Landscape: String Theory and the Illusion of Intelligent Design*. New York: Hachette, 2006.

Sutton, Christine. Review of *Einstein's Universe* (BBC2, March 14, 7:05 p.m.). *New Scientist* 81/1145 (March 8, 1979): 784.

Swinburne, Richard. *The Existence of God*. Oxford: Clarendon, 1979.

Talbott, William. "Bayesian Epistemology." In *The Stanford Encyclopedia of Philosophy*, edited by Edward N. Zalta, Winter 2016. https://plato.stanford.edu/archives/win2016/entries/epistemology-bayesian.

Tanzella-Nitti, G. "The Two Books Prior to the Scientific Revolution." *Annales Theologica* 18 (2004): 51–83.

Tarnas, Richard. *The Passion of the Western Mind*. London: Pimlico, 1991.

Taylor, Geoffrey Ingram. "Interference Fringes with Feeble Light." *Proceedings of the Cambridge Philosophical Society* 15/1 (1909): 114–15. http://www.gsjournal.net/Science-Journals/Historical%20Papers-Mechanics%20/%20Electrodynamics/Download/2659.

Taylor, Sean V., et al. "Searching Sequence Space for Protein Catalysts." *Proceedings of the National Academy of Sciences USA* 98 (2001): 10596–601.

Tegmark, Max. "Infinity Is a Beautiful Concept—and It's Ruining Physics." *Discover*, February 20, 2015. https://www.discovermagazine.com/the-sciences/infinity-is-a-beautiful-concept-and-its-ruining-physics#.W_ZbZZNKhE5.

———. "Is 'The Theory of Everything' Merely the Ultimate Ensemble Theory?" *Annals of Physics 270/1* (1998): 1–51.

———. "The Multiverse Hierarchy." In *Universe or Multiverse?* edited by Bernard Carr, 99–126. Cambridge: Cambridge Univ. Press, 2007.

———. *Our Mathematical Universe: My Quest for the Ultimate Nature of Reality*. New York: Knopf, 2014.

———. "What Scientific Idea Is Ready for Retirement? Infinity." *Edge*, February 12, 2014. https://www.edge.org/response-detail/25344.

Thaxton, Charles B., and Walter L. Bradley. "Information and the Origin of Life." In *The Creation Hypothesis: Scientific Evidence for an Intelligent Designer*, edited by J. P. Moreland, 193–97. Downers Grove, IL: InterVarsity, 1994.

Thaxton, Charles, Walter L. Bradley, and Roger L. Olsen. *The Mystery of Life's Origin: Reassessing Current Theories*. New York: Philosophical Library, 1984.

Thayer, H. S. *Newton's Philosophy of Nature: Selections from His Writings*. New York: Macmillan, 1974.

Thomson, G. P. "Experiments on the Diffraction of Cathode Rays." *Proceedings of the Royal Society of London, Series A, Containing Papers of a Mathematical and Physical Character* 117/778 (February 1, 1928): 600–609. http://www.ymambrini.com/My_World/History_files/GP%20Thomson%20orginal%20paper.pdf.

Tipler, Frank. J. "Energy Conditions and Spacetime Singularities." *Physical Review D 17/10* (1978): 2521–28.

Tokuriki, Nobuhiko et al. "The Stability Effects of Protein Mutations Appear to Be Universally Distributed," *Journal of Molecular Biology* 369/5 (June 22, 2007): 1318–32, https://doi.org/10.1016/j.jmb.2007.03.069.

Tokuriki, Nobuhiko, and Dan S. Tawfik. "Stability Effects of Mutations and Protein Evolvability." *Current Opinion in Structural Biology* 19/5 (2009): 596–604.

Tompa, Peter, and George D. Rose. "The Levinthal Paradox of the Interactome." *Protein Science* 20/12 (2011): 2074–79.

Torrance, Thomas F. *Divine and Contingent Order*. Oxford: Oxford Univ. Press, 2000.

Tour, James. "Animadversions of a Synthetic Chemist." *Inference* 2/2, May 2016. https://inference-review.com/article/animadversions-of-a-synthetic-chemist.

———. "An Open Letter to My Colleagues." *Inference* 3/2 (2017).

———. "Time Out." *Inference* 4/4 (2019).

Trimble, Virginia. "Anybody but Hubble!" July 8, 2013. arXiv:1307.2289.

———. "Extragalactic Distance Scales: H_0 from Hubble (Edwin) to Hubble (Hubble Telescope)." *Space Science Reviews* 79, no. 3/4 (1997): 793–834.

———. "H_0: The Incredible Shrinking Constant, 1925–1975." *Publications of the Astronomical Society of the Pacific* 108/730 (December 1996): 1073–82. doi:10.1086/133837.

Tyson, Neil deGrasse. In *Cosmos: A Spacetime Odyssey*. "Episode 3: When Knowledge Conquered Fear." http://investigacion.izt.uam.mx/alva/cosmos2014_01-1.pdf, Episode 3, pages 34–50. See also: https://www.fox.com/watch/cebd37976d6172a938ec77621fb35699/.

Ugural, A. C., and S. K. Fenster. "Hooke's Law and Poisson's Ratio." In *Advanced Strength and Applied Elasticity*, 66–71. 4th ed. Prentice Hall, 2003.

Ungureanu, James. *Science, Religion, and the Protestant Tradition: Retracing the Origins of Conflict*. Pittsburgh, PA: Univ. of Pittsburgh Press, 2019.

Vaidman, Lev. "Quantum Theory and Determinism." *Quantum Studies: Mathematics and Foundations* 1/1–2 (2014): 5–38.

Van Till, Howard. *The Fourth Day*. Grand Rapids, MI: Eerdmans, 1986.

Van Till, Howard, Davis Young, and Clarence Menninga. *Science Held Hostage*. Downers Grove, IL: InterVarsity, 1988.

Venema, Dennis, and Dan Kuebler. "Biological Information and Intelligent Design." *BioLogos*, June 9, 2016. https://biologos.org/articles/biological-information-and-intelligent-design.

Venema, Dennis, and Scott McKnight. *Adam and the Genome: Reading Scripture After Genetic Science*. Grand Rapids, MI: Brazos, 2017.

Vessot, R. F. C., et al. "Test of Relativistic Gravitation with a Space-Borne Hydrogen Maser." *Physical Review Letters* 45 (1980): 2081–84.

Viereck, G. S. *Glimpses of the Great*. New York: Macauley, 1930.

Vilenkin, Alexander. "The Beginning of the Universe." *Inference* 1/4, October 2015.

———. "Creation of Universes from Nothing." *Physics Letters 117B* (1982): 25–28.

———. *Many Worlds in One: The Search for Other Universes*. New York: Hill and Wang, 2006.

———. "Quantum Cosmology." February 11, 1993. arXiv:gr-qc/9302016.

———. "Quantum Cosmology and the Initial State of the Universe." *Physical Review D* 37/4 (February 15, 1988): 888–97. doi:10.1103/PhysRevD.37.888.

———. "Quantum Origin of the Universe." *Nuclear Physics B* 252 (January 1, 1985): 141–52.

Vilenkin, Alexander, and Masaki Yamada. "Tunneling Wave Function of the Universe." *Physical Review D* 98/6 (August 6, 2018): 066003. doi:10.1103/PhysRevD.98.066003.

Voltaire. "Epître à l' auteur du livre des *Trois imposteurs*." In *OEuvres complètes de Voltaire*, ed. Louis Moland. 1768; Paris: Garnier, 1877–85.

Wagner, Andreas. *The Arrival of the Fittest*. New York: Penguin, 2014.

Wald, George. "The Origin of Life." *Scientific American* 191 (1954): 44–53.

Wallace, David. "The Quantum Measurement Problem: State of Play." December 3, 2007. arXiv:0712.0149.

Wallerstein, George, et al. "Synthesis of the Elements in Stars: Forty Years of Progress." *Reviews of Modern Physics* 69/4 (October 1, 1997): 995–1084.

Weinberg, Steven. *The First Three Minutes: A Modern View of the Origin of the Universe*. New York: Basic Books, 1993.

Wells, Jonathan. "Membrane Patterns Carry Ontogenetic Information That Is Specified Independently of DNA." *BIO-Complexity* 2014/2 (April 2016): 1–28.

West, John G. *Darwin's Corrosive Idea: The Impact of Evolution on Attitudes About Faith, Ethics, and Human Uniqueness*. Seattle: Discovery Institute Press, 2016.

White, Andrew D. *A History of the Warfare of Science with Theology in Christendom*. New York: Appleton, 1896.

Whitehead, Alfred North. *Science and the Modern World*. New York: Free Press, 1925.

Wilczek, Frank. "The Cosmic Asymmetry Between Matter and Antimatter." *Scientific American* 243 (December 1980): 82–90.

Willard, Dallas. "Knowledge and Naturalism." In *Naturalism: A Critical Analysis*, edited by William Lane Craig and J. P. Moreland. London: Routledge, 2000.

Willford, J. N. "Sizing up the Cosmos: An Astronomers Quest." *New York Times*, March 12, 1991, B9.
Wiltshire, David L. "An Introduction to Quantum Cosmology." In *Cosmology: The Physics of the Universe*, edited by B. A. Robson, N. Visvanathan, and W. S. Woolcock, 473–531. Singapore: World Scientific, 1996.
Wolf, Huri I., and Eugene V. Koonin. "On the Origin of the Translation System and the Genetic Code in the RNA World by Means of Natural Selection, Exaptation, and Subfunctionalization." *Biology Direct* 2 (2007): 1–25.
Woodward, Joe. "The End of Evolution." *Alberta Report*, December 1996, 33.
Yenter, Timothy. "Samuel Clarke." In *The Stanford Encyclopedia of Philosophy*, edited by Edward N. Zalta, Winter 2017. https://plato.stanford.edu/archives/win2017/entries/clarke.
Yockey, H. P. "A Calculation of the Probability of Spontaneous Biogenesis by Information Theory." *Journal of Theoretical Biology* 67/3 (August 7, 1977): 377–98.
———. *Information Theory and Molecular Biology.* Cambridge: Cambridge Univ. Press, 1992.
Young, Thomas. "Bakerian Lecture: Experiments and Calculations Relative to Physical Optics." *Philosophical Transactions of the Royal Society* 94 (1804): 1–16. doi:10.1098/rstl.1804.0001.
Zilsel, Edgar. "The Genesis of the Concept of Physical Law." *Philosophical Review* 51/3 (1942): 245–79.

Credits and Permissions

The inclusion of any figures, illustrations, photographs, diagrams, charts, or other types of images in this book should not be construed as an endorsement of the ideas and arguments contained in this book on the part of any copyright holders or creators of those images, other than the author of the book himself. All illustrations not listed here © 2020 Ray Braun Design.

Figure 1.1a Richard Dawkins photograph courtesy of Mike Cornwell [CC BY-SA 2.0], via Wikimedia Commons.

Figure 1.1b Lawrence Krauss photograph courtesy of: Jvangiel [CC BY-SA 3.0], via Wikimedia Commons.

Figure 1.2 Allan Sandage photograph courtesy of the Carnegie Institution for Science.

Figure 1.3 Dean Kenyon photograph used with permission from Dean Kenyon.

Figure 1.4a John William Draper photograph courtesy of John Sartain [Public domain], via Wikimedia Commons.

Figure 1.4b Andrew Dickson White photograph courtesy of Robinsons - Ithaca, New York [Public domain], via Wikimedia Commons.

Figure 1.5 Peter Hodgson photograph E/13/3/1 used with permission from the President and Fellows of Corpus Christi College, Oxford.

Figure 1.6 Johannes Kepler photograph courtesy of unidentified painter [Public domain], via Wikimedia Commons.

Figure 1.7 William of Ockham photograph courtesy of Andrea di Bonaiuto [Public domain], via Wikimedia Commons.

Figure 2.2 Title page of John Ray's book [Public domain], via Wikimedia Commons.

Figure 2.3 Robert Boyle photograph © The Royal Society.

Figure 2.4 Isaac Newton photograph courtesy of Godfrey Kneller [Public domain], via Wikimedia Commons.

Figure 2.6 *Principia* frontispiece image. Reproduced by kind permission of the Syndics of Cambridge University Library. File name: Adv.b.39.2, p. i (frontispiece).

Figure 3.1 David Hume photograph courtesy of Allan Ramsay [Public domain], via Wikimedia Commons.

Figure 3.2 Immanuel Kant photograph by Johann Gottlieb Becker [Public domain], via Wikimedia Commons.

Figure 3.3 William Paley photograph courtesy of National Portrait Gallery, London.

Figure 3.4 Charles Darwin photograph by Henry Maull and John Fox courtesy of Beao [Public domain], via Wikimedia Commons.

Figure 3.5 Pierre Laplace photograph by Biblioteca de la Facultad de Derecho y Ciencias del Trabajo Universidad de Sevilla (Flickr: 1029006) [CC BY-SA 2.0], via Wikimedia Commons].

Figure 3.6 Karl Marx photograph by author unknown [Public domain], via Wikimedia Commons.

Figure 3.7 Sigmund Freud photograph by Max Halberstadt [Public domain], via Wikimedia Commons.

Figure 4.2 Edgar Allan Poe photograph by Edwin H. Manchester [Public domain], via Wikimedia Commons.

Figure 4.3a Harlow Shapley photograph courtesy of Harvard University News Office, via AIP Emilio Segrè Visual Archives, Physics Today Collection.

Figure 4.3b Heber Curtis photograph courtesy of Rockefeller University.

Figure 4.4 Edwin Hubble photograph by Johan Hagemeyer (1884-1962) [Public domain], via Wikimedia Commons.

Figure 4.5 Henrietta Leavitt photograph by Author Unknown [Public domain], via Wikimedia Commons.

Figure 4.7 Edwin Hubble photograph courtesy of HUB 1042 (4) Edwin Powell Hubble Papers, The Huntington Library, San Marino, California.

Figure 4.8 Vesto Slipher photograph courtesy of the Special Collections Research Center, University of Chicago Library.

Figure 4.12 Hubble spiral galaxy photograph courtesy of Carnegie Institution for Science.

Figure 4.13 Hubble spiral galaxy photograph courtesy of Carnegie Institution for Science.

Figure 5.1 Albert Einstein photograph by Ferdinand Schmutzer [Public domain], via Wikimedia Commons.

Figure 5.3 Illustration © 2020 Ray Braun Design; adapted by permission of Fred Heeren.

Figure 5.4 Aleksandr Friedmann photograph by Author Unknown [Public domain], via Wikimedia Commons.

Figure 5.5 Georges Lemaitre photograph courtesy of the Catholic University of Louvain, Georges Lemaitre Archives.

Figure 5.6 Einstein at telescope photograph courtesy of The Edwin Powell Hubble papers, The Huntington Library, San Marino, California (HUB 1057).

Figure 5.7 Sir Arthur Eddington photograph by George Grantham Bain Collection, Library of Congress Prints and Photographs Division Washington, DC [Public domain], via Wikimedia Commons.

Figure 5.8 Gold, Bondi, and Hoyle photograph by permission of the Master and Fellows of St. John's College, Cambridge.

Figure 5.10 Wilson and Penzias photograph courtesy AIP Emilio Segrè Visual Archives, Physics Today Collection.

Figure 5.12 Illustration © 2020 Ray Braun Design; adapted by permission from an original by John Wiester.

Figure 5.13 Illustration © 2020 Ray Braun Design; original public domain image from NASA.

Figure 5.14 Robert Jastrow photograph courtesy AIP Emilio Segrè Visual Archives, Physics Today Collection. Used with permission.

Figure 6.1 Stephen Hawking photograph courtesy of David Montgomery/Getty Images.

Figure 6.2 George Ellis photograph courtesy of David Monniaux [CC BY-SA 3.0], via Wikimedia Commons.

Figure 6.4 Alan Guth photograph courtesy of Justin Knight

Figure 7.1 Fred Hoyle photograph courtesy of AIP Emilio Segrè Visual Archives, Clayton Collection.

Figure 7.3 Illustration © 2020 Ray Braun Design; the graph appears in Barnes, "The Fine-Tuning of the Universe for Intelligent Life," 537.

Figure 7.6 Paul Davies photograph courtesy Christopher Michel.

Figure 7.7 John Polkinghorne photograph courtesy of Louis Monier/Gamma-Rapho/ Getty Images.

Figure 8.2 Sir Roger Penrose photograph courtesy of Festival della Scienza [CC BY-SA 2.0], via Wikimedia Commons.

Figure 8.3 William Dembski photograph courtesy of Laszlo Bencze.

Figure 9.1 Watson and Crick photograph courtesy of A. Barrington Brown / Photo Researchers, Inc.

Figure 9.4 Illustration © 2020 Ray Braun Design; adapted by permission. Courtesy of I. L. Cohen of New Research Publications.

Figure 9.6 Claude Shannon photograph courtesy of the Estate of Francis Bello / Photo Researchers, Inc.

Figure 9.7 Illustration © 2020 Ray Braun Design; adapted by permission from an original drawing by Fred Heeren.

Figure 9.9 Charles Thaxton photograph courtesy of Charles Thaxton.

Figure 9.10 Peter Lipton photograph printed by permission from Howard Guest.

Figure 10.3 Murray Eden photograph courtesy of MIT Museum.

Figure 10.4 Gerd Muller photograph courtesy Susanne Müller [CC BY-SA 4.0], via Wikimedia Commons.

Figure 10.5 Michael Denton photograph courtesy of Brian Gage.

Figure 10.6 David Berlinski photograph courtesy of Nicholas DeSciose.

Figure 10.7 Douglas Axe photograph courtesy of Biola University.

Figure 11.1 Christian de Duve photograph courtesy of Ingbert Gruttner, The Rockefeller University.

Figure 11.2 Sean Carroll photograph courtesy of Sgerbic [CC BY-SA 4.0], via Wikimedia Commons.

Figure 11.6 Charles Sanders Peirce photograph courtesy of National Oceanic and Atmospheric Administration/Department of Commerce / Public domain.

Figure 12.1 Michael Shermer photograph by Byrd Williams; uploaded by User:Loxton. [CC BY 3.0], via Wikimedia Commons.

Figure 12.2 Ernan McMullin photograph courtesy of the University of Notre Dame Archives.

Figure 12.4 William Lane Craig photograph courtesy of C. Bertuzzi. Used with permission.

Figure 12.5 J. P. Moreland photograph courtesy of J. P. Moreland. Used with permission.

Figure 13.4 Michael Polanyi photograph courtesy of National Portrait Gallery, London.

Figure 13.5 Luke Barnes photograph; used with permission.

Figure 15.1 Thomas Nagel photograph © Sorel: Courtesy of NYU Photo Bureau.

Figure 15.2 John Sutherland photograph courtesy of The Royal Society [CC BY-SA 3.0], via Wikimedia Commons.

Figure 15.4c dGRN figures © Copyright 2008 National Academy of Sciences, U.S.A. Figure 1D, 1E, and 7, Oliveri, P.; Tu, Q.; and Davidson, E. H., "Global Regulatory Logic for Specification of an Embryonic Cell Lineage," Proceedings of the National Academy of Sciences USA, 105 (2008): 5955-62. Use of PNAS material does not imply any endorsement by PNAS or the National Academy of Sciences or the authors. Used with permission.

Figure 16.4 Bruce Gordon photograph courtesy of Michael Tims / Houston Baptist University.

Figure 16.5 Leonard Susskind photograph courtesy of Jeff Singer/Redux.

Figure 16.6 Robin Collins photograph courtesy of Messiah College.

Figure 17.3 Erwin Schrodinger photograph courtesy of Carl Wolf / National Portrait Gallery, Smithsonian Institution.

Figure 17.6a John Wheeler photograph courtesy of AIP Emilio Segrè Visual Archives.

Figure 17.6b Bryce DeWitt photograph courtesy of Ric Spencer, in the University of North Carolina at Chapel Hill Photographic Laboratory Collection #P0031, North Carolina Collection Photographic Archives, The Wilson Library, University of North Carolina at Chapel Hill.

Figure 19.2 Max Tegmark photograph courtesy of Justin Knight Photography.

Figure 19.3 Ludwig Boltzmann photograph courtesy of University of Chicago Library.

Figure 20.1 Isaac Newton painting by Enoch Seeman © National Portrait Gallery, London.

Figure 20.3 Neil deGrasse Tyson photograph by Thor Nielsen / NTNU (Norwegian University of Science and Technology [CC BY SA 2.0], via Wikimedia Commons.

Figure 20.4 *Principia* image (Prop. 12); Reproduced by kind permission of the Syndics of Cambridge University Library. File name: Adv.b.39.1, p. 418.

Figure 21.1 Stephen Hawking photograph courtesy of Mackenzie Stroh/Contour/Getty Images.

Figure 21.2 Alvin Plantinga photograph courtesy of Jonathunder [CC BY-SA 3.0], via Wikimedia Commons.

Figure 21.3 Robert Koons photograph courtesy of LifeTouch, Inc.

Index

Page references followed by *fig* indicate an illustrated figure.